LEARNSMART® >

The market leading **adaptive study tool** proven to strengthen memory recall, increase class retention and boost grades.

> Moves students beyond memorizing

> Allows instructors to align content with their goals

> Allows instructors to spend more time teaching higher level concepts

SMARTBOOK™ >

The first—and only—adaptive reading experience designed to transform the way students read.

> Engages students with a personalized reading experience

> Ensures students retain knowledge

LEARNSMART PREP™

An adaptive course preparation tool that quickly and efficiently helps students prepare for college level work.

> Levels out student knowledge

> Keeps students on track

LEARNSMART ACHIEVE™

A learning system that continually adapts and provides learning tools to teach students the concepts they don't know.

> Adaptively provides learning resources

> A time management feature ensures students master course material to complete their assignments by the due date

WWW.LEARNSMARTADVANTAGE.COM

About the Cover

This photograph by Michael Collier shows a beautifully stark, sparsely populated landscape north of Capitol Reef National Park in central Utah. The area has an arid climate and receives less than 20 cm (eight inches) of rain every year, sometimes in cloudbursts like the one captured in the photograph.

During such a storm, roads become impassable, and the nearby, aptly named Muddy Creek swells with flashfloods. Over time, rainfall has eroded the tilted Jurassic sedimentary layers in the foreground, gradually lowering their surface. The two mountains in the center of the picture—Hebes, the sloping mountain in front of the rain, and Cedar Mountain, behind the rain—are capped with horizontal layers of igneous rock. These rocks formed when magma was injected between the layers four million years ago, when the surface stood about a kilometer higher than it does today. These igneous rocks are much more resistant to erosion than the reddish and white sedimentary layers; as a result, the two mountains remain standing, while the surrounding surface has been lowered. The present-day scene represents a single snapshot into a continuously evolving landscape, reflecting the interplay between the land surface, hydrosphere, atmosphere, and biosphere.

Michael Collier's photography is featured in textbooks, magazines, and a series of beautiful books showing landscapes from the air. His books have been about various national parks, climate change, the San Andreas Fault, Alaskan glaciers, and other geographic locations.

STEPHEN J. REYNOLDS
Arizona State University

ROBERT V. ROHLI
Louisiana State University

JULIA K. JOHNSON
Arizona State University

PETER R. WAYLEN
University of Florida

MARK A. FRANCEK
Central Michigan University

exploring physical geography

CYNTHIA C. SHAW
Lead Illustrator, Art Director

McGraw Hill Education

EXPLORING PHYSICAL GEOGRAPHY

Published by McGraw-Hill, a business unit of The McGraw-Hill Companies, Inc., 1221 Avenue of the Americas, New York, NY 10020. Copyright © 2015 by The McGraw-Hill Companies, Inc. All rights reserved. Printed in the United States of America. No part of this publication may be reproduced or distributed in any form or by any means, or stored in a database or retrieval system, without the prior written consent of The McGraw-Hill Companies, Inc., including, but not limited to, in any network or other electronic storage or transmission, or broadcast for distance learning.

Some ancillaries, including electronic and print components, may not be available to customers outside the United States.

This book is printed on acid-free paper.

2 3 4 5 6 7 8 9 0 DOW/DOW 1 0 9 8 7 6 5 4

ISBN 978–0–07–809516–0
MHID 0–07–809516–6

Senior Vice President, Products & Markets: *Kurt L. Strand*
Vice President, General Manager, Products & Markets: *Marty Lange*
Vice President, Content Production & Technology Services: *Kimberly Meriwether David*
Managing Director: *Thomas Timp*
Brand Manager: *Michelle Vogler*
Director of Development: *Rose M. Koos*
Product Developer: *Jodi Rhomberg*
Director of Digital Content: *Andrea M. Pellerito, Ph.D.*
Marketing Manager: *Matthew Garcia*
Director, Content Production: *Terri Schiesl*
Content Project Manager: *Sandy Wille*
Buyer: *Nicole Baumgartner*
Designer: *Matthew Backhaus*
Cover Image: *Front: Rain Storm above Muddy Creek near Cedar Mountain, Utah* © *Michael Collier*
Back: © *Robert Daly / age fotostock*
Senior Content Licensing Specialist: *Lori Hancock*
Layout: *Stephen J. Reynolds, Julia K. Johnson, Cynthia C. Shaw, and Lachina Publishing Services*
Compositor: *Lachina Publishing Services*
Typeface: *9/10.5 Avenir LT Std*
Printer: *R. R. Donnelley*

All credits appearing on page or at the end of the book are considered to be an extension of the copyright page.

Library of Congress Cataloging-in-Publication Data

Cataloging-in-Publication Data has been requested from the Library of Congress.

The Internet addresses listed in the text were accurate at the time of publication. The inclusion of a website does not indicate an endorsement by the authors or McGraw-Hill Education, and McGraw-Hill Education does not guarantee the accuracy of the information presented at these sites.

BRIEF CONTENTS

CONTENTS

CHAPTER 4:
ATMOSPHERIC MOISTURE   108

CHAPTER 5:
WEATHER SYSTEMS AND SEVERE WEATHER   142

CHAPTER 6:
ATMOSPHERE-OCEAN-CRYOSPHERE INTERACTIONS   180

CHAPTER 7:
CLIMATES AROUND THE WORLD   212

CHAPTER 16:
SOILS   502

CHAPTER 17:
ECOSYSTEMS AND
BIOGEOCHEMICAL CYCLES   534

TELLING THE STORY . . .

WE WROTE *EXPLORING PHYSICAL GEOGRAPHY* so that students could learn from the book on their own, freeing up instructors to teach the class in any way they want. I (Steve Reynolds) first identified the need for this type of book while I was a National Association of Geoscience Teachers' (NAGT) distinguished speaker. As part of my NAGT activities, I traveled around the country conducting workshops on how to infuse active learning and scientific inquiry into introductory college science courses, including those with upwards of 200 students. In the first part of the workshop, I asked the faculty participants to list the main goals of an introductory science course, especially for nonmajors. At every school I visited, the main goals were similar to those listed below:

- to engage students in the process of scientific inquiry so that they learn what science is and how it is conducted,
- to teach students how to observe and interpret landscapes and other aspects of their physical environment,
- to enable students to learn and apply important concepts of science,
- to help students understand the relevance of science to their lives, and
- to enable students to use their new knowledge, skills, and ways of thinking to become more informed citizens.

I then asked faculty members to rank these goals and estimate how much time they spent on each goal in class. At this point, many instructors recognized that their activities in class were not consistent with their own goals. Most instructors were spending nearly all of class time teaching content. Although this was one of their main goals, it commonly was not their top goal.

Next, I asked instructors to think about why their activities were not consistent with their goals. Inevitably, the answer was that most instructors spend nearly all of class time covering content because (1) textbooks include so much material that students have difficulty distinguishing what is important from what is not, (2) instructors needed to lecture so that students would know what is important, and (3) many students have difficulty learning independently from the textbook.

In most cases, textbooks drive the curriculum, so my coauthors and I decided that we should write a textbook that (1) contains only important material, (2) indicates clearly to the student what is important and what they need to know, and (3) is designed and written in such a way that students can learn from the book on their own. This type of book would give instructors freedom to teach in a way that is more consistent with their goals, including using local examples to illustrate geographic concepts and their relevance. Instructors would also be able to spend more class time teaching students to observe and interpret landscapes, atmospheric phenomena, and ecosystems, and to participate in the process of scientific inquiry, which represents the top goal for many instructors.

COGNITIVE AND SCIENCE-EDUCATION RESEARCH

To design a book that supports instructor goals, we delved into cognitive and science-education research, especially research on how our brains process different types of information, what obstacles limit student learning from textbooks, and how students use visuals versus text while studying. We also conducted our own research on how students interact with textbooks, what students see when they observe photographs showing landscape features, and how they interpret different types of scientific illustrations, including maps, cross sections, and block diagrams that illustrate evolution of environments. *Exploring Physical Geography* is the result of our literature search and of our own science-education research. As you examine *Exploring Physical Geography*, you will notice that it is stylistically different from most other textbooks, which will likely elicit a few questions.

A How Does Wind Transport Sediment?

Wind is generated by differences in air pressure and at times is strong enough to transport material, but only relatively small and lightweight fragments, like sand and clay. Transport of these materials by the wind is most efficient in dry climates, where there is limited vegetation to bind materials together and hold them on the ground.

Wind is capable of transporting sand and finer sediment, as well as lightweight plant fragments and other materials lying on the surface. It generally moves material in one of three ways and can deposit sediment in various settings, some of which are shown in the photographs below.

Most materials on Earth's surface are not moved by the wind because they are too firmly attached to the land (such as rock outcrops), are too large or heavy to be moved, or are both.

Wind

Dust in Suspension

Bouncing Grains

Stationary Grains

Rolling Grains

09.11a1

Wind can pick up and carry finer material, such as dust, silt, and salt. This mode of transport is called *suspension*, and wind can keep some particles in the air for weeks, transporting them long distances.

If wind velocity is great enough, it can roll or slide grains of sand and silt and other loose materials across the ground.

Very strong winds can lift sand grains, carry them short distances, and drop them. This process is akin to bouncing a grain along the surface and is called *saltation*.

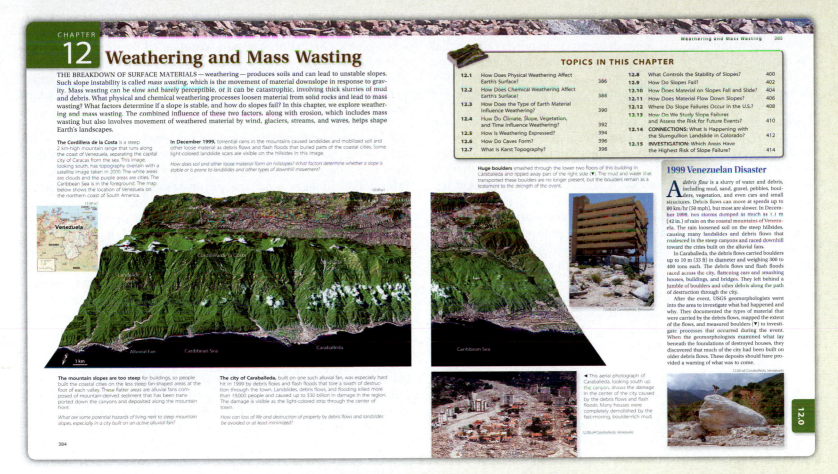

Exploring Physical Geography promotes inquiry and science as an active process. It encourages student curiosity and aims to activate existing student knowledge by posing the title of every two-page spread and every subsection as a question. In addition, questions are dispersed throughout the book. Integrated into the book are opportunities for students to observe patterns, features, and examples before the underlying concepts are explained. That is, we employ a *learning-cycle approach* where student exploration precedes the introduction of geographic terms and the application of knowledge to a new situation. For example, chapter 12 on slope stability, pictured above, begins with a three-dimensional image of northern Venezuela and asks readers to observe where people are living in this area and what natural processes might have formed these sites.

Wherever possible, we introduce terms after students have an opportunity to observe the feature or concept that is being named. This approach is consistent with several educational philosophies, including a learning cycle and just-in-time teaching. Research on learning cycles shows that students are more likely to retain a term if they already have a mental image of the thing being named (Lawson, 2003). For example, this book presents students with maps showing the spatial distribution of earthquakes, volcanoes, and mountain ranges and asks them to observe the patterns and think about what might be causing the patterns. Only then does the textbook introduce the concept of tectonic plates.

Also, the figure-based approach in this book allows terms to be introduced in their context rather than as a definition that is detached from a visual representation of the term. We introduce new terms in italics rather than in boldface, because boldfaced terms on a textbook page cause students to immediately focus mostly on the terms, rather than build an understanding of the concepts. The book includes a glossary for those students who wish to look up the definition of a term to refresh their memory. To expand comprehension of the definition, each entry in the glossary references the pages where the term is defined in the context of a figure.

WHY ARE THE PAGES DOMINATED BY ILLUSTRATIONS?

Physical geography is a visual science. Geography textbooks contain a variety of photographs, maps, cross sections, block diagrams, and other types of illustrations. These diagrams help portray the spatial distribution and geometry of features in the landscape, atmosphere, oceans, and biosphere in ways words cannot. In geography, a picture really is worth a thousand words.

Exploring Physical Geography contains a wealth of figures to take advantage of the visual and spatial nature of geography and the efficiency of figures in conveying geographic concepts. This book contains few large blocks of text — most text is in smaller blocks that are specifically linked to illustrations. Examples of our integrated figure-text approach are shown throughout the book. In this approach, each short block of text is one or more complete sentences that succinctly describe a geographic feature, geographic process, or both of these. Most of these text blocks are connected to their illustrations with leader lines so that readers know exactly which feature or part of the diagram is being referenced in the text block. A reader does not have to search for the part of the figure that corresponds to a text passage, as occurs when a student reads a traditional textbook with large blocks of text referencing a figure that may appear on a different page. The short blocks are numbered if they should be read in a specific order.

This approach is especially well suited to covering geographic topics, because it allows the text to have a precise linkage to the geographic location of the aspect being described. A text block discussing the Intertropical Convergence Zone in Costa Rica can have a leader that specifically points to the location of this feature. A cross section of atmospheric circulation can be accompanied by short text blocks that describe each part of the system and that are linked by leaders directly to specific locations on the figure. This allows the reader to concentrate on the concepts being presented, not deciding what part of the figure is being discussed.

The approach in *Exploring Physical Geography* is consistent with the findings of cognitive scientists, who conclude that our minds have two different processing systems, one for processing pictorial information (images) and one for processing verbal information (speech and written words). This view of cognition is illustrated in the figure below. Cognitive scientists also speak about two types of memory: *working memory*, also called short-term memory, holds information that our minds are actively processing, and *long-term memory* stores information until we need it (Baddeley, 2007). Both the verbal and pictorial processing systems have a limited amount of working memory, and our minds have to use much of our mental processing space to reconcile the two types of information in working memory. For information that has both pictorial and verbal components, as most geographic information does, the amount of knowledge we retain depends on reconciling these two types of information, on transferring information from working memory to long-term memory, and on linking the new information with our existing mental framework. For this reason, this book integrates text and figures, as in the example shown here.

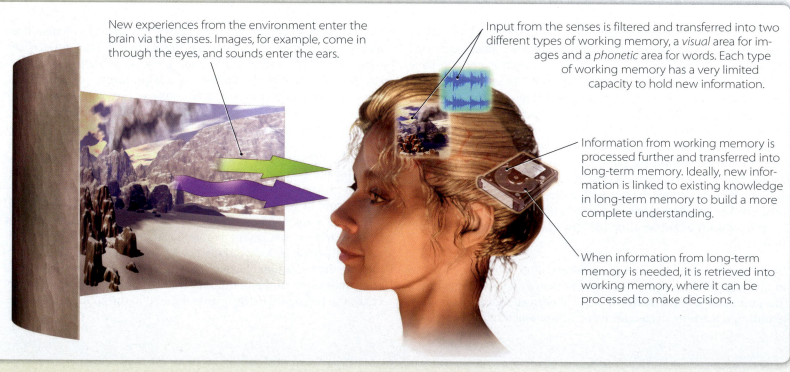

New experiences from the environment enter the brain via the senses. Images, for example, come in through the eyes, and sounds enter the ears.

Input from the senses is filtered and transferred into two different types of working memory, a *visual* area for images and a *phonetic* area for words. Each type of working memory has a very limited capacity to hold new information.

Information from working memory is processed further and transferred into long-term memory. Ideally, new information is linked to existing knowledge in long-term memory to build a more complete understanding.

When information from long-term memory is needed, it is retrieved into working memory, where it can be processed to make decisions.

WHY ARE THERE SO MANY FIGURES?

This textbook contains more than 2,600 figures, which is two to three times the number in most introductory geography textbooks. One reason for this is that the book is designed to provide a concrete example of each process, environment, or landscape feature being illustrated. Research shows that many college students require concrete examples before they can begin to build abstract concepts (Lawson, 1980). Also, many students have limited travel experience, so photographs and other figures allow them to observe places, environments, and processes they have not been able to observe firsthand. The numerous photographs, from geographically diverse places, help bring the sense of place into the student's reading. The inclusion of an illustration for each text block reinforces the notion that the point being discussed is important. In many cases, as in the example on this page, conceptualized figures are integrated with photographs and text so that students can build a more coherent view of the environment or process.

Exploring Physical Geography focuses on the most important geographic concepts and makes a deliberate attempt to eliminate text that is not essential for student learning of these concepts. Inclusion of information that is not essential tends to distract and confuse students rather than illuminate the concept; thus, you will see fewer words. Cognitive and science-education research has identified a redundancy effect, where information that restates and expands upon a more succinct description actually results in a decrease in student learning (Mayer, 2001). Specifically, students learn less if a long figure caption restates information contained elsewhere on the page,

such as in a long block of text that is detached from the figure. We avoid the redundancy effect by including only text that is integrated with the figure.

The style of illustrations in *Exploring Physical Geography* was designed to be more inviting to today's visually oriented students who are used to photo-realistic, computer-rendered images in movies, videos, and computer games. For this reason, many of the figures were created by world-class scientific illustrators and artists who have worked on award-winning textbooks, on Hollywood movies, on television shows, for *National Geographic,* and in the computer-graphics and gaming industry. In most cases, the figures incorporate real data, such as satellite images, weather and climatological data, and aerial photographs. Our own research shows that many students do not understand cross sections and other subsurface diagrams, so nearly every cross section in this book has a three-dimensional aspect, and many maps are presented in a perspective view with topography. Research findings by us and other researchers (Roth and Bowen, 1999) indicate that including people and human-related items on photographs and figures attracts undue attention, thereby distracting students from the features being illustrated. As a result, our photographs have nondistracting indicators of scale, like dull coins and plain marking pens. Figures and photographs do not include people or human-related items unless we are trying to (1) illustrate how geographers study geographic processes and features, (2) illustrate the relevance of the processes on humans, or (3) help students connect and relate to the human dimension of the issue.

WHY DOES THE BOOK CONSIST OF TWO-PAGE SPREADS?

This book consists of two-page spreads, most of which are further subdivided into sections. Research has shown that because of our limited amount of working memory, much new information is lost if it is not incorporated into long-term memory. Many students keep reading and highlighting their way through a textbook without stopping to integrate the new information into their mental framework. New information simply displaces existing information in working memory before it is learned and retained. This concept of cognitive load (Sweller, 1994) has profound implications for student learning during lectures and while reading textbooks. Two-page spreads and sections help prevent cognitive overload by providing natural breaks that allow students to stop and consolidate the new information before moving on.

Each spread has a unique number, such as 6.10 for the tenth topical two-page spread in chapter 6. These numbers help instructors and students keep track of where they are and what is being covered. Each two-page spread, except for those that begin and end a chapter, contains a *Before You Leave This Page* checklist that indicates what is important and what is expected of students before they move on. This list contains learning objectives for the spread and provides a clear way for the instructor to indicate to the student what is important. The items on these lists are compiled into a master *What-to-Know List* provided to the instructor, who then deletes or adds entries to suit the instructor's learning goals and distributes the list to students before the students begin reading the book. In this way, the *What-to-Know List* guides the students' studying.

Two-page spreads and integrated *Before You Leave This Page* lists offer the following advantages to the student:

- Information is presented in relatively small and coherent chunks that allow a student to focus on one important aspect or geographic system at a time.
- Students know when they are done with this particular topic and can self-assess their understanding with the *Before You Leave This Page* list.

- Two-page spreads allow busy students to read or study a complete topic in a short interval of study time, such as the breaks between classes.
- All test questions and assessment materials are tightly articulated with the *Before You Leave This Page* lists so that exams and quizzes cover precisely the same material that was assigned to students via the *What-to-Know* list.

The two-page spread approach also has advantages for the instructor. Before writing this book, the authors wrote most of the items for the *Before You Leave This Page* lists. We then used this list to decide what figures were needed, what topics would be discussed, and in what order. In other words, *the textbook was written from the learning objectives*. The *Before You Leave This Page* lists provide a straightforward way for an instructor to tell students what information is important. Because we provide the instructor with a master *What-to-Know* list, an instructor can selectively assign or eliminate content by providing students with an edited *What-to-Know* list. Alternatively, an instructor can give students a list of assigned two-page spreads or sections within two-page spreads. In this way, the instructor can identify content for which students are responsible, even if the material is not covered in class. Two-page spreads provide the instructor with unparalleled flexibility in deciding what to assign and what not to cover. It allows this book to be easily used for one-semester and two-semester courses.

CONCEPT SKETCHES

Most items on the *Before You Leave This Page* list are by design suitable for student construction of concept sketches. Concept sketches are sketches that are annotated with complete sentences that identify geographic features, describe how the features form, characterize the main geographic processes, and summarize histories of landscapes (Johnson and Reynolds, 2005). An example of a concept sketch is shown to the right.

Concept sketches are an excellent way to actively engage students in class and to assess their understanding of geographic features, processes, and history. Concept sketches are well suited to the visual nature of geography, especially cross sections, maps, and block diagrams. Geographers are natural sketchers using field notebooks, blackboards, publications, and even napkins, because sketches are an important way to record observations and thoughts, organize knowledge, and try to visualize the evolution of landscapes, circulation in the atmosphere and oceans, motion and precipitation along weather fronts, layers within soils, and biogeochemical cycles. Our research data show that a student who can draw, label, and explain a concept sketch generally has a good understanding of that concept.

In a warm front, warm air moves across the surface, displacing cold air. The warm air is less dense than cold and so rises over the cold air, producing stratiform clouds.

Close to the surface position of the front, raindrops can pass through the thin wedge of cold air, remaining as raindrops.

WARM FRONT

WARM AIR

Stratus Clouds

COLD AIR

If the warm air rises so high that it is at freezing temperatures, precipitation can start as snow that reaches all the way to the ground.

If the ground below the warm front is below freezing, the raindrops freeze as they encounter cold objects on the surface, producing freezing rain.

Farther back from the surface position, raindrops have to fall through a thicker amount of cold air, and so freeze on the way down, producing sleet.

REFERENCES CITED

Baddeley, A. D. 2007. *Working memory, thought, and action.* Oxford: Oxford University Press, 400 p.

Johnson, J. K., and Reynolds, S. J. 2005. Concept sketches—Using student- and instructor-generated annotated sketches for learning, teaching, and assessment in geology courses. *Journal of Geoscience Education*, v. 53, pp. 85–95.

Lawson, A. E. 1980. Relationships among level of intellectual development, cognitive styles, and grades in a college biology course. *Science Education*, v. 64, pp. 95–102.

Lawson, A. 2003. *The neurological basis of learning, development & discovery: Implications for science & mathematics instruction.* Dordrecht, The Netherlands: Kluwer Academic Publishers, 283 p.

Mayer, R. E. 2001. *Multimedia learning.* Cambridge: Cambridge University Press, 210 p.

Roth, W. M., and Bowen, G. M. 1999. Complexities of graphical representations during lectures: A phenomenological approach. *Learning and Instruction*, v. 9, pp. 235–255.

Sweller, J. 1994. Cognitive Load Theory, learning difficulty, and instructional design. *Learning and Instruction*, v. 4, pp. 295–312.

HOW IS THIS BOOK ORGANIZED?

Two-page spreads are organized into 18 chapters that are arranged into five major groups: (1) introduction to Earth, geography, and energy and matter; (2) atmospheric motion, weather, climate, and water resources; (3) introduction to landscapes, earth materials, sediment transport, plate tectonics, and tectonic processes (e.g., volcanoes and earthquakes); (4) processes, such as streamflow and glaciation, that sculpt and modify landscapes; and (5) soils, biogeography, and biogeochemical cycles. The first chapter provides an overview of geography, including the scientific approach to geography, how we determine and represent location, the tools and techniques used by geographers, and an introduction to *natural systems*—a unifying theme interwoven throughout the rest of the book. Chapter 2 covers energy and matter in the Earth system, providing a foundation for all that follows in the book.

The second group of chapters begins with an introduction to atmospheric motion (chapter 3), another theme revisited throughout the book. It features separate two-page spreads on circulation in the tropics, high latitudes, and mid-latitudes, allowing students to concentrate on one part of the system at a time, leading to a synthesis of lower-level and upper-level winds. Chapter 3 also covers air pressure, the Coriolis effect, and seasonal and regional winds. This leads naturally into chapter 4, which is a thorough introduction to atmospheric moisture and the consequences of rising and sinking air, including clouds and precipitation. Chapter 5 follows with a visual, map-oriented discussion of weather, including cyclones, tornadoes, and other severe weather. The next chapter (chapter 6), unusual for an introductory geography textbook, is devoted entirely to interactions between the atmosphere, oceans, and cryosphere. It features sections on ocean currents, sea-surface temperatures, ocean salinity, and a thorough treatment of ENSO and other atmosphere-ocean oscillations. This leads into a chapter on climate (chapter 7), which includes controls on climate and a climate classification, featuring a two-page spread on each of the main climate types, illustrated with a rich blend of figures and photographs. These spreads are built around globes that portray a few related climate types, enabling students to concentrate on their spatial distribution and control, rather than trying to extract patterns from a map depicting all the climate types (which the chapter also has). The climate chapter also has a data-oriented presentation of climate change. This second part of the book concludes with chapter 8, which presents the hydrologic cycle and water resources, emphasizing the interaction between surface water and groundwater.

The third part of the book focuses on landscapes and tectonics. It begins with chapter 9, a visually oriented introduction to understanding landscapes, starting with familiar landscapes as an introduction to rocks and minerals. The chapter has a separate two-page spread for each family of rocks and how to recognize each type in the landscape. It presents a brief introduction to weathering, erosion, and transport, aspects that are covered in more detail in later chapters on geomorphology. Wind transport, erosion, and landforms are integrated into chapter 9, rather than being a separate, sparse-content chapter that forcibly brings in non-wind topics, as is done in other textbooks. It also covers relative and numeric dating and how we study the ages of landscapes. It is followed by chapter 10 on plate tectonics and regional features. Chapter 10 begins with having students observe large-scale features on land and the sea floor, as well as patterns of earthquakes and volcanoes, as a lead-in to tectonic plates. Integrated into the chapter are two-page spreads on continental drift, paleomagnetism, continental and oceanic hot spots, evolution of the modern oceans and continents, the origin of high elevations, and the relationship between internal and external processes. The last chapter in this third part (chapter 11) presents the processes, landforms, and hazards associated with volcanoes, deformation, and earthquakes. It also explores the origin of local mountains and basins, another topic unique to this textbook.

The fourth group of chapters concerns the broad field of geomorphology—the form and evolution of landscapes. It begins with chapter 12, a more in-depth treatment of weathering, mass wasting, and slope stability. This chapter also has two-page spreads on caves and karst topography. Chapter 13 is about streams and flooding, presenting a clear introduction to drainage networks, stream processes, different types of streams and their associated landforms and sediment, and how streams change over time. It ends with sections on floods, calculating stream discharges, some examples of devastating local and regional floods, and the many ways in which streams affect people. Chapter 14 covers glaciers and glacial movement, landforms, and deposits. It also discusses the causes of glaciation and the possible consequences of melting of ice sheets and glaciers. Chapter 15 covers the related topic of coasts and changing sea levels. It introduces the processes, landforms, and hazards of coastlines. It also covers the consequences of changing sea level on landforms and humans.

The fifth and final group of chapters focuses on the biosphere and begins with chapter 16, which explores the properties, processes, and importance of soil. This chapter covers soil characterization and classification, including globes showing the spatial distribution of each main type of soil. It ends with a discussion of soil erosion and how soil impacts the way we use land. Chapter 17 provides a visual introduction to ecosystems and biogeochemical cycles. It addresses interactions between organisms and resources within ecosystems, population growth and decline, biodiversity, productivity, and ecosystem disturbance. The last part of chapter 17 covers the carbon, nitrogen, phosphorus, and sulfur cycles, the role of oxygen in aquatic ecosystems, and invasive species. The final chapter in the book, chapter 18, is a synthesis chapter on biomes. It discusses factors that influence biomes and then contains a two-page spread on each major biome, with maps, globes, photographs, and other types of figures to convey where and why each biome exists. It includes a section on sustainability and ends with a synthesis that portrays biomes in the context of many topics presented in the book, including energy balances, atmospheric moisture and circulation, climate types, and soils.

TWO-PAGE SPREADS

Most of the book consists of *two-page spreads*, each of which is about one or more closely related topics. Each chapter has four main types of two-page spreads: opening, topical, connections, and investigation.

Opening Two-Page Spread

Opening spreads introduce the chapter, engaging the student by highlighting some interesting and relevant aspects and posing questions to activate prior knowledge and curiosity.

Topical Two-Page Spread

Topical spreads comprise most of the book. They convey the geographic content, help organize knowledge, describe and illustrate processes, and provide a spatial context. The first topical spread in a chapter usually includes some aspects that are familiar to most students, as a bridge or scaffold into the rest of the chapter. Each chapter has at least one two-page spread illustrating how geography impacts society and commonly another two-page spread that specifically describes how geographers study typical problems.

Connections Two-Page Spread

The next-to-last two-page spread in each chapter is a *Connections spread* designed to help students connect and integrate the various concepts from the chapter and to show how these concepts can be applied to an actual location. *Connections* are about real places that illustrate the geographic concepts and features covered in the chapter, often explicitly illustrating how we investigate a geographic problem and how geographic problems have relevance to society.

Investigation Two-Page Spread

Each chapter ends with an *Investigation* spread that is an exercise in which students apply the knowledge, skills, and approaches learned in the chapter. These exercises mostly involve virtual places that students explore and investigate to make observations and interpretations and to answer a series of geographic questions. Investigations are modeled after the types of problems geographers investigate, and they use the same kinds of data and illustrations encountered in the chapter. The Investigation includes a list of goals for the exercises and step-by-step instructions, including calculations and methods for constructing maps, graphs, and other figures. These investigations can be completed by students in class, as worksheet-based homework, or as online activities.

DIGITAL RESOURCES

McGraw-Hill offers various tools and technology products to support *Exploring Physical Geography*, 1st edition.

McGraw-Hill ConnectPlus™ (**www.mcgrawhillconnect.com**) is a web-based assignment and assessment platform that gives students the means to better connect with their coursework, with their instructors, and with the important concepts that they will need to know for success now and in the future. The following resources are available in Connect:

- Auto-graded assessments
- LearnSmart, an adaptive diagnostic and learning tool
- SmartBook™, an adaptive reading experience
- Powerful reporting against learning outcomes and level of difficulty
- The full textbook as an integrated, dynamic eBook which you can also assign
- Base Map and Google Earth exercises
- McGraw-Hill Tegrity®, which digitally records and distributes your lectures with a click of a button
- Instructor Resources such as an Instructor's Manual, Power-Point® images, and Test Banks.

With ConnectPlus, instructors can deliver assignments, quizzes, and tests online. Instructors can edit existing questions and write entirely new problems; track individual student performance—by question, by assignment, or in relation to the class overall—with detailed grade reports; integrate grade reports easily with learning management systems such as WebCT and Blackboard®; and much more. By choosing Connect®, instructors are providing their students with a powerful tool for improving academic performance and truly mastering course material. Connect allows students to practice important skills at their own pace and on their own schedule. Importantly, students' assessment results and instructors' feedback are all saved online, so students can continually review their progress and plot their course to success.

LEARNSMART®

McGraw-Hill LearnSmart® is available as an integrated feature of McGraw-Hill Connect. It is an adaptive learning system designed to help students learn faster, study more efficiently, and retain more knowledge for greater success. LearnSmart assesses a student's knowledge of course content through a series of adaptive questions. It pinpoints concepts the student does not understand and maps out a personalized study plan for success. This innovative study tool also has features that allow instructors to see exactly what students have accomplished and a built-in assessment tool for graded assignments.

Visit **www.learnsmartadvantage.com** to discover for yourself how the LearnSmart diagnostic ensures that students will connect with the content, learn more effectively, and succeed in your course.

SMARTBOOK™

McGraw-Hill Smartbook™ is the first and only adaptive reading experience available for the higher education market. Powered by an intelligent diagnostic and adaptive engine, SmartBook facilitates the reading process by identifying what content a student knows and doesn't know through adaptive assessments. As the student reads, the reading material constantly adapts to ensure that the student is focused on the content he or she needs the most to close any knowledge gaps. Visit the following site for a demonstration: **http://www.learnsmartadvantage.com**.

Tegrity® is a service that makes class time available all the time by automatically capturing every lecture in a searchable format for students to review when they study and complete assignments. With a simple one-click start and stop process, you capture all computer screens and corresponding audio. Students replay any part of any class with easy-to-use, browser-based viewing on a

PC or Mac. Educators know that the more students can see, hear, and experience class resources, the better they learn. With Tegrity Campus, students quickly recall key moments by using Tegrity Campus's unique search feature. This search helps students to efficiently find what they need, when they need it, across an entire semester of class recordings. Help turn your students' study time into learning moments immediately supported by your lecture. To learn more about Tegrity, watch a two-minute Flash demo at **http://tegritycampus.mhhe.com**.

Customizable Textbooks:

Create what you've only imagined. Introducing **McGraw-Hill Create**™ —a new, self-service website that allows you to create custom course materials—print and eBooks—by drawing upon McGraw-Hill's comprehensive, cross-disciplinary content. Add your own content quickly and easily. Tap into other rights-secured third-party sources as well. Then, arrange the content in a way that makes the most sense for your course. You can even personalize your book with your course name and information. Choose the best format for your course: color print, black-and-white print, or eBook. The eBook is now viewable on an iPad! And when you are finished customizing, you will receive a free PDF review copy in just minutes! Visit McGraw-Hill Create at **www.mcgrawhillcreate.com** today and begin building your perfect book.

CourseSmart eBook

CourseSmart is a new way for faculty to find and review eBooks. It's also a great option for students who are interested in accessing their course materials digitally and saving money. *CourseSmart* offers thousands of the most commonly adopted textbooks across hundreds of courses. It is the only place for faculty to review and compare the full text of a textbook online, providing immediate access without the environmental impact of requesting a print exam copy. At *CourseSmart*, students can save up to 50% off the cost of a print book, reduce their impact on the environment, and gain access to powerful Web tools for learning, including full text search, notes and highlighting, and e-mail tools for sharing notes between classmates.

To review complimentary copies or to purchase an eBook, go to **www.coursesmart.com**.

ACKNOWLEDGMENTS

Writing a totally new type of introductory geography textbook would not be possible without the suggestions and encouragement we received from instructors who reviewed various drafts of this book and its artwork. We are especially grateful to people who contributed entire days either reviewing the book or attending symposia to openly discuss the vision, challenges, and refinements of this kind of new approach. Our colleagues Paul Morin and Mike Kelly contributed materials in various chapters, for which we continue to be grateful.

This book contains over 2,600 figures, several times more than a typical introductory geography textbook. This massive art program required great effort and artistic abilities from the illustrators and artists who turned our vision and sketches into what truly are pieces of art. We are especially appreciative of Cindy Shaw, who was lead illustrator, art director, and a steady hand that helped guide a diverse group of authors. For many figures, she extracted data from NOAA and NASA websites and then converted the data into exquisite maps and other illustrations. Cindy also fine-tuned the authors' layouts, standardized illustrations, and prepared the final figures for printing. Chuck Carter produced many spectacular pieces of art, including virtual places featured in the chapter-ending Investigations. Susie Gillatt contributed many of her wonderful photographs of places, plants, and creatures from around the world, photographs that helped us tell the story in a visual way. She also color corrected and retouched most of the photographs in the book. We also used visually unique artwork by Daniel Miller, David Fierstein, and Susie Gillatt. Suzanne Rohli performed magic with GIS files, did the initial work on the glossary, and helped in many other ways. We were ably assisted in data compilation and other tasks by geography students Emma Harrison, Abeer Hamden, Peng Jia, and Javier Vázquez, and by Courtney Merjil. Terra Chroma, Inc., of Tucson, Arizona, supported many aspects in the development of this book, including funding parts of the extensive art program and maintenance of the ExploringPhysicalGeography.com website.

Many people went out of their way to provide us with photographs, illustrations, and advice. These helpful people included Susie Gillatt, Vladimir Romanovsky, Paul McDaniel, Lawrence McGhee, Charles Love, Cindy Shaw, Sandra Londono, Lynda Williams, Ramón Arrowsmith, John Delaney, Nancy Penrose, Dan Trimble, Bixler McClure, Michael Forster, Vince Matthews, Ron Blakey, Doug Bartlett, Ed DeWitt, Phil Christensen, Scott Johnson, Peg Owens, Emma Harrison, Skye Rodgers, Steve Semken, and David Walsh.

We used a number of data sources to create many illustrations. Reto Stöckli of the Department of Environmental Sciences at ETH Zürich and NASA Goddard produced the Blue Marble and Blue Marble Next Generation global satellite composites. We are very appreciative of the NOAA Reanalysis Site, which we used extensively, and for other sites of the USDA, NASA, USGS, NRCS, and NPS.

We have treasured our interactions with the wonderful Iowans at McGraw-Hill Higher Education, who enthusiastically supported our vision, needs, and progress. We especially thank our current and previous publishers Michelle Vogler, Ryan Blankenship, and Marge Kemp for their continued encouragement and excellent support. Jodi Rhomberg and Sandy Wille skillfully and cheerfully guided the development of the book during the entire publication process, making it all happen. Lori Hancock helped immensely with our ever-changing photographic needs. We also appreciate the support, cooperation, guidance, and enthusiasm of Thomas Timp, Marty Lange, Kurt Strand, Matt Garcia, Lisa Nicks, David Hash, Traci Andre, Tammy Ben, and many others at McGraw-Hill who worked hard to make this book a reality. Jeff Lachina and Matthew Orgovan of Lachina Publishing Services were a pleasure to work with while helping prepare the book for internal review and final printing. Kevin Campbell provided thorough copy editing and also compiled the index. Angie Sigwarth and Rose Kramer provided excellent proofreading that caught small gremlins before they escaped. Our wonderful colleague Gina Szablewski expertly directed the development of LearnSmart materials and provided general encouragement.

Finally, a project like this is truly life consuming, especially when the author team is doing the writing, illustrating, photography, near-final page layout, media development, and development of assessments, teaching ancillaries, and the instructor's website. We are extremely appreciative of the support, patience, and friendship we received from family members, friends, colleagues, and students who shared our sacrifices and successes during the creation of this new vision of a textbook. Steve Reynolds thanks the ever-cheerful, supportive, and talented Susie Gillatt; John and Kay Reynolds; and our mostly helpful book-writing companions, Widget, Jasper, and Ziggy. Julia Johnson thanks Annabelle Louise and Hazel Johnson, and the rest of her family for enthusiastic support and encouragement. Steve and Julia appreciate the support of their wonderful colleagues at ASU and elsewhere.

Robert Rohli is grateful to his wife Suzanne, a geographer herself, for her patient and unflagging assistance with so many aspects of this project. In addition, their son, Eric, and daughter, Kristen, also contributed in various direct and indirect ways. Their support and enthusiasm, and the encouragement of so many other family members and friends, particularly Bob's and Suzanne's parents, was an important motivator. Rohli also feels deep appreciation for so many dedicated mentors who stimulated his interest in physical geography while he was a student. These outstanding educators include John Arnfield, David Clawson, Carville Earle, Keith Henderson, Jay Hobgood, Merrill Johnson, Ricky Nuesslein, Kris Preston, John Rayner, Jeff Rogers, Rose Sauder, and many others. And finally, Rohli thanks the many students over the years whose interest in the world around them makes his job fun.

Peter Waylen thanks his wife, Marilyn, for her continued unstinting support and encouragement in this and all his other academic endeavors. He would also like to acknowledge geographers who have been very influential in guiding his satisfying and rewarding career, the late John Thornes, Ming-ko Woo, César Caviedes, Bill Courtwright, and Ole Humlum. He also thanks Germán Poveda for the stream of stimulating new ideas, including Daisy World with a hydrologic cycle. Peter thanks his coauthors, especially Steve Reynolds and Julia Johnson, for providing the opportunity to participate in this novel and exciting project.

Mark Francek wishes to thank his wife, Suezell, who from the onset, said, "You can do this!" despite his initial doubts about being able to find the time to complete this project. His five kids and two grandkids have also been supportive, making him smile and helping him not to take his work too seriously. Mark's academic mentors over the years, including Ray Lougeay, Lisle Mitchell, Barbara Borowiecki, and Mick Day, have instilled in him a love of field work and physical geography. He also thanks the hundreds of students he taught over the years. Their eagerness to learn has always pushed him to explore new academic horizons. Finally, Mark appreciates working with all his coauthors. He marvels at their patience, kindness, and academic pedigrees.

Cindy Shaw, lead illustrator, is grateful to John Shaw and Ryan Swain, who were of enormous help with final art-file preparation. She particularly appreciates the support of her ever-patient husband, Karl Pitts, who during the project adapted to her long working hours and a steady diet of take-out food. As a scientist, he was always interested and happy to bounce ideas around and clarify any questions. Finally, Cindy thanks all the authors for being a pleasure to work with.

All the authors are very grateful for the thousands of students who have worked with us on projects, infused our classrooms with energy and enthusiasm, and provided excellent constructive feedback about what works and what doesn't work. We wrote this book to help instructors, including us, make students' time in our classes even more interesting, exciting, and informative. Thank you all!

REVIEWERS

Special thanks and appreciation go out to all reviewers. This book was improved by many beneficial suggestions, new ideas, and invaluable advice provided by these reviewers. We appreciate all the time they devoted to reviewing manuscript chapters, attending focus groups, surveying students, and promoting this text to their colleagues.

We would like to thank the following individuals who wrote and/or reviewed learning goal–oriented content for **LearnSmart**.

Florida Atlantic University, Jessica Miles
Northern Arizona University, Sylvester Allred
Roane State Community College, Arthur C. Lee
State University of New York at Cortland, Noelle J. Relles
University of North Carolina at Chapel Hill, Trent McDowell
University of Wisconsin—Milwaukee, Gina Seegers Szablewski
University of Wisconsin—Milwaukee, Tristan J. Kloss
Elise Uphoff

Special thanks and appreciation go out to all reviewers, focus group and Symposium participants. This first edition (through several stages of manuscript development) has enjoyed many beneficial suggestions, new ideas, and invaluable advice provided by these individuals. We appreciate all the time they devoted to reviewing manuscript chapters, attending focus groups, reviewing art samples, and promoting this text to their colleagues.

PHYSICAL GEOGRAPHY REVIEWERS

Antelope Valley College, Michael W. Pesses
Arizona State University, Bohumil Svoma
Austin Peay State University, Robert A. Sirk
Ball State University, David A. Call
California State University–Los Angeles, Steve LaDochy
California State University–Sacramento, Tomas Krabacker
College of Southern Idaho, Shawn Willsey
College of Southern Nevada, Barry Perlmutter
Eastern Washington University, Richard Orndorff
Eastern Washington University, Jennifer Thomson
Florida State University, Holly M. Widen
Florida State University, Victor Mesev
Frostburg State University, Phillip P. Allen
Frostburg State University, Tracy L. Edwards
George Mason University, Patricia Boudinot
Las Positas College, Thomas Orf
Lehman College, CUNY, Stefan Becker
Long Island University, Margaret F. Boorstein
Mesa Community College, Steve Bass
Mesa Community College, Clemenc Ligocki
Metro State, Kenneth Engelbrecht
Metro State, Jon Van de Grift
Minnesota State University, Forrest D. Wilkerson
Monroe Community College, SUNY, Jonathon Little
Moorpark College, Michael T. Walegur
Normandale Community College, Dave Berner
Northern Illinois University, David Goldblum
Oklahoma State University, Jianjun Ge
Oregon State University, Roy Haggerty
Pasadena City College, James R. Powers
Rhodes College, David Shankman
Samford University, Jennifer Rahn
San Francisco State University, Barbara A. Holzman
South Dakota State University, Trisha Jackson

South Dakota State University, Jim Peterson

Southern Illinois University–Edwardsville, Michael J. Grossman

Southern Utah University, Paul R. Larson

State University of New York at New Paltz, Ronald G. Knapp

Texas State University–San Marcos, David R. Butler

The University of Memphis, Hsiang-te Kung

Towson University, Kent Barnes

United States Military Academy, Peter Siska

University of Calgary, Lawrence Nkemdirim

University of Cincinnati, Teri Jacobs

University of Colorado–Boulder, Jake Haugland

University of Colorado–Colorado Springs, Steve Jennings

University of Georgia, Andrew Grundstein

University of Missouri, C. Mark Cowell

University of Nevada–Reno, Franco Biondi

University of North Carolina–Charlotte, William Garcia

University of Oklahoma, Scott Greene

University of Saskatchewan, Dirk de Boer

University of Southern Mississippi, David Harms Holt

University of Tennessee, Derek J. Martin

University of Tennessee–Knoxville, Julie Y. McKnight

University of Wisconsin–Eau Claire, Christina M. Hupy

University of Wisconsin–Eau Claire, Joseph P. Hupy

University of Wisconsin–Eau Claire, Garry Leonard Running

University of North Dakota, Paul Todhunter

Weber State University, Eric C. Ewert

PHYSICAL GEOGRAPHY FOCUS GROUP AND SYMPOSIUM PARTICIPANTS

Ball State University, Petra Zimmermann

Blinn College, Rhonda Reagan

California State University–Los Angeles, Steve LaDochy

Georgia State University, Leslie Edwards

Indiana Purdue University–Indianapolis (IUPUI), Andrew Baker

Kansas State University, Doug Goodin

Mesa Community College, Steven Bass

Minnesota State University, Ginger L. Schmid

Northern Illinois University, Lesley Rigg

Northern Illinois University, Mike Konen

South Dakota State University, Bruce V. Millett

Texas A&M University, Steven Quiring

University of Alabama, Amanda Epsy-Brown

University of Colorado–Boulder, Peter Blanken

University of North Carolina–Greensboro, Michael Lewis

University of Oklahoma, Scott Greene

University of Wisconsin–Oshkosh, Stefan Becker

ABOUT THE AUTHORS

STEPHEN J. REYNOLDS

Stephen J. Reynolds received an undergraduate degree from the University of Texas at El Paso, and M.S. and Ph.D. degrees in geosciences from the University of Arizona. He then spent ten years directing the geologic framework and mapping program of the Arizona Geological Survey, completing a new *Geologic Map of Arizona*. Steve currently is a professor in the School of Earth and Space Exploration at Arizona State University, where he has taught various courses about regional geology, earth resources, evolution of landscapes, field studies, and teaching methods. He was president of the Arizona Geological Society and has authored or edited nearly 200 maps, articles, and reports on the evolution of Western North America. He also coauthored several widely used textbooks, including the award-winning *Exploring Geology*. His current science research focuses on regional geology, geomorphology, and resources of the Southwest, including northern Mexico. For a dozen years, he has done science-education research on student learning in college science courses, especially the role of visualization. He was the first geoscientist with his own eye-tracking laboratory, where he and his students have researched student learning, including the role of textbooks and other educational materials. Steve is known for innovative teaching methods, has received numerous teaching awards, and has an award-winning website. As a National Association of Geoscience Teachers (NAGT) distinguished speaker, he traveled across the country presenting talks and workshops on how to infuse active learning and inquiry into large introductory geology classes. He is commonly an invited speaker to national workshops and symposia on active learning, visualization, and teaching.

ROBERT V. ROHLI

Robert Rohli received a B.A. in geography from the University of New Orleans, an M.S. degree in atmospheric sciences from The Ohio State University, and a Ph.D. in geography from Louisiana State University (LSU). He is currently in his 14th year at LSU, where his time is divided between faculty director of the LSU Residential Colleges Program and professor of geography. Previously, he served as assistant professor of geography at Kent State University (KSU) and as regional climatologist at the Southern Regional Climate Center. His teaching and research interests are in physical geography, particularly synoptic and applied meteorology/climatology, atmospheric circulation variability, and hydroclimatology. He has taught Physical Geography, Climatology, Meteorology, Physical Climatology, World Climates, Methods in Synoptic Climatology, Applied Meteorology, Analysis of Spatial Data, Water Resources Geography, and others. Major themes in his teaching include the systems approach to physical geography, collaboration among students from different disciplines in producing group research projects, and development of applied problem-solving skills. He has been an active supporter of undergraduate education initiatives, including living-learning communities, the Communication across the Curriculum program, improved teaching assessment methods, and outreach activities—especially those that promote geography. He has published more than 35 refereed research articles, mostly on topics related to synoptic or applied climatology, and over 20 loosely refereed manuscripts, encyclopedia articles, proceedings papers, and technical reports. He has also coauthored *Climatology*, a widely used textbook, and *Louisiana Weather and Climate*.

JULIA K. JOHNSON

Julia K. Johnson is currently a full-time faculty member in the School of Earth and Space Exploration at Arizona State University. Her M.S. and Ph.D. research involved structural geology and geoscience education research. She teaches introductory geoscience to more than 1,500 students per year, both online and in person, and supervises the associated in-person and online labs. She also coordinates the introductory geoscience teaching efforts of the School of Earth and Space Exploration, helping other instructors incorporate active learning and inquiry into large lecture classes. Julia coordinated an innovative project focused on redesigning introductory geology classes so that they incorporated more online content and asynchronous learning. This project was very successful in improving student performance, mostly due to the widespread implementation of concept sketches and partly due to Julia's approach of decoupling multiple-choice questions and concept-sketch questions during exams and other assessments. Julia is recognized as one of the best science teachers at ASU and has received student-nominated teaching awards and very high teaching evaluations in spite of her challenging classes. Her efforts have dramatically increased enrollments. She coauthored the widely used *Exploring Geology* textbook and publications on geology and science-education research, including an article in the *Journal of Geoscience Education* on concept sketches. She is the lead author of *Observing and Interpreting Geology*, an innovative laboratory manual in which all learning is built around a virtual world. She also developed a number of websites used by students around the world, including the *Visualizing Topography* and *Biosphere 3D* websites.

PETER R. WAYLEN

Peter Waylen is Robin and Jean Gibson Professor of Geography at the University of Florida. He holds a B.Sc. in geography from the London School of Economics, England, and a Ph.D. from McMaster University, Canada. He has also served as assistant professor at the University of Saskatchewan, visiting associate professor at the University of Waterloo, Canada, Hartley Visiting Research Fellow at the University of Southampton, England, and visiting scholar in the Department of Engineering Hydrology, University College Galway, Ireland. His teaching and research interests are in the fields of hydrology and climatology, particularly the temporal and spatial variability of risks of such hazards as floods, droughts, freezes, and heat waves, and the way in which these vary in the long run, driven by global-scale phenomena like ENSO. He has worked throughout Anglo- and Latin America, and several parts of Africa. He teaches Introductory Physical Geography, Principles of Geographic Hydrology, and Models in Hydrology, and was selected University of Florida Teacher of the Year in 2002. His research is principally interdisciplinary and collaborative with colleagues and students. It has been supported variously by the Natural Research Council of Canada, NSF, NOAA, NASA, and the Inter-American Institute for Global Change Research. Results appear in over 70 geography, hydrology, and climatology refereed outlets and book chapters. He served as chair of the Department of Geography from 2004–11.

MARK A. FRANCEK

Mark Francek is a geography professor at Central Michigan University (CMU). Before coming to CMU in 1988, Francek earned his doctorate in geography from the University of Wisconsin-Milwaukee, his master's in geography from the University of South Carolina, and his bachelor's degree in geography and psychology from the State University College at Geneseo, New York. He has teaching and research interests in earth science education, physical geography, and soil science. Mark has pedaled twice across America and teaches biking geography field classes in and around the Great Lakes region and the Appalachian Mountains. He has authored and coauthored more than 30 scholarly papers, funded in part by the NSF and the State of Michigan, and has presented his research at numerous national and state conferences. At CMU, Francek has served as acting director of the Environmental Studies Program and director of the Science and Technology Residential College. He has received state and national teaching awards, including the CMU Teaching Excellence Award, the Carnegie Foundation for the Advancement of Teaching Michigan Professor of the Year, the Presidents Council State Universities of Michigan Distinguished Professor of the Year, the National Council for Geographic Education Distinguished Teaching Award, and the National Earth Science Teachers of America Service Award. His "Earth Science Sites of the Week" Listserv, which highlights the best earth-science websites and animations, is distributed through several Listservs and reaches thousands of K–16 educators from around the world.

Illustrators and Artists

CINDY SHAW

Cindy Shaw holds a B.A. in zoology from the University of Hawaii–Manoa as well as a master's in education from Washington State University, where she researched the use of guided illustration as a teaching and learning tool in the science classroom. Now focusing on earth science, mapping, and coral reef ecology, she writes and illustrates for textbooks and museums, and develops ancillary educational materials through her business, Aurelia Press. Her kids' novel, *Grouper Moon*, is used in many U.S. and Caribbean science classrooms and is making a real impact on shaping kids' attitudes toward fisheries conservation. Currently landlocked in Richland, Washington, Cindy escapes whenever possible to travel, hike, and dive the reefs to field-sketch and do reference photography for her projects.

CHUCK CARTER

Chuck Carter has worked in the artistic end of the science and entertainment industries for more than 30 years. He helped create the popular computer game *Myst* in 1992. Chuck worked on more than two dozen video games as an artist, art director, computer graphics supervisor, and group manager. He has a decades-long relationship with *National Geographic* as an illustrator and helping launch National Geographic Online. Carter worked as a digital matte painter for science fiction shows like Babylon5, Crusade, and Mortal Kombat, as well as art and animation for motion rides like Disney's Mission to Mars and Paramount's Star Trek: the Experience. His illustration clients include *Wired* magazine, *Scientific American*, and numerous book publishers. He is co-founder of Eagre Interactive Inc.

SUSIE GILLATT

Susie Gillatt grew up in Tucson, Arizona, where she received a bachelor of arts degree from the University of Arizona. She has worked as a photographer and in different capacities in the field of video production. She is president of Terra Chroma, Inc., a multimedia studio. Initially specializing in the production of educational videos, she now focuses on scientific illustration and photo preparation for academic books and journals. Many of the photographs in this book were contributed by Susie from her travels to experience different landscapes, ecosystems, and cultures around the world. For her own art, she especially enjoys combining photography with digital painting and exploring the world of natural patterns. Her award-winning art has been displayed in galleries in Arizona, Colorado, and Texas.

exploring physical
geography

The Nature of Physical Geography

THE EARTH HAS A WEALTH of intriguing features, from dramatic mountains to intricate coastlines and deep ocean trenches, from lush, beautiful valleys to huge areas of sparsely vegetated sand dunes. Above the surface is an active, ever-changing atmosphere with clouds, storms, and variable winds. Occupying all these environments is life. In this chapter and book, we examine the main concepts of physical geography, along with the tools and methods that physical geographers use to study the landscapes, oceans, climate, weather, and ecology of Earth.

The large globe spanning these two pages is a computer-generated representation of Earth, using data collected by several satellites. On land, brown colors depict areas of rock, sand, and soil, whereas green areas have a more dense covering of trees, bushes, and other vegetation. Oceans and lakes are colored blue, with greenish blue showing places where the water is shallow or where it contains mud derived from the land. Superimposed on Earth's surface are light-colored clouds observed by a different satellite, one designed to observe weather systems.

What are all the things you can observe from this portrait of our planet? What questions arise from your observations?

Most questions that arise from observing this globe are within the domain of *physical geography*. Physical geography deals with the landforms and processes on Earth's surface, the character and processes in oceans and other bodies of water, atmospheric processes that cause weather and climate, and how these various aspects affect life, and much more.

01.00.a2 Santorini, Greece

Natural hazards, including volcanic eruptions and earthquakes, are a major concern in many parts of the world. In the Greek Island of Santorini (◄), people live on the remains of a large volcano that was mostly destroyed in a huge eruption 3,600 years ago, an eruption that probably gave rise to the story of Atlantis.

What occurs during a volcanic eruption? Do all volcanoes erupt in the same way, and how can we recognize a volcano in the landscape?

01.00.a3 Morocco

The Sahara Desert, on the opposite side of the Mediterranean Sea from Greece, has a very different climate. Here, a very dry environment results, forming huge areas covered by sand dunes (▲) with almost no vegetation.

What do the features of the landscape—the landforms—tell us about the surface processes that are forming and affecting the scenery? What causes different regions to have different climates, some that are hot and dry, and others that are cold and wet? Is the climate of the Sahara somehow related to the relative lack of clouds over this area, as shown on the globe?

TOPICS IN THIS CHAPTER

01.00.a4 Tibet

Water is the most important resource on the planet, and Earth's temperatures allow water to occur in three states of matter — solid, liquid, and vapor. Examine this photograph (◄) and identify all the ways in which water is expressed on the surface and in the atmosphere. Is some water likely present but not visible? Geographers are concerned with where resources are, what causes a resource to be where it is, how to locate more of this resource, and how to reconcile the inevitable economic, environmental, and cultural trade-offs involved in using a resource.

How does water occur in the atmosphere, how is its presence expressed, and what is its role in severe weather? How does water occur and move on Earth's surface and what landforms result from running water?

01.00.a5 Indonesia

Oceans cover three-fourths of Earth's surface. Ocean temperatures, currents, and salinity all play a major role in global weather, climate, and the livability of places, even for those far from the coast. The oceans and nearby lands (▲) represent important habitats for plants and animals, which can be greatly impacted by human activities.

How do satellites help us measure the temperature, salinity, and motion of the oceans, and how do changes in any of these factors affect plants and animals that live in or near the sea?

01.00.a1

The Ancient and Modern Discipline of Geography

Geographers seek to understand the Earth. They do this by formulating important and testable questions about the Earth, employing principles from both the natural and social sciences. Geographers use these principles to portray features of the Earth using maps and technologically intensive tools and techniques that are distinctly geographical. Geographers synthesize the diverse information revealed by these tools to investigate the interface between the natural and human environment. The study of the spatial distribution of natural features and processes occurring near Earth's surface, especially as they affect, and are affected by, humans, is physical geography.

The ancient discipline of geography is especially relevant in our modern world, partly because of the increasing recognition that many problems confronting society involve complex interactions between natural and human dimensions. Such problems include the complex spatial distribution and depletion of natural resources; contamination of air, water, and soils; susceptibility of areas to natural disasters, including earthquakes, volcanic eruptions, flooding, and landslides; formation of and damage caused by hurricanes, tornadoes, and other severe weather; the current and future challenges of global environmental change, and the environmental implications of globalization. The topics and questions introduced on these pages provide a small sample of the aspects investigated by physical geographers and are discussed more fully in the rest of the book. We hope you enjoy the journey learning about our fascinating planet.

1.0

1.1 What Is Physical Geography?

PHYSICAL GEOGRAPHY IS THE STUDY of spatial distributions of phenomena across the landscape, processes that created and changed those distributions, and implications for those distributions on people. Geography is both a natural and social science. Geographers think broadly, emphasizing interconnections and complex issues, solving complicated problems such as resource management, environmental impact assessment, disease diffusion, and urban planning. Although many such occupations do not have the title of *geographer*, they require a geographic perspective. Let's have a closer look at what the geographic perspective entails.

A What Approach Do Geographers Use to Investigate Important Issues?

Geographers approach problems from different perspectives than other natural and social scientists. Specifically, geographers think *spatially*, meaning they emphasize the setting, such as location, in addressing problems, and *holistically*, integrating ideas from a wide variety of the natural and social sciences. In many ways, it is not *what* is studied that makes it geography, but instead *how* it is studied. The decision of whether to drill for oil in Alaska's Arctic National Wildlife Refuge (ANWR) is a complicated issue that can be best understood using the geographic approach.

1. This figure (▶) shows a three-dimensional perspective of the central part of ANWR, looking south with the ice-covered Arctic Ocean in the foreground. ANWR is well known for its abundant caribou and other Arctic animals. Before reading on, examine this scene and think about all the information you would need if you wanted to understand how drilling for oil and gas might impact the caribou.

2. To understand this issue, you might ask a series of questions. Where do the caribou live? Since they migrate seasonally, where are they at different times of the year? What do they eat, where are these foods most abundant, and what factors control these abundances? Where is water available, and how much rain and snow do different parts of the region receive? Is the precipitation consistent from year to year? When is the mating season, and where do the mothers raise their young?

01.01.a1

3. You could also ask questions about the subsurface oil reserves. Where is the oil located, and what types of facilities will be required to extract and transport the oil? How much land will be disturbed by such activities, and how will this affect the caribou?

4. The issues of ANWR nicely illustrate why we would use a geographic approach. Most of the questions we asked here had a *spatial component*, as indicated by the word "where" and could be best answered with some type of map. The questions also have an explicit or implicit societal component, such as how development could affect the traditional way of life of the native people of the region.

01.01.a2

01.01.a3 ANWR, AK

5. The *spatial perspective* allows us to compare the locations of the physical, environmental, economic, political, and cultural attributes of the issue. On this map (◀), ANWR is the large area outlined in orange. Its size is deceptive since Alaska is huge (by far the largest state in the U.S.). For comparison, ANWR is only slightly smaller than the state of South Carolina.

6. Directly to the west of ANWR is the Prudhoe Bay oil field, the largest oil field in North America. Not all of ANWR is likely to contain oil and natural gas, and an assessment of the oil resources by the U.S. Geological Survey (USGS) identified the most favorable area as being near the coast. To properly consider the question about oil drilling, we would want to know where this favorable area is, how much land will be disturbed by drilling and subsequent activities, when these disturbances will occur, and how these compare with the location of caribou at different times of the year, especially where they feed, mate, and deliver their young.

7. The *holistic perspective* allows us to examine the interplay between the environment and the aesthetic, economic, political, and cultural attributes of the problem. Most of ANWR is a beautiful wilderness area (▲), as well as being home to caribou, various plants and animals, and native people.

B How Does Geography Influence Our Lives?

Observe this photograph, which shows a number of different features, including clouds, snowy mountains, slopes, and a grassy field with horses and cows (the small, dark spots). For each feature you recognize, think about what is there, what its distribution is, and what processes might be occurring. Then, think about how these factors influence the life of the animals and how it would influence you if this were your home.

01.01.b1 Henry Mtns., UT

The snow-covered mountains, partially covered with clouds, indicate the presence of water, an essential ingredient for life. The mountains have a major influence on water in this scene. Melted snow flows downhill toward the lowlands, to the horses and cows. The elevation and shape of the land influence the spatial distribution and type of precipitation (rain, snow, and hail) and the pattern of streams that develop to drain water off the land.

The horses and cows roam on a flat, grassy pasture, avoiding slopes that are steep or barren of vegetation. The steepness of slopes reflects the strength of the rocks and soils, and the flat pasture resulted from loose sand and other materials that were laid down during flooding along a desert stream. The distribution of vegetation is controlled by steepness of slopes, types of soils and other material, water content of the soil, air temperatures, and many other factors, all of which are part of physical geography. The combined effect of such factors in turn affect, and are affected by, the human settlements in the area to make every place, including this one, distinctive and unique.

01.01.b2

300 m

A better view of the spatial distribution of the green pasture is provided by this aerial photograph (a photograph taken from the air, like from a plane). This view of the pasture and adjacent areas reveals the shape of the pasture, and we could measure its length, width, and area. Such measurements would help us decide how many horses and cows the land could support.

01.01.b3

300 m

Geographers calculate various measures of the landscape, like the steepness of slopes, and then overlay this information on the original map or image. In the figure above, red shading shows the steepest slopes, along and below the pinkish cliff. Yellow and green indicate less steep slopes, and relatively flat areas are unshaded. Such a map would help us decide which areas could be new pastures.

01.01.b4

5 km

This image shows the shape of the land across the region, including the mountains (the pasture is on the left part of the map). Colors indicate the average amount of precipitation with green showing the highest amounts. As we might predict, the mountains receive more rain and snow.

Before You Leave This Page Be Able To

✓ Describe the geographic approach.

✓ Describe some examples of information used by physical geographers and how these types of information could influence our lives.

1.1

1.2 How Do We Investigate Geographic Questions?

PHYSICAL GEOGRAPHERS STUDY DIVERSE PROBLEMS, ranging from weather systems and climate change to ocean currents and landscape evolution. The types of data required to investigate each of these problems are equally diverse, but most geographers try to approach the problem in a similar, objective way, guided by spatial information, and relying on various geographic tools. Geography utilizes approaches from the natural and social sciences, blending them together in a geographic approach. Like other scientists, geographers pose questions about natural phenomena and their implications, propose a possible explanation (hypothesis) that can be tested, make predictions from this hypothesis, and collect data needed to critically evaluate whether the hypothesis passes the tests.

A How Do Geographers Approach Problems?

Geographers ask questions like the following:

- Where is it?
- Why is it where it is?
- How did it get where it is?
- Why does it matter where it is?
- How does "where it is" influence where other things are and why they are there?

The conceptual basis of these questions lies in the notion that the *location* of something affects, and is a product of, other features or processes in both the natural and human environment, and of interactions between the natural and human environments. Natural and human phenomena are constantly changing and constantly impacting other features in new ways, influencing aspects like site selection and risk of natural hazards. To address such complex issues, we use a variety of tools and methods, such as maps, computer-simulation models, aerial photographs, satellite imagery, statistical methods, and historical records. The figure to the right illustrates some aspects to consider.

01.02.a1

QUESTION	CONCEPT	METHOD
Where?	Location	**Observation:** In the field, maps, satellite images, aerial photographs
How did it get there in its form?	Natural environment / Human environment → Changes over time; movement, diffusion	**Analysis:** Field work, lab methods, computer modeling, quantitative analysis, surveys
What is the significance?	Impacts on other aspects of the natural and human environment	**Interpretation and Synthesis:** Reach conclusions based on data from available tools

B What Is the Difference Between Qualitative and Quantitative Data?

Geographers approach problems in many ways, asking questions about Earth processes and collecting data that help answer these questions. Some questions can be answered with qualitative data, but others require quantitative data, which are numeric and are typically visualized and analyzed using data tables, calculations, equations, and graphs.

01.02.b1 Augustine Island, AK

01.02.b2 Augustine Island, AK

01.02.b3 Augustine Island, AK

When Augustine volcano in Alaska erupts, we can make various types of observations and measurements. Some observations are *qualitative*, like descriptions, and others are measurements that are *quantitative*. Both types of data are essential for documenting natural phenomena.

Qualitative data include descriptive words, labels, sketches, or other images. We can describe this picture of Augustine volcano with phrases like "contains large, angular fragments," "releases steam," or "the slopes seem steep and unstable." Such phrases can convey important information about the site.

Quantitative data involve numbers that represent measurements. Most result from scientific instruments, such as this thermal camera that records temperatures on the volcano, or with measuring devices like a compass. We could also collect quantitative measurements about gases released into the air.

C   How Do We Test Alternative Explanations?

Science proceeds as scientists explore the unknown—making observations and then systematically investigating questions that arise from observations that are puzzling or unexpected. Often, we try to develop several possible explanations and then devise ways to test each one. The normal steps in this *scientific method* are illustrated below, using an investigation of groundwater contaminated by gasoline.

Steps in the Investigation

Observations

Questions Derived from Observations

Proposed Explanations and Predictions from Each Explanation

4. One explanation is that the buried tank is the source of contamination.

5. Another explanation is that the buried tank is not the source of the contamination. Instead, the source is somewhere else, and contamination flowed into the area.

Results of Investigation

Conclusions

01.02.c1

1. Someone makes the *observation* that groundwater from a local well near an old buried gasoline tank contains gasoline. The first step in any investigation is to make observations, recognize a problem, and state the problem clearly and succinctly. Stating the problem as simply as possible simplifies it into a more manageable form and helps focus our thinking on its most important aspects.

2. The observation leads to a *question*—Did the gasoline in the groundwater come from a leak in the buried tank? Questions may be about what is happening currently, what happened in the past, or, in this case, who or what caused a problem.

3. Scientists often propose several explanations, referred to as *hypotheses*, to explain what they observe. A hypothesis is a causal explanation that can be tested, either by conducting additional investigations or by examining data that already exist.

---

6. We develop *predictions* for each explanation. A prediction for the explanation in number 4 might be that the tank has some kind of leak and should be surrounded by gasoline. Also, if the explanation in number 4 is true, the type of gasoline in the tank should be the same as in the groundwater. Next, we plan some way to *test* the predictions, such as by inspecting the tank or analyzing the gasoline in the tank and groundwater.

7. To study this problem, an early step is to compile all the necessary data. This might include maps showing the location of water wells, locations of gas stations and other possible sources of gasoline, and the direction of groundwater flow. In our case, investigation discovered no holes in the tank or any gasoline in the soil around the tank. Records show that the tank held leaded gasoline, but gasoline in the groundwater is unleaded. We compare the results of any investigation with the predictions to determine which possible explanation is most consistent with the new data.

8. Data collected during the investigation support the conclusion that the buried tank is not the source of contamination. Any explanation that is inconsistent with data is probably incorrect, so we pursue other explanations. In this example, a nearby underground pipeline may be the source of the gasoline. We can devise ways to evaluate this new hypothesis by investigating the pipeline. We also can revisit the previously rejected hypothesis if we discover a new way in which it might explain the data.

9. When approaching such problems, the goal is to collect data, assemble information, and draw conclusions without letting our personal bias interfere with carrying out good science. We want to reach the explanation that best explains all the data. Few things are ever "proved" in science, some can be "disproved," but generally we are left to weigh the pros and cons of several still-viable explanations. We choose the one that, based on the data, is most likely to be correct.

Before You Leave This Page Be Able To

☑ Summarize some of the aspects commonly considered using a geographic approach.

☑ Explain the difference between qualitative and quantitative data, providing examples.

☑ Explain the logical scientific steps taken to critically evaluate an explanation.

1.3 How Do Natural Systems Operate?

EARTH HAS A NUMBER OF SYSTEMS in which matter and energy are moved or transformed. These involve processes of the solid Earth, water in all its forms, the structure and motion of the atmosphere, and how these three domains (Earth, water, and air) influence life. Such systems are *dynamic*, responding to any changes in conditions, whether those changes arise internally *within* the system or are imposed externally, from *outside* the system.

A What Are the Four Spheres of Earth?

Earth consists of four overlapping spheres—the atmosphere, biosphere, hydrosphere, and lithosphere—each of which interacts with the other three spheres. The atmosphere is mostly gas, but includes liquids (e.g., water drops) and solids (e.g., ice and dust). The hydrosphere represents Earth's water, and the lithosphere is the solid Earth. The biosphere includes all the places where there is life—in the atmosphere, on and beneath the land, and on and within the oceans.

01.03.a1

The *atmosphere* is a mix of mostly nitrogen and oxygen gas that surrounds Earth's surface, gradually diminishing in concentration out to a distance of approximately 100 km, the approximate edge of outer space. In addition to gas, the atmosphere includes clouds, precipitation, and particles such as dust and volcanic ash. The atmosphere is approximately 78% nitrogen, 21% oxygen, less than 1% argon, and smaller amounts of carbon dioxide and other gases. It has a variable amount of water vapor, averaging about 1%.

The *biosphere* includes all types of life, including humans, and all of the places it can exist on, above, and below Earth's surface. In addition to the abundant life on Earth's surface, the biosphere extends about 10 km up into the atmosphere, to the bottom of the deepest oceans, and downward into the cracks and tiny spaces in the subsurface. In addition to visible plants and animals, Earth has a large population of diverse microorganisms.

The *hydrosphere* is water in oceans, glaciers, lakes, streams, wetlands, groundwater, moisture in soil, and clouds. Over 96% of water on Earth is saltwater in the oceans, and most fresh water is in ice caps, glaciers, and groundwater, not in lakes and rivers.

The *lithosphere* refers generally to the solid upper part of the Earth, including Earth's crust and uppermost mantle. Water, air, and life extend down into the lithosphere, so the boundary between the solid Earth and other spheres is not distinct, and the four spheres overlap.

B What Are Open and Closed Systems?

Many aspects of Earth can be thought of as a system—a collection of matter, energy, and processes that are somehow related and interconnected. For example, an air-conditioning system consists of some mechanical apparatus to cool the air, ducts to carry the cool air from one place to another, a fan to move the air, and a power source. There are two main types of systems: *open systems* and *closed systems*.

An *open system* allows matter and energy to move into and out of the system. A tree, like these aspen (◄), is an open system, taking in water and nutrients from the soil, extracting carbon dioxide from the air to make the carbon-rich wood and leaves, and expelling oxygen as a by-product of photosynthesis, fueled by externally derived energy from the Sun.

A *closed system* does not exchange matter, or perhaps even energy, with its surroundings. The Earth as a whole (►) is fundamentally a closed system with regard to matter, except for the escape of some light gases into space, the arrival of occasional meteorites, and the exit and return of spacecraft and astronauts.

01.03.b1

01.03.b2

C How Do Earth Systems Operate?

Systems consist of matter and energy, and they respond to internally or externally caused changes in matter and energy, as a tree responds to a decrease in rain (matter) or colder temperatures during the winter (energy). Systems can respond to such changes in various ways, either reinforcing the change or counteracting the change.

System Inputs and Responses

1. One of Earth's critical systems involves the interactions between ice, surface water, and atmospheric water. This complex system, greatly simplified here (▶), remains one of the main challenges for computer models attempting to analyze the causes and possible consequences of climate change.

01.03.c1

2. Liquid water on the surface evaporates (represented by the upward-directed blue arrows), becoming water vapor in the atmosphere. If there is enough water vapor, small airborne droplets of water accumulate, forming these low-level clouds.

3. Under the right conditions, the water freezes, becoming snowflakes or hail, which can fall to the ground. Over the centuries, if snow accumulates faster than it melts, the snow becomes thick and compressed into ice, as in glaciers.

4. The water molecules in snow and ice can return directly to the atmosphere via several processes.

5. If temperatures are warm enough, snow and ice can melt, releasing liquid water that can accumulate in streams and flow into the ocean or other bodies of surface water. Alternatively, the meltwater can evaporate back into the atmosphere. Melting also occurs when icebergs break off from the glacier.

6. The movement of matter and energy carried in the various forms of water is an example of a *dynamic system*—a system in which matter, energy, or both, are constantly changing their position, amounts, or form.

Feedbacks

7. The system can respond to changes in various ways, which can either reinforce the effect, causing the overall changes to be more severe, or partially or completely counteract the effect, causing changes to be less severe. Such reinforcements or inhibitors are called *feedbacks*.

8. In our example, sunlight shines on the ice and water. The ice is relatively smooth and light-colored, reflecting much of the Sun's energy upward, into the atmosphere or into space. In contrast, the water is darker and absorbs more of the Sun's energy, which warms the water.

01.03.c2

9. If the amount of solar energy reaching the surface, or trapped near the surface, increases, for whatever reason, this may cause more melting of the ice. As the front of the ice melts back, it exposes more dark water, which absorbs more heat and causes even more warming of the region. In this way, an initial change (warming) triggers a response that causes even more of that change (more warming). Such a reinforcing result is called a *positive feedback*.

10. The warming of the water results in more evaporation, moving water from the surface to the atmosphere, which in turn may result in more clouds. Low-level clouds are highly reflective, so as cloud cover increases they intercept more sunlight, leading to less warming. This type of response does not reinforce the change but instead dampens it and diminishes its overall effect. This dampening and resultant counteraction is called a *negative feedback*.

11. As this overly simplified example illustrates, a change in a system can be reinforced by positive feedbacks or stifled by negative ones. Both types of feedbacks are likely and often occur at the same time, each nudging the system toward opposite behaviors (e.g., overall warming or overall cooling). Feedbacks can leave the system largely unchanged, or the combined impact of positive and negative feedbacks can lead to a stable but gradually changing state, a condition called *dynamic equilibrium*.

Before You Leave This Page Be Able To

☑ Describe Earth's four spheres.

☑ Explain what is meant by open and closed systems.

☑ Sketch and explain examples of positive and negative feedbacks.

1.3

1.4 What Are Some Important Earth Cycles?

MATTER AND ENERGY MOVE within and between each of the four spheres. A fundamental principle of all natural sciences is that energy and matter can be neither created nor destroyed, but only transferred from one form to another—the *First Law of Thermodynamics*. A second principle is that energy and matter tend to become dispersed into a more uniform spatial distribution—the *Second Law of Thermodynamics*. As a result, matter and energy are stored, moved, dispersed, and concentrated as part of natural *cycles*, where material and energy move back and forth among various sites within the four spheres.

A What Is Cycled and Moved in the Atmosphere?

Atmospheric processes involve the redistribution of *energy* and *matter* from one part of the atmosphere to another. Moving air masses have *momentum*, which can be transferred from one object to another.

Energy

01.04.a1

Storage and transfer of energy are the drivers of Earth's climate and weather. Energy can be moved from one part of the atmosphere to another, such as by air currents associated with storms (▲). Also, energy is released or extracted from the local environment when water changes from one state of matter to another, such as from a liquid to a gas.

Matter

01.04.a2

Water in all of its forms, along with other matter, moves globally, tending to disperse, but other factors prevent an even spatial distribution. As a result, some regions like here in the western U.S. are more humid and cloudy than others, as shown in this satellite image (▲) of water vapor (blue is more, brown is less). Also, water cycles between vapor, liquid, and solid states.

Momentum

01.04.a3

Moving air masses have mass and velocity, so they have *momentum*, which is defined as mass times velocity. A dense, fast-moving dust storm has more momentum than a gentle breeze in dust-free air. Winds near the surface are slowed by interactions with trees, hills, buildings, etc. Winds aloft are faster (▲) and can transfer their momentum downward.

B How Are Matter and Energy Moved in the Hydrosphere?

Many processes in the Earth occur as part of a cycle, a term that describes the movement of matter and energy between different sites in Earth's surface, subsurface, and atmosphere. The most important of these is the *hydrologic cycle*, which involves local-to-global-scale storage and circulation of water and associated energy near Earth's surface.

1. Water vapor in the atmosphere can form drops, which can remain suspended in the air as clouds or can fall to the ground as *precipitation*. Once on the surface, water can coat rocks and soils, and it causes these solids to decompose and erode.

2. When winter snows don't melt completely, as is common at higher elevations and at polar latitudes, ice accumulates in glaciers, which are huge, flowing fields of ice. Glaciers transport sediment and carve the underlying landscape.

3. Moving water on the surface can flow downhill in *streams*, encountering obstacles, like solid rock and loose debris. The flowing water and the material it carries breaks these obstacles apart and picks up and transports the pieces. Flowing water is the most important agent for sculpting Earth.

6. Water evaporates from the oceans, surface waters on land, and from soils and plants, returning to the atmosphere and completing the hydrologic cycle. Phase changes between a solid, liquid, and gas involve energy transformations.

Precipitation

Streams

Evaporation

Groundwater Flow

Ocean Currents

01.04.b1

5. Water can sink into the ground and travel through cracks and other empty spaces in rocks and soils, becoming *groundwater*. Groundwater can react chemically with rocks through which it flows, dissolving or depositing material. It typically flows toward lower areas, where it may emerge back on Earth's surface as springs.

4. The uppermost part of oceans is constantly in motion, partly due to friction (and transfer of momentum) between winds in the atmosphere and the surface of the oceans. Winds in the oceans cause waves that erode and shape shorelines.

How Does the Rock Cycle Affect Materials of the Lithosphere?

Matter and energy are also stored and moved on Earth's surface and subsurface. The *rock cycle*, summarized below and discussed in more detail later, describes the movement of matter and energy on and below Earth's surface at timescales from seconds to billions of years, involving such processes as erosion, burial, melting, and uplift of mountains.

1. *Weathering:* Rock is broken apart or can be altered by chemical reactions when exposed to sunlight, rain, wind, plants, and animals. This weathering creates loose pieces of rock called *sediment*.

8. *Uplift:* Rock may be uplifted back to the surface where it is again exposed to weathering.

2. *Erosion and Transport:* Sediment is then stripped away by *erosion*, and moved by gravity, glaciers, flowing water, or wind.

3. *Deposition:* After transport, the sediment is laid down, or deposited when it reaches the sea or at any point along the way, such as beside the stream.

4. *Burial and Formation of Rock:* Sediment is eventually buried, compacted by the weight of overlying sediment, and perhaps cemented together by chemicals in the water to form a harder rock.

01.04.c1

7. *Solidification:* As magma cools, either at depth or after being erupted onto the surface, it begins to crystallize and solidify into rock.

6. *Melting:* A rock exposed to high temperatures may melt and become molten, forming magma.

5. *Deformation and Related Processes:* Rock can be subjected to strong forces that squeeze, bend, and break the rock, causing it to deform. The rock might also be heated and deformed so much that it is changed into a new kind of rock.

D **What Cycles and Processes Are Important in the Biosphere?**

The biosphere includes life and all of the places it exists. It overlaps the atmosphere, hydrosphere, and lithosphere, extending from more than 10 km above the surface to more than 10 km beneath sea level, both on the seafloor and within Earth's subsurface. Life interacts with the other three spheres, forming a number of important cycles, several of which are described later.

Plants exchange gases with the atmosphere. They extract carbon dioxide (CO_2) from the atmosphere and use the carbon for their leaves, stems, roots, spines, and other leafy or woody parts. Plants also release oxygen, a key ingredient in life. Life on Earth is currently the main source of the oxygen and CO_2 in the atmosphere.

Life interacts with the hydrologic cycle. Plants take in water from the rocks and soil, which may have arrived from lakes and other bodies of surface water, or directly from the atmosphere, as during precipitation. Plants then release some water back into the environment. The Sun is the ultimate source of energy for photosynthesis, as well as movement of matter and energy in the atmosphere and most movement of material on Earth's surface.

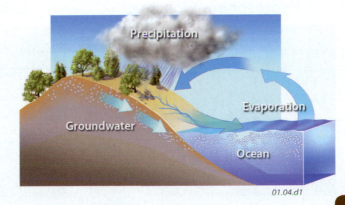
01.04.d1

Life also interacts with aspects of the rock cycle. Plants help break down materials on Earth's surface, such as when a plant's roots push open fractures in rocks and soil. Plants help stabilize soils, inhibiting erosion, by slowing down the flow of water, allowing it to remain in contact with rocks and soil longer. This increases the rate at which weathering breaks down materials. Humans have altered the surface of Earth by removing vegetation that would compete with crops, villages, and cities.

Water, nutrients, and other materials are stored and cycled through the biosphere, at local to global scales. We use the term *biogeochemical cycle* to indicate that plants, animals, and bacteria, in addition to chemical and physical processes, are involved in the cycling of a chemical substance through different parts of the environment. For example, the movement of carbon between the biosphere, atmosphere, hydrosphere, and lithosphere is a biogeochemical cycle known as the *carbon cycle*.

Before You Leave This Page Be Able To

✓ Describe some examples of transfer of energy, matter, and momentum in the atmosphere.

✓ Sketch and describe the hydrologic cycle, rock cycle, and how the biosphere exchanges matter with the other spheres.

1.4

1.5 How Do Earth's Four Spheres Interact?

ENERGY AND MATTER MOVE between the land, water, atmosphere, and biosphere—between the four spheres. There are various expressions of these interactions, many of which we can observe in our daily lives. In addition to natural interactions, human activities, such as the clearing of rain forests, can affect interactions between the spheres. Changes in one component of one sphere can cause impacts that affect components of other spheres.

A What Are Some Examples of Energy and Matter Exchanges Between Two Spheres?

The four spheres interact in complex and sometimes unanticipated ways. As you read each example below, think of other interactions—observable in your typical outdoor activities—that occur between each pair of spheres.

Atmosphere-Hydrosphere

01.05.a1 Indonesia

The Sun's energy evaporates water from the ocean and other parts of the hydrosphere, moving the water molecules into the atmosphere. The water vapor can remain in the atmosphere or can condense into tiny drops that form most clouds. Under the right conditions, the water returns to the surface as precipitation.

Atmosphere-Lithosphere

01.05.a2 Mount Veniaminof, AK

Active volcanoes emit gases into the atmosphere, and major eruptions release huge quantities of steam, sulfur dioxide, carbon dioxide, and volcanic ash. In contrast, weathering of rocks removes gas and moisture from the atmosphere. Precipitation accumulates on the land, where it can form standing water, groundwater, or erosion-causing runoff.

Atmosphere-Biosphere

01.05.a3 Indonesia

Plants and animals utilize precipitation from the atmosphere, and some plants can extract moisture directly out of the air without precipitation. Large-scale circulation patterns in the atmosphere are a principal factor in determining an area's climate, and the climate directly controls the types of plants and animals that inhabit a region.

Hydrosphere-Lithosphere

01.05.a4 Glacier Parkway, Alberta, Canada

Channels within a stream generally bend back and forth as the water flows downhill. The water is faster and more energetic in some parts of the stream than in others, and so erodes into the streambed and riverbank. In less energetic sections, sediment will be deposited on the bed, like the gravel in this photograph. Earth's surface can be uplifted or dropped down, as during an earthquake, and the resulting changes can influence the balance of erosion and deposition.

Hydrosphere-Biosphere

01.05.a5 Belize

Oceans contain a diversity of life, from whales to algae, and everything in between. Coral reefs represent an especially life-rich environment, formed when living organisms extract materials dissolved in or carried by seawater, to produce the hard parts of corals, shells, and sponges. At greater ocean depths, where waters are colder, shells and similar biologic materials dissolve, transferring material back to the seawater.

Lithosphere-Biosphere

01.05.a6 Near Badlands National Park, SD

The clearest interaction between the lithosphere and biosphere is the relationship between plants and soils. The type of soil helps determine the type of plants that can grow, and in turn depends on the type of starting materials (rocks and sediment), the geographic setting of the site (e.g., slope versus flat land), climate, and other factors. Plants remove nutrients from the soil but return material back to the soil through roots and annual leaf fall, or plant death and decay.

B To What Extent Do Humans Influence Interactions Between the Spheres?

Anyone who has flown in an airplane or spent some time using Google Earth® appreciates the amazing amount of human influence on the landscape. The intent of development is almost always to improve the human condition, but the complex chain reaction of impacts that cascade through the system can cause unintended and often harmful impacts elsewhere in one or more of the four spheres, as illustrated in the examples below. Some consequences of human impacts are not felt immediately but only appear much later, after the activity has continued for many years.

01.05.b1

01.05.b2 Grand Coulee Dam, WA

01.05.b3 Phoenix, AZ

Humans clear forests, a critical part of the biosphere, to provide lumber and grow food. In addition to the loss of habitat for plants and animals, deforestation reduces the amount of CO_2 that can be extracted out of the atmosphere and stored in the carbon-rich trunks, branches, and leaves of plants. Removing plant cover also causes increased runoff, which enhances soil erosion and leads to the additional loss of plant cover—an unintended consequence and a positive feedback.

Over 80,000 dams exist in the U.S., providing water supplies, generating electricity, protecting towns from flooding, and providing recreational opportunities. Dams also alter the local water balance by interrupting the normal seasonal variations in flows of water and by capturing silt, sand, wood, and other materials that would normally go downstream. Construction and filling of the reservoir disrupts ecosystems, displaces people, and threatens or destroys plant and animal communities.

Local warming of the atmosphere occurs near cities because of normal urban activities (lighting, heating, etc.) and because many urban materials, like dark asphalt, capture and store more heat than natural open space. Heat is also released from car exhausts and industrial smokestacks. Non-natural drainage systems cause rapid accumulation and channeling of water. Development infringes on natural plant and animal communities, disturbs or covers soil, and alters erosion rates.

Geography in Our Modern World

The science of geography has ancient origins, arising from the need of early civilizations for maps for navigation, planning, and other purposes. The examples above illustrate current societal issues that should also be considered from a spatial perspective—the percentage decrease in the area of rain forests, the outline of areas that will be flooded by construction of a dam, or the patterns of population growth and resulting changes in local temperatures. Understanding location and spatial distributions is as important in our modern world as it was in ancient times, because these factors are crucial in understanding the environment or determining possible sites for any human activity. For example, what spatial factors should be considered when planning a new subdivision or a new business? Geographic advantages can be the difference between success and failure, and an understanding of natural and human environments around the world is as important as ever. The map shown here depicts the average precipitation in the lower 48 states, with purple and blue designating the highest average annual precipitation, red and orange indicating the lowest precipitation, and green and yellow designating intermediate precipitation. How would you describe some of the main patterns? Where are the highest and the lowest precipitation amounts? Why are these wet or dry areas located where they are? What are some implications of the spatial distribution of high-precipitation versus low-precipitation regions? How might these variations in precipitation influence agriculture or the water supplies of the growing desert cities of the U.S. Southwest? Geographers address these and many other types of questions.

01.05.t1

Miles 200 400 600

Before You Leave This Page Be Able To

✓ Provide an example of an interaction between each pair of spheres.

✓ Describe examples of how humans can affect the natural system in each of the four spheres.

✓ Describe why geographic factors are important when considering environmental issues or when evaluating potential sites for a new agricultural area or business.

1.5

1.6 How Do We Depict Earth's Surface?

EARTH'S SURFACE DISPLAYS various features, including mountains, hillslopes, and river valleys. We commonly represent such features on the land surface of an area with a *topographic map* or *shaded-relief map*, each of which is useful for certain purposes. Some maps allow us to visualize the landscape and navigate across the land, whereas others permit the quantitative measurement of areas, directions, and steepness of slopes.

A How Do Maps Help Us Study Earth's Surface?

Maps are the primary way we portray the land surface. Some maps depict the shape and elevation of the land surface, whereas others, like a soil map, represent the materials on that surface. Views and maps of SP Crater in northern Arizona provide a particularly clear example of the relationship between the land surface and different types of maps.

01.06.a1

1. This perspective view has aerial photography superimposed over topography (shape of the land). What features do you observe in the topography? Which areas are high in elevation? What are the most distinctive features?

2. The area has distinct, cone-shaped hills surrounded by broad, less steep areas. The hills are small volcanoes, which form when fragments of molten rock are ejected into the air and settle around a volcanic vent.

5 km

3. In the center of the area is a nearly black feature, which is a solidified *lava flow* formed when fluid magma erupted onto the surface in the last 5,000 years. The volcano at the southern end of the lava flow is named *SP Crater*, and is well known to many physical geographers.

4. Examine other features in the scene. Note the light-gray areas in the upper left parts of the image, and the linear features that cut across the gray rocks. Different materials are forming different types of landscapes. This entire area has a relatively dry climate, with few trees to obscure the landform features.

01.06.a2 SP Crater, northern AZ

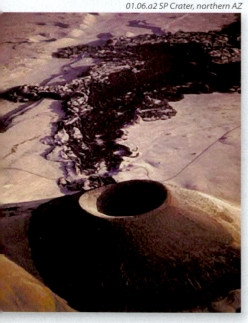

01.06.a3 SP Crater, northern AZ

◄ This aerial photograph shows SP Crater and the dark lava flow that erupted from the base of the volcano. Note that the slopes of the volcano are much steeper than those of the surrounding land.

▲ This photograph, taken from the large crater south of SP Crater, shows the crater (on the left) and several other volcanoes. The view is toward the north. Try to match some of the features in this photograph with those shown in the larger perspective view. In starting such activities, it is always best to begin with a known feature (in this case SP Crater).

Before You Leave These Pages Be Able To

✓ Describe how shaded relief and topographic maps depict the surface.

✓ Describe what contours on a topographic map represent, how the shapes of contours reflect the shapes of features, and how contour spacing indicates the steepness of a slope.

✓ Sketch and describe the meaning of the terms *elevation*, *depth*, *relief*, and *slope*; include how we express steepness of slope.

▼ A *shaded-relief map* emphasizes the shape of the land by simulating light and dark shading on the hills and valleys. The individual hills on this map are volcanoes. The area is cut by straight and curving stream valleys that appear as gouges in the landscape. Simulated light comes from the upper left corner of the image.

01.06.a4

Shaded-Relief Map

SP Crater

2 km

▼ A *topographic map* shows the elevation of the land surface with a series of lines called *contours*. Each contour line follows a specific elevation on the surface. Standard shaded-relief maps and topographic maps depict the shape of the land surface but give no specific and direct information about what lies beneath.

01.06.a5

Topographic Map

Most topographic maps show every fifth contour with a darker line, to emphasize the broader patterns and to allow easier following of lines across the map. These dark lines are called *index contours*.

Adjacent contour lines are widely spaced where the land surface is fairly flat (has a gentle slope).

Contour lines are more closely spaced where the land surface is steep, such as on the slopes of the volcano. Note how the shapes of the contours reflect the shapes of the different volcanoes.

B How Do We Refer to Differences in Topography?

Earth's surface is not flat and featureless, but instead has high and low parts. Topography is steep in some areas but nearly flat in others. We use common terms to refer to the height of the land and the steepness of slopes.

Elevation, Depth, and Relief

1. The height of a feature above sea level is its *elevation*. Scientists describe elevation in *meters* or *kilometers* above sea level, but some maps and most signs in the U.S. list elevation in feet.

2. Beneath water, we talk about *depth*, generally expressing it as depth below sea level. We use *meters* for shallow depths and *kilometers* for greater ones.

01.06.b1

3. We also refer to the height of a feature above an adjacent valley. The difference in elevation of one feature relative to another is *topographic relief*. Like elevation, we measure relief in meters or feet; we refer to rugged areas as having *high relief* and to flatter areas as having *low relief*.

Slope and Gradient

4. One way to represent the topography of an area, especially the steepness of the land surface, is to envision an imaginary slice through a terrain, like this one through SP Crater (▶). The dark line shows the change in elevation across the land surface, and is a *topographic profile*. Cliffs and slopes that drop sharply in elevation are *steep* slopes, whereas topography that is less steep is referred to as being *gentle*, as in a gentle slope.

W   E
26°

01.06.b2 SP Crater, northern AZ

5. We describe steepness of a slope in *degrees* from horizontal. The eastern slope of SP Crater has a 26-degree slope (26° slope). We also talk about *gradient*—a 26° slope drops 480 meters over a distance of one kilometer, typically expressed as 480 m/1,000 m or simply as 0.48.

1.6

1.7 What Do Latitude and Longitude Indicate?

IMAGINE TRYING TO DESCRIBE the location of an "X" on a featureless sphere. What system would you devise to convey the location? If the sphere did not have any markings or seams, we would need to first establish a frame of reference—a place on the sphere from which to reference the location of the X. For these reasons, we have devised systems of imaginary gridlines on the Earth. These are referenced as angles from the known points within or on the Earth. The most commonly used imaginary gridlines are *latitude* and *longitude*, which are displayed on many maps and are provided by the location capabilities of many cellular phones.

A How Do We Represent Locations on a Globe?

If you were trying to convey the location of the X on the sphere, or the location of a city on our nearly spherical planet, a good place to begin visualizing the problem is to establish a framework of imaginary gridlines. Another important aspect is to consider how lines and planes interact with a sphere.

Parallels

We could draw lines that circle the globe, each staying the same distance from the North or South Pole. The lines are parallel to one another and remain the same distance apart, and so are called *parallels*. In addition, these lines are parallel to imaginary cuts through the Earth, perpendicular to Earth's spin axis (which goes through the North and South Poles). The parallel that is halfway between the North and South Pole is the *equator*.

01.07.a1

If we traveled along one of these lines (i.e., along a parallel), we would stay at the same distance from the pole as we encircled the planet. In other words, our position in a north-south framework would not change.

Meridians

Lines that encircle the globe from North Pole to South Pole are called *meridians*. Meridians do not stay the same distance apart and are not parallel. Instead, meridians are widest at the equator and converge toward each pole. A meridian would be the path you would travel if you took the most direct route from the North Pole to the South Pole, or from south to north.

01.07.a2

The term *meridian* comes from a Latin term for midday because the Sun is along a meridian (i.e., is due south or north) at approximately noon. The terms A.M. (for before noon) and P.M. (for after noon) are also derived from this Latin term (e.g., post meridiem).

Great Circles

The intersection of a plane and a sphere is a curved, circular line that encircles the sphere. If the plane is constrained to pass through the center of the sphere, we call the resulting intersection a *great circle*. A great circle also represents the shortest distance between two points on a sphere and so is the path airlines travel over long distances.

01.07.a3

A great circle divides the sphere into two equal halves. The equator is a great circle, separating the Earth into two hemispheres—the *Northern Hemisphere* north of the equator and the *Southern Hemisphere* south of the equator. A north-south oriented great circle is used to separate the *Western Hemisphere*, which includes North and South America, from the *Eastern Hemisphere*, which includes Europe, Asia, Africa, and Australia. Antarctica, over the South Pole, and the Arctic Ocean, over the North Pole, each straddle the great circle between the Eastern Hemisphere and Western Hemisphere.

Small Circles

An imaginary plane that intersects a sphere at any other angle, not going through Earth's center, is called a *small circle*. As is obvious from the small circle illustrated in this figure, a small circle does not divide the globe into equal halves.

01.07.a4

Note that all parallels, which are oriented east-west, are small circles, except the equator, which is a great circle (it divides the planet into two equal halves).

In contrast, each north-south meridian, when paired with its counterpart on the other side of the globe, forms a great circle. Any such pair of meridians divides the globe into two equal halves. When viewed together, parallels and meridians divide the planet into a grid of somewhat rectangular regions. Such regions encompass greater area near the equator than near the poles, due to the convergence of meridians toward the poles.

B What Are Latitude and Longitude?

If you were a pilot flying from New York City to Moscow, Russia, how would you know which way to go? Our imaginary grid of parallels and meridians provides a precise way to indicate locations using latitude and longitude, which are expressed in degrees. Fractions of a degree are expressed as decimal degrees (e.g., 9.73°) or as minutes and seconds, where there are 60 minutes (indicated by ') in a degree and 60 seconds (") in a minute (e.g., 9° 43' 48").

This map (▶) illustrates the nature of the problem. If we want to navigate from New York to Moscow, we can tell from this map that we need to go a long way to the east and some amount to the north. These directions, although accurate, would not be good enough to guide us to Moscow. We need to more precisely specify the locations of each place and then figure out the shortest flight path. Fortunately, we can find on the Internet that the location of New York City, as given by latitude and longitude, is 40.7142° N, 74.0064° W. The location of Moscow is 55.7517° N, 37.6178° E. Now if we only knew what these numbers signify!

01.07.b1

Before you go any further, note the yellow line on this map, which represents the shortest route between New York and Moscow. It does not look like the shortest route on this flat, two-dimensional map, but it is indeed the shortest route on the three-dimensional globe. What type of path do you think this flight route follows? It is a great circle.

Latitude

The *latitude* of a location indicates its position north or south of the equator. Lines of latitude are *parallels* that encircle the globe east-west.

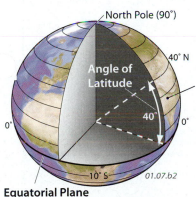

01.07.b2

The angle created by drawing lines connecting the position of an object on the Earth's surface to the center of the Earth, and then to the equator defines the number of degrees of *latitude* of the object's position. In the Northern Hemisphere, latitude is expressed as degrees north. In the Southern Hemisphere, latitude is expressed as degrees south or as negative degrees.

01.07.b3

Parallels of latitude run east-west around the Earth. The zero line of latitude is the equator, with the values increasing to 90 at the north and south poles. There are ten million meters from the equator to the North (or South) Pole, so one degree of latitude is approximately 111 km (69 miles).

Longitude

The *longitude* of a location indicates its east-west position. Lines of longitude are *meridians* that encircle the globe north-south.

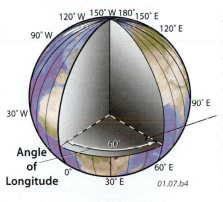

01.07.b4

As a starting point, a zero-degree meridian is defined as the north-south line that passes through Greenwich, U.K.—this is called the *Prime Meridian*. The angle created by the object's position, the center of the Earth, and the Prime Meridian defines that object's *longitude*, given as degrees east or west of the Prime Meridian. Meridians west of the Prime Meridian often are expressed as negative degrees.

01.07.b5

Meridians of longitude run north-south. They are widest at the equator (where a degree of longitude is also about 111 km) and converge at higher latitudes until they meet at the poles. Starting at the zero meridian through Greenwich, values increase to 180° E and 180° W at the *International Date Line*, an imaginary line that runs through the middle of Pacific Ocean (not shown; on the opposite side of the globe).

In addition to the equator, there are a few lines of latitude that are especially important. These include the *Tropic of Cancer* and *Tropic of Capricorn*, which are 23.5° north and south of the equator, respectively. Also important are the *Arctic Circle* and *Antarctic Circle*, which are 66.5° north and south of the equator (23.5° away from the corresponding pole). As discussed later, the 23.5° angle is how much the Earth's axis is tilted with respect to the Sun.

01.07.b6

Before You Leave This Page Be Able To

✔ Sketch and explain what is meant by a parallel, meridian, great circle, and small circle.

✔ Sketch and explain the meaning of latitude and longitude, indicating where the zero value is for each measurement.

1.8 What Are Some Other Coordinate Systems?

WE USE OTHER SYSTEMS, besides latitude and longitude, to describe location. These include the Universal Transverse Mercator (UTM) system, the State Plane Coordinate System (SPCS), and the Public Land Survey System (PLSS). Each is very useful for certain applications, and some are used to specify the location of real-estate properties appearing on legal documents associated with purchasing a house. Therefore, they are relevant to most citizens, even those who are not geographers.

A How Do We Use the UTM System?

Maps can show large regions, even the entire world. The main considerations for displaying large regions arise mostly from the fact that we live on a three-dimensional world (a sphere) and flat maps are two-dimensional. One solution to this challenge is the *Universal Transverse Mercator* (UTM) system, a method of identifying locations across the nonpolar part of the Earth. UTM is the most useful method of location for people who frequently hike or camp, or for people, like forest rangers and wildlife researchers, who work outdoors in nonurban settings.

1. The UTM system slices the nonpolar region of Earth into 60 north-south zones, each 6° of longitude wide. The slices are numbered from 1 to 60, with numbers increasing eastward from the International Date Line. A slice comprises two *UTM zones,* one in the Northern Hemisphere and another in the Southern Hemisphere. For example, most of Florida is in UTM zone 17 N, whereas the southern tip of South America is mostly in UTM Zone 19 S. What is the UTM zone for the place where you grew up or go to school?

01.08.a1

UTM ZONE NUMBERS

2. The slices are further subdivided into grid zones, each 20° of latitude long, as shown by the rectangles on this map. The purpose of UTM zones is to ensure that location is accurately portrayed in the middle of each division, with the understanding that there will be increasing distortion away from the middle. Due to large distortions that occur in the UTM system near the poles, UTMs are typically only used between 80° N and 80° S latitudes (we do not use UTM within 10° of the poles).

3. For a location within a grid zone, we specify coordinates as *eastings* and *northings*. Eastings are a measure of the number of meters east or west of the central meridian for that zone. Northings are a measure of the position north or south of the equator. The map below shows the aerial photograph of the horse and cow pasture shown earlier in this chapter, but this time with a UTM grid labeled with eastings (along the bottom of the map) and northings (along the left side of the map).

4. The advantage of the UTM system is that it is a "square" grid system measured in meters rather than degrees, so it is convenient for measuring direction and distance. Note how useful this grid and UTM system would be if you were riding around trying to record the location of each horse in the pasture. Two horses (not visible here) are grazing at an easting of 495250 and a northing of 4214100; can you determine about where these horses are? Are they in the pasture?

01.08.a2

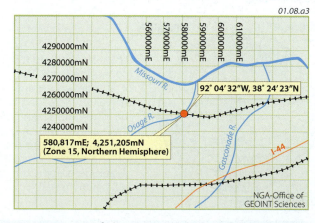

01.08.a3

580,817mE; 4,251,205mN
(Zone 15, Northern Hemisphere)

92° 04' 32"W, 38° 24' 23"N

NGA-Office of GEOINT Sciences

5. We can specify locations using several systems, and convert from one location system to another. The map above shows the position of a site expressed in both latitude-longitude (commonly called "lat-lon") and UTM coordinates. There are Internet sites that allow easy conversion from lat-lon to UTM and vice versa. To go from UTM to lat-lon, you have to specify the UTM zone, which can be determined using the large map near the top of this page.

B | How Do We Describe Locations Using the State Plane Coordinate System?

The State Plane Coordinate System (SPCS) is a third system for mapping, used only in the U.S. Its name derives from the system ignoring the curvature of the Earth, treating the surface as a "plane," and being based on a reference frame that is optimized for each state. As a result, the system can use X-Y coordinates to represent positions, simplifying land surveys and calculations of distances and areas. Another advantage is that the projection was chosen based on the geographic orientation of the state or section of the state, to minimize distortion for that area.

In the SPCS, most states are subdivided into two or more zones called *state plane zones*; some states are a single zone. Alaska has 10 zones and Hawaii has five zones. The boundaries of the zones generally are east-west or north-south, but are not straight, following local county boundaries (trying to keep a county within a single zone).

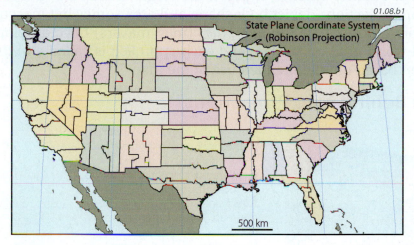

01.08.b1

State Plane Coordinate System (Robinson Projection)

500 km

States that are elongated east-west, such as Tennessee, use different map approaches to generate the state plane coordinates than states like Illinois that are elongated north-south. The goal is to customize the drawing of the map so as to minimize the distortion that is always present when trying to show features of a spherical Earth on a flat piece of paper. So local U.S. maps, such as for flood zones, roads, or property delineation are likely to use the SPCS. If you buy a house in the U.S., the legal documents will likely use SPCS to specify the location of the property, perhaps accompanied by a survey in UTM.

C | How Do We Describe Location Using the Public Land Survey System?

The Public Land Survey System (PLSS) is another system used in the U.S. for describing the location of lands and for subdividing larger land parcels into smaller ones. When you hear someone refer to a "section of land" or a "quarter-section," they are talking about PLSS. The PLSS is also called the township-range system because these terms are used to describe areas of land.

The Public Land Survey System was designed specifically for public lands, such as those administered by the U.S. Department of the Interior, and as a result is most widely used in states where there are federal lands (▼). It is not used in many eastern states, where there is little land that is not privately owned, and in Texas, which has much state-owned land.

PLSS is defined for different regions, in each based around some initial point. From this point, a *Principal Meridian* extends both north and south and a *Base Line* extends both east and west. Beginning at the Principal Meridian, the land is subdivided into six-mile-wide, north-south strips of land called *ranges*. Beginning at the Base Line, the land is subdivided into six-mile-wide, east-west strips of land called *townships*.

01.08.c2

States Included in the PLSS

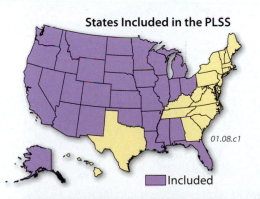

01.08.c1

■ Included

Each square of the township-range grid generally is six miles in an east-west direction and six miles in a north-south direction, so it is 36 mi² in area. Each grid square is further subdivided into 36 sections that are each one square mile in area. Township and range lines and section boundaries are included on many topographic maps.

Each one-square-mile section can be further divided into quarters, eighths, and even smaller subdivisions. The rectangle in the southeastern corner of Section 14 would be described as being in the eastern half of the southeast quarter of Section 14, Township 2 South and Range 3 East. This is abbreviated: E1/2 SE1/4, S. 14, T2S, R3E.

1.8

1.9 How Do Map Projections Influence the Portrayal of Spatial Data?

EARTH IS NOT FLAT, so a flat map cannot portray all locations accurately. An ideal map would preserve directions, distances, shapes, and areas, but it is not possible to preserve all four of these accurately. Instead, either the *shape* of features on a map, such as country outlines, is preserved or the *area* of features is preserved, but never both at the same time. Many map projections depict both shape and area somewhat inaccurately, as a trade-off, so that neither will be shown more inaccurately. *Cartographers* (map makers) have developed different ways of projecting our three-dimensional world onto a flat map, and each approach is called a *map projection*. The particular type of projection is chosen based on the intended use of the map.

A What Is the Rationale Behind Map Projections?

A *map projection* is a mathematical algorithm used to represent places on a three-dimensional spherical Earth onto a flat map. Imagine shining a light through a partially transparent globe and observing the image projected on the back wall (▶).
This is what a map projection does, but

01.09.a1

in a quantitative, numeric way. While many projections exist, the best projection for a given map will introduce the least distortion for the key areas being shown. Whenever a map is made, some distortion is introduced by the projection. It is impossible to avoid distorting either shapes or areas, or doing some distortions of each.

Some map projections attempt to preserve shapes, and are called *conformal*. If shapes are preserved (▶), directions may be preserved but areas are distorted and scale will vary across the map. These imperfections get worse for maps that show larger areas.

In *equal-area* projections, areas are preserved but shape is distorted (▶). Compass directions cannot be shown correctly, so such a map should not be used for navigation. If the proper projection is selected for a given application, distortions are minimized for the aspect (e.g., shape) that is most important and for the region of most interest.

01.09.a2

01.09.a3

B What Are the Major Types of Projections and What Advantages Does Each Offer?

Sinusoidal

Perhaps the easiest projection to visualize conceptually is to imagine peeling an orange and slicing it in a few strategic places to allow it to be flattened without buckling (▼). *Sinusoidal* projections work on this same premise. If the map can be interrupted so that areas of lesser significance for a given application are not shown, then less distortion exists in the areas that are shown. Straight, parallel lines remain so, and have their correct length. Meridians become progressively longer toward the edges of each lobe of the map. While areas are preserved, shape distortion increases near the edges of each lobe.

01.09.b2

01.09.b1

The shape distortion problem in such projections can be mitigated by increasing the number of central meridians around which accuracy is preserved (▲). However, this comes at the expense of having more areas of interrupted coverage. Notice how the central meridians are straightest and appear at right angles to the parallels at the equator. These are the areas that are depicted most accurately for this type of projection. The most common type of map using this projection strategy is called a *Goode projection*.

Cylindrical

In *cylindrical projections*, the globe is transformed to a flat page by projecting a globe outward onto a cylinder. The projection starts at a line, called the *standard line*, where the globe touches the cylinder, usually at the equator. These types of map projections have no distortion at the standard line (equator), but distortion becomes worse with increasing distance from the standard line. The resulting maps portray parallels of latitude as straight lines with the same length as the equator (that is, distorted in length) and depicts meridians also as straight lines intersecting the parallels at right angles.

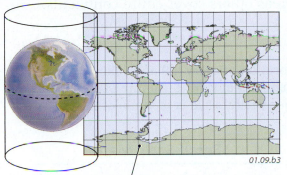

01.09.b3

◄ Cylindrical projections depict compass directions as straight lines, so they are excellent for navigation. However, because the meridians are depicted (falsely) as being parallel to each other, east-west exaggeration of distances is severe, particularly in high latitudes. To allow these maps to be conformal (preserve shapes), north-south distances are stretched to match the east-west exaggeration. This makes high-latitude areas greatly exaggerated in size, but they retain shapes.

High-latitude distortion increases to such an extent that the poles cannot be shown.

The most familiar type of cylindrical projection is the Mercator map, which became an important tool in the Age of Exploration. The part of the map at the right, which is a *Mercator projection*, portrays Greenland as being larger than the conterminous U.S. Is this true?

01.09.b4

Some maps blend aspects of a cylindrical map with other types of projections. The *Robinson projection* (►) is a commonly used projection of a world map, especially in textbooks. It does not fully preserve areas or shapes, instead representing a compromise between equal-area and conformal projections. The meridians curve gently, and the parallels are straight lines horizontally across the map. A feature it shares with cylindrical maps is severe distortion near the poles.

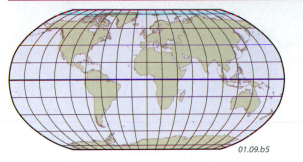

01.09.b5

Conical

Conical projections involve conceptualizing a cone over the globe, usually with the apex of the cone vertically above the pole. No distortion occurs along the arc where the globe touches the cone—the standard line, usually a parallel of latitude. If the cone slices through the globe and intersects the surface along two arcs (usually parallels of latitude), the projection is called *polyconical*. In either case, distortion increases with distance away from these arcs.

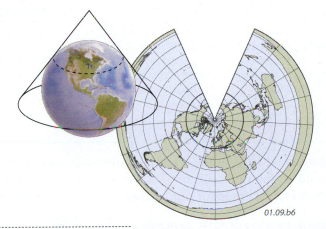

01.09.b6

In conical projections, parallels are concentric circles and meridians are lines radiating from the center of curvature of the parallels. This family of projections is neither equal area nor conformal.

Conical projections can only show areas within a single, complete hemisphere (since regions that curve underneath the globe cannot be projected). This type of map works best when the area mapped is small in latitudinal extent. Notice how poorly this projection performs for showing a large area (◄).

Planar

In *planar projections*, the plane onto which the map is projected touches the globe in a single point, which becomes the center of the map. Distortion increases away from this point, and any straight line from this point is a line of true direction. Again, only one full hemisphere can be shown on such a map.

The pole is the focus in the *polar stereographic projection*. Scale becomes exaggerated toward the equator, but all lines connecting the shortest distance between two points on the sphere (great circles) are shown as straight lines. Planar projections are useful, therefore, for air navigation.

01.09.b7

1.9

1.10 How Do We Use Maps and Photographs?

MAPS ARE AMONG OUR MOST IMPORTANT TOOLS for depicting and analyzing spatial information, whether we are interested in environmental issues or election results. Cartographers generate different kinds of maps that are designed to show Earth's landscape features, its weather and climate, and the distribution of plants, animals, or many other types of variables. Some cover small areas of Earth's surface, whereas others cover entire continents.

A How Much Area Do Maps Portray?

If we are hiking across the landscape, we want a detailed map that shows the location of every hill and valley. If we are interested in global climate change, we may want a map showing average temperatures for the entire planet. We use the general term *scale* to describe how much area the map shows. More specifically, scale is the ratio of the distance on a map to the actual distance (in the same units) on Earth.

Large-Scale Maps

01.10.a1

SCALE 1:24 000

◄ This topographic map shows hills and lakes that formed as glacial features in Kettle Moraine State Forest in central Wisconsin. We can convey the scale of the map in three ways. First, we can report the scale with *words*—on the original version of this map (reduced here to fit on the page), one cm on the map equals 24,000 cm on the surface. Second, we can report this same information as a *ratio* of a distance on the map to the actual distance on the ground, which is called the map's *representative fraction;* for the original version of this map the representative fraction was 1:24,000, as reported on the map. Third, most maps include some type of visual *bar scale.*

The original scale of this map, 1:24,000, is the typical scale used in the U.S. for topographic maps, with one inch equaling 2,000 feet. A map like this, which shows a local area, has a large representative fraction. It would require a relatively large map to show a large area—it is called a *large-scale map.*

Small-Scale Maps

01.10.a2

50 km

The map above shows state forests in Wisconsin. With this type of regional map, we can portray a relatively large area with a relatively small map—such a map has a small representative fraction and is a *small-scale map.*

B How Are Maps Made?

Originally, topographic maps were produced by sending a team of surveyors out in the field and having them map the area, drawing lines on paper maps, and taking notes. Today, such maps can be produced directly from laser and radar measurements from orbiting spacecraft or from pairs of photographs taken from slightly different perspectives.

1. Aerial photographs are typically taken from a plane or satellite as it flies across the terrain. The onboard, downward-pointing camera takes photographs at specific intervals in such a way that there is some overlap between the area captured by two successive photographs. The perspective of the camera is slightly different between the two photographs in the same way that our two eyes simultaneously have a slightly different perspective of the same scene. Test this concept by looking around, closing one eye at a time, and noticing how objects shift slightly in position relative to one another. The apparent shift is related to their difference in distance from us.

01.10.b1 Arabia Terra, Mars

2. Two aerial photographs that have overlapping coverage and slightly different perspectives (◄) are called a *stereo pair,* usually generated by two images taken seconds apart, while the plane or satellite is moving. In other cases, both images are taken at the same time but by two different cameras placed a specific distance apart. When the two photographs of a stereo pair are placed at a proper distance side by side, a tool called a *stereoscope* enables us to see the scene in 3D, with the hills appearing to stand in relief above valleys. Such stereo pairs can be used in the construction of a topographic map, and many of the earliest topographic maps were originally constructed using stereo pairs.

C How Can Maps Be Used for Reporting Information?

Sometimes we make new maps in the field, such as by using surveying equipment to construct a topographic map that depicts the shape of topography. In most cases, we use existing maps, like the ones shown previously, and mark on the map the location of things we observe, such as the locations of glacial features or certain types of trees. In either case, this type of map actually produces new knowledge and is therefore a form of *primary data*.

The procedure is to visit the field site with an appropriately detailed map or aerial image, upon which observations can be plotted. Such maps and images upon which data are plotted are called *base maps*. The base map can be a large-scale topographic map or a detailed aerial photograph, like the one shown to the right. Observations and other information are plotted directly on the base map or on a partially transparent overlay. Alternatively, locations can be determined with a hand-held GPS device where the coordinates are saved and later mapped by various mapping programs.

This aerial photograph shows different materials on the surface of several levels (elevations) along a desert river channel. The gray part, bounded by the dashed red lines, represents the active channel and related areas that are flooded during most years. The lower terrace is slightly higher in elevation above the channel, and is flooded less frequently. The upper terrace is high enough to avoid any flooding. This map was produced by walking through the field area and drawing on the aerial photograph the boundaries between different areas. This map would be useful for determining flooding potential and other types of land-use planning.

01.10.c1 Queen Creek, AZ

D How Can Maps Be Used to Analyze and Interpret the Environment?

Preexisting maps become the basis for various interpretations. For example, the annotated aerial photograph above could be used to plan the locations of a subdivision, especially deciding where *not* to build. A preexisting map that is used for providing the input for answering some other question is known as a *secondary data source*.

01.10.d1

◄ Many maps contain a combination of data and interpretations. Examine this map of the Arctic National Wildlife Refuge (ANWR), an area discussed at the start of this chapter, and identify aspects that are data versus those aspects that represent some type of interpretation. Then continue reading below.

Data – The locations of features on this map would be considered data. These include the outline of the coastline, the boundary between Alaska and Canada, the locations of rivers and roads, and the outline of ANWR.

Interpretations – Other aspects of the map are interpretations, which commonly represent an expert's opinion of a situation. On this map of ANWR, interpretations include the migration routes of caribou (the large black arrows) and the locations where caribou give birth to their calves (calving areas, in green). To produce these interpretations, wildlife experts observed the habits of caribou and then generalized these to the interpretations shown. This map, consisting of data and interpretations, would be considered a secondary data source. It might be used to determine which areas are permissible for drilling.

◄ Surface weather maps likewise are a combination of data and interpretations. The edge of the continent and the outlines of states are clearly data. Weather maps also show analysis of the location of areas of relatively high atmospheric pressure (H), low atmospheric pressure (L), and weather fronts—lines interpreted to mark the boundary between air of very different temperatures and humidity. The triangles and semicircles point in the direction of air movement, another interpretation. Such maps help us predict today's likelihood of having precipitation and high winds.

Before You Leave This Page Be Able To

☑ Explain the difference between large-scale and small-scale maps and when you would use each.

☑ Describe a stereo pair and how it could be used to make a map.

☑ Describe how we can use maps to record new information or analyze and interpret existing information about the environment.

1.10

1.11 How Do We Use Global Positioning Systems and Remote Sensing?

THE GLOBAL POSITIONING SYSTEM (GPS) and remote sensing have greatly increased the accuracy of geographic field studies and given geographers new methods for performing geographic analyses. GPS helps geographers define spatial relationships among Earth's surface features, and a wide variety of remote sensing techniques help geographers define regional patterns and monitor changing environmental conditions.

A What Is GPS?

Most people think of GPS as a navigation system in our cars, or embedded in our cellular telephones, or a handheld device used for location and guidance while hiking across the countryside. GPS provides the accurate position on Earth's surface including latitude, longitude, elevation, and even how fast we are traveling. This information comes from a series of satellites orbiting Earth that send radio signals to ground-based receivers, like the ones on our dashboards, in our phones, or in our hands when we go hiking.

01.11.a1

The U.S. government launches, controls, and monitors a constellation of 24 satellites orbiting in six different planes around Earth (◄). Two generations of satellites currently operate in the GPS constellation (►). A third type will soon be deployed to improve the accuracy and reliability of satellite signals.

The time required for a radio signal from a satellite to reach a receiver on Earth is related to its distance to the receiver. A GPS receiver "knows" where each satellite is located in space at the instant when the GPS unit receives the signal. Calculating the distances from four or more satellites allows the GPS unit to calculate its own position, commonly with a precision and accuracy of several meters (for a handheld GPS unit). Higher precision can be achieved by occupying a single site for a long interval of time and then averaging the measurements.

01.11.a2

B How Do We Use GPS to Study Geographic Features?

GPS is used in a variety of applications from tracking wildlife migration or package delivery, to improving ocean and air travel. Even farmers use GPS to harvest crops and improve yield. Geographers use GPS for a variety of activities, including monitoring changes in the environment, collecting more accurate field data when surveying or mapping, and making decisions about how to best prevent or address natural disasters. Geographers employ two types of GPS devices, the familiar handheld GPS and the Differential GPS (DGPS).

01.11.b1

A handheld GPS device is a navigation tool for finding a location. These instruments operate on the same principles as all other GPS devices in that they receive radio signals from orbiting satellites that contain information about the position and distance of the satellite. GPS works best outside and with a clear view of the sky, but it can operate with reduced accuracy in forested areas, deep canyons, and other settings where parts of the sky, and therefore view of the satellites, is partially blocked.

Geographers use handheld GPS mostly for field work, including mapping the locations of landscape features, determining locations of water and soil samples, and inventorying populations of plants and animals.

Differential GPS (DGPS) is the same as GPS but with a correction signal added to improve the precision and accuracy. Accuracy is enhanced because the correction signal performs an independent check of each GPS satellite's signal. DGPS can provide accuracy of less than several meters.

Geographers use DGPS when precision is important, such as in surveys of changes in the land surface over timescales of decades or to gauge the erosion effects of a recent hurricane on a shoreline and its communities.

01.11.b2

C What Is Remote Sensing?

The term *remote sensing* refers to techniques used to collect data or images from a distance, including the processing of such data, and the construction of maps using these techniques. Remote sensing can be carried out using a helicopter, airplane, satellite, balloon, ship, or other vehicles, or it can be performed with instruments fixed on the land surface. The instrument-carrying vehicle or site is called the *platform*, and the instrument that collects the images and other data is the *sensor*. There are two general types of remote-sensing systems: *passive systems* and *active systems*.

Passive Remote Sensing

01.11.c1 New Orleans, LA after Hurricane Katrina

In passive remote sensing, the sensor points at the area of interest and records whatever light, heat, or other energy is naturally coming from that region. Aerial photography and most satellite images, like the one to the left, are recorded by passive sensors. The sensors are tuned to collect specific types and wavelengths of energy, such as infrared, visible, and ultraviolet energy. Most sensors collect an array of similar frequencies.

Active Remote Sensing

In active remote sensing, an energy source, usually on the same platform as the sensor, directs a beam of energy downward or sideways toward the area of interest. Such energy can include radar, as shown here, microwaves, laser light, or other types of energy. The sensor then measures how much of this energy returns to the platform and whether this energy has been modified by its interaction with the surface or atmosphere.

Radar Waves

01.11.c2

D What Types of Remote Sensing Are Used By Geographers?

Geographers use a variety of remote sensing techniques, measuring various types of energy, to study Earth's atmosphere, hydrosphere, lithosphere, and biosphere. Geographers also document and investigate patterns in land use, vegetation cover, erosion rates, extent of pollution, atmospheric water content and circulation, and ocean temperatures.

Visible and Near Infrared (IR)

01.11.d1 Washington, DC

Aerial photographs typically record visible light reflected off an area, but some photographs and many satellite images also record adjacent bands of infrared energy (near-IR). On near-IR images, vegetation commonly is depicted with a reddish tone, as in this image of Washington, D.C.

Thermal Infrared (IR)

01.11.d2 Providence, RI

Objects also emit energy at thermal infrared wavelengths, either from the internal heat of an object or from heat initially gained from the Sun. This image shows thermal-IR derived temperatures of Providence, Rhode Island, with lighter colors showing hotter areas in the city.

Microwave

01.11.d3

Images from microwave-sensing satellites and ground-based stations provide us with weather images in nightly newscasts. Microwaves can penetrate clouds and haze, providing a clear view of the ground at all times. They also can measure the height of the sea surface, as shown here.

Multispectral

01.11.d4 Nili Patera, Mars

Some multipurpose satellites collect data at multiple wavelengths of energy, therefore the name *multispectral*. Multispectral data are used for studying natural hazards, inventorying plant communities, tracking forest fires, and observing landscapes on other planets, as shown here.

Radar, Sonar, and Lidar

01.11.d5 Mount St. Helens, WA

Radar, sonar, and a newer technique called lidar all involve emitting waves of a certain wavelength and then measuring how much is reflected back to the sensor and the time required for the various beams to return. These data allow us to map the surface, like volcanic features.

Before You Leave This Page Be Able To

✓ Explain what GPS is, how it works, and how we can use it to investigate geographic problems.

✓ Summarize what remote sensing is, explaining the difference between passive and active systems.

✓ Describe five types of remote sensing data and one example of how each type is used.

1.11

1.12 How Do We Use GIS to Explore Spatial Issues?

MAPS ARE USED FOR REPORTING OBSERVATIONS and making interpretations from previously collected observations, and they can also be analyzed to create new maps. Maps created from aerial photographs, satellite imagery, and field observations can be stored in computer databases called *geographic information systems (GIS)*, where a variety of information can then be combined quickly and efficiently to examine relationships among the different features. For example, we might be interested in comparing the distribution of plant communities with the type of soil, average precipitation, and percentage of nights when it freezes. Modern geographic analysis using GIS is data-rich and computer-intensive, providing the clearest way to explore many types of spatial relationships.

A What Are Geographic Information Systems?

The first step required to analyze geographic problems is to decide what key data sets are needed to best understand the issues. Then, maps, images, and data are imported into a GIS database. Maps, aerial photographs, and satellite images need to be matched to standard geographic coordinates, such as UTM. The process of taking an image or map and matching it to standard coordinates is called *georeferencing* or *rectification*.

Once data are in the GIS, we can overlay different maps to compare different data sets. We commonly call each map a GIS *layer* because in the GIS computer we can arrange the different maps (layers) one on top of another, as shown in the figure to the right. Examining multiple kinds of data in this way would be useful, for example, in selecting a site for some type of facility.

The Region

Locations Within 10 km of Interstate Highway (Green)

Undeveloped Land (White)

Agriculturally Non-productive Soils (White)

Non-conserved Lands (Pink)

Best Location for Airport (Blue Column)

01.12.a1

Suppose city planners want to find the best location for a new airport. They first need to determine what already exists in the region so they examine a satellite image or aerial photograph (the top layer in the GIS layers shown here). Priority is given to sites that are within 10 km of an interstate highway (second layer from top). The site must contain a large area of undeveloped land (third layer), does not have soil that could be productively farmed (fourth layer), is not designated as wetland (fifth layer), and other considerations, such as who owns the property and whether the site has unstable slopes and soils (not shown). Without a GIS, several individual maps would have to be compared by hand to determine the best possible sites. A GIS can store each of these maps digitally and allow the user to identify any locations that meet the specified criteria. The digital format also allows for easy updating, and inquiries can be run on the potential sites to glean further information, such as current land costs.

Another popular use of GIS is for route optimization. If you have ever used Google Maps® or a navigation system on your phone or in your car, you've made use of this feature of GIS. In the example below, GIS is used to show a route that minimizes the distance that a shipping company must travel to deliver packages to locations marked with an "X" on a given day. Similar inquiries can be used to optimize routes for garbage pickup, hurricane evacuation, or school bus stops.

01.12.a2

The site selection and routing applications of GIS can be used together. For example, a power company may want to minimize the distance that power lines must be run, while still keeping the lines as far as possible from heavily populated and low-lying areas. Also, the power lines ideally are hidden from view as much as possible by local topography and forests.

GIS can be used in combination with remote sensing to classify areas on images. This can be very useful in determining land use and land cover for environmental applications. In its simplest form, areas could be classified as being urban, forested, agricultural, undeveloped, or water. More elaborate classifications could be used to identify types of wetlands and forests (▼) or other important aspects.

01.12.a3 Mermentau River Basin, LA

Land Cover Classes
- Bare Land
- Cultivated Land
- Deciduous Forest
- Estuarine Emergent Wetland
- Evergreen Forest
- Grassland
- High Intensity Developed
- Low Intensity Developed
- Mixed Forest
- Palustrine Aquatic Bed
- Palustrine Emergent Wetland
- Palustrine Forested Wetland
- Palustrine Scrub/Shrub Wetland
- Scrub/Shrub
- Unconsolidated Shore
- Water
- Mermentau River Basin

Another key advantage of GIS is that it allows direct comparison of changes in an area over time, whether those changes are in land use, vegetation cover, population density, or some other variable. By comparing earlier or later images of the same area, rates of urban sprawl, deforestation, or wetland loss could be documented, allowing us to consider possible remedies in an informed manner.

B What Kinds of Calculations Are Possible with GIS?

Spatial analysis is a cornerstone of geography, and GIS can evaluate the spatial distributions within the data, highlight correspondences among different variables, and automate the identification of properties of spatial distributions.

If the locations of observed features can be considered as points, such as stations that reported precipitation in the past 48 hours or sites in a stream where dissolved oxygen is low, physical geographers often want to know whether those features (1) occur in clusters, (2) are spaced approximately evenly apart (i.e., regular distribution), or (3) are distributed randomly across the landscape. *Point-pattern analysis* can answer this question objectively, and GIS can complete the geostatistical calculations automatically. Analysis proceeds on the principle of laying an imaginary grid over the study area and determining whether the object of interest is distributed significantly less evenly (i.e., clustered) or more evenly (i.e., regular) across the grid cells than would be expected in a random distribution.

Types of Spatial Distributions

1. Geographers are interested in the *distributions* of objects, whether of plants or animals or the locations of gas stations. There are a number of different types of distributions, three of which are shown here (▶). The name of each type conveys the overall character of the distribution, but there are usually exceptions (called *outliers*) to the general patterns.

Clustered   Regular   Random

01.12.b1

4. Objects can have *random* distribution, such as the trees on the right side of this figure. Some trees occur together, but not consistently. Point-pattern analysis indicates that the trees are neither clustered nor distributed regularly. The random arrangement fails to provide an obvious explanation of tree location.

2. Objects can have a *clustered* distribution, with objects tending to occur together rather than being dispersed widely apart. Point-pattern analysis would conclude that the trees on the left side of the diagram are clustered. We may conclude that individual trees prefer proximity to a feature present in only part of the area, such as next to ponds, streams, and other sources of water. In many cases, the explanation for a clustered distribution is not so obvious as in this case.

3. Objects can also have a more *regular* distribution, more or less evenly spaced apart and not spatially clustered. Point-pattern analysis would suggest a regular distribution of trees in this middle location. Perhaps individual trees, when away from lakes and streams, are more likely to survive if spread out as far as possible from other trees, increasing access to precipitation.

Calculations on Lines and Polygons

In addition to analyzing the distributions of points, GIS can calculate distances and angles between points, as when a navigation system calculates the distance to a destination. GIS can also be used to quantify the extent of areas, represented as polygons, such as the park shown here. GIS can calculate areas enclosed within polygons, identify the center of an irregularly shaped polygon, or calculate the perimeter of any area, like this park.

Dimensions of Park

100 m

150 m

01.12.b2

Rainfall Amounts

3.2 cm   2.0 cm

???

2.7 cm   1.6 cm

Spatial Interpolation

There are statistical methods of estimating a data value at any location if we have values for nearby locations. Such analysis, commonly called *spatial interpolation*, is done by comparing the location to known values of other points at different distances. For example, the amounts of rainfall, in centimeters, during a storm were measured at four weather stations, but we need to estimate the amount received in a nearby location that lacks a weather station. There are various statistical strategies, with names like *kriging* and *inverse-distance methods*. A common approach is to weigh the known values at other points by the distance from known points to the unknown point.

Identification of Buffer Zones

01.12.b3

We may not want two things in close proximity, like a school next to earthquake faults that are most active (◀). To solve such a problem, we need to compare the locations of schools with faults capable of causing earthquakes. GIS can contain all of these types of information, and can outline a *buffer zone* near such faults, within which school construction would be prohibited. Or a buffer zone may be drawn around a hurricane to indicate the zone of hurricane-force winds. These buffered areas can then be overlain on other maps, such as one of population density, to identify the number of people whose homes are affected by the hurricane.

Before You Leave This Page Be Able To

✓ Explain what GIS is, including the concept of map overlay.

✓ Give several examples of the types of spatial distributions tested in point pattern analysis.

✓ Explain the types of functional calculations that GIS can do on spatial data.

1.12

1.13 What Is the Role of Time in Geography?

WE LIVE ON A GLOBE THAT ROTATES, causing locations on the surface to pass from day to night and back again. Not everyone witnesses sunrise at the same time, because the Sun rises at different times in different locations. Some ideas from geography, especially the concept of longitude, help us understand these differences and describe time so that society can operate in a more orderly manner. Most of us think of time as the hours, minutes, and seconds on a clock, but much longer units of time are used when considering Earth's long history.

A How Do We Define Time Globally?

Some units of time, like a year or length of day, arise from natural progressions of the Earth as it orbits around the Sun in a year and completes a full daily rotation in 24 hours. Other measurements of time are locally based, so in the 1800s the world had to agree on an international system for defining time, based on the *Prime Meridian* and the *International Date Line*.

Prime Meridian is defined as the 0° longitude measurement on the Earth, passing through the British Royal Observatory in Greenwich, U.K. This location was also chosen as the reference point for world time, a time called *Greenwich Mean Time* (GMT) or *Coordinated Universal Time* (UTC). Time anywhere in the world is referenced relative to time at Greenwich.

The globe shown here has meridians spaced equally apart, so that there are 24 zones centered on the lines, one for each of the 24 hours in a day. If you could instantaneously travel from one meridian to the next, there would be a one-hour time difference. If political and other considerations did not intervene, the distribution of time zones could precisely follow lines of longitude, each 15° apart.

The *International Date Line* (IDL) is defined as the 180° measurement of longitude in the Pacific Ocean—the meridian on the exact opposite side of the Earth from Greenwich. Segments of the IDL have been shifted east and west to accommodate the needs of some nations in the Pacific, so that travel and trade among those islands is less problematic.

If you cross the International Date Line, you cross into a different calendar day. Traveling westward across the IDL puts you one day ahead of the date (or day) relative to immediately east of the IDL; this is described as "gaining a day." Moving eastward across the line you move to one day earlier, so we say you "lose a day."

B How Are Time Zones Defined?

The world is divided into 24 time zones, based loosely on longitude. This map color-codes these 24 time zones, most of which have irregular boundaries because they follow natural or political boundaries or try to keep some population center in a single zone. The boundaries between the four time zones covering the contiguous U.S. are mostly drawn along state or county boundaries or natural features.

For most latitudes of the Earth, the Sun shines on all 360° of longitude sometime during the course of a 24-hour day, covering 15° of longitude of new territory each hour. The Earth is divided into 24 time zones, each about 15° of longitude wide. Areas within a time zone adopt the same time, and there is a one-hour jump from one time zone to the next. If you are in one time zone, the time zone to the west is one hour earlier, and the time zone to the east is one hour later.

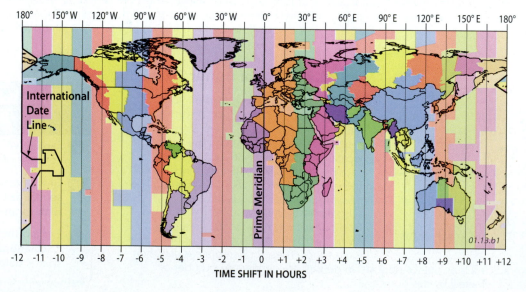

In most of the U.S. and Canada, clocks during the summer are set to one hour later for much of the year. This *Daylight Savings Time* (DST) provides daylight for an extra hour during the evening and one less hour of daylight in the morning. Some areas, like Saskatchewan and most of Arizona, do not observe DST, remaining instead on "standard time."

C | How Do We Refer to Rates of Events and Processes?

Many aspects of physical geography involve the rates of processes, such as how fast a hurricane is moving toward a coastline or how fast water in a river is flowing. We calculate the rates of such processes in a similar way to how we calculate the speed of a car or a runner. Some rates, like wind velocities inside a tornado, are very rapid, whereas other rates, like the uplift of a mountain range, are very slow. For scientific work, units are metric, so we talk about millimeters per year, kilometers per hour, or similar units of distance and time.

Calculating Rates

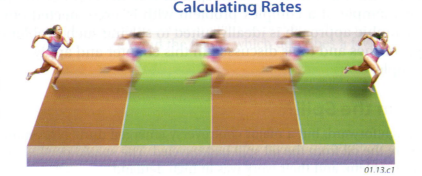

1. A runner (▶) provides a good reminder of how to calculate rates. A rate is how much something changed divided by the time required for the change to occur. Generally, we are referring to how much distance something travels in a given amount of time.

2. If this runner sprinted 40 meters in 5 seconds, the runner's average speed is calculated as follows:

$$distance/time = 40\ m/5\ s = 8\ m/s$$

3. What is the runner's average speed if she runs 400 meters in 80 seconds? Go ahead, try it. We don't need to provide you the answer to this one.

01.13.c1

Relatively Rapid Earth Processes

4. Some Earth processes are relatively rapid, occurring within seconds, minutes, or days. Relatively rapid natural processes include the velocity of the atmospheric jet stream (100s km/hr), speeds of winds inside a severe storm (100s km/hr), motion of the ground during earthquakes (5 km/s), movement of an earthquake-generated wave (tsunami) across the open ocean (100s km/hr), and the catastrophic advance of an explosive volcanic eruption (100s km/hr).

5. Hurricane Sandy, a huge and incredibly destructive storm that occurred in October 2012, is an example of a rapid natural process. Sandy originated as a tropical storm but migrated up the East Coast of the U.S. until it turned inland and struck New Jersey. While in the tropics, Sandy had winds estimated at 185 km/hr (115 mi/hr), but it had weakened considerably by the time the storm came ashore. The storm killed nearly 300 people along its path and caused damages of over $70 billion, mostly in New Jersey and New York.

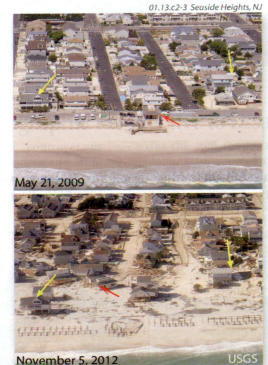

01.13.c2-3 Seaside Heights, NJ

May 21, 2009

November 5, 2012   USGS

6. The two photographs above show Seaside Heights, N.J., before and after the storm. The yellow and red arrows point to the same houses in both photographs. Although the photographs were taken several years apart, nearly all the damage occurred within a 24-hour period. In this short time, the shape of the coastline was extensively rearranged, houses were destroyed, and the entire neighborhood was covered in a layer of beach sand washed in by the waves.

Relatively Slow Earth Processes

7. Other Earth processes are very slow, occurring over decades, centuries, or millions of years. Natural processes that are fairly slow include movement of groundwater (m/day), motion of continents (cm/yr), and uplift and erosion of the land surface (as fast as mm/yr, but typically much slower). Although these processes are relatively slow, the Earth's history is long (4.55 billion years), so there is abundant time for slow processes to have big results, such as uplift of a high mountain range.

01.13.c4 Superstition Mtns., AZ

8. Observe this photograph taken along a canyon wall and ask yourself how long each feature took to form. You do not need to arrive at any answers, just the questions. The tan, brown, and yellowish rocks are all volcanic rocks, formed from molten rock and volcanic ash erupted from an ancient volcano.

9. Several questions about the rates of processes come to mind. Each layer in the volcanic rocks may represent a single pulse of eruption and could have accumulated rapidly, in minutes or hours. How long did it take to form all the layers? The landscape currently is being eroded, and this process has been occurring in this area for millions of years. How long will it take for the large brown blocks to fall or slide off the lower cliff? Some of these questions are about the present (how fast is erosion occurring), some are about the past (eruptions), and others are about the future (the blocks). The easiest questions to answer are usually about the present, and the hardest ones are about the past or the future.

Before You Leave This Page Be Able To

☑ Explain what GMT is and where the starting point is.

☑ Describe why we have time zones, and what influences time-zone boundaries and width.

☑ Explain how we calculate rates, giving some examples of relatively fast and slow natural processes.

1.15 What Might Happen If This Location Is Deforested?

YOU HAVE BEEN EMPLOYED by a county planning commission. You are asked to assess any possible impacts of logging (removing trees) of a mountainside in the area under your jurisdiction. To address the problem, you rely on your broad perspective and skills in the use of maps, satellite-image interpretation, physical geographic principles, and the scientific method. Then you will communicate your findings to policy makers and politicians on the commission.

Goals of This Exercise:

- Develop a list of observations about the area, accompanied by ideas of possible impacts of deforestation.
- Produce a list of questions you have about how different features and other aspects might be impacted by deforestation and how these might affect logging activities.
- Develop a list of the types of information you would need to answer your most important questions.

This three-dimensional perspective shows an area with a variety of terrains and climate. Make some observations about the landscape and record your observations on a sheet of paper or on the worksheet. As you study this landscape, record any questions you have about the impact of deforestation on the atmosphere, hydrosphere, lithosphere, and biosphere. The questions are what are important here, not the answers. You will be able to answer most or all of your questions by the end of the course. Include in your questions any you have about the potential impacts of deforestation on humans. Listed around the figure are some of the features and other aspects you should consider, but there are others not listed. Think broadly and be creative. We have listed a few questions to help you begin.

Patterns of atmospheric circulation, including prevailing wind direction and precipitation patterns:

Streams, lakes, and glaciers:

Slopes: How might the steepness of slopes interact with deforestation? Are steeper slopes more difficult or easier to log? What happens to soil if we remove vegetation on a steep slope? How about on a more gentle slope? Does erosion occur faster or slower on steep slopes?

Local climate: How cold is this place normally? Does most precipitation occur as snow or as rain, and how does this vary with the time of year? Is the precipitation spread out evenly over the year, or does it mostly occur during some seasons? If the winds are blowing from the water to the land, how do the mountains affect this moist air?

Farmland:

Types of Rocks, Sediment, and Soil:

Procedures

1. Examine the scene and make observations about the landscape. In your notes, write down what you consider to be the most important observations about the area.

2. Think about how different features or aspects could be affected by deforestation, or how that feature or aspect could affect the proposed logging. Write down the most important questions that arise from your considerations. For example, what are the implications if the area has a relatively wet climate compared to a drier climate? How would removal of the trees affect erosion along the stream? Consider any other factors that might be important with regard to the issue but might not be obvious in the scene. For example, does the economy of this region depend on logging?

3. List any types of information you would like to know about this location to answer your questions or further clarify the problem. When considering how logging could affect erosion of the slopes, for example, we would want to know whether the plan was to log steep slopes, gentle slopes, or both.

Water supplies from the streams, lake, and groundwater:

Impact on how sunlight reaches the bare ground:

Optional Activities

Your instructor may have you complete one or more of the additional activities listed below. For these activities, type your answers into a word-processing document and submit this to your instructor in the manner in which she or he indicates (paper copy versus electronic copy).

Nearby cities and towns:

4. List all the layers you would you want to have in a GIS database to study the issue and any remote sensing techniques that would be needed to collect data.

5. For two aspects that you consider during this investigation, generate a research question that addresses an important point capable of being answered using evidence in the form of data and observations. List the types of data you would need to answer these two questions and how you might collect them (field studies, satellites, etc.).

Coastal environment, which features a current flowing parallel to the coast:

6. For one of your two research questions, develop a well-conceived hypothesis—a statement, not a question, that explains what you suspect the answer to your question will be. Think about the various tools that a physical geographer uses. Which tools would you use to test your specific hypothesis?

Birds and other animals on land; fish in the lake, streams, and ocean; creatures on the seafloor:

01.15.a1

1.15

Energy and Matter in the Atmosphere

ALMOST ALL NATURAL SYSTEMS on Earth derive their energy from the Sun, but not all areas receive the same amount of sunlight. Instead, the amount of energy reaching Earth's surface varies from region to region and from time to time. The interaction of energy with the Earth's atmosphere and surface determine the climate, weather, and habitability of an area.

02.00.a2 Namibia, Africa

02.00.a3 Philippines

Sunlight warms the land and oceans, which in turn warm our atmosphere (◄), making some regions, such as this spectacular desert in Namibia, warmer than others.

What type of energy is in sunlight, and does all of the Sun's energy make it to Earth's surface?

The Sun rises and sets each day, except in some polar places where the Sun shines 24 hours a day during the summer (▶). In other places, on the opposite pole of the planet, there is total darkness during the same 24 hours.

What causes variations in the number of daylight hours, both from place to place and from season to season?

Many regions have seasons, changing from the warm days of summer to the cold, snowy times of winter (▶).

What causes the change from season to season, and do all areas experience summer at the same time?

02.00.a4 Sapporo, Japan

Tropical areas like Indonesia (▶) do not have a distinct summer and winter but may have a rainy season and a dry season.

Why do some regions experience summer and winter but others do not?

02.00.a5 Banda Island, Indonesia

02.00.a6

Antarctica, during its winter, has a dramatic thinning of the overlying ozone layer in the atmosphere, shown here in purple (◄).

What is ozone, what causes this thinning, and why is there so much global concern about this phenomenon when ozone makes up less than 0.001% of all the gas in the atmosphere?

Arctic Ocean

Asia

Pacific Ocean

Australia

Antarctica

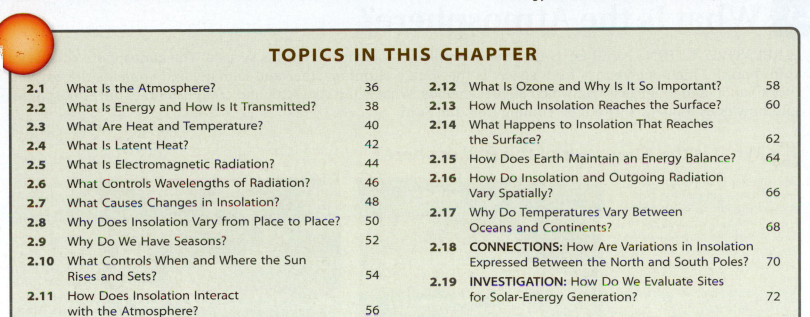

TOPICS IN THIS CHAPTER

Ancient people, such as the builders of Stonehenge (▶) 4,000 years ago, used changes in the position of the Sun over time to schedule important activities, such as the planting of crops.

What causes the seasons, and why do temperatures change from season to seasons?

02.00.a7 Stonehenge, Wiltshire, England

02.00.a8 West-Central IL

02.00.a1

Clouds, as in the thunderstorm above (▲), consist of small drops of water and ice crystals. The water to make the drops and crystals evaporated from the surface using energy from the Sun.

How much energy is needed to cause evaporation, and where does that energy ultimately go?

02.00.a9 Earth and Moon

Astronauts on the NASA Apollo Mission took this famous picture (▲) of "Earthrise" over the Moon.

Where does the light coming from the Earth in the photograph originate, and does the loss of this light affect Earth's overall energy balance?

2.0

What Is the Atmosphere?

A RELATIVELY THIN LAYER OF GAS — the *atmosphere* — surrounds Earth's surface. The atmosphere shields us from harmful high-energy rays from space, is the source of our weather and climate, and contains the oxygen, water vapor, and other gases on which all life depends. What is the character and composition of the atmosphere, and how does it interact with light coming from the Sun?

A What Is the Character of the Atmosphere?

Examine this view of Earth, taken from a spacecraft orbiting high above the Earth. As viewed from space, Earth is dominated by three things: the blue oceans and seas, the multicolored land, and clouds. If you look closely at the very edge of the planet, you can observe a thin blue fringe that is the atmosphere. From this perspective, the atmosphere appears to be an incredibly thin

02.01.a1

02.01.a2

As viewed from the ground, the atmosphere mostly appears as a blue sky with variable amounts of clouds, which are com- monly nearly

layer that envelopes our planet, separating us from the dark vastness of space. Clouds, which are so conspicuous in any image of Earth taken from space, mostly circulate within the lower atmosphere, bringing rain and snow. The winds that move the clouds are also within the atmosphere.

white or some shade of gray. During sunset and sunrise, the sky can glow reddish or orange. The colors of the sky and clouds are due to the way sunlight interacts with matter in the atmosphere.

B What Is the Structure of the Atmosphere?

The atmosphere extends from the surface of Earth upward for more than 100 km, with some characteristics of the atmosphere going out to thousands of kilometers. The atmosphere is not homogeneous in any of its attributes, but instead has different layers that vary in temperature, air pressure, and the amount and composition of gases. Each layer has the term "sphere" as part of its name, referring to the way each layer successively wraps around the Earth with a roughly spherical shape. Examine the figure below and then read the text from the bottom left, starting with the lowest and most familiar part of the atmosphere.

4. The top layer of the atmosphere is the *thermosphere*, derived from the Greek word for heat because this layer, surprisingly, can become very hot (more than 1,500°C) as gas particles intercept the Sun's energy. It is the altitude where the spectacular *auroras* (e.g., "Northern Lights") originate from interactions of solar energy and energetic gas molecules.

3. Above the stratosphere is the *mesosphere*, where "meso" is Greek for "middle," in reference to this layer being in the middle of the atmosphere. The mesosphere starts at 50 km, the top of the stratosphere, and goes up to more than 80 km (~50 miles) in altitude. The upper part of the mesosphere is very cold (−85°C, −120°F), and is considered by many scientists to be the coldest place within the Earth system. It is within the mesosphere that most small meteors burn up, producing the effect called "shooting stars." Radio waves from Earth bounce off this layer and the thermosphere, allowing us to hear radio stations from far away.

2. The next layer up is the *stratosphere*, beginning at an altitude of about 10 km above sea level, at about the elevation of Earth's highest peaks. The name is derived from a Latin term for spreading out, referring to its layered (not mixed) character. The name also helps convey that temperatures are stratified, varying from cooler lower altitudes to warmer upper ones. The lowest part of the stratosphere is an altitude at which many commercial jets fly because the air offers less resistance to motion, allowing appreciable fuel savings.

1. The lowest layer of the atmosphere is the one with which we surface-dwellers interact. It contains the air we breathe, clouds, wind, rain, and other aspects of weather. This layer is the *troposphere*, with the name "tropo" being derived from a Greek word for turning or mixing, in reference to the swirling motion of clouds, wind, and other manifestations of weather.

02.01.b1

5. On the right side of this figure are scattered bright dots that represent gas molecules in the atmosphere. The molecules are infinitely smaller and more abundant than shown here. Note that the molecules are more abundant lower in the atmo- sphere and become much less abundant upward. Over 70% of the mass of the atmosphere is in the lowest 10 km, that is, within the troposphere. The mesosphere and thermosphere contain only a few tenths of a percent of the atmosphere's mass.

C   What Is the Composition of the Atmosphere?

The atmosphere is not completely homogeneous in its vertical, horizontal, or temporal composition, but chemists and atmospheric scientists have estimated its average composition, as represented in the graph and table below.

The atmosphere is held in place by the balance between gravity (which is directed downward and keeps the gases close to the surface) and a buoyancy force (which is directed upward and exists because material tends to flow toward the vacuum of outer space). Near the surface, the greater weight of the overlying atmosphere results in more molecules being tightly packed close to the surface of the Earth and a rapid thinning of the number of molecules with distance upward in the atmosphere, away from the surface. This effect of gravity also influences the composition of the atmosphere, with a higher proportion of heavier gases low in the atmosphere (in the troposphere) and a higher proportion of lighter gases, such as hydrogen, higher up in the atmosphere.

02.01.c1

Composition of Atmosphere (by volume)	
Gas	**%**
Nitrogen (N_2)	78.084
Oxygen (O_2)	20.946
Argon (Ar)	0.934
Water vapor (H_2O)	0.400
Carbon dioxide (CO_2)	0.039
Neon (Ne)	0.002
Helium (He)	0.001
Less than 0.001% each	
Methane (CH_4)	
Krypton (Kr)	
Hydrogen (H_2)	
Nitrous oxide (N_2O)	
Xenon (Xe)	
Ozone (O_3)	
Nitrogen dioxide (NO_2)	

As shown by this diagram and table, the two dominant gases are nitrogen (78%) and oxygen (21%), followed by argon. Several other gases, such as water vapor, carbon dioxide (CO_2), methane (CH_4), nitrous oxide (N_2O), and ozone (O_3) play significant roles in global climate through their interaction with energy emitted by the Sun and re-emitted by the Earth.

The atmosphere also contains various types of solids and liquids called *aerosols*, such as dust, industrial pollutants, and tiny drops of liquid from volcanic eruptions. Aerosols play an important role in the energy balance of the Earth and aid in the formation of clouds and precipitation.

D   How Does the Atmosphere Interact with Energy from the Sun?

Gas in the atmosphere interacts with visible light and other energy radiated by the Sun, as well as energy reflected from and radiated from Earth's surface. An understanding of the possible types of interactions helps explain what we see every day, such as colors, as well as the underlying causes of weather and climate. Here we discuss four types of interactions: transmission, reflection, absorption, and scattering. The discussion will emphasize light, but the principles are applicable to other forms of radiant energy, such as ultraviolet energy.

Transmission—An object can be entirely or mostly transparent to light and other forms of radiant energy, in the way that a clear glass sphere permits most light to pass through. Allowing such energy to pass through is called *transmission*. The atmosphere is largely transparent to visible light, allowing it to pass through and illuminate Earth's surface. For some types of radiant energy the atmosphere is only partially transparent, because certain molecules interact with some of the incoming radiation. Energy from the Sun is efficiently transmitted through space, which is nearly a vacuum, lacking many molecules that could interact with the energy.

Absorption—Objects retain some of the energy that strikes them, and this process of retention is called *absorption*. Objects have varying degrees of absorption. This dark, dull-textured sphere is highly absorbent, retaining most of the light that strikes it, accounting for the sphere's dark color (not much light coming back off the surface). Energy that is absorbed by an object can be released back in another form (like heat from the dark ball). The giving out of radiant energy is called *emission*.

02.01.d1

Reflection—Instead of passing through an object, some light can bounce off the object, the process of *reflection*, as illustrated by light reflecting off a polished metal sphere. Not all light and energy bounce off real objects, so reflectivity can be thought of as a continuum from objects being perfectly reflective to having no reflectivity. For visible light, objects that are white or made of shiny, polished metal are highly reflective, whereas dark, rough objects have low reflectivities. In the summer sun, a white car, which reflects much of the Sun's heat, is cooler than a dark car that is not so reflective. If an object, like a leaf, preferentially reflects green light, we see the object as green.

Scattering—Most objects reflect some light, but in a way that disperses the energy in various directions, *scattering* the light. The shiny but rough sphere shown here has a highly reflective surface, but the roughness causes light to be scattered in various directions. Such scattering occurs when light and other energy strike the land, and it is also caused by certain types of aerosols and gases, like water vapor, in the atmosphere. Gases in the atmosphere preferentially scatter blue light, which spreads out in all directions through the atmosphere, causing the dominantly blue color of the sky.

Before You Leave This Page Be Able To

✓ Describe what the atmosphere is, including its average composition.

✓ Sketch and describe the layers in the atmosphere.

✓ Describe four ways that matter interacts with light and other radiant energy.

2.1

2.2   What Is Energy and How Is It Transmitted?

THE TRANSMISSION OF ENERGY, and the interactions between energy and matter, define the character of our planet and control weather, climate, and the distribution of life, including humans. Here, we examine the fundamentals of energy, including what it is, where it comes from, and how it is moved from one place to another.

A   What Is Energy?

All matter contains *energy*, which is the capability of an object to do work, such as pushing or pulling adjacent objects, changing an object's temperature, or changing the state of an object, as from a liquid to a gas. How is such energy expressed at the scale of atoms and molecules (combinations of atoms)?

1. Energy is expressed at an atomic level by *motions* of atoms, molecules, and their constituent parts. These motions include changes in position and in-place vibrations and rotations. The more motion the atoms and molecules display, the more energy the system contains. Atoms and molecules in a gas can move at over 200 m/s, or about 500 miles/hour. These fast-moving objects collide with adjacent ones, causing atmospheric pressure in a similar fashion to the way in which air molecules hold out the walls of an inflated balloon.

02.02.a1

Less Motion = Cooler

More Motion = Hotter

3. Energy can also be tied up *within* the atomic or molecular structure of matter. When we heat a liquid, we impart energy to the constituent molecules, causing some of the molecules to move apart, escaping as vapor. But these molecules also carry stored energy, which can be released when the molecules recombine into a liquid. This type of energy is called *potential energy*, because it is not being expressed directly, but could potentially be released. Where potential energy is related to a change in the *state of matter*, such as from a liquid to a gas, as in the example presented here, it is called *latent energy* (latent means hidden).

2. The temperature of an object is a measure of the average energy level (motions) of its molecules and atoms. The molecules of all objects on Earth, regardless of temperature, are moving, some more than others. This type of energy, due to motions of objects, is called *kinetic energy*.

B   How Does Energy Relate to the State of Matter?

In our everyday world, matter exists in three different forms or *states*—solid, liquid, and gas. The energy content of the material determines which of these states of matter dominates at any time and place.

1. When matter is in a *solid state*, the constituent atoms and molecules are bound together, such as in the structured internal architecture of a crystal. The energy levels (motions) of the atoms and molecules are low enough that the solid can withstand the vibrations and other motions without coming apart. As a solid is heated up by an external energy source, like the Sun, the motions become more intense.

02.02.b1

2. If enough energy is added to a solid, the motions begin to break the bonds that hold the solid together. The material begins to melt, turning into a *liquid*, which is a collection of mobile atoms and molecules that more or less stay together but are not held into the rigid form of a solid. For example, energy provided by the environment increases the vibrations of the water molecules in an ice crystal until the bonds holding the ice crystal together begin to disintegrate and the ice melts.

3. If even more energy is added, such as by placing the liquid water on a hot burner, the added energy causes even more energetic motions of water molecules in the liquid, allowing more and more molecules to break free of the liquid and enter the air as a *gas*—water vapor. As a result, atoms and molecules in their gaseous state are more energetic than in their solid or liquid equivalents. Not only are the molecules in a gas intensely vibrating and rotating, molecules can now move freely through the air at high speeds—having a large amount of kinetic energy.

C   What Occurs During the Processes of Warming and Cooling?

When something changes temperature, we say that it is warming or cooling, but what actually occurs during a change in temperature? For warming or cooling to occur, energy must be transferred from one object to another, such as from a flame to the air around it.

02.02.c1

02.02.c2

When an object increases in temperature, it is *warming*. Warming occurs where an object *gains energy* from the surroundings, resulting in the increase in temperature. You gain energy when you feel the warming benefit of a campfire (▶). There is no need to touch the flames to feel the warming effect. Energy is transmitted from the fire to the surroundings. When you heat up water in a pan on the stove, energy is transferred from the burner to the pan and then from the pan to the water. If you add enough energy, the water molecules become so warm and energetic that they start to escape the pan during the process of boiling.

The opposite of warming is *cooling*. Cooling is the *loss of energy* to the surroundings. When you are out in cold weather, you lose energy to the cold air, causing you to feel cold. A similar process occurs when you add warm water to ice cubes in a glass dish (◀). Initially, the water is warmer than the ice, but once the water and ice are in contact, molecular motion in the water is transmitted to the cooler ice. This transfer of energy adds more energy to the ice, eventually melting it. As the surrounding water loses energy to the ice, it cools, resulting in cold water. Cooling indicates a loss of energy, whereas warming indicates a gain of energy.

D   What Are the Four Dominant Types of Energy Transfer?

Heat is thermal energy transferred from higher-temperature to lower-temperature objects. Heat transfer, also called *heat flux* or *heat flow*, results when two adjacent masses have different temperatures. The four mechanisms of heat transfer are *conduction, radiation, convection,* and *advection.* An understanding of the various mechanisms of energy transfer is crucial in understanding weather, climate, evolution of landscapes, and many of our daily activities.

Conduction — A water-filled pan on a burner gets hot as thermal energy is transferred by direct contact between the burner and pan, and the pan and water. Heat transfer by direct contact is *conduction*, which involves transferring thermal energy from the warmer object (with more energy) to the cooler one (with less energy). Energy can only be transferred from the more energetic one to the less energetic ones. Molecules are most densely packed in solids and least densely packed in gases, so conduction is a more important mode of transfer in solids, whereas gases conduct energy much less efficiently.

02.02.d1

Advection — Moving a pan full of hot water away from the stove also transfers heat from one place to another. Energy transfer by the *horizontal movement* of a material, such as moving the pan sideways off the burner, is called *advection*. Fog along many coastlines results from warm air flowing sideways and mixing with cooler air over the coast. This horizontal transfer of energy via the moving air is advection.

Radiation — A hot burner on a stove can warm your hands a short distance away. Such warming occurs because heat from the burner radiates through the air, a process called *radiation* or *radiant heat transfer*. Radiation is energy transmission by means of electrical and magnetic fields. All objects constantly emit various kinds and quantities of this particular type of electrical and magnetic radiation, which is known as *electromagnetic radiation*. This form of energy is even capable of passing through a vacuum.

Convection — Energy is conducted through the base of the pot and into the lowest layer of water contacting the pot. These molecules move more, requiring more room to do so, which causes the volume occupied by the warmer water to increase and its *density* (mass per unit volume) to decrease. Being less dense than the overlying cooler water, the warm water rises. When the rising water reaches the surface, it cools and flows back down the sides. This type of vertical heat transfer by flow of a gas, a liquid, or a weak solid is *convection*. If the material flows around a circular path, as in the pan, we use the term *convection cell*.

Before You Leave This Page Be Able To

✓ Describe what constitutes the energy in an object.

✓ Describe how energy relates to states of matter.

✓ Describe the four major mechanisms of energy transfer and provide an example of each.

2.2

2.3 What Are Heat and Temperature?

THE TERMS HEAT AND TEMPERATURE are used every day, but what do they actually mean? *Temperature* is a measure of the object's internal kinetic energy—the energy contained within molecules that are moving, and *heat* is thermal energy transferred from one object to another. Moving molecules drive many processes in the Earth-ocean-atmosphere system, such as evaporation, precipitation, and erosion.

A   What Is Sensible Heat?

1. The term heat is used in two ways. Scientists use *heat* to refer to the transfer of thermal energy from a warmer object to a cooler one or to the energy that is transferred in this way. The amount of heat is specified in a unit called a *Joule*, a measure of work or energy. Two common examples illustrate heat nicely.

2. What happens when you hold a cup of hot tea? Your hand feels heat coming from the cup. You are feeling the transfer of thermal energy from the cup to your hand. For this to happen, water molecules were heated and made to move. Once in the cup, the moving molecules collided with the inside of the cup, warming it. Then, that heat is transferred through the cup and against your hand via conduction—from the burner to the bottom of the kettle, to the water, to the cup, and finally to your hand. Conduction and convection both help distribute heat within the kettle and the cup.

02.03.a1

3. What do you think happens when you hold a cold glass of ice water? In this case, the molecules in your hand are more energetic than the ones in the cold drink, so heat is transferred from your hand to the glass. Your hand feels cold because you are losing thermal energy to the cold glass.

4. This type of heat, which changes the temperature of two objects through exchange, is called *sensible heat* because we can sense it. But do we sense the actual temperature of an object or just the heat gain or loss? Try this experiment: Find a metal object and a wooden or plastic object in the same place. Place your hand on each and observe what you feel. Go do it, and then come back and continue reading. There, you no doubt sensed that the metal felt colder than the wood or plastic, but both have been in the room for a while and so are exactly the same temperature. This experiment shows that we sense heat gain and loss more than the actual temperature. In your experiment, metal conducted heat away from your hand faster, and so felt colder, but it wasn't.

B   What Is Temperature and How Do We Measure It?

Temperature is a quantitative measure of the average kinetic energy of molecules in an object—in other words, the hotness or coldness. Measurement of temperature of an object, whether a solid, liquid, or gas, involves the transfer of sensible heat from the object to some type of measuring device, usually a thermometer. Official temperatures are measured in a variety of ways, depending on the accuracy that is required and the location of the measurement.

02.03.b1

02.03.b2

02.03.b3

02.03.b4

The mercury-in-glass thermometer is the most familiar tool to measure temperature. Mercury is a convenient element because many of its physical properties remain consistent over the range of temperatures experienced on Earth. As the mercury's temperature increases, it expands and fills more of the tube. When it cools, the mercury contracts and withdraws down the tube.

Infrared thermometers calculate the temperature of a solid or liquid surface by pointing the sensor at the surface as it measures a range of wavelengths of energy emitted by that surface. Equations then relate wavelength to energy and energy to temperature. Infrared thermometry is a form of remote sensing, convenient when the surface is too far away, or too dangerous, to be measured directly.

When we need to measure sudden and slight temperature changes very precisely, we use special thermometers in which differences in energy content cause a thermoelectric response that can be wired to a computerized data recorder. The temperature can be calculated using specific equations that derive the amount of heat as a function of the amount of electrical current and the resistance of the electrical circuit.

Temperatures in the atmosphere are usually measured by weather balloons. These include instrument packages called *radiosondes* that measure a range of variables at various heights as the balloon ascends. Temperature is measured using thermoelectric principles. Wind, humidity, and other variables are also measured and relayed to the ground via radio signal.

C How Does the Fahrenheit Scale Relate to the Celsius and Kelvin Scales?

Most Americans are familiar with the *Fahrenheit temperature scale*, in which 32° represents the freezing point of water and 70° is a comfortable temperature. Nearly all nations except the U.S. use the *Celsius scale*, and scientists use the Celsius scale or a related scale called the *Kelvin scale*. We typically compare the scales with reference to the temperatures at which water freezes or boils, called the *freezing point* and *boiling point*, respectively.

In the Fahrenheit scale (°F), the boiling point of water is 212° (at sea level, but lower at higher elevations). In the Celsius scale (°C), the boiling point is 100° (at sea level).

Typical room temperature is 77°F, which is equivalent to 25°C.

The freezing point of pure water is 32°F, which is equivalent to 0°C. The Celsius scale was calibrated to the freezing and boiling points of water, with 100°C separating the two, a convenient scale.

The two scales correspond to one another at −40° (that is, −40°F equals −40°C).

Water Boils at Sea Level (100°C, 212°F)

Typical Room Temperature (25°C, 77°F)

Water Freezes (0°C, 32°F)

°C = °F (−40°C, −40°F)

02.03.c1

Water Boils at Sea Level (100°C, 373 K)

Typical Room Temperature (25°C, 298 K)

Water Freezes (0°C, 273 K)

02.03.c2

Both the Celsius and the Fahrenheit scales are "arbitrary" in the sense that zero degrees doesn't mean that there is a lack of internal energy. Likewise, a doubling of the Fahrenheit temperature from 40° to 80° does not mean that there is twice as much internal energy. In scientific calculations, we need a temperature scale that allows us to relate changes in internal energy to the absolute amount of heat gained or lost by a system.

The Kelvin temperature scale (K) was devised as an "absolute" temperature scale to remedy these problems. In the Kelvin system, 0 K corresponds to the temperature at which no internal energy exists and all molecular motion theoretically ceases. This temperature is known as *absolute zero*, and is −273°C or −460°F. Doubling the internal energy of molecules would double their speed and be associated with a doubling of the Kelvin temperature.

In the Kelvin system, water freezes at 273 K and boils at 373 K. Converting from Celsius to Kelvin temperature is easy:

$$K = C + 273$$
or
$$C = K - 273$$

Conversions between Fahrenheit and Kelvin can be made by converting first to Celsius and then to the other scale. Note that we do not use a degree symbol in reporting temperatures with the Kelvin scale.

Between the boiling and freezing points of water, there are 100 Celsius degrees but 180 Fahrenheit degrees. So one Fahrenheit degree is only 100/180 (or 5/9) of a Celsius degree. This fact, along with the different "starting point" (i.e., the freezing point of water) forms the basis of converting between Fahrenheit and Celsius. To convert from Fahrenheit to Celsius, we must first subtract 32 degrees from the Fahrenheit temperature to allow for the fact that the starting point is offset by 32 degrees in the two systems. Then we must multiply by 5/9 to allow for the differences in the value of a degree on each scale. The equations for converting back and forth are as follows:

$$C = 5/9 \times (F - 32)$$

$$F = (C \times 9/5) + 32$$

D How Many Stations Report Temperature?

This globe (►) shows part of the worldwide distribution of weather stations that have temperature data sets for at least many decades. The distribution is uneven, with most stations being on land, with relatively few stations in oceanic areas (on islands). Most stations are concentrated in densely populated areas, especially in the lowlands of more developed regions, such as the eastern U.S. Other regions, such as the center of South America (the Amazon rain forest), have very few stations to represent rather large areas. In recent decades, remote-sensing techniques have allowed truly global temperature coverage, but these data sets are not available as far back in time.

02.03.d1

Before You Leave This Page Be Able To

✓ Explain what temperature is.

✓ Describe the strengths and weaknesses of the various temperature scales.

✓ Convert back and forth between the scales.

✓ Explain how various instruments to measure temperature work.

✓ Characterize the distribution of temperature-monitoring stations.

2.3

2.4 What Is Latent Heat?

WATER OCCURS IN ALL THREE PHYSICAL STATES—solid, liquid, and gas—at temperatures commonly found on Earth. Although the chemical structure of water remains unchanged from state to state, the three states, also called *phases*, are differentiated by the physical spacing of the water molecules. Considerable quantities of energy, contained as *latent heat*, are involved in these changes of state, and act as moderators of global climate.

A What Are the Forms of Latent Heat?

The chemical substance *water* consists of two hydrogen atoms bonded to a single oxygen atom, with a chemical formula of H_2O. The change in state between any two of these phases requires the addition of energy or involves the release of energy, depending in which direction the change is occurring (e.g., liquid to solid versus solid to liquid).

1. When ice is placed in warmer surroundings, like an ice cube on a kitchen counter, energy from its environment flows into the ice, increasing the internal motions of water molecules in the solid, crystalline structure. At first the ice heats up (increases in temperature), but once it reaches a certain temperature (its melting point) it begins to melt. *Melting* requires energy to be added (as shown by the blue-to-red color of the arrow for melting).

2. During melting, energy input into the system is stored (absorbed) in water molecules of the liquid—as *latent heat*. The latent heat associated with melting is called the *latent heat of fusion*. If enough energy is removed for the liquid water to be converted back into ice during *freezing*, the latent heat stored in the water molecules is released back into the surroundings as heat (also called *latent heat of fusion*). For freezing to continue, this released energy must be dispersed into the surroundings. The warm flow of air coming from the back or bottom of a household freezer is the heat being dispersed from the cooling system.

02.04.a1

3. Conversion from a liquid phase to a vapor phase, *evaporation*, also requires an input of energy from the surroundings, as when you boil water in a pan. During evaporation, energy added to the liquid breaks the bonds holding the water molecules together, allowing molecules to escape as a gas. The liberated gas molecules are moving fast and so have increased kinetic energy and carry energy stored as *latent heat*. When the gas molecules are cooled and recombine into a liquid, the process of *condensation*, the latent heat is released back into the surroundings. The latent heat associated with evaporation and condensation is the *latent heat of vaporization*.

4. Water can also go directly from the solid state to a vapor, the process of *sublimation*. Sublimation requires energy from the surroundings and stores latent heat in the gas molecules—the *latent heat of sublimation*. The reverse process, converting water vapor directly into ice, is called *deposition* and is the main way snowflakes form. Deposition releases the latent heat back into the environment, but when it is snowing it is cold enough that we would not easily notice any addition of latent heat to the cold air.

B What Happens to Temperature During Melting and Boiling?

1. An interesting thing happens to temperature when we heat ice and convert it to liquid water and then water vapor, as shown by the graph to the right. This graph plots the energy input into the system (measured in a unit of energy called a kilojoule) versus the resulting temperature if we start with a kilogram of ice. The process begins in the lower left corner, with ice at −20°C, well below its melting temperature (0°C).

2. The initial input of energy into the system causes ice to increase in temperature, represented by the first inclined, red part of the line. The increase in energy is expressed as sensible heat.

3. When the temperature reaches 0°C, the melting point, the ice starts to melt. The temperature does not change during melting (as shown in the short, horizontal brown line). Instead, all the increase in energy is going into breaking the bonds and is stored as latent heat.

4. Once all the ice is melted, the increase in energy again causes an increase in temperature (sensible heat), as shown by the second inclined red line. The temperature of the water increases until it reaches the boiling point (100°C).

5. As the water starts to boil, it does not increase in temperature (long, horizontal brown line). All increase in energy is used to convert liquid into the vapor phase and is stored as latent heat.

6. Once all the water has become vapor, further heating causes the steam to increase temperature (inclined, red line). In this entire process, more energy was used to change states (brown horizontal lines) than was used to increase temperature (the inclined, red lines). Of the total amount of energy used, less than 20% went to change temperature and more than 80% was used to change state!

02.04.b1

C   What Does Latent Heat Do to the Surroundings?

Latent energy added to the water molecules or released by the water molecules allows the phase change to proceed, but it also impacts the temperature of the environment. More than five times the energy is involved in evaporating or condensing water (the latent heat of vaporization) than in raising the temperature of the same mass of water from the freezing point to the boiling point. The large quantities of latent heat have a huge role in many aspects of our world, including changes in atmospheric temperature.

Water: Vapor, Liquid, and Ice

This graph shows conditions under which each phase occurs. For reference, a pressure of 1.0 bar is the average atmospheric pressure at sea level, and 25°C is standard room temperature.

02.04.c1

Ice occurs at low temperatures, whereas liquid and water vapor are favored by higher temperatures. Higher pressure acts to hold the water molecules within the liquid rather than allowing them to escape into the air. If we cool vapor, we get liquid at higher pressures or ice or at lower ones, causing the formation of clouds or precipitation, both of which consist of liquid (drops of water) and ice.

Phases, such as vapor, that require more energy to form are called *high-energy states*, whereas less energetic ones are *low-energy states*.

Cooling and Heating the Air

When water in the atmosphere changes state, it releases or takes in thermal energy, heating or cooling the surrounding air. This diagram illustrates the change in air temperature during phase changes. Red arrows indicate that phase change in this direction causes the surrounding air to heat up. Blue arrows indicate that the surrounding air must provide heat to the phase change, and so the air cools.

02.04.c2

Energy must be released into the surroundings for water to go from a higher-energy state to a lower-energy molecular state. Heat is released (red arrows) when water vapor forms droplets or ice crystals, or when liquid water freezes—this warms the surrounding air.

Heat is taken in from the surroundings (blue arrows) when ice melts or water evaporates, or ice sublimates. Any of these processes cool the air.

D   What Are Examples of Latent Heat in the Environment?

02.04.d1

1. In 2004 Hurricane Ivan deposited an average of 15 cm of rainfall over Duval County and the city of Jacksonville in northeastern Florida. When we consider the amount of rainfall, the area of the county, and the typical amount of latent heat released during condensation, we calculate that this storm, in this one county, released latent heat equivalent to much more than all the electrical energy consumed by the world since electricity was harnessed for human use. Many such storms occur each year and cover even larger areas.

3. A cold beverage in a can, bottle, or glass warms as a result of conduction from the cold container and its warmer surroundings, like a table, but it warms even more quickly because condensation releases latent heat on the outside of the container.

02.04.d3

2. Severe freezes are a serious threat to the citrus industry in the southern and southwestern U.S. Sustained subfreezing temperatures may ruin the crop and even destroy the tree. To prevent extensive losses, grove owners spray the trees with water, which freezes on the crop. At first sight the weight of the accumulated ice only seems to add to the damage by breaking off fruit and limbs, but each kilogram of water that freezes on the trees releases enough energy to the fruit and plant to prevent cold temperatures from destructively freezing water within the cells of the plant.

02.04.d2 Florida

Before You Leave This Page Be Able To

☑ Explain the relationship between latent heat and changes in the state of water.

☑ Compare the quantities of energy involved in the latent heats of fusion and vaporization, compared to changes in temperatures.

☑ Sketch and describe how latent heat affects the environment, providing some examples.

2.4

2.5 What Is Electromagnetic Radiation?

ELECTROMAGNETIC RADIATION is one of the fundamental forces of nature. It dominates our daily interactions with the world, determining the color of objects, the character of the air we breathe, and the physical characteristics of the water we drink. Electromagnetic radiation is essential to the operation of weather and our climate system and to all lifeforms on the Earth's surface, including us.

A What Are Some Common Examples of Electromagnetic Radiation?

Electromagnetic radiation (EMR) is all around us, although we may not be aware of all of its manifestations. Its most obvious expression is as light, which is the only kind of EMR that we can observe with our eyes. EMR is also what causes sunburns, is the heat we feel from a heat lamp, and is expressed in all sorts of wireless communications, including TV, radio, and WiFi.

Visible light is one type of electromagnetic radiation (EMR). Visible light is itself composed of different types of EMR, which we can observe by using a prism (▶) to spread out visible light into its rainbow colors. On following pages, we explain what causes different colors.

02.05.a1

02.05.a2

There are many types of EMR that are not visible to us. These include radio waves (like WiFi), X-rays, and microwaves (◀) that cook our food. These three types of EMR cannot be observed, but are used daily.

B What Is the Character of Electromagnetic Radiation?

Electromagnetic radiation consists of energy radiated from charged particles and manifested as interacting electrical and magnetic fields. For most applications, it is useful to think of electromagnetic radiation as a series of waves that, if unconfined, radiate out in all directions. When observed in certain ways, such as at microscopic distances and exceptionally short timescales, EMR appears to be composed of particles. Here, we will treat EMR as waves.

1. To visualize the rather abstract concept of EMR, think of it as a series of waves (▶) of electrical and magnetic energy that are moving from one place to another, in this case, from left to right. Like any waves, some parts are higher and some are lower. The direction in which a wave is moving is the *direction of propagation*. The wave shown here is propagating from left to right. To envision how such a wave moves, think about what happens when you shake the end of a rope or string. The rope curves into the series of waves, which move from your hand outward to the end of the rope.

Direction of Wave Propagation

Amplitude

Motion of Material Within Wave

Wavelength

02.05.b1

2. In this type of wave, the motion of any part of the wave is mostly up and down, parallel to the small arrows on this figure—the rope stays in your hand, so it is not moving away from you, although the waves are. Motion within the wave (up and down) is perpendicular to the direction of propagation of the wave (left to right).

3. In describing such waves, whether in a rope or EMR, the term *amplitude* refers to the height of the wave, from trough to crest. The greater the difference in height between the top and bottom of the wave, the greater the amplitude. Surfers like large amplitude waves.

4. We use *wavelength* to describe the distance between two adjacent crests (tops) or two adjacent troughs (bottoms). The longer the distance between two adjacent crests or troughs, the greater the wavelength. If we know the speed of the wave, we can describe the *frequency*—the number of waves per second passing a point.

5. Electromagnetic waves travel through electric and magnetic fields, which can be envisioned as two mutually perpendicular planes. In this figure (▶), the electric field, containing the electrical component of the wave, is vertical (dark gray), whereas the magnetic field is horizontal (colored blue with the orange horizontal arrows).

Plane of Electrical Component

Electrical Component of Wave

Magnetic Component of Wave

Plane of Magnetic Component

02.05.b2

6. The electrical and magnetic waves have the same wavelength, and the crests of the two types of waves are the same distance along the direction of wave propagation (i.e., the crests line up). Motion in both types of waves is perpendicular to the direction the wave propagates, as described in the previous figure. Electromagnetic waves get their name from these linked electrical and magnetic waves that move in unison.

C How Is Electromagnetic Energy Generated and Transmitted?

Electromagnetic radiation is generated by changes associated with charged particles—atoms and molecules produced by vibrations within the particle, by changes in the energy level of electrons, and by fusion of particles, as occurs in the Sun. The Sun is the dominant source of energy for Earth, emitting EMR at a variety of wavelengths.

02.05.c1

1. Atoms have a tiny central core—the *nucleus*—that is so much smaller than the entire atom that it cannot be shown here. The red and yellow spheres are electrons. Groups of electrons travel around the nucleus at different distances, called *electron shells.* Each shell has a different level of energy, increasing away from the nucleus. If an electron in an outer, higher energy shell drops into an inner, lower energy shell, it must emit the extra energy as an EMR wave, which radiates out in all directions (shown here as one direction for simplification).

02.05.c2

2. Electrons within atoms and the bonds (shown here as springs) within molecules vibrate back and forth, changing positions slightly. This motion emits EMR, and the faster such motions occur, the higher the frequency of the EMR that is emitted.

02.05.c3

3. The Sun is huge and extremely hot. It is nearly all composed of hydrogen and helium, the two lightest chemical elements. Through a series of steps, protons and neutrons join together to produce larger particles, such as helium—the process of *fusion*, which releases huge amounts of EMR that radiate outward.

4. Earth is approximately 150 million km from the Sun, and the nearly complete vacuum of space separating the two objects eliminates the possibility of conduction, convection, and advection, all of which require a medium for the transfer to occur (there is no medium in a vacuum). The Sun's immense mass means that very little material can escape the Sun's gravitational pull. Therefore, almost all energy comes to Earth via radiation, which can pass through a vacuum, as illustrated by the fact that light from a flashlight (▶) can easily pass through sealed flasks, one containing air and other containing a vacuum.

02.05.c4

D How Much Energy Does an Object Emit?

Scientists have discovered a number of important quantitative relationships, some of which are considered laws of nature, including those that govern the production and transfer of energy. One important relationship is the Stefan-Boltzmann Law, which relates an object's temperature to the amount of EMR it emits.

1. The Stefan-Boltzmann Law is represented graphically by this figure, which plots the *amount of emitted energy* as a function of temperature in Kelvin (K), for a specific type of object that absorbs all the energy that strikes it. We are simplifying this discussion a bit, which your instructor will recognize, but we don't think you'll mind.

TEMPERATURE (°C AND K)

Y-axis: ENERGY EMITTED (1000s W/m²), 0 to 700
Top axis (°C): -273°, -73°, 127°, 327°, 527°, 727°C
Bottom axis: 0, 200, 400, 600, 800, 1,000 K

02.05.d1

2. The graph starts at a temperature of zero Kelvin (−273°C), and at this temperature the amount of emitted energy is zero. All molecular motion ceases at zero Kelvin—which is why it is called *absolute zero*—so at this temperature no radiation would be emitted. No motion means no EMR.

3. With an increase in temperature (moving to the right on the graph), the amount of energy emitted increases, slowly at first. What this indicates is that as an object increases in temperature, the object emits more energy (in the form of EMR). In other words, hot objects emit more radiation than do cold ones. Since temperature represents the amount of motion associated with the atoms and molecules within a material, the graph means that the amount of EMR increases as the amount of atomic scale motion increases.

4. At higher temperatures, the amount of energy emitted increases dramatically as a function of temperature, much faster than it did at lower temperatures. Compare how much the amount of emitted energy increased from 200 to 400 K with how much it increases from 800 to 1,000 K. The shape of the curve indicates that hot objects emit much larger quantities of energy than cooler ones.

5. Using the equation from which this graph was derived, and factoring in some other aspects not discussed here, we can use an object's surface temperature to predict how much energy it should emit. The Earth's average temperature is 283 K, whereas the Sun's average surface temperature is 6,000 K (way off the graph)—we would predict the Sun emits much more energy than the Earth. From the calculations, the Sun should emit more than 200,000 times more energy than Earth!

Before You Leave This Page Be Able To

✓ Explain what EMR is and how it is generated.

✓ Explain how the surface temperature of an object is related to the quantity of EMR it emits.

2.5

2.6 **What Controls Wavelengths of Radiation?**

ELECTROMAGNETIC RADIATION takes many different forms, each of which have very different properties, determined by their wavelengths. Variations in wavelength explain the existence of different colors and warming of the Earth due to climate change. How are differences in the wavelength of electromagnetic energy (EMR) expressed in our world and in the Solar System?

A What Range of Wavelengths Does an Object Emit?

Not all molecules in an object vibrate or move at the same speed, so an object emits energy with some variation in the wavelengths and amounts of energy it emits. Some molecules vibrate or move faster, emitting EMR with shorter wavelengths and higher energies. Other molecules vibrate or move more slowly, resulting in EMR with longer wavelengths and lower energies. Temperature is a measure of the *average energy content* of all molecules in an object.

Molecules in a cube-shaped container (▶) are color coded by *energy level*; red is lowest, purple is highest, and other colors are in between. Molecules possessing the average energy level (i.e., temperature of the object) are in olive green. Observe the relative proportions of the different energy levels (that is, the different amounts of each color of molecule). Which are most abundant and which are least abundant?

02.06.a1

02.06.a2

This graph (◀) shows the number of molecules for each class of energy in the square object we just examined. The height of each column represents the number of molecules emitting energy at that wavelength.

The peak of the graph represents the wavelengths and energy levels that are most common. Columns to the left of the peak have lower energy and longer wavelengths than the peak, whereas those to the right have higher energy and shorter wavelengths. Taking into account all the molecules, there will be an average energy level emitted by this material at this temperature (near the peak).

B What Type of Energy Is Emitted from an Object?

The rate of motion of molecules within an object is related to the object's overall temperature. There is also a relationship between temperature and the wavelength of energy emitted by that object.

1. Examine the figure below, which shows three blocks of the same material but at different temperatures, with the blue block being cool, the brown block being warmer, and the red block being the hottest. Coming off each block are arrows depicting the amount and wavelength of EMR being emitted. Observe how the amounts and wavelengths relate to temperature.

Cool   Warm   Hot

02.06.b1

2. If an object is relatively cool, the motion of its constituent atoms and molecules will be relatively slow. This in turn causes the rate of energy shifts to be less frequent, resulting in energy with a longer wavelength.

3. As the temperature of an object increases, it not only emits more EMR (as shown by more arrows for the warm and hot blocks in the figure), but the wavelength of the EMR becomes shorter at higher temperatures. In a hot object, the atoms and molecules are moving relatively fast, and they change energy states more often, producing EMR with a shorter wavelength. If you shake a rope rapidly, you make shorter waves than if you shake it slowly. Go try it!

4. Whether an object is cool or hot, it emits a range of wavelengths of EMR, as described in the section above. However, scientists discovered a numeric relationship relating the dominant wavelength of EMR to temperature, a relationship known as *Wien's Law*.

5. The red curve on this graph (▶) represents Wien's Law, plotting the dominant wavelength of EMR emitted as a function of temperature. The curve is highest on the left, indicating that an object with a low temperature emits EMR with very long wavelengths.

02.06.b2

6. As the temperature of an object increases, the wavelength decreases very rapidly at first and then more slowly toward higher temperatures. Wien's Law indicates that cold objects emit much longer wavelength energy that hot objects. Using this graph or the equation associated with Wien's Law, we can use the temperature of an object to then predict the dominant wavelength of EMR it emits. For the Sun's very high temperatures (off the curve to the right), we predict energy wavelengths of 0.5 μm (micrometers). The Earth's much cooler temperature yields longer wavelengths of about 10 μm.

C What Is the Electromagnetic Radiation Spectrum?

The various objects in the universe emit a diversity of wavelengths of electromagnetic radiation. EMR varies from relatively long-wavelength, low-energy radio waves, which bombard our bodies every second of every day, to short-wavelength, high-energy X-rays and gamma rays, which are not so benign. The different wavelengths (and therefore frequencies) of electromagnetic radiation are arranged in what is called the *electromagnetic spectrum*, shown below. Inspect this figure and then read the associated text. In this figure, the longer wavelengths are at the top.

1. The longest wavelengths of EMR are *radio waves*. These include the waves that carry signals from typical FM and AM radio stations, and also include even longer radio waves called VLF (for very low frequency), with wavelengths of up to 100 km. VLF waves can penetrate significant depths of water and so are used for communication with submarines.

2. *Microwaves* are another form of EMR, with shorter wavelengths, higher frequencies, and higher energies than radio waves. Microwave ovens use a specific frequency of microwave that energetically excites (heats) water molecules.

3. Next on the scale is *infrared energy* (IR). Although we cannot see IR, our skin is sensitive to it, so we often think of infrared radiation as heat. Infrared energy is incredibly important on Earth, playing a key role in keeping our planet a hospitable temperature. There are several types of IR, including *thermal-IR*, which is close to microwaves in wavelength, and *near-IR*, which is near to visible light, the next entry on the spectrum.

4. *Visible light* occupies a relatively narrow part of the spectrum. It varies from red colors at long wavelengths to violet colors at short ones. Orange, yellow, green, and blue are in between red and violet.

5. Next to, and with shorter wavelengths (higher frequencies) and more energy than visible light, is *ultraviolet light* (UV). Ultraviolet is more energetic than visible light and is known to cause skin cancers and possible genetic mutations.

6. The shortest wavelength (highest frequency and therefore most energetic) waves are *X-rays* and *gamma rays*. These are potentially harmful. The energy in X-rays is used in medical technology as it will pass through soft tissue, but not bone.

02.06.c2

7. The different colors of visible light are different wavelengths of energy, ranging between 0.4 μm (violet) and 0.7 μm (red). The component wavelengths of visible light can be seen when the light is split by an optical prism or in a rainbow (▲).

8. The human eye is sensitive to radiation of the wavelengths between violet and red, and this is why this portion of the EMR spectrum is known as "visible light." The Sun's wavelengths are concentrated at 0.5 μm, where we predicted them based on Wien's Law. This wavelength is in the middle of visible light, coinciding with blue light. It is not a coincidence that the human eye developed to detect EMR of the wavelengths that carry most energy from the Sun.

02.06.c1

9. The Sun produces huge quantities of energy per square meter of its surface, as indicated by the yellow area on the graph to the right. It also produces most of its energy near 0.5 μm, centered on the wavelength of visible light; these wavelengths are called *shortwave radiation*. The Sun emits a very low proportion of its energy below 0.1 μm and above 1.0 μm in wavelength.

10. The Earth emits low quantities of energy, as indicated by the relatively small blue area on the graph, which is here greatly exaggerated relative to the curve for the Sun. Most of that energy is emitted back to space at wavelengths of around 10 μm, called *longwave radiation*. As we will discover, the fact that Earth receives its energy at short wavelengths, and returns it at wavelengths 20 times longer, is of fundamental importance to maintaining temperatures on Earth favorable for life.

02.06.c3

Before You Leave This Page Be Able To

☑ Describe why an object emits a range of EMR wavelengths.

☑ Describe the relationship among molecular motions, wavelengths of emitted EMR, and temperature.

☑ Summarize the electromagnetic spectrum, indicating the relative order of different types of EMR on the spectrum.

2.6

2.7 What Causes Changes in Insolation?

THE ENERGY TRANSMITTED from the Sun to Earth, called incoming solar radiation, or *insolation*, has varied only slightly during the short time for which we have accurate measurements from satellites. How much energy do we receive from the Sun, and why does it vary at all?

A How Much Energy Is Transferred from the Sun to the Earth?

1. To help us envision how much energy Earth receives, imagine a thin circular disc (▶) that is the same diameter as Earth, making it just large enough to intercept all energy that would fall on Earth's more complex, spherical shape.

2. According to both measurements and calculations, a relatively consistent amount of energy is radiated on this disc every second. This relatively consistent amount of energy—equivalent to 1,366 watts per square meter (W/m²)—is called the *solar constant*. This equates to only about one two-billionth of the power emitted from the Sun, since the Sun's energy goes out in all directions, not just toward Earth. Each planet would have a different solar constant due to differences in their distance from the Sun, with planets farther from the Sun receiving much less energy.

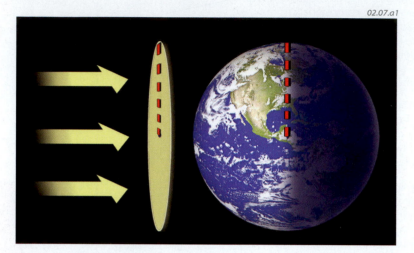

02.07.a1

4. During the course of a day, this amount of energy is spread out over the entire surface area of the planet and its atmosphere. A sphere has four times more surface area than a circle of the same radius, so the energy must be spread out across four times more area for the sphere compared

3. However, each place on Earth's surface does not get 1,366 W/m² of power from the Sun all the time. Any place experiencing night receives zero, whereas equatorial locations experiencing clear skies during the summer may get 600–700 W/m² at noon. Something happens to the rest of the insolation between the time it reaches the atmosphere and the time it reaches the surface.

to the disc. Dividing 1,366 W/m² by four yields 341.5 W/m². Geographers and solar scientists actually use 345 W/m² (slightly more than ¼ of 1,366) for the average insolation to the top of the atmosphere because Earth is not exactly a sphere.

B How Does the Amount of Insolation Change During a Year?

The amount of insolation varies during the year because the Earth is closer to the Sun at certain times of the year, but not the times you would expect—Earth is closer to the Sun during the Northern Hemisphere winter.

1. Although the average Earth-Sun distance is 150 million km, Earth's orbit is slightly elliptical rather than circular. The orbit is so nearly circular that it is virtually impossible to notice the elliptical shape in an accurately drawn view looking perpendicular to the orbital plane (▶). However, Earth is slightly closer to the Sun at some times of year than others, with the Earth-Sun distance varying about 3% during the course of the year.

2. The date of closest approach is called the *perihelion*, where "peri" is a Greek word for "around" and "helio" refers to the Sun. Perihelion currently occurs in early January, but the exact date varies a bit each year, and moves forward by one day each 65 years. Over several hundred thousand years it may change season completely, causing changes in our climate.

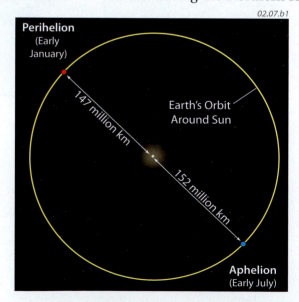

02.07.b1

Perihelion (Early January)

147 million km

Earth's Orbit Around Sun

152 million km

Aphelion (Early July)

4. This variation in Earth-Sun distance, from perihelion to aphelion and back again, constantly changes the average amount of insolation Earth's atmosphere receives. When the Earth is closest to the Sun (during the Northern Hemisphere winter) it intercepts more of the insolation. Changes in the amount of insolation striking the Earth on any given day can be calculated knowing the solar constant and the Earth-Sun distance on that day. From these calculations, the amount of energy is greatest at perihelion, decreases for half the year as Earth approaches aphelion, and then increases the next half of the year on the way to perihelion. The total amount of variation in energy related to Sun-Earth distance is about 7%.

3. The date of farthest approach is the *aphelion*, where "apo" comes from a Greek word meaning "from." The aphelion is, perversely for Northern Hemisphere dwellers, early in July. In other words, we are farthest from the Sun during our summer—the seasons cannot be explained by our changing distance from the Sun. As with perihelion, the actual date of aphelion varies slightly from year to year. For now and the immediate future, it occurs in early July.

C Does the Sun's Output of Energy Change Over Years, Decades, and Centuries?

Like most natural systems, the Sun's activity varies, sometimes increasing and other times decreasing. The Sun exhibits cycles that repeat over the course of a decade or so, and it has longer fluctuations that occur over multiple decades. These cycles influence the Sun's total output of energy, called the *total solar irradiance* (or simply TSI).

02.07.c1   02.07.c2   02.07.c5

1. This top pair of images (▶) shows the Sun at two different times, approximately 9 years apart. In the left image, the Sun has a number of dark spots, termed *sunspots*. These are places that are slightly cooler than the rest of the Sun, and so show up darker. In the image to the right, the Sun lacks any sunspots. The number of sunspots varies from year to year.

2. This second pair of images shows the Sun at the same two times as in the top images, but displays emissions at ultraviolet wavelengths, rather than visible light (▶). They indicate that the Sun is much more active and emitting more overall energy when there are more sunspots. So more sunspots means more energy output from the Sun. Data collected by satellites since 1978 show small (<1%) monthly changes in the output of solar energy and a correlation between sunspot activity and TSI.

3. Another expression of the Sun's activity are bursts of intense energy and matter directed into space, termed *solar flares* (▲). The bright regions producing solar flares in the image above are so large they could contain the Earth many times over. Solar flares pose a hazard to people and equipment in space and can affect wireless communications on Earth.

02.07.c3   02.07.c4

4. Scientists have been observing and counting the number of sunspots over time, calculating the average number present at any time, a value called the *sunspot number*. This graph (▶) plots monthly averages of sunspot number as a function of time since 1950. From the data, there is an approximately 11-year cycle of sunspot activity, where the number of sunspots starts out at some minimum number and then increases to a maximum number and then decreases again. This full cycle from minimum to maximum and back to minimum is called a *solar cycle* (or sunspot cycle). Solar cycles are numbered sequentially from the 1700s (the first accurate sunspot data) to the present—we are currently in solar cycle 24.

02.07.c6

5. As you can observe from this graph, the maximum number of sunspots (the heights of the "peaks" on this graph) varies from cycle to cycle. The higher the sunspot number, the more active the Sun. The duration of a cycle can also be more than or less than 11 years. The length of a cycle can help solar scientists predict the maximum sunspot number, and therefore solar activity, for the next cycle.

6. Sunspots have been counted for several centuries, revealing some important patterns (▶). This plot shows sunspot observations from the 1600s. The data are much more accurate for recent times, compared to the early observations, but the broad patterns shown are accepted by most scientists. Red points are early, less accurate measurements, blue data are more recent, and the black line is a smoothed average.

400 YEARS OF SUNSPOT OBSERVATIONS

02.07.c7

7. There have been several times, such as the *Maunder Minimum*, when there were few or no sunspots for many years in a row. The Maunder Minimum occurred during one of the coldest periods in the last 1,000 years, a time called the *Little Ice Age*. In contrast, our modern warming has occurred during a time of very high solar activity, as recorded by numerous sunspots.

Before You Leave This Page Be Able To

✓ Explain the solar constant.

✓ Sketch, label, and explain Earth's perihelion and aphelion and their effect on insolation on Earth.

✓ Explain how sunspot activity varies over time and why this might be important for understanding Earth's climate.

2.7

2.8 Why Does Insolation Vary from Place to Place?

DESPITE A FAIRLY CONSISTENT supply of energy from the Sun, considerable differences in the quantity of insolation are experienced between the poles and equator, and also over several timescales—most noticeably, changes between seasons and between night and day. On Earth, variations in insolation are mostly related to latitude.

A What Controls the Insolation Reaching the Atmosphere and Earth's Surface?

The amount of sunlight varies from time to time and from place to place. In the summer, sunlight feels more intense than it does in winter. Also, there are huge ranges in temperature between the warm, tropical regions of the world and the cold, polar regions. A main factor controlling the maximum possible amount of insolation is position on the Earth, specifically the latitude,

02.08.a1

1. When thinking about insolation, an important aspect to consider is the orientation of energy with respect to the surface being irradiated. In the figure to the right, a flat surface is illuminated by two identical lights, each with an energy equal to the amount of solar energy striking the top of the atmosphere (345 W/m²). The left light is shining directly down on the surface, so the surface receives the same amount of energy per square meter as the light emits (345 W/m²).

2. The second light is angled 45° relative to the surface, so when its light strikes the surface the light is spread out over a larger area. As a result, the amount of energy per area is less, and the area would be less brightly lit. For an angle of 45°, the energy is spread out over an area 1.41 times longer than for the left light, so the resulting energy is less, at 244 W/m². The angle between the direction of energy emission and the surface is called the *angle of incidence* and is 90° for the left light and 45° for the right one.

3. The geometric relationship shown in the previous figure applies to the entire Earth because of its spherical shape. The figure to the right shows how equal rays of solar energy strike different parts of the Earth, varying in angle of incidence from a position along the equator to ones nearer the poles. Each orange bar is of equal width and represents insolation equal to the solar constant.

02.08.a2

6. In contrast to the equator, insolation arriving above polar regions strikes Earth's atmosphere at a very low angle of incidence, much lower than the angled light in the figure above. As a result, the energy represented by the ray (the solar constant) spreads over a relatively large surface area (the small red box). Insolation striking that part of the atmosphere, as measured in energy amount per area, is considerably dispersed, diminished to perhaps only 20% of the solar constant.

5. In the mid-latitude regions north or south of the equator, insolation strikes the outside of Earth's atmosphere at a slightly oblique angle and is distributed over a wider area than at the equator. Thus the amount of energy is distributed over a larger area compared to the equator, and the energy per area is less. The area shown (on the outside of the atmosphere) receives about 60% of the solar constant, more if the area is closer to the equator or less if it is farther away—that is, the amount of insolation varies as a function of latitude.

4. Some insolation strikes the outside of Earth's atmosphere perpendicularly, like a light pointed directly toward a surface. This part of the atmosphere is receiving the maximum insolation possible, or an amount of energy equivalent to the solar constant, at the "top" of the atmosphere, 1,366 W/m².

7. In addition to these geometric effects, the atmosphere further influences the amount of insolation that reaches Earth's surface, depending on the angle of the incoming energy. As shown in this figure (▶), energy that comes directly down at the surface, like a vertical light, passes through the atmosphere in the shortest distance possible. Some energy is reflected or absorbed by the atmosphere.

02.08.a3

8. Energy coming at a lower angle of incidence must pass through more of the atmosphere to reach the surface. This increases the amount of atmospheric attenuation and causes less energy to reach the surface. This is a further reason why early morning and late afternoon sunlight does not feel as intense as that at noon, and why the winter Sun, which is low in the sky, does not feel as intense as the summer Sun, which is higher in the sky.

9. When viewing the entire Earth (▶), latitude strongly influences the severity of atmospheric attenuation. Near the equator at noon, the Sun's energy is coming at a high angle of incidence, and experiences relatively less attenuation.

02.08.a4

10. In the mid-latitudes, the energy is more oblique to the Earth, so must pass through more of the atmosphere. This results in more atmospheric attenuation, resulting in less energy reaching Earth's surface.

11. In polar and high-latitude regions, the angle of incidence is very low, and the Sun's energy must pass through even more of the atmosphere. The amount of atmospheric attenuation is greater, so less energy passes through to the surface.

B   What Is the Zenith Angle and How Does It Affect Insolation?

1. The angle made by two lines, one drawn from the observer to the Sun at its position and the other drawn directly over the observer's head, is called the *zenith angle*. Note that the zenith angle is measured from vertical, regardless of the slope of the land. When the shadow of an object is directly underneath the object (▼), the zenith angle is 0°.

Zenith Angle   SUN

02.08.b1

02.08.b2

2. This figure (►) shows the Sun at four different positions, corresponding to four different zenith angles (at noon). The 10 thin orange rays of sunlight represent equal amounts of energy (the solar constant). The brackets on the ground indicate the width over which that same amount of sunlight strikes the surface.

0°	30°	60°	90°
10 m (100%)	11.5 m (87%)	20 m (50%)	0 m (0%)

02.08.b3

3. In this example, if the zenith angle is 0°, as happens near the equator and in the tropics, the 10 rays of sunlight intersect the Earth's surface over a width of 10 meters. The light is perpendicular to the surface, so the area receives the maximum amount (100%) of energy possible.

4. In a subtropical latitude, with a zenith angle of 30°, the 10 rays of sunlight are spread out over a wider area, so the surface receives less insolation per square meter compared to a site at the equator, but not much less (87%).

5. If the zenith angle is larger, like 60° in a high-latitude location, the 10 rays of sunlight are spread out over 20 m (twice the area as at the equator), so any area receives only 50% of the sunlight per square meter that the equator does.

6. For a location directly at Earth's North or South Pole (the Sun position on the far right), the zenith angle can be 90°, and none of the 10 rays of sunlight hit the ground. Brrrrrrrrrrrrrrr!

C   How Does Insolation Vary by Time and Location on Earth's Surface?

The amount of sunlight that reaches the surface varies greatly with latitude, due to the reinforcing effects of zenith angle and atmospheric attenuation. These variations in insolation result in huge ranges in temperature between the warm, tropical regions of the world and the cold, polar regions.

02.08.c1 Raja Ampat, Indonesia

02.08.c2 Antarctica

02.08.c3 Hermosa, CO

The least variation in insolation through the year occurs in equatorial regions, which do not experience summer versus winter. The length of daylight in equatorial regions is always around 12 hours, and zenith angles (at noon) are always small. The Sun is directly overhead in March and September and nearly so the rest of the year, so temperatures remain warm throughout most of the year.

Polar regions experience the maximum variation in insolation. During the summer this region has 24 hours of sunlight, allowing fully lit outdoor excursions (▲), but during the winter there is no sunlight. Even during the summer, large zenith angles prevent insolation from increasing too much. The combination of months of darkness and low insolation causes polar areas to have very cold average temperatures.

Areas between the tropical and polar regions, such as those in the mid-latitudes, experience an intermediate amount of variability in insolation and so typically have distinct seasons. The mid-latitudes have some sunlight and darkness every day, with longer days and lower zenith angles during the relatively warm summer. The winter has shorter days and larger zenith angles, so it is colder than the summer.

Before You Leave This Page Be Able To

☑ Sketch, label, and explain the relationship between latitude and the amount of insolation, including atmospheric effects.

☑ Sketch, label, and explain zenith angle and how it influences the amount of insolation an area receives.

☑ Summarize how zenith angle and variations in length of day are expressed in tropical, polar, and mid-latitude locations.

2.8

2.9 Why Do We Have Seasons?

DURING THE YEAR MOST LOCATIONS progress through different seasons, from warmer summers to cooler winters and back again. The progression from season to season accompanies changes in the position of the Sun, such as from higher in the sky during the summer to lower in the sky during the winter. Except at the equator, the durations of daylight and darkness also vary, with longer days during the summer and longer nights during the winter. What causes most parts of Earth to have different seasons? Why do tropical areas not have summer and winter?

A What Causes the Seasons?

The annual "march of the seasons" indicates variations in the amount of sunlight received at different latitudes during the course of a year. The main cause of these variations is that Earth's axis of rotation, about which our planet spins once during a 24-hour day, is tilted relative to the plane in which we orbit the Sun, the *orbital plane.*

1. Earth's axis of rotation is tilted relative to our planet's orbital plane, as shown by the figure below, which is a sideways perspective of Earth's nearly circular orbit. The axis remains fixed in orientation with respect to the orbital plane and the stars throughout the year—the tilt does not change during a year. As Earth orbits the Sun, the Northern Hemisphere faces in the direction of the Sun during some times of the year, while the Southern Hemisphere faces the Sun at other times. Observe the relationship between the orientation of the rotation axis and the direction to the Sun.

7. The seasons therefore are controlled by the relationship between the tilt of Earth's rotation axis and the orbit of Earth around the Sun. Each season is bounded by a solstice and an equinox, which in turn represent days of either a maximum or minimum amount of apparent tilt relative to the Sun.

2. During December, the North Pole and Northern Hemisphere face away from the Sun. The Northern Hemisphere therefore receives less direct insolation and so experiences the cooler temperatures of winter. Rotation about the axis brings most parts of the hemisphere into the sunlight for part of the day.

6. In late June, the North Pole and Northern Hemisphere more directly face the Sun and so start to experience summer. The *June Solstice* (Summer Solstice in the Northern Hemisphere) marks the date when the North Pole's rotation axis is pointed most directly toward the Sun. At the same time, the South Pole and Southern Hemisphere face away from the Sun and start to experience winter.

September 21

23.5°

(COMPONENTS NOT TO SCALE)

December 21

June 21

23.5°

23.5°

23.5°

02.09.a1

March 21

3. In contrast, the South Pole and Southern Hemisphere more directly face the Sun and receive more direct insolation. The Southern Hemisphere therefore has its summer (at the same time the Northern Hemisphere has its winter). The *December Solstice* (Winter Solstice in the Northern Hemisphere, Summer Solstice in the Southern Hemisphere) is the date when the North Pole (marking Earth's rotation axis) is pointed most away from the Sun.

4. As Earth orbits around the Sun, the rotation axis no longer points directly away from the Sun. In March (and later again in September), Earth's axis is pointing sideways relative to the Sun—neither the North Pole nor the South Pole is inclined toward the Sun. In this position, neither hemisphere receives more insolation than the other, and so both experience the more moderate temperatures of spring and fall. The *March Equinox* and *September Equinox* are times when the axis is exactly sideways and the durations of daylight and darkness are equal (equinox is Latin for "equal night").

5. Between March and June, Earth's continued orbit causes the North Pole and Northern Hemisphere to increasingly face the Sun and to warm up as summer approaches. During this time, the South Pole and Southern Hemisphere begin to face away from the Sun, cooling down on the way to winter.

8. In thinking about the seasons, it is important to remember that Earth's orbit is nearly circular, as shown in this view looking straight down on Earth's orbital plane (▶). In this figure, as in the one above, the sizes of the Earth and Sun are greatly exaggerated relative to the size of the orbit, and the Earth is shown much larger than it actually is relative to the Sun.

9. The nearly circular geometry of Earth's orbit reinforces the fact that seasons are not caused in any way by differences in distance from the Sun. In fact, Earth is slightly closer to the Sun during the northern winter (perihelion) and slightly farther away during the northern summer (aphelion), opposite to what would be required to explain the seasons.

02.09.a2

10. In each globe, Earth's rotation axis is represented by the small purple rod protruding from the North Pole (pointing toward the viewer and a little toward the left). Note that the axis points in the same direction throughout the year—it points toward Polaris, the North Star (not shown). Over thousands of years, the axis varies in orientation, not always pointing at Polaris. The amount of tilt varies by a degree either way (22.5° to 24.5°) over 40,000 years, but we are most interested here in the current tilt of 23.5°, an angle that will reappear throughout this book.

B What Factors Determine the Temporal and Spatial Variations in Insolations?

Earth's rotational axis is tilted relative to the plane in which we orbit the Sun (the orbital plane). The tilt of the axis is currently 23.5° from vertical to the orbital plane, an angle that is reflected in important geographic features on our planet, including how we define the Tropic of Cancer, Tropic of Capricorn, Arctic Circle, and Antarctic Circle.

The Tropics

1. These two globes show the Earth on the solstices—days when one of the poles is most exposed to the Sun and the other pole most faces away. During the December Solstice, the noon Sun is no longer directly overhead at the equator but instead is 23.5° to the south, directly over the *Tropic of Capricorn*. On this day, the equator has a zenith angle of 23.5°.

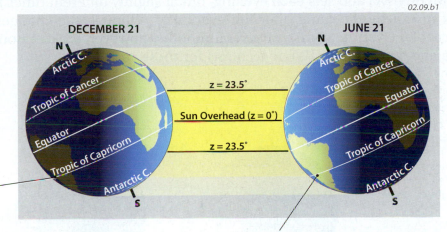

02.09.b1

4. The rest of the year, the overhead Sun migrates between the Tropics of Cancer and Capricorn, delivering relatively intense and constant insolation between these latitudes, causing a warm region called the *tropics*. In contrast, mid-latitude and polar regions receive a highly variable supply of energy and exhibit more variability during the year—that is, they exhibit more *seasonality*.

2. During the December Solstice, the Southern Hemisphere receives the maximum amount of insolation it will receive all year, marking the start of the southern summer. Also on this day, the Northern Hemisphere receives the least amount of insolation for the year, marking the beginning of the northern winter.

3. During the June Solstice, Earth is on the opposite side of the Sun, and the North Pole faces toward the Sun by its maximum amount. The overhead Sun, and the position of maximum insolation, have migrated to a latitude 23.5° north of the equator, the *Tropic of Cancer*. On the day of the June Solstice, the Northern Hemisphere receives the most insolation of the year, shifting into summer, while the Southern Hemisphere receives the least amount and moves into winter.

The Polar Circles

5. These two globes show the same two days as those above, the solstices. During the December Solstice, the northward limit of sunlight is the Arctic Circle, located at a latitude of 66.5° N, or 23.5° south of the North Pole. Along the Arctic Circle, the Sun will barely appear on the horizon at noon time and then disappear for a very long night. All places north of that latitude receive no insolation on this day—they have 24 hours of darkness.

02.09.b2

8. At this time, the area within the Arctic Circle (66.5° N) is fully illuminated. Note that the same angles keep reoccurring—23.5° and 66.5° (which is 90° minus 23.5°). These values, for the two tropics (Cancer and Capricorn) and for the two polar circles (Arctic and Antarctic), are the same as the tilt angle of the rotation axis. The tilt of Earth's axis controls the locations of the tropics and polar circles.

6. Note that the entire Antarctic Circle, in the Southern Hemisphere, is illuminated, but with very oblique sunlight (high zenith angles). As the Earth completes its daily rotation, the South Pole and nearby areas remain in the sunlight all 24 hours—these places have 24 hours without darkness.

7. During the June Solstice, locations in the Southern Hemisphere now face away from the Sun. The southern limit of sunlight is the Antarctic Circle, at latitude 66.5° S. Here, the Sun will appear on the horizon for a brief moment at noon.

City	Latitude	Equinox	Dec. Sol.	June Sol.
Saskatoon, Canada	52.0° N	52.0°	75.5°	28.5°
Jacksonville, FL	30.5° N	30.5°	54.0°	7.0°
San José, Costa Rica	10.0° N	10.0°	13.5°	33.5°
Quito, Ecuador	0.0°	0.0°	23.5°	23.5°

9. Using the 23.5° angle and an area's latitude, we can calculate the zenith angle for any place on Earth. Examine this table (◄), which lists zenith angles for the equinoxes and both solstices for several cities. Note that the zenith angle varies by 47° from solstice to solstice—47° is two times 23.5°.

Before You Leave This Page Be Able To

☑ Sketch and explain how the position of the Earth with respect to the Sun at various times of the year explains the seasons.

☑ Explain the solstices and equinoxes in terms of Earth-Sun position, seasons, and locations of tropics and polar circles.

2.9

2.10

What Controls When and Where the Sun Rises and Sets?

THE SUN RISES EACH MORNING and sets each evening, but at slightly different times from day to day. Also, the Sun does not rise or set in exactly the same direction every day, although the changes from day to day are so gradual as to be unnoticeable. Over the course of several months, however, we can notice significant changes in when and where sunrise and sunset occur. What accounts for these variations?

A Why Does the Sun Rise and Set?

1. At any moment in time, half of the Earth's surface area is sunlit, experiencing day, and half is in the darkness of night. The dashed line encircling the world separates the lighted and dark halves and is called the *circle of illumination*. When viewed straight on, as in these figures, the circle of illumination appears as a line, but it has a curved shape from any other perspective. The colored areas on these globes are time zones.

2. In the left globe, North America and western South America are on the side of the Earth hidden from the Sun, and so it is night. At the same time, eastern South America and Africa are on the side facing the Sun and so are in daylight. With a rotating Earth, this view is only an instantaneous snapshot of which areas are in sunlight and which are in darkness.

TIME 1

TIME 2
(4 Hours Later)

Circles of Illumination

02.10.a1

4. The circle of illumination, the boundary between day and night, moves westward across the surface as the planet rotates eastward. As this occurs, the Sun has not changed position—the Earth has simply rotated. It finishes a complete rotation in 24 hours.

3. In the right globe, the Earth has rotated an additional four hours—the globe rotates to the right when viewed in this perspective, or counterclockwise when viewed from above the North Pole. At this later time, South America and eastern North America have rotated into the sunlight (it is morning), but the western U.S. is still in the last remaining hours of night.

B Why Does the Length of Daylight Vary Through the Year?

In most parts of the world, there are significant differences between the length of daylight from season to season. In such regions, days are noticeably shorter during the winter than during the summer. In accordance with this, nights are longer during the winter and shorter during the summer. In contrast, at the equator, there are always 12 hours of day and 12 hours of night, irrespective of the time of year.

1. These globes show the circle of illumination at three different times of year: the December Solstice, an equinox, and the June Solstice. To help us visualize the circle of illumination, the three larger globes are depicted as if being observed by the small figure next to the corresponding small globes. The axial tilt remains fixed in orientation as Earth orbits the Sun, but here it is portrayed from different perspectives.

2. During an equinox (the large, center globe), the tilt axis is oriented neither toward nor away from the Sun. So, the pattern of light and dark is symmetrical with respect to the equator and other lines of reference. It takes the same amount of time for every location to rotate in and out of sunlight. At equinox, every location on Earth has 12 hours of sunlight and darkness.

02.10.b1

DECEMBER SOLSTICE

EQUINOX

Circle of Illumination

JUNE SOLSTICE

Arctic Circle

Equator

Tropic of Capricorn

Circle of Illumination

Equator

Tropic of Cancer

Equator

Antarctic Circle

Circle of Illumination

3. At other times of the year, Earth's axis appears tilted toward or away from the Sun, and so the circle of illumination is not symmetrical relative to lines of reference, such as the tropics. In the left globe, representing the December Solstice, any line of latitude in the Southern Hemisphere, such as the Tropic of Capricorn, is more in sunlight than in darkness. Therefore, days are longer and nights are shorter. The opposite is true for any latitude in the Northern Hemisphere, which is more in the dark than in the light, causing nights to be longer than days.

5. The variation in the lengths of day versus night from season to season increases with increasing latitude, either north or south of the equator. Such seasonal changes in day length are absent at the equator and greatest at the poles.

4. During the June Solstice and adjacent months, the opposite is true—more of the Northern Hemisphere is in sunlight than is in darkness. As a result, days are longer than nights in the Northern Hemisphere during this time (the northern summer). In contrast, more of the Southern Hemisphere is in darkness, so days are shorter and nights are longer during this time (the southern winter).

C Why Do Arctic Areas Sometimes Have 24 Hours of Sunlight or Darkness?

The most extreme variations in the lengths of daytime and nighttime occur in the highest latitudes, including the Arctic region around the North Pole and the Antarctic region around the South Pole. North of the Arctic Circle and south of the Antarctic Circle, summer days can have 24 hours of straight daylight. During the winter, it can remain dark for all 24 hours, night after night. Either condition can last for months. How is this so?

02.10.c1 Arctic Circle

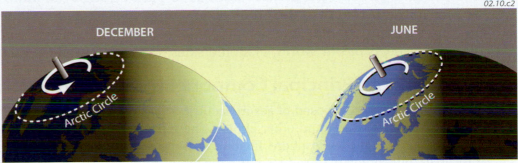
02.10.c2

This image combines different photographs to show the path of the Sun during several hours at a location north of the Arctic Circle. The Sun remains low in the sky and dips toward the horizon at midnight, but never actually sets—this location has 24 hours of sunlight during the middle of summer. The low Sun means that insolation striking the land is spread out and so is relatively weak, and it has a long and attenuated path through the air.

The figure above shows sunlight on the north polar region during the December Solstice. All the area within the Arctic Circle is in darkness (it is on the side opposite to the Sun). As the Earth rotates about its axis, the entire area within the circle remains out of the sunlight—24 hours of darkness. In days following the Solstice, sunlight begins to creep into the Arctic Circle, so less of the Arctic has 24-hour nights.

The opposite situation occurs during the June Solstice, shown here. Note that the entire Arctic Circle faces the Sun and will remain in sunlight as the Earth rotates about its axis. On days before and after the solstice (when the North Pole less directly faces the Sun), areas just inside the Arctic Circle would not be in constant sunlight, and so would have slightly less than 24 hours of sunlight and minutes to hours of nighttime.

D What Controls the Time and Direction of Sunrise and Sunset?

Except at the equator, the times of sunrise and sunset shift slightly from day to day, but typically by less than a few minutes every day. From month to month, however, we notice significant differences in the times of sunrise and sunset, and therefore in the duration of day and night. The changes in time are accompanied by gradual changes in the direction from which the Sun rises and the direction in which it sets.

1. This figure depicts where and when the Sun rises and sets at a location at 45° N latitude, which is halfway between the equator and the North Pole. Observe this figure and note how the locations of sunrise and sunset change by date.

2. The December and June dates, marking the two solstices, show the two extremes. The locations and times of sunrise and sunset fall between these two extremes for all other dates. Either equinox is halfway between the solstices. Each day, the Sun's path defines a circle, which lies on a plane that is inclined at an angle that is equal to the site's latitude (45° in this case).

02.10.d1

3. At the December Solstice, the Northern Hemisphere faces away from the Sun the maximum amount, and so the Sun is as low in the sky as it ever gets. The Sun rises and sets as far south as any day of the year, rising in the southeast and setting in the southwest. This is the shortest day and longest night of the year.

4. At either equinox, neither pole faces the Sun, and so there are 12 hours of daylight and 12 hours of darkness at every location on Earth. Everywhere on Earth, the Sun rises due east of the site and sets due west on an equinox. If you wanted to determine directions without a compass, and it happened to be the date of an equinox, you could precisely determine an east-west direction by drawing a line from the direction of sunrise to the direction of sunset, as was done by ancient cultures. Also, you could determine your latitude by measuring the zenith angle at noon on this day—your latitude is the zenith angle.

5. At the June Solstice, the Northern Hemisphere faces the Sun, and so the Sun is the highest in the sky it will be all year. The Sun rises and sets as far north as any day of the year, rising in the northeast and setting in the northwest. This is the longest day and the shortest night of the year.

Before You Leave This Page Be Able To

☑ Sketch, label, and explain why the length of day and night vary during the year, and which dates have the longest and shortest daylight hours.

☑ Sketch, label, and explain why a polar region can have 24 hours of daylight or 24 hours of darkness.

☑ Sketch, label, and explain where the sun rises and sets, and has the highest and lowest zenith angles, for different times of the year.

2.10

How Does Insolation Interact with the Atmosphere?

INSOLATION REACHES THE EARTH but has to pass through the atmosphere before it reaches us. The atmosphere does not transmit all of the Sun's energy; some wavelengths of energy are partially or completely blocked by atmospheric components, such as gas molecules. The interactions between insolation and the atmosphere explain many aspects of our world, like blue skies, red sunsets, and even the existence of life.

A What Are the Principal Components of the Atmosphere?

The atmosphere is composed of gas molecules, especially nitrogen and oxygen, and small solid particles, including dust, and drops of water and other liquids. These particles and drops are together called *aerosols*, to convey that they are suspended in the atmosphere, floating with the moving air. Most gases and aerosols are produced by natural processes, but some are introduced into the atmosphere by activities of humans.

02.11.a1 Iceland

02.11.a2 Mount Etna, Italy

02.11.a3

Gases—The atmosphere is more than 99% nitrogen (N_2) and oxygen (O_2), with lesser amounts of argon (Ar), and water vapor (H_2O), as in the steam shown above. It contains trace amounts of other molecules, such as carbon dioxide (CO_2), methane (CH_4), nitrous oxide (N_2O), and sulfur dioxide (SO_2).

Solid Particles—If tiny enough, solid particles can be suspended in the atmosphere. Such particles include volcanic ash (▲), and wind-blown dust, salt, and pollen. They also include soot and smoke from natural and human-caused fires. Some small particles are produced by chemical reactions in the air.

Drops of Liquid—The atmosphere contains drops of water, with much smaller amounts of other liquids. Most water drops are tiny enough to remain suspended in the air as in clouds (▲). If small drops combine or otherwise grow, they may become too heavy to remain suspended, falling as rain.

B How Do Atmospheric Components Affect Insolation?

Insolation, like all types of electromagnetic radiation (EMR), can be affected by material through which it passes. As solar energy attempts to pass through the atmosphere, it interacts in various ways with the different atmospheric components. The types of interaction that occur depend on the size and physical nature of the component (e.g., solid versus a gas), the wavelength of the energy (blue versus red light, for example), and other factors.

02.11.b2

Reflection and Absorption

Reflection—Some atmospheric components can partially *reflect* incoming insolation, such as by this snowflake. Reflected energy can be returned directly into space or can interact with other atmospheric components and remain in the atmosphere.

02.11.b1
Snowflake

Absorption—An atmospheric component, such as this gas molecule, can instead *absorb* the energy, converting the incoming electromagnetic energy into kinetic energy expressed as motions of the molecule.

Gas Molecule

Scattering

Scattering—Insolation can be *scattered* by atmospheric components (▶), which send the energy off in various directions. Some processes of scattering affect shorter wavelengths of EMR more than longer ones; for example blue light is scattered more than red light. Other processes affect all wavelengths of solar energy equally.

Sky Color—As insolation enters the atmosphere, blue and violet light are preferentially scattered, and this scattered light causes us to see the sky as blue. The remaining light that passes through gives the Sun a yellowish color. When sunlight passes through the atmosphere at a low angle, as during sunrise and sunset, most colors, except orange and red, have been scattered out. Scattering of the remaining orange and red light produces the familiar orange and red glow.

C   How Do Different Layers of the Atmosphere Interact with Insolation?

If the atmosphere consisted of gases that did not interact with insolation, we would expect that its temperature would decline with distance from Earth. However, atmospheric temperatures exhibit some surprising changes with altitude, as a direct result of the interactions between insolation and atmospheric components, specifically gases.

1. Observations from high-flying planes, balloons, and rockets reveal that the atmosphere is divided vertically into four layers distinguished by their thermal properties. These layers are, from bottom to top, the *troposphere, stratosphere, mesosphere,* and *thermosphere.* In the troposphere and mesosphere, temperatures decline with increasing altitude, as expected, displaying a *normal temperature gradient.* Temperatures in the stratosphere and thermosphere, however, actually increase with altitude—a *temperature inversion* (reverse gradient). Why?

2. The unexpected change in temperature gradient in some layers is due to the fact that insolation arrives at the outside of the atmosphere in a variety of wavelengths of the electromagnetic (EM) spectrum, from very short ones (X-rays) to very long ones (radio waves). Most energy is in visible-light wavelengths, with significant amounts of adjacent wavelengths of ultraviolet (UV) and infrared (IR) energy. Some wavelengths of energy interact with certain atmospheric molecules, transferring energy as they do so. Different interactions occur in different layers, accounting for the differences in temperature gradient, and therefore the distinctions among the different layers. The four atmospheric layers are separated by distinct breaks called *pauses.*

3. The very shortest and most energetic of EM radiation from the Sun are X-rays and gamma rays, with wavelengths approaching the size of gas molecules. These incoming wavelengths of EMR are effectively intercepted by the few molecules of nitrogen (N_2) and oxygen (O_2) in the uppermost parts of the atmosphere—the *thermosphere.* The greatest number of interceptions (and transfer of energy) occurs at the first opportunity for the gamma and X-rays to encounter these gases, at the outermost parts of the thermosphere. This causes the outer thermosphere to warm up, but so few molecules exist at such levels that you would freeze to death instantly, even at temperatures approaching 1,200°C. A progressively smaller proportion of these energetic rays penetrates lower in the thermosphere, so fewer energy exchanges occur, and temperatures decline downward across the thermosphere.

4. The next layer down, the *mesosphere,* possesses no particular properties to intercept wavelengths of insolation, so it displays a normal temperature gradient (temperature decreases upward). The boundary at the top of the mesosphere is the *mesopause.*

5. The *stratosphere* has relatively high concentrations of the trace gas *ozone* (O_3), which effectively absorbs UV wavelengths. The same principle prevails as in the thermosphere, with the greatest amount of absorption occurring near the top of the stratosphere, which is therefore relatively warm. Progressively less absorption of incoming UV occurs downward, resulting in a decrease in temperatures downward—a temperature inversion. The top of the stratosphere is the *stratopause.*

6. Conditions in the troposphere can be very complex with the presence of various sublayers of air and the influence of clouds, but temperatures usually decrease upward (a normal temperature gradient). The top of the troposphere is the *tropopause.*

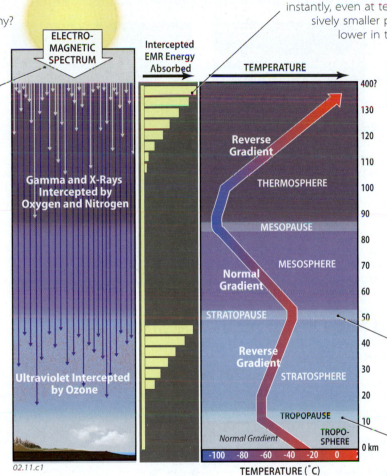

02.11.c1

7. The greatest amounts of the Sun's emitted energy is at wavelengths of visible light, along with UV and IR. (▼) Interactions between solar energy and various components of the atmosphere (such as N_2) cause some wavelengths of energy to be intercepted through absorption, scattering, or reflection in the atmosphere. Therefore, the spectrum of EM energy that reaches the surface is different in detail from that which enters the top of the atmosphere, as depicted by the curve showing the amount of insolation at sea level by this graph. Certain wavelengths of EMR are absorbed by certain components, such as water (H_2O), oxygen (O_2), and carbon dioxide (CO_2), as expressed by dips in the curve for insolation received at sea level.

02.11.c2

Before You Leave This Page Be Able To

✓ Summarize the principal components of the atmosphere.

✓ Sketch and explain reflection, absorption, and scattering, and how they explain the color of the sky.

✓ Sketch, label, and explain the four layers in the atmosphere, how they interact with insolation, and how these interactions explain the different temperature gradients.

2.11

What Is Ozone and Why Is It So Important?

OZONE IS AN ESSENTIAL GAS in the atmosphere, shielding life on the surface from deadly doses of ultraviolet radiation from the Sun. In the past several decades, there has been major concern about the loss of ozone in our atmosphere, particularly a seasonal decrease in ozone above Antarctica. How can a gas that generally constitutes less than one molecule in every 10 million in the air be so important, and why does the Antarctic region experience severe ozone loss?

A   What Is Ozone and Where Does It Occur?

Ozone is a molecule composed of three oxygen atoms bonded together (O_3), instead of the much more common arrangement of two oxygen atoms in a molecule of oxygen gas (O_2). More than 90% of ozone occurs in the stratosphere, but ozone also occurs in lesser amounts in the troposphere and mesosphere.

This graph (▶) plots the concentration of ozone in the atmosphere as a function of altitude above Earth's surface. Note that concentrations are very low overall, measured in parts per million (ppm—the number of ozone molecules for every one million molecules of atmosphere). The maximum concentration is in the middle of the stratosphere, a zone called the *ozone layer*, but even here the ozone concentration is only 6 to 8 ppm. Ozone concentrations decrease rapidly into the overlying mesosphere.

02.12.a1

02.12.a2 Los Angeles, CA

Ozone is a component of *smog* (◀), a type of air pollution common in larger cities, especially under certain atmospheric conditions. The term smog was derived from air pollution that looks like smoke and fog.

Ozone is also produced naturally due to ionization of oxygen gas molecules in the air near lightning. In such settings, we can sense ozone's distinctive smell, and this is the origin of the term *ozone*, after the Greek verb for "to smell."

02.12.a3

There is a slight increase in concentrations low in the troposphere, just above the surface. Whereas ozone in the ozone layer (stratospheric ozone) is beneficial to life, ozone near the surface (*ground-level ozone*) is a harmful air pollutant, produced by sunlight striking hydrocarbons, such as car exhaust, in the air.

B   How Is Ozone Produced and Destroyed in the Stratosphere?

1. Electromagnetic radiation (EMR) with a wavelength of visible light or shorter (e.g., ultraviolet) has the ability to affect the chemical bonds in a compound, a process called *photodissociation*. Such processes occur throughout the atmosphere, but particularly in the stratosphere, where (1) much of the EMR arriving from the Sun has not yet been absorbed and (2) the concentration of molecules becomes high enough to allow photodissociation to proceed. This figure shows how photodissociation forms and destroys ozone.

2. Oxygen in the atmosphere is mostly oxygen gas (O_2), composed of two oxygen atoms bonded together. These molecules absorb EMR from the Sun, including several types of ultraviolet (UV) energy. They are strongly affected by an energetic type of UV called UV-C, which has relatively short wavelengths. UV-C can break apart the two oxygen atoms, liberating a free (unbonded) oxygen atom (O).

3. The freed oxygen atom can quickly bond with another free oxygen atom, forming a new molecule of oxygen gas.

4. Alternatively, a freed oxygen atom can combine with an existing oxygen gas molecule to form a molecule of ozone (O_3). This process is the way ozone is produced in the atmosphere.

5. Ozone molecules are capable of absorbing a longer wavelength form of ultraviolet called UV-B, in addition to UV-C. UV-B has lower energy than UV-C, but more of it reaches Earth's surface. UV-B is the wavelength of ultraviolet radiation that causes sunburn and contributes to skin cancer. It also produces vitamin D in our bodies when it interacts with our skin.

6. UV-C and UV-B can photodissociate an ozone molecule, leaving a molecule of oxygen gas and a free oxygen atom. These can recombine with other atoms and molecules to form another molecule of ozone, so the process of formation and destruction of ozone is a cycle, with ozone and oxygen gas continuously being broken part, combined, and broken apart again. If the rates of formation and destruction of ozone molecules are equal, the ozone concentration will remain constant. Variations in atmospheric conditions, however, change the relative rates, causing concentrations to change over time and from place to place.

02.12.b1

C   What Is the Distribution of Ozone in the Atmosphere?

These two globes show the amount of ozone in the atmosphere at two different months, as measured from a satellite. In general, ozone concentrations are relatively low (green) over the equator, and are higher (yellows and oranges) in middle and high latitudes. The difference between the two globes shows that ozone concentrations change with the season, responding to changes in the patterns of insolation. Ozone amounts are expressed as Dobson units (DU).

OZONE (DU)
100   500

April 1, 2001

October 1, 2001

02.12.c1

In April, the highest concentrations are in higher latitude regions of the Northern Hemisphere.

In October, the highest values are in high latitudes of the Southern Hemisphere, but there is a huge area with very low concentrations centered over the South Pole (at the bottom of the globe). This low exists because total darkness during the Antarctic winter lasts for several months, during which time there is no insolation to form ozone, but ozone continues to be chemically destroyed (see below).

02.12.c2

This figure shows ozone concentrations in the main part of the stratosphere. Ozone is most concentrated (red zones on the figure) in the lower middle stratosphere—the ozone layer. The highest concentrations are in high latitudes, and below an altitude of about 25 km. In equatorial regions, the zone of maximum ozone is at higher altitudes, typically more than 25 km. Production of ozone is greatest in the tropical equatorial regions, which receive the most insolation, including UV, but a very slow-moving circulation pattern in the stratosphere moves the ozone upward and laterally toward the poles, accounting for the higher concentrations there.

D   What Is Causing Depletion of Ozone and Formation of the "Ozone Hole"?

Ozone is critical to life because it shields us from dangerous UV-C and UV-B radiation. In the past half century, we became aware that our protective ozone shield was being depleted by human activities, especially the production of chlorofluorocarbons (CFC), chemicals released from aerosol cans, air conditioning units, refrigerators, and polystyrene. As a result, production of CFCs was limited by international agreement, the Montreal Protocol, adopted in the 1990s.

Sept. 16, 2011

02.12.d1

1. This image shows ozone concentrations in the Southern Hemisphere as measured by a NASA satellite, with the lowest amounts in purple. The image, taken in September 2011, shows 10.6 million square miles of the Antarctic region with severe ozone depletion. This depletion is described as "thinning of the ozone layer."

2. Most ozone is destroyed by natural processes, but humans have introduced chemicals that accelerate such losses. Chlorofluorocarbons (CFCs) contain halogens, elements like chlorine and bromine that easily bond with another element or molecule. Halogens can break apart ozone by attracting one of the oxygen atoms away, forming a new molecule, such as one with chlorine and oxygen (ClO).

02.12.d2

3. The ClO molecule is short lived, and the chlorine atom breaks away to combine with other O atoms, typically outcompeting O_2 for bonding with O, thereby breaking apart ozone or keeping it from forming. On average, a single Cl may be responsible for the destruction of up to 10,000 ozone molecules. More than 70% of halogens in the atmosphere were introduced by humans.

4. Halogens are regarded as a major factor in the depletion of the ozone layer in the Southern Hemisphere, shown in the top globe. The bottom globe shows the Northern Hemisphere at an equivalent time of year (spring in both places). Why is there thinning of the ozone layer over Antarctica but less so over the North Pole?

5. The contrasting geography of the two polar regions plays a major role. Antarctica, in the center of the top globe, is a continent surrounded by ocean. This arrangement produces a zone of rapid circumpolar winds, which effectively exclude the import of ozone from nonpolar areas.

6. In contrast, the Arctic is an ocean surrounded by irregularly shaped continents (Eurasia and North America). The alternating pattern of oceans and continents induces far greater north-south movement and complex wind patterns, which encourage the exchange and mixing of gases, including ozone, into and out of the northern polar atmosphere.

October 1, 2001
April 1, 2001
02.12.d3

Before You Leave These Pages Be Able To

✓ Explain what ozone is, where it is, and how it protects us.

✓ Describe the natural and human-related processes that contribute to ozone formation and destruction.

2.12

2.13 How Much Insolation Reaches the Surface?

NOT ALL INSOLATION reaches Earth's surface. Much of it is intercepted (absorbed, scattered, or reflected) by the atmosphere. Measurements and models allow us to account for the destination of insolation globally. Approximately 69% of the energy arriving at the top of the atmosphere remains in the Earth's system, of which 20% is stored (absorbed) in the atmosphere and 49% is absorbed by, and heats, the Earth's surface—the rest (31%) is lost back into space. This accounting of insolation is termed the *global shortwave-radiation budget*.

A How Is Insolation Intercepted in the Atmosphere?

02.13.a1

The atmosphere absorbs radiation of various wavelengths, including ultraviolet (UV) light, an energetic shortwave EMR emitted by the Sun. The absorption of this harmful energy in the atmosphere allows life to exist on Earth's surface. UV is absorbed chiefly by oxygen gas, ozone, and nitrogen gas. As shown in this graph, the shortest-wavelength, highest-energy UV (UV-C) is absorbed first, followed by longer-wavelength, lower-energy UV-B. Nearly all UV-A, the longest-wavelength, least-energetic UV, is transmitted to the surface. When a wavelength of EMR passes through a substance, such as the atmosphere, without being intercepted, we say that the substance is "transparent to the energy." The atmosphere is nearly transparent to UV-A.

02.13.a2

Clouds can absorb, scatter, and reflect insolation, bouncing it back higher into the atmosphere and even into space. Clouds in the lower part of the troposphere, *low-level clouds*, can be highly reflective, like the ones shown here. The dark underside of the clouds indicates that little light is transmitted.

02.13.a3

Clouds higher in the atmosphere, such as these wispy *cirrus clouds*, are typically less reflective, but still reflect about 50% of insolation striking them. These *high-level clouds* occur in the middle to upper troposphere and are typically composed of ice crystals, unlike low-level clouds, which contain mostly water drops.

B What Happens to Insolation That Reaches Earth's Surface?

Earth's surface is diverse in its topography, rocks, soil, bodies of water, types of vegetation, and amount of human development. Each type of surface material has certain colors and other characteristics, so some surfaces reflect relatively little insolation and absorb much heat, while others reflect most insolation and absorb less heat.

1. The Apollo astronauts saw the Earth rise above the Moon's horizon (▼). The human eye is only sensitive to EMR in the visible (0.4–0.7 μm) wavelengths, and the Earth emits most of its energy at wavelengths 20 times longer than this (infrared). What the astronauts saw was visible light originating from the Sun and reflecting from the Earth. All objects reflect some insolation, and the percentage of insolation that is reflected by an object is termed its *albedo*. Albedo varies from 0% for a theoretical black object that reflects no light to 100% for a perfectly white, smooth one; no common objects have these end-member values. The average albedo for the entire Earth-ocean-atmosphere system is 31%.

2. Different land surfaces have different albedos. When humans use the land surface, they generally change the albedo compared to its natural state, modifying the energy balance, with local and perhaps regional implications. Examine the albedo values for different types of land cover and think how changes in the land cover would affect the energy balance.

02.13.b1

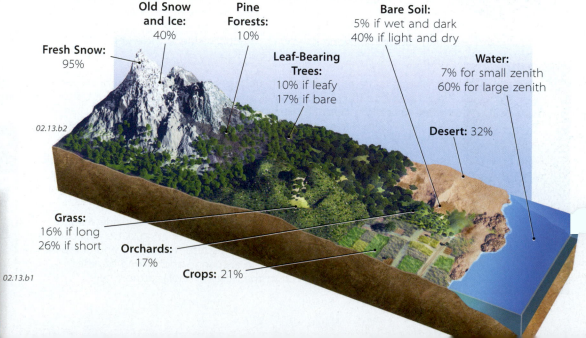

02.13.b2

Old Snow and Ice: 40%

Pine Forests: 10%

Bare Soil: 5% if wet and dark 40% if light and dry

Fresh Snow: 95%

Leaf-Bearing Trees: 10% if leafy 17% if bare

Water: 7% for small zenith 60% for large zenith

Desert: 32%

Grass: 16% if long 26% if short

Orchards: 17%

Crops: 21%

C   What Percentage of Insolation Goes Where?

We can measure and model how much insolation is intercepted in the air versus how much reaches the surface. Some insolation that reaches the surface is reflected upward into the atmosphere, and some of this reflected energy goes all the way back into space. Most energy is absorbed by the water, land, and vegetation. The figure below shows the global shortwave-radiation budget—how much insolation goes where, a key component of the global energy balance.

1. We begin by considering the total amount of insolation arriving at the top of the atmosphere as being represented as 100%. Then, we can examine what percentages of this total amount end up where. The left side of this figure depicts reflection and scattering of insolation, whereas the right side represents absorption.

2. *Scattering* by various air components (gas molecules, dust, etc.) returns 7% of total insolation back to space. Such scattering causes blue skies, red sunsets, and red sunrises, and is what illuminates the surface in shady areas and during heavy overcast days. Scattering decreases the amount of insolation that reaches the surface.

3. The amount of energy *reflected by clouds* is variable and controlled by many complex factors. Different types of clouds have different albedos, and some lower-level clouds are obscured below other, higher clouds. The location of a cloud is also important—tropical clouds receive more direct overhead sunlight and so have more available insolation to reflect, whereas low clouds over the poles receive little direct light, or no light at all during winter. Also, the amount of the world covered by clouds varies, depending on the season and other aspects of weather. Considering all the factors, clouds on average reflect 20% of the planet's total insolation back to space.

4. Chemical constituents of the atmosphere, such as O_2 and N_2 gas, absorb EMR of various wavelengths, amounting to 17% of the total insolation. Events on Earth's surface may cause considerable changes in the number of aerosols in the troposphere. Volcanic eruptions emit gases, volcanic ash, and other aerosols into the atmosphere, and these effects persist for several years. These additions increase the amount of absorption in the atmosphere.

5. Clouds absorb only a small percentage of insolation, about 3% of total insolation. The complex interaction between reflection, scattering, and absorption by various layers of clouds makes their impact on climate difficult to quantify—the net effect of clouds is one of the most difficult aspects to represent in global climate models. Some types of clouds tend to hold in heat, warming the surface, whereas others have an overall cooling effect.

INSOLATION 100%

Planetary Albedo 31%

Absorption by Dust and Molecules 17%

Scattering 7%

Energy Absorbed by Atmosphere 20%

Energy Reflected by Clouds 20%

Energy Reflected by Surface 4%

Absorportion by Clouds 3%

Energy Absorbed by Ground 49%

02.13.c1

7. So of the total insolation, 20% is converted to sensible heat or latent heat after absorption in the atmosphere, causing an increase in temperature or a change in state (mostly evaporation of water), respectively. A total of 49% of the insolation is retained by the land, as what is called *ground heat*. The remaining 31% is reflected back into space, accounting for the remainder of the 100% of insolation (20% + 49% + 31% = 100%).

6. Earth's surface albedo is fairly low but varies widely. Snow is highly reflective (high albedo), but many rocks are dark (low albedo) and rough (increasing the amount of scattering). Water generally has a lower albedo than land, and more than 70% of the Earth is covered with water. Due to reflection, scattering, and absorption in the atmosphere, only 53% of insolation reaches Earth's surface. But 4% of this energy is reflected back to space once it hits the surface, leaving 49% to be absorbed.

Changing the Global Radiation Budget

02.13.c2 Luzon Island, Philippines

In June 1991 Mount Pinatubo in the Philippines erupted (◄), throwing volcanic ash and gases up to 34 km into the troposphere and stratosphere. The ash circled the Earth, reducing insolation by about 1 watt per square meter (W/m^2), and cooling global temperatures 0.5 C°. The image to the right shows, in red, the amount of dust circulating around the tropical regions of the world the following December (more than a half year later). The eruption was also accompanied by a decrease in global vegetation, as measured by satellite, suggesting extra absorption of insolation by volcanic ash. The effects lasted for about two years until the ash either fell out of the atmosphere or was washed out by precipitation.

02.13.c3

Before You Leave This Page Be Able To

✓ Summarize how the atmosphere intercepts insolation.

✓ Describe albedo, give five examples, and summarize the importance for Earth's energy balance.

✓ Sketch, label, and explain the interactions of insolation with the atmosphere and surface, noting the percentages of insolation lost and stored in various ways.

2.13

2.14

What Happens to Insolation That Reaches the Surface?

APPROXIMATELY HALF OF INSOLATION is transmitted to Earth's surface, and this energy is variably reflected, absorbed, and re-emitted. Earth absorbs energy of short wavelengths, including insolation, but re-emits it at longer wavelengths. Certain *greenhouse gases* in the Earth's atmosphere interact with this outgoing long-wavelength radiation, complicating the return of this energy to space and helping keep our planet at a hospitable temperature.

A How Does Insolation Interact with the Surface?

Insolation that reaches Earth's surface is at short wavelengths, centered on the visible spectrum. It also includes ultraviolet (UV) and infrared (IR) wavelengths adjacent to the visible spectrum, commonly called *near-UV* and *near-IR*, respectively. Such electromagnetic radiation (EMR), with wavelengths less than 4 μm, is called *shortwave radiation*. By contrast, Earth emits energy at longer wavelengths, or *longwave radiation*. Shortwave radiation that reaches Earth's surface can be converted into other forms of energy.

Shortwave Radiation Converted to Sensible Heat

1. Some energy that strikes Earth's surface is *absorbed* by molecules, increasing their temperature, an increase in sensible heat. The heat stored in the land and water is called ground heat. The amount of heating of the surface, and the flux of ground heat, depends on all the factors that influence the distribution of insolation, such as latitude, season, length of day, and cloud cover. It is also influenced by whether insolation strikes land or water, by moisture and mineral content of the surface, and other factors.

02.14.a1

2. As Earth's land and water transfer sensible heat to the adjacent atmosphere, the air warms. Heating of Earth's land and water therefore warms the adjacent atmosphere—actually more than direct insolation does. The land and water surfaces heat up and cool down slowly. It takes both a while to exchange heat with the air, so warmest surface temperatures of the day typically occur hours after noon. For the same reason, coldest temperatures typically occur just before sunrise.

Shortwave Radiation Converted to Latent Heat

3. Energy striking Earth's surface can also be converted to latent heat as ice melts. As the short-wave energy of the insolation converts ice to liquid, molecules in the liquid begin to carry the associated latent heat of fusion. The water can flow away from the site of melting, or can remain and refreeze, releasing the latent heat to the local environment.

02.14.a2

4. Shortwave energy from insolation can also cause water to evaporate, forming water vapor. Molecules in the vapor carry the *latent heat of vaporization* into the atmosphere. When the vapor condenses into water drops, such as in clouds or as precipitation, the molecules give back this latent heat to the surrounding atmosphere, warming it (sensible heat). Through this process, shortwave radiation is converted first to latent heat and then to sensible heat. Similar processes include conversion of ice to vapor (sublimation) and vapor to ice (deposition).

Shortwave Radiation Converted to Longwave Radiation

5. Shortwave radiation that strikes the surface can be absorbed by materials it encounters, such as rocks, soils, wood, and water, increasing the motions of their constituent molecules. These materials emit some of this energy and are at temperatures appropriate for the energy emissions to be at long wavelengths (according to Wien's Law). In this way, shortwave radiation from the Sun is absorbed by the surface and then radiated as longwave radiation.

02.14.a3

6. The emitted longwave radiation can return to space or be intercepted by clouds, gas molecules, or solid aerosols. Some of the energy absorbed by these atmospheric components is then re-emitted as longwave radiation in all directions—into space, into other parts of the atmosphere, or back toward Earth's surface. Longwave energy that is emitted downward is commonly called *counter-radiation*, and helps keep the Earth at temperatures suitable for life.

B How Do Greenhouse Gases Interact with Electromagnetic Energy?

The atmosphere contains certain components, most notably water molecules, that absorb some wavelengths of *outgoing longwave radiation* (OLR). This energy is then emitted via longwave radiation to the surrounding atmosphere or even back down to Earth, keeping the planet warmer than it would be otherwise. Many people use the term *greenhouse effect* to refer to this warming influence, such as the way glass allows in sunlight but traps heat to keep a greenhouse warm. Atmospheric components that exhibit this behavior are called *greenhouse gases*.

1. The figure below shows absorption (vertical axis of graph) of EMR at various wavelengths (horizontal axis), by individual greenhouse gases. It also shows the amount of radiation that is scattered by gases and particles (second graph from bottom). Peaks on each curve show where the gas absorbs a certain wavelength of energy. A value of 1 indicates very high absorption or scattering at that wavelength. The top graph shows the composite effect for all gases in the atmosphere.

2. Characteristic wavelengths of incoming visible light and adjacent wavelengths of UV are barely impacted by atmospheric gases. As a result, insolation at these wavelengths is mostly transmitted through the atmosphere, illuminating and heating Earth's surface.

3. Insolation at shorter wavelengths, such as UV, is nearly all absorbed by atmospheric gases, especially by oxygen gas, ozone, and scattering by particles. These atmospheric gases and particles shield us from this dangerous radiation.

4. The two shaded curves on the bottom of the diagram show the wavelengths of energy emitted by the Sun (yellow) versus the Earth (blue). The majority of insolation entering the top of Earth's atmosphere is in the form of visible light (0.4–0.7 μm), whereas Earth emits its radiation back to space in the form of longwave radiation, mostly as *thermal infrared*, at wavelengths between about 8 and 20 μm. There is a clear separation between the wavelengths of *incoming shortwave radiation* and those of *outgoing longwave radiation*. The curves for individual gases show that each greenhouse gas intercepts some amounts of certain wavelengths of outgoing longwave radiation.

5. There is a band of frequencies at 8–13 μm that pass through the atmosphere without much loss from scattering and other effects. This is expressed as the large trough in the top curve, which shows the total effect of absorption and scattering for the entire atmosphere. The trough represents a part of the thermal IR spectrum where most energy is transmitted through the atmosphere—this region of the longwave spectrum is called the *atmospheric window* because it allows most longwave radiation out through the atmosphere.

6. Water vapor (H_2O) is an abundant, and the most important, greenhouse gas. It absorbs a wide range of wavelengths of OLR, including most wavelengths greater than 15 μm. The quantity of atmospheric water vapor varies from place to place and from time to time. It is typically higher over the oceans than over land. Water vapor and clouds help the surface stay warm by emitting extra counter-radiation downward, moderating changes in temperature between day and night.

7. Carbon dioxide (CO_2) is an atmospheric gas present in trace amounts, measured in parts per million. Quantities of CO_2 in the atmosphere have increased as the world warmed from the last glacial advance. Some CO_2 released into the atmosphere is from natural sources, such as volcanic eruptions and the natural decay of vegetation.

8. Oxygen and ozone intercept only some wavelengths of OLR. The previously noted ability of these gases to effectively absorb *incoming* UV radiation can be seen clearly by high peaks on the far left side of this diagram.

9. Nitrous oxide (N_2O) is a by-product of many industrial chemical processes, particularly production of fertilizers. It absorbs some energy at thermal-IR wavelengths.

10. Methane (CH_4), sometimes known as "marsh gas," occurs naturally as the result of the decay of organic material. Methane can also be released from oil and gas production or when frozen ground thaws or the seafloor warms up. It absorbs some thermal-IR energy.

2.14

2.15 How Does Earth Maintain an Energy Balance?

SIXTY-NINE PERCENT OF INSOLATION received at the outside of the Earth's atmosphere is available for sensible, ground, and latent heating. Ultimately all of this energy must be returned to space as longwave radiation in order to attain a balance between incoming and outgoing radiation. A greater loss to space would cool the global system, and a smaller loss would increase global temperatures. Just as there is a shortwave radiation budget, there is a budget of *global outgoing longwave radiation*. By interacting with outgoing longwave radiation (OLR), greenhouse gases help maintain Earth's hospitable temperature. Earth would be a very different planet if the greenhouse effect operated in a different manner or if the amount of greenhouse gases was different.

Sensible and Latent Heat Flux from Earth's Surface

1. There are various ways that Earth's surface and atmosphere transfer longwave energy. This page examines losses via the transfer of sensible and latent heat, and the facing page deals with losses through emission of longwave radiation. Read both pages in a counterclockwise order.

2. Of the 100 units of insolation (shortwave) that enter the top of Earth's atmosphere, 31 units are reflected or scattered by the atmosphere directly to space. That leaves 69 units within the surface and atmosphere.

3. Gas molecules, clouds, and various particles absorb 20 units, or 20% of insolation. For now, remember that this energy must be somehow released back into space.

4. The other 49 units of insolation are transmitted through the atmosphere. These units are then absorbed by Earth's surface, including land, oceans, and other bodies of water. These 49 units must somehow also escape the surface, or else the surface would keep heating up indefinitely as the Sun continued to transmit shortwave energy.

5. If 49 units of insolation reach Earth's surface, this is more than twice the amount (20 units) that is absorbed in the atmosphere. An implication of this is that the Sun heats Earth's surface more than it does the atmosphere, and in turn, the surface heats the atmosphere (by 29 units). Warming of the atmosphere from below in this way is one reason why air temperatures generally decrease upward with increasing altitude.

6. As the atmosphere is heated from below, the warmer air near the surface starts to rise upward, inducing convection in the troposphere. Approximately 7 units of energy are transmitted, mostly by convection, to the adjacent air as sensible heat. This flow of energy is called the *sensible heat flux*.

7. Most of the Earth is covered by ocean, and many land areas include lakes, wetlands, and heavily vegetated regions, so much of the energy reaching the surface goes to *latent heat flux* (melting ice, evaporating water, and transpiration from plants). Melting of ice only transfers energy between different parts of the surface (ice sheet to sea, for example), so it does not directly impact the atmospheric energy budget—but evaporation does. As the warm air rises convectively, it carries aloft the recently evaporated water vapor into the ever cooler air at higher altitudes. Eventually the moist air cools sufficiently to condense into water drops and form clouds, which then release the latent heat into the atmosphere. Almost half of all the energy reaching the surface of Earth (23 of 49 units) is returned to the atmosphere in this way.

Reflected and Scattered From Planet (31 Units)

Sun

31 Units

(100 Units)

Absorbed by Atmosphere (20 Units)

Insolation Reaching Surface - (49 Units)

Sensible Heat Flux (7 Units)

Latent Heat Flux (23 Units)

02.15.a1

8. The combined contributions of sensible and latent heat carry about 30 of the 49 units of the shortwave radiation stored at the surface into the atmosphere. Note that the numbers on these two pages add up to slightly more than 100 units (100%) because values have been rounded to whole numbers. When carried out with more precise numbers, it all adds up to 100 units.

Longwave Energy Flux From Surface and Atmosphere

13. We started with 100 units of insolation, so these 69 units plus the 31 units of shortwave insolation reflected (the planetary albedo) provide a perfect balance of input and output of energy to and from the Earth's land-ocean-atmosphere system. Keep in mind that these values are average annual values for the globe. Any individual place is unlikely to experience such a balance. Circulation of the atmosphere and oceans transfers energy from places that have an excess relative to the global average to those areas that have a relative deficit.

12. A total of 69 units of longwave energy go back into space: 57 units emitted by the atmosphere and the 12 units emitted directly from the surface.

11. The 57 units in the atmosphere are eventually emitted to space, in the form of longwave radiation.

10. At this point, the atmosphere has 50 units of energy—20 units from insolation initially absorbed by the atmosphere and 30 units it received from the surface via sensible and latent heat flux. This heat energy cannot be transmitted to space directly via conduction or convection because space is essentially a vacuum. Instead the atmosphere releases this energy by emitting longwave radiation in all directions: upward to space, sideways to other parts of the atmosphere, and downward to the surface. There is a back-and-forth exchange of radiant energy between the atmosphere and surface, largely controlled by the greenhouse effect. As a result of these interactions, there is a net flux of 7 units upward. These 7 units join the 50 units already in the atmosphere for a total of 57 units.

Emitted Directly to Space (12 Units)

69 Units

Emitted Directly by Atmospheric Molecules (57 Units)

Longwave Energy Flux (Net Flux 7 Units)

02.15.a2

9. Of the 49 units of shortwave radiation that reach Earth's surface, 12 units are emitted directly from the surface (water and land) to space as longwave radiation, without significant interactions with the atmosphere. This is possible because of the *atmospheric window* that allows certain IR wavelengths to radiate upward through the atmosphere with only minimal losses to absorption, reflection, and scattering.

A World without the Greenhouse Effect

Imagine two Earths, identical in all other ways (solar constant and planetary albedo) except the presence of a greenhouse effect. In the figure shown here, the lower globe has greenhouse gases in its atmosphere, whereas the upper globe does not.

From surface and satellite observations, average global surface temperatures on Earth currently are about +15°C (+59°F). We can estimate the average surface temperature of the imaginary planet using the Stefan-Boltzmann Law, relating the energy emission and temperature. Using this law, the surface temperature of the imaginary world is predicted to be −18°C (−9°F), significantly colder than the currently observed temperature of Earth.

EARTH WITHOUT ATMOSPHERE

Forecasted Surface Temp = -18 °C

EFFECT OF ATMOSPHERE 15°C - (-18°C) = +33°C

Observed Surface Temp = 15 °C

EARTH WITH ATMOSPHERE

02.15.t1

Some of the difference between these two estimates must be caused by greenhouse gases, which are keeping Earth's current temperatures warmer, thereby allowing water to exist in a liquid state, a key factor in supporting life as we know it. Calculating the actual contribution (in degrees of warming) of greenhouse gases is too complex to pursue here, because such calculations involve many other factors.

From these rough calculations, we can see why there is concern over the impacts of increased concentrations of greenhouse gases due to human-related emissions of carbon dioxide (CO_2), methane (CH_4), and nitrous oxide (N_2O). We address the changing concentrations of greenhouse gases and the broader topic of *climate change* in a later chapter on climate.

Before You Leave This Page Be Able To

✓ Sketch, label, and explain the nature of the Earth's overall energy balance, including the longwave radiation balance.

✓ Explain the linkage between the longwave radiation balance and the "greenhouse effect" on Earth.

2.15

2.16 How Do Insolation and Outgoing Radiation Vary Spatially?

FLOWS OF ENERGY into and out of Earth's system vary spatially, as a function of latitude, land or sea, cloud cover, and many other factors. The pattern of insolation also changes over several timescales, from daily rotation of the planet to the longer changes in season, causing spatial and temporal imbalances—zones of surplus energy and zones with an energy deficit, relative to the planetary average. These energy imbalances provide the driving force for global weather and climate.

A How Does Insolation Vary Spatially?

Insolation striking the top of the atmosphere, when averaged over a year, shows a smooth gradient, from higher amounts over the equator to much lower amounts at the poles. A more complex pattern emerges if we examine how much insolation actually reaches the surface during the course of a year, as shown in the figure below.

1. At the broadest scale, latitude controls insolation. The highest amounts are in low latitudes, and the lowest amounts are near the poles. Purples on this globe are as low as 120 W/m², whereas orange and red are more than 250 W/m².

2. Although insolation is strong in the tropics it is reduced somewhat due to absorption, reflection, and scattering by the abundant clouds that characterize these same regions.

3. The highest values are in subtropical deserts.

4. Universally low values mark the Antarctic and Arctic, which have days to months of total darkness. The lengthy summer days do not make up for the dark winters.

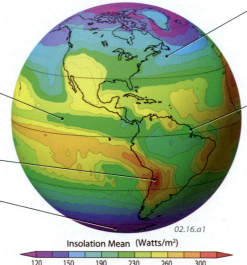

02.16.a1

Insolation Mean (Watts/m²)

120 150 190 230 260 300

5. The mid-latitudes are between 30° and 60° latitude, between the lines shown on this globe. At these latitudes, there is a relatively steep gradient, with the amount of annual insolation decreasing toward the pole.

6. There is a band of somewhat lower insolation just north of the equator, indicating a region of frequent thick clouds, which raises the local albedo of these regions.

7. There is a marked contrast between the gradual variation in insolation patterns over oceans, versus the irregular patchwork of variations over continents. The more irregular character over land is because of its variable land covers and elevations, which in turn influence the albedo, cloud cover, and other factors.

B How Does Outgoing Longwave Radiation Vary Spatially?

Using satellites, we can measure the amount of longwave radiation that leaves the Earth system (land, oceans, and atmosphere). The figure below shows the total amount of outgoing longwave radiation (OLR), averaged over an entire year. The picture for any single day is more complex, with more irregular weather-related patterns. The picture also varies in detail from hour to hour or week to week, because of local variations in storms, cloud cover, and other variables.

1. The quantity of energy emitted from an object depends on its surface temperature. Latitude strongly controls annual insolation, so it also controls larger patterns of both temperature and OLR.

2. The highest amounts of OLR are in the tropics, which receive more insolation than other regions. Some of this "extra" energy is directly returned as OLR from the surface and from the atmosphere.

3. The pattern of high OLR in the tropics is complicated over land, such as a zone of lower OLR along the western side of South America. This is due to a combination of a high mountain range (the Andes) and a cold ocean current (the Humboldt Current) along the west coast, both of which reduce the temperature, and reduced temperatures affect OLR.

02.16.b1

OLR Mean (Watts/m²)

150 180 220 250 280

4. The Arctic and Antarctic differ from one another, although it is difficult to see from this perspective. The oceanic Arctic emits more OLR, appearing warmer, than the continental Antarctic. A similarly low OLR is associated with land in Greenland, compared to the adjacent ocean, which emits more OLR because it is warmer.

5. Conversely, some of the highest continental values of OLR are between 20° and 30° latitude, on either side of the equator. These are the global deserts, such as the Sahara in North Africa. In addition to having high temperatures, the absence of water vapor allows OLR to escape through the atmosphere.

6. The patterns in this globe and the one above change with the season, as the latitude of maximum insolation shifts north and south.

How Does the Balance Between Incoming and Outgoing Radiation Vary Spatially?

Tropical regions receive much more insolation than do mid-latitude or polar regions, and this results in an unequal distribution of energy as a function of latitude—tropical regions have more Sun-induced energy than the poles. Energy flows from regions of relative surplus toward regions with relative deficits.

Radiative Balances

1. This graph shows the average amount of insolation at the top of the atmosphere versus the amount of outgoing energy (OLR) emitted, as a function of latitude, from the equator on the left to the North Pole on the right. A graph constructed from the equator to the South Pole would be similar.

2. Like a bank account, the difference between incoming and outgoing radiation is a location's *radiation balance*. On average, regions near the equator, tropics, and subtropics receive more energy from insolation than they emit in OLR—they have a positive balance or a *radiation surplus*. The region of surplus extends as far north (and south) as about 35° latitude, well past the Tropics of Cancer and Capricorn. The amount of insolation and OLR barely change across the tropics (left part of the top graph).

3. Poleward of the tropics, insolation declines rapidly, while the amount of outgoing radiation decreases more gradually.

02.16.c1

4. The two curves cross around 35° latitude, represented by the vertical dashed line. At this location, the amount of outgoing radiation (OLR) equals the amount of incoming radiation, on average—at this latitude, there is a radiation balance of zero.

5. Poleward of 35°, regions emit more OLR than the shortwave radiation that they receive from insolation—they have a *radiation deficit*. Insolation continues declining rapidly at mid latitudes, reaching very low values in the polar regions. The energy balance becomes increasingly negative toward the poles.

6. Energy is transferred from areas of surplus to those of deficit, toward the poles.

Large-Scale Patterns of Radiative Surplus and Deficit

7. This globe shows the mean annual radiation balance based on satellite observations. Orange, red, and dark pink represent zones of surplus, whereas yellow, green, and blue are regions of deficit. The white dotted lines show 35° N and 35° S. What patterns do you notice?

8. As expected, the regions with the highest energy surplus, depicted in red and orange, are concentrated along the equator, extending out across the tropics on both sides of the equator.

9. The patterns on land are locally more complex than the patterns in the ocean, suggesting the complicating influence of variations in elevation and the resulting complex patterns of clouds, snow, wind directions, etc.

02.16.c2

10. The regions with the most severe energy deficits are near the poles. These regions lose much more energy to space than they gain from low-angle sunlight during the summer, and they receive no insolation during the winter.

11. The mid-latitudes, between the tropics and the polar circles, exhibit a slight radiative deficit, represented here by a light pink color. They gain energy transferred poleward from the tropics.

12. More than half of the planet's surface area has an energy surplus, and less than half has an energy deficit. The total amount of surplus energy equals the amount of energy deficit, and the planet is in an overall energy balance. As we will explore later, the transfer of energy from areas of radiative surplus to those of deficit drive much of our wind patterns and weather.

Before You Leave This Page Be Able To

✓ Sketch and explain the global patterns of insolation and outgoing longwave radiation.

✓ Sketch and explain global patterns of surplus and deficit energy, using a graph or map, and explain how these are expressed in large-scale patterns in the regional distribution of areas of energy excess and deficits.

2.16

2.17 Why Do Temperatures Vary Between Oceans and Continents?

WATER EXHIBITS VERY DIFFERENT thermal properties from those displayed by the rocks and soil of land. These differences in thermal properties cause oceans and land to warm and cool at different rates, leading to significant temperature variations between oceans and land. Such differences help explain major patterns of global temperature and climate.

A   How Do Water and Earth Materials Respond to the Same Changes in Energy?

Heat Capacity

1. In evaluating how water, rocks, and other materials heat up or cool down, an important consideration is how much energy is needed to heat up an object, and how much heat that object can retain. A physical attribute called *heat capacity* expresses how much heat is required to change a volume's temperature by one Kelvin.

HEAT CAPACITY

Less   More

02.17.a1

SPECIFIC HEAT

Land   Water

Less   More

02.17.a2

Specific Heat

3. To compare the inherent thermal responses of different materials, irrespective of how much of the material is present, we use a property called *specific heat capacity*, or simply *specific heat*. Specific heat is the amount of energy needed to increase a kilogram mass of a substance by 1 K (or 1 C°).

2. The heat capacity of an object is determined by the kind of material in the object, such as rock versus water, and by the size of the object. The larger block above has a higher heat capacity than the smaller block, as long as both are composed of similar materials.

4. The specific heat of water is four times that of most rocks and materials. This means that it takes four times more energy to heat water than it takes to heat an equivalent mass of rock.

Thermal Responses of Water Versus Other Earth Materials

5. Due to their differences in specific heat, oceans (water) heat up during the day differently than land. If the same amount of insolation strikes water and land, the land will increase in temperature 4 degrees (K or C°) for every degree the water increases. As a result, land warms up much faster than the ocean, under the same environmental conditions, but some of this difference is offset by increased losses of heat from the land to the air.

02.17.a3

1°C   4°C

Water   Land

02.17.a4

Cools Faster

Cools Slowly

Water   Land

6. At night, the water and land both lose energy to the cool night air. For the same energy loss, land cools more than does water. Also, the land became hotter during the day, and hot objects radiate more energy than cool ones, so the land loses energy faster than the water. As a result, land cools off much faster than water at night.

Depth of Heating, Cooling, and Mixing

7. Another factor in how water and land respond to changes in insolation and air temperatures is how deeply heat is able to enter each material. Nearly all water allows at least some transmission, so shortwave radiation can penetrate to depths of tens of meters or more. In contrast, rocks and soil are largely opaque, so insolation is confined to the surface, and heat must move downward into the land by conduction, which it does by only a meter or two during the day.

SUN

Water   02.17.a5   Land

8. Some materials, such as water, are relatively mobile, which allows them to flow and mix. Under calm ocean conditions (left column), limited mixing causes surface waters warmed by the Sun to remain near the surface, so there is a strong temperature contrast with depth. Surface winds induce waves (center column), resulting in turbulence, which carries warm waters downward, mixing them with cooler waters at a depth. Salt water (right column) is more dense than fresh water, so any waters that are saltier than normal, such as from partial evaporation, can sink, causing mixing of the water column. Mixing allows heat to be carried deeper into the water column (much faster than heat is conducted) and brings up cooler water that gives off energy to the atmosphere more slowly (because it is cool). As a result, mixed water heats up more slowly than does land, which experiences almost no vertical mixing.

02.17.a6

Calm   Turbulent   Salty

Unmixed Water Column   Mixed Water Column   Mixed Water Column

Latent Heat

9. Water has another unique capacity relative to land—it can store abundant energy as latent heat. Insolation that strikes water is transformed into one of three different fluxes of energy. Some energy goes into heating the water (ground heat), some heats the air (sensible heat in the atmosphere), but a large amount goes into latent heat produced by evaporation. In contrast, insolation striking land goes almost entirely into sensible heat in the atmosphere and into ground heat. Land contains some water, but lesser amounts of its insolation goes into latent heat.

02.17.a7

Distribution of Continents and Oceans

10. Another factor that influences the global energy budget, and the balance of energy that falls on land versus the oceans, is the difference between the Northern and Southern Hemispheres. As can be observed on any map or globe, the Northern Hemisphere has the majority of the planet's landmasses, whereas the Southern Hemisphere is dominated by oceans. As a result, an equal amount of insolation striking both hemispheres will result in more latent heat being generated in the Southern Hemisphere than in the Northern Hemisphere. In December, when the Southern Hemisphere more directly faces the Sun, more insolation will fall on water than during June.

02.17.a8

B How Do Temperatures Reveal Thermal Differences Between Ocean and Land?

1. These various factors, from specific heat to latent heat, cause land and water to respond very differently to insolation and to the change from day to night. Land, with its relatively low specific heat, limited mixing, and limited amount of latent heating, warms up more quickly than water and reaches higher temperatures. At night, land's higher daytime temperatures cause it to lose heat more rapidly than does water. This keeps the night warm for a while, but eventually the cool night air dominates.

2. Water, with its high specific heat, partial transparency, and ability to mix vertically, heats up more slowly and does not reach as high a temperature. Also, much insolation is converted into latent heat that is transferred to the atmosphere via evaporation and condensation, so this energy is not available to heat the body of water. Large bodies of water therefore experience smaller temperature variations and more moderate temperatures overall, compared to land. Land areas adjacent to the water can partly experience the moderate temperatures caused by the unique thermal properties of water.

02.17.b1

Mean Annual Temperature

02.17.b2

3. These three globes show the yearly average temperature, average January temperature, and average July temperature, all for 2011. Observe the temperatures shown on each globe and use concepts presented in this chapter to try to explain the main patterns. Aspects to consider include variations in insolation due to latitude and clouds, land-sea contrasts, and the distribution of continents. These globes show data for the different times of year, so they express seasonal variations. In each globe, red and orange are hotter, blue and purple are colder, and yellow and green are intermediate in temperature. The overall average temperatures (◀) range from less than −50°C in Antarctica to locally greater than +20°C in some tropical and subtropical regions.

Average January Temperature

02.17.b3

4. Average January temperatures (◀) go from less than −50°C near the North Pole to more than +25°C in the tropics.

5. Average July temperatures (▶) are higher in the Northern Hemisphere than in the Southern Hemisphere.

Average July Temperature

02.17.b4

Before You Leave This Page Be Able To

☑ Sketch, label, and explain all the factors that cause land and ocean at the same latitudes to heat and cool differently, identifying how these factors affect temperatures.

☑ Summarize some of the main patterns in the global distribution of average, maximum, and minimum temperatures.

2.17

How Are Variations in Insolation Expressed Between the North and South Poles?

VARIATIONS IN INSOLATION, both as a function of latitude and from season to season, help explain many aspects of our world — average temperatures, hours of daylight, type of climate and weather, the type of landscape, and the overall livability of a place. For a transect down the west coasts of the Americas, from the Arctic to the Antarctic, we examine the average monthly amounts of insolation, length of day, and temperature, as a way to connect concepts in this chapter with actual places. Examine the photographs, graphs, and text for each place, and think about what explains the patterns for that place and the variations from one place to the next.

Northern Alaska and Canada

02.18.a1 ANWR, AK

The North Pole is located in the Arctic Ocean, but parts of Alaska and Canada are north of the Arctic Circle (66.5° N). Compare the graphs to the right, which show the monthly variation in insolation (first graph), number of daylight hours (bar graph) and average temperatures (line graph). Note that during parts of winter there is no daylight.

CAMBRIDGE BAY, NUNAVUT, CANADA

02.18.a2

02.18.a3

Pacific Northwest

02.18.a4 British Columbia, Canada

The northwestern part of the mainland U.S. and adjacent parts of British Columbia, Canada, straddle the famous 49th parallel (49° N latitude). They are squarely in the mid-latitudes. Like most of the places on this page, the region is near the ocean, so its temperature variations are moderated somewhat by the waters of the adjacent ocean.

VANCOUVER, BRITISH COLUMBIA, CANADA

02.18.a5

02.18.a6

Baja California

02.18.a7 Baja, Mexico

Baja California, part of Mexico, is a desert peninsula bordered by the Pacific Ocean to the west and the Gulf of California to the east. La Paz, the capital of Baja California Sur, is at a latitude of 24° N, just north of the Tropic of Cancer (23.5° N). Note that the graphs for Baja and the two previous places display a maximum centered on June to August (summer).

LA PAZ, BAJA CALIFORNIA SUR, MEXICO

02.18.a8

02.18.a9

Ecuador

02.18.a10 Ecuador

The South American country of Ecuador is named for its position on the equator. It is on the west coast of the continent and contains parts of the Andes, Amazon rain forest, and the Galápagos Islands, famous for their active volcanoes and unusual animals. Note the pattern on the length-of-day bar graph and the minor variation in monthly insolation for the equator.

GALAPAGOS ISLANDS

02.18.a11

02.18.a12

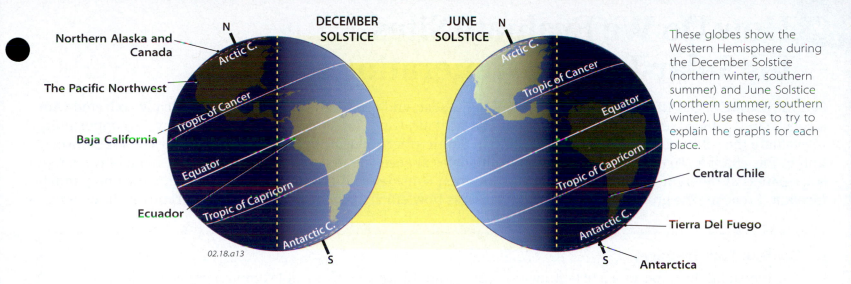

N   DECEMBER SOLSTICE   JUNE SOLSTICE   N

Northern Alaska and Canada — Arctic C.
The Pacific Northwest
Tropic of Cancer
Baja California
Equator
Ecuador — Tropic of Capricorn
Antarctic C.
S

Arctic C.
Tropic of Cancer
Equator
Tropic of Capricorn — Central Chile
Antarctic C. — Tierra Del Fuego
S — Antarctica

02.18.a13

These globes show the Western Hemisphere during the December Solstice (northern winter, southern summer) and June Solstice (northern summer, southern winter). Use these to try to explain the graphs for each place.

Central Chile

02.18.a14 Santiago, Chile

About halfway down the western coast of South America is the central part of Chile. The country's capital of Santiago is at a latitude of 33.5° S, south of the Tropic of Capricorn. It is inland and higher than the coast. Note that in the Southern Hemisphere, the graphs of insolation, hours of daylight, and temperatures now have central troughs rather than peaks.

SANTIAGO, CHILE
02.18.a15   02.18.a16

Tierra Del Fuego

02.18.a17 Ushaia, Argentina

Tierra del Fuego is the southernmost tip of South America, in southern Chile and Argentina. Ushuaia, Argentina, at a latitude of 54.8° S, is called the southernmost city in the world. It is not within the Antarctic Circle, so every day has some daylight and darkness. The southern tip of South America, surrounded by cold seas, is a frigid, stormy place.

USHUAIA, ARGENTINA
02.18.a18   02.18.a19

Antarctica

02.18.a20 Antarctica

Antarctica, centered over the South Pole, consists of vast plains of ice with some impressive mountain ranges. The main part of the continent is within the Antarctic Circle, so it has months of darkness. The South Pole has continuous darkness from April to September (the southern winter), but 24 hours of sunlight from October to February. It is always a very cold place.

SOUTH POLE
02.18.a21   02.18.a22

Before You Leave This Page Be Able To

☑ Sketch the general patterns of average insolation, number of daylight hours, and temperatures for three sites: one in the Northern Hemisphere, one in the Southern Hemisphere, and one near the equator. Explain the patterns in terms of latitude.

2.18

How Do We Evaluate Sites for Solar-Energy Generation?

SOLAR ENERGY IS A RENEWABLE SOURCE of light, heat, and electricity. Solar energy is typically collected with a solar panel, which can generate electricity or can heat air, water, or some other fluid. You have an opportunity to evaluate the solar-energy potential of five sites in South America, using concepts you have learned about insolation in this chapter. To do this, you will use zenith angles at each site to determine how directly sunlight will strike solar panels at different times of the year. You will then consider some regional factors that influence how much insolation reaches the ground. From these data, assess how efficient each site will be at generating solar energy.

Goals of This Exercise:

- Understand the importance of latitude in constraining the maximum possible amount of insolation striking the surface, to help evaluate the solar energy potential of a site.
- Use factors controlling regional amounts of insolation, along with characteristics of the geographic setting of a site, to evaluate how much insolation is likely to actually arrive at Earth's surface, the key factor in the solar-energy potential of each site.

1. The globe below is a view centered on the continent of South America, and it shows the South Pole and adjacent parts of the Southern Hemisphere. It locates five possible sites you will consider for their solar energy potential: (1) the Galápagos Islands, part of the country of Ecuador; (2) Macapá, a small settlement in Brazil; (3) La Serena, a coastal town of Chile; (4) near Mar Chiquita, in central Argentina, and (5) Ushuaia, the capital of the Tierra del Fuego Province of southernmost Argentina. Some characteristics of each site are described below.

Site	Latitude	Zenith Angle		
		Equinox	Dec. Sol.	June Sol.
Galápagos	0°	0°	23.5° S	23.5° N
Macapá	0°	0°	23.5° S	23.5° N
La Serena	30° S	30°	7.5°	53.5°
Mar Chiquita	30° S	30°	7.5°	53.5°
Ushuaia	~60° S	60°	36.5	83.5°

2. The Galápagos are a series of islands in the Pacific Ocean, far west of South America. They are known for their exotic animal species, made famous by the visits of Charles Darwin. Although surrounded by the ocean, many parts of the islands receive little rain and are similar to a desert.

3. La Serena is a popular tourist site along the Pacific coast of Chile. It has a cool, desert climate, caused in part by a cold ocean current (the Humboldt Current) that limits the amount of moisture in the air. It is on the fringe of the Atacama Desert, the driest place on Earth (the tan strip along the coast to the north of 30° S on the globe).

4. Ushuaia, the southernmost city in the world, is near Cape Horn, the southern tip of South America. Cape Horn is surrounded on three sides by ocean, some of which is coming from near Antarctica and is very cold. As a result, Ushuaia has a relatively cold and humid climate. It is close to 55° S latitude, but for this exercise we will consider it to be at 60° S to simplify our calculations.

7. From the latitude of each site we can calculate the zenith angle at each site for an equinox and both solstices. These results are presented in the table above.

6. Macapá is a small community in Brazil, near the mouth of the Amazon River, just inland from the sea. It is next to the Amazon River, the world's largest river, and is part of the huge Amazon rain forest, which stretches from the Atlantic Ocean westward all the way to the foothills of the Andes. As part of the rain forest, Macapá is a humid, rainy place.

5. Mar Chiquita is a small inland lake on the plains of central Argentina, far from the oceans but at a relatively low elevation. The proposed site is north of the lake, far enough that the lake does not significantly affect the local climate. This part of Argentina is east of the Andes, the huge mountain range along the western coast of South America. The Andes block moisture from the west, causing the plains of Argentina to be relatively dry.

02.19.a1

Procedures

1. Read descriptions of each site and consider how the geographic setting might impact the favorability of the site for solar-energy production.

2. For each site, use the zenith angles to look up the maximum percentage of solar energy that is theoretically available.

3. Use a graph of insolation as a function of latitude and month to evaluate how the maximum amount of insolation for each site varies during the year.

4. List the pros and cons of building a solar-energy facility at each site.

5. OPTIONAL EXERCISE: Your instructor may provide you with simple formulas for calculating zenith angle from latitude and have you determine zenith angles and solar favorability for other sites, some near where you live.

8. In considering the various sites, an important factor is that the solar panel will be horizontal. So we can use zenith angle as a measure of whether the sunlight is coming in perpendicular to the solar panel, the optimum orientation. Although this orientation works fine near the equator, at other latitudes solar panels are inclined at some other angle so as to maximize the amount of sunlight (keeping the panel perpendicular to the direction of sunlight, wherever possible).

02.19.a2

9. Use the zenith angle for each site to determine the maximum percentage of energy that can be produced by the solar panel for that zenith angle, relative to the maximum amount it could produce if it were perfectly perpendicular to the Sun. Write down your results to guide your considerations in evaluating each site.

Percentage of Solar Energy Available	
Zenith Angle	Percent
0°	100%
10°	98%
20°	94%
30°	87%
40°	76%
50°	64%
60°	50%
70°	34%
80°	17%
90°	0%

10. Use this graph (▶) to estimate the quantity of insolation arriving at the top of the atmosphere above each location, during the months of the solstices (December and June) and the equinoxes (March and September). The graph is in W/m². During which month does each site theoretically (ignoring atmospheric effects) receive the most and the least insolation? Which site or sites receive the most and the least insolation overall? Record your results.

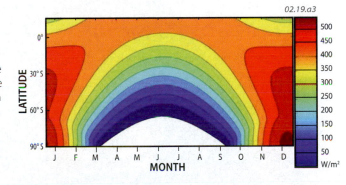

02.19.a3

11. Since the numbers on this graph represent the amount of insolation reaching the top of the atmosphere, we must now consider how the climate of each site might decrease this amount, such as from excessive cloudiness. Use the globes below to complete this step. Finally, combine your zenith calculations with results from the graph to the left and the globes below to make a list showing the pros and cons of each site. Choose the best site and be able to defend your conclusions.

Downward Shortwave Radiation Flux This globe shows the amount of shortwave radiation (insolation) that actually reaches the surface. Red, orange, and yellow show higher amounts of insolation reaching the surface, whereas blue and purple show low amounts. The generally higher amounts of insolation in the tropics versus the low amounts near the South Pole show the effects of latitude. The deviations from the broad pattern, such as the green low over the Amazon, are due to processes in the atmosphere that reflect, absorb, or scatter insolation. Use what you know about the geographic setting of each site to propose what processes explain the pattern on this globe.

02.19.a4

Outgoing Longwave Radiation Flux (OLR) This globe shows the amount of OLR emitted by Earth's surface. Such radiation is not useful for generating solar energy because it is longwave and, therefore, less energetic than shortwave, and because it is going upward, not down. It is a useful indication, however, of processes going on in the atmosphere because it suggests, for example, the abundance of water vapor (a greenhouse gas) and clouds, which can absorb longwave radiation before it exits the atmosphere, thereby reducing OLR. On this globe, orange represents high amounts of OLR, whereas purple and blue represent much smaller amounts. What do the patterns indicate, and what are the implications of both globes for solar energy?

02.19.a5

2.19

Atmospheric Motion

MOTION OF THE EARTH'S ATMOSPHERE has one of the greatest influences on human lives, controlling climate, rainfall, weather patterns, and long-range transportation. It is driven largely by differences in insolation, with influences from other factors, including topography, land-sea interfaces, and especially rotation of the planet. These factors control motion at local scales, like between a mountain and valley, at larger scales encompassing major storm systems, and at global scales, determining the prevailing wind directions for the entire planet. All of these circulations are governed by similar physical principles, which explain wind, weather patterns, and climate.

Large-scale patterns of atmospheric circulation are shown here for the Northern Hemisphere. Examine all the components on this figure and think about what you know about each. Do you recognize some of the features and names? Two features on this figure are identified with the term "jet stream." You may have heard this term watching the nightly weather report or from a captain on a cross-country airline flight.

What is a jet stream and what effect does it have on weather and flying?

Prominent labels of H and L represent areas with relatively higher and lower air pressure, respectively.

What is air pressure and why do some areas have higher or lower pressure than other areas?

Distinctive wind patterns, shown by white arrows, are associated with the areas of high and low pressure. The winds are flowing outward and in a clockwise direction from the high, but inward and in a counterclockwise direction from the low. These directions would be reversed for highs and lows in the Southern Hemisphere.

Why do wind patterns develop around areas of high and low pressures, why do these patterns spiral, and why are some spirals clockwise and others counterclockwise?

North of the equator, prevailing winds (shown with large gray arrows) have gently curved shapes. For most of human history, transportation routes depended on local and regional atmospheric circulation. These winds were named "trade winds" because of their importance in dictating the patterns of world commerce. The trade winds circulate from Spain southwestward, causing Christopher Columbus to land in the Bahamas rather than the present U.S.

What causes winds blowing toward the equator to be deflected to the west?

Prevailing winds from the north and south converge near the equator. This zone of convergence, called the *Intertropical Convergence Zone (ITCZ)*, is a locus of humid air and stormy weather.

What causes winds to converge near the equator, and why does this convergence cause unsettled weather?

Polar Cell

Polar Easterlies

60N

Polar Front Jet Stream

L

H

30N

Northeast Trade Winds

EQUATOR

TOPICS IN THIS CHAPTER

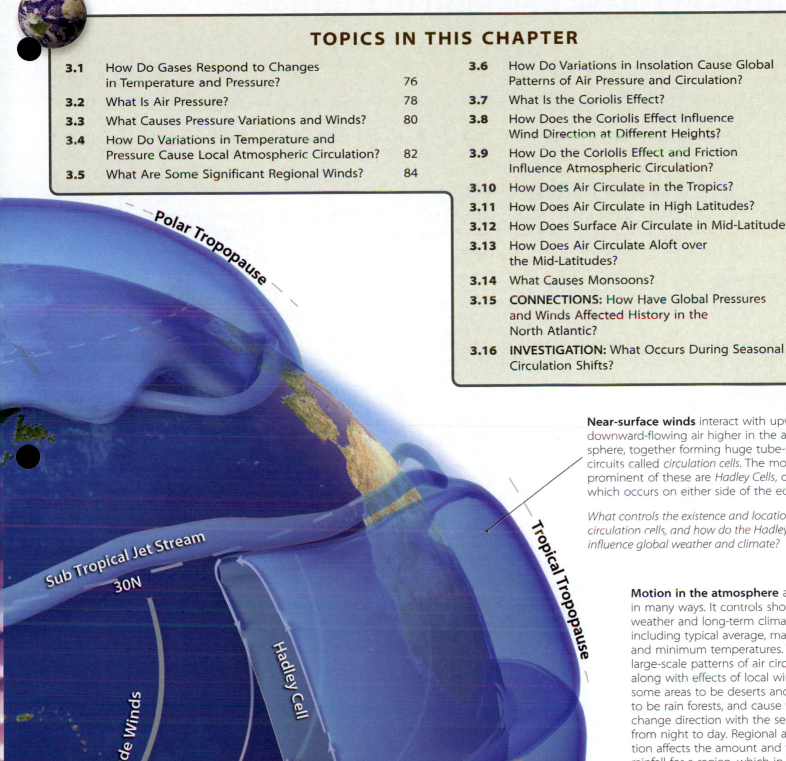

Near-surface winds interact with upward- and downward-flowing air higher in the atmosphere, together forming huge tube-shaped air circuits called *circulation cells*. The most prominent of these are *Hadley Cells*, one of which occurs on either side of the equator.

What controls the existence and location of circulation cells, and how do the Hadley Cells influence global weather and climate?

Motion in the atmosphere affects us in many ways. It controls short-term weather and long-term climate, including typical average, maximum, and minimum temperatures. The large-scale patterns of air circulation, along with effects of local winds, cause some areas to be deserts and others to be rain forests, and cause winds to change direction with the seasons and from night to day. Regional air circulation affects the amount and timing of rainfall for a region, which in turn controls the types of soils, vegetation, agriculture, and animals situated in an area. Winds determine which areas of the U.S. are more conducive to wind-power generation than others. The result of these global, regional, and local atmospheric motions is a world in which the tropics are not too hot, the polar areas are not too cold, and no areas have too little moisture for life.

03.00.a1

3.0

3.1 How Do Gases Respond to Changes in Temperature and Pressure?

THE ATMOSPHERE CONSISTS LARGELY OF GASES, with lesser amounts of liquids, such as drops of water, and solids, such as dust and ice. By nature, gases expand easily or contract in volume in response to changes in temperature and pressure. Variations in temperature and resulting changes in pressure are the main drivers of motion in the atmosphere.

A How Does a Gas Behave When Heated or Cooled?

The quantity of insolation entering the atmosphere exhibits considerable spatial variability, especially as a function of latitude, and temporal variations on both daily and seasonal time scales. These variations in insolation in turn lead to differences in temperatures. How do the gases in the atmosphere respond to changes in temperature?

03.01.a1 Namibia, Africa

1. Consider what happens when we want to make a hot-air balloon rise (◄). Typically, a propane-powered burner heats ambient air, causing the air to expand in volume. This increase in volume inflates the balloon. Since the same amount of gas now occupies a much larger volume, the *density* of the heated air is less than the density of the surrounding air, so the balloon rises. So, as air increases in temperature, it tends to increase in volume and become less dense.

2. The figure below shows how a quantity of gas responds to either an increase in temperature (heating) or a decrease in temperature (cooling). The starting condition is represented by the cube of gas on the left.

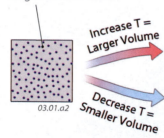

Increase T = Larger Volume

Decrease T = Smaller Volume

03.01.a2

3. An increase in the temperature of a gas means more energetic molecules, so a larger volume is needed to accommodate the same amount of gas.

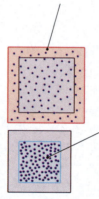

4. If a gas cools, the molecules within it have less kinetic energy (motions) and can therefore be packed into a smaller volume. The gas has a higher density and will tend to sink.

5. This example shows that temperature and volume of a gas are directly related—in fact they are *proportional* if pressure is held constant. Such a proportional relationship means that if temperature is doubled, volume doubles too. If temperature decreases by half, volume does too. This specific relationship is called *Charles's Law*, which is one of the fundamental laws governing the behavior of gases, and it explains why a hot air balloon rises.

B What Happens When a Gas Is Compressed?

If a gas is held at a constant temperature but forced to occupy a smaller *volume*, the *pressure* of the gas increases. Pressure is proportional to the number of collisions of the molecules. If the same gas fills a larger volume, the pressure decreases. In both cases, if we instead change the pressure, the volume of the gas will adjust accordingly. A material, like a gas, that can be compressed, is said to be *compressible*.

1. Molecules of gas in the sealed container in the left canister are under pressure, represented by the two weights resting on top. At some temperature, the molecules have a fixed amount of energy, and some of the moving molecules are hitting the moveable lid, resisting the downward force of the attached weight.

2. Removing a weight reduces the downward pressure on the gas. However, the gas retains its same average energy level (temperature) and therefore exerts the same upward force on the moveable lid as before. The upward force from the gas molecules exceeds the downward force of the weight and so raises the lid, increasing the volume occupied by the gas. In this way, a decrease in pressure results in an increase in volume, if the gas does not change temperature.

3. Increasing the downward pressure by adding weight on the original canister causes the lid to slide down. This increase in pressure causes a decrease in volume. As the gas is compressed into a smaller volume, the number of the molecules impacting the lid increases. When this upward force from the gas molecules equals the downward force from the weight, the lid stops moving, and the volume and pressure of the gas stop changing.

4. The relationship between pressure and volume of a gas, under conditions of constant temperature, is *inversely proportional*—if pressure increases, volume decreases. If pressure decreases, volume increases. Either pressure or volume can change, and the other factor responds accordingly, changing in the opposite direction by a proportional amount. That is, if the volume is cut in half, the pressure doubles. If the volume doubles, the pressure is cut in half. This inversely proportional relationship between pressure and volume, under constant temperature, is called *Boyle's Law*.

Less Pressure = More Volume

More Pressure = Less Volume

03.01.b1

C   How Are Temperatures and Pressures Related?

Since Charles's Law relates volume to temperature, and Boyle's Law relates volume to pressure, we might suspect that we can relate temperature and pressure. Combining Charles's Law and Boyle's Law leads to the *Ideal Gas Law*, which relates temperature, pressure, and density (mass divided by volume). Basic aspects of the Ideal Gas Law help explain the processes that drive the motion of matter and associated energy in the atmosphere.

03.01.c1

1. We can represent the Ideal Gas Law with a figure, with words, or with an equation. We begin with this figure (▶), which expresses the two sides of the equation. On one side of the equation (the left in this figure) is pressure. On the right side of the equation are density and temperature. What the Ideal Gas Law states is that if we increase a variable on one side of the equation (like increasing pressure), then one or both of the variables on the other side of the equation have to change in the same direction—density or temperature have to also change, or perhaps both do.

2. Examine this figure and envision changing any one of the three variables (pressure, density, and temperature), and consider how the other two variables would respond to satisfy the visual equation.

3. What happens if pressure increases? If temperature does not change, then density must increase. If pressure increases but density does not change, then temperature has to increase. Alternatively, temperature and density can both change. This three-way relationship partly explains why temperatures are generally warmer and the air is more dense at low elevations, where the air is compressed by the entire weight of the atmosphere, than at higher elevations, where there is less air. Higher pressure often results in higher temperatures.

4. What does the relationship predict will happen if a gas is heated to a higher temperature? If the density does not change, the pressure exerted by the gas on the plunger must increase. If the pressure does not change, the density must decrease. This is because density and temperature are on the same side of the equation, so an increase in one must be matched by a decrease in the other—if the other side of the equation (pressure) does not change. The relationship indicates that heated air can become less dense, which allows it to rise, like in a hot air balloon.

5. The Ideal Gas Law can also be expressed by the equation to the right:

$$P = R \rho T$$

where P is pressure, R is a constant, ρ is density (shown by the greek letter rho), and T is temperature. Note how this equation roughly corresponds to the figure above.

D   How Can Differences in Insolation Change Temperatures, Pressures, and Density in the Atmosphere?

The way gas responds to changes in temperature and pressure is the fundamental driver of motion in the atmosphere. Since temperature changes are largely due to insolation, we can examine how insolation affects the physical properties of gas and how this drives atmospheric motion.

03.01.d1

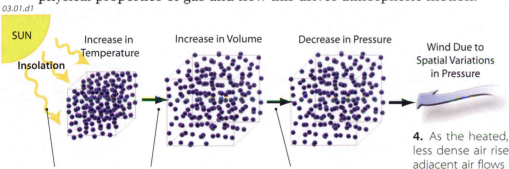

5. In this way, the response of gas to changes in temperature, pressure, and density (or volume), as expressed by the gas laws, is the primary cause of motion in the atmosphere. Variations in insolation cause changes in temperature, pressure, and density, which in turn cause air to move within the atmosphere.

1. The Sun is the major energy source for Earth's weather, climate, and movements of energy and matter in the atmosphere and oceans. In the figure above, insolation strikes Earth's surface (land or water), which in turn heats a volume of gas in the overlying atmosphere.

2. The increase in temperature results in expansion of the gas because of the increased kinetic energy of the molecules in the gas, which is an increase in volume. If the same number of gas molecules occupy more volume, the density of the air decreases (the air becomes less dense).

3. The increase in volume can result in a decrease in pressure (less frequent molecular collisions). As a result, the air mass is now less dense than adjacent air that was heated less. The more strongly heated and expanded air rises because it is less dense relative to surrounding air (which was not heated as much and so is more dense).

4. As the heated, less dense air rises, adjacent air flows into the area to replace the rising air. The end result is a vertical and lateral movement of air within the atmosphere—vertical motion within and above the rising air, and lateral motion of surrounding air toward the area vacated by the rising air.

Before You Leave This Page Be Able To

☑ Explain why a gas under constant pressure expands and contracts with changes in temperature.

☑ Explain why a gas at a constant temperature expands or contracts with changes in pressure.

☑ Sketch and summarize the Ideal Gas Law and how the three gas laws explain motion in the atmosphere due to variations in insolation.

3.2 What Is Air Pressure?

PRESSURE OF GASES WITHIN THE ATMOSPHERE is highly variable, both vertically and laterally. These variations in pressure define the structure of the atmosphere and also determine the nature and direction of atmospheric motions. If one place in the atmosphere has higher pressure than another place, this imbalance of pressure (and therefore also atmospheric mass) must be evened out, causing the air to flow. How do we describe and measure pressure, and how do we use these measurements to understand or even predict the flow of air?

A What Is Pressure?

Pressure is an expression of the force exerted on an area, usually from all directions. In the case of a gas, pressure is related to the frequency of molecular collisions, as freely moving gas molecules collide with other objects, such as the walls of a container holding the gas. It is such collisions that keep a balloon, soccer ball, or bicycle tire inflated.

1. Molecules of gas in a sealed glass container move rapidly in random directions, and some strike the walls of the container. The force imparted by these collisions is pressure. The more collisions there are, the more pressure is exerted on the walls of the container.

03.02.a1

2. If we push down on the lid of the container, the same number of molecules are confined into a smaller space. Lower parts of the container walls are now struck by a greater number of the more closely packed gas molecules, so the pressure is greater. Decreasing the volume of a gas increases its pressure, consistent with Boyle's Law.

3. What happens if we put a weight (▶) on top of the lid (center container) and then either cool or heat the gas in the container?

03.02.a2

4. If we cool the container by placing it in ice, the molecules become less energetic and so strike the walls and lid of the container less often—the gas pressure decreases and the lid moves down.

5. If we instead heat the container, the gas molecules become more energetic and strike the walls and lid of the container more often—the gas pressure increases and lifts the lid.

6. The equation to the right illustrates what pressure measures and the units we use to describe it. The units of pressure are used throughout this book in describing weather, climate, and the flow of water.

$$\text{Pressure} = \frac{\text{Force}}{\text{Area}} = \frac{\text{Mass} \cdot \text{Acceleration}}{\text{Area}} = \frac{\text{kg m sec}^{-2}}{\text{m}^2} = \frac{\text{Newtons}}{\text{m}^2} = \text{Pascals}$$

100 Pascals = 1 millibar (mb)

7. Pressure is a force exerted on a given surface area.

8. According to Newton's second law, force is the product of mass and acceleration.

9. The unit of mass is the kilogram (kg), acceleration is in meters/second/second (m/s²), and area is in square meters (m²).

10. A force of one kg m/s² is called a *Newton*. Pressure, measured in Pascals, is an expression of the number of Newtons of force exerted on a square meter of surface.

11. The air pressure at Earth's surface is many Pascals, so we express pressures in a larger, related unit called a *bar*, or in *millibars* (1/1000 of a bar).

B How Is Air Pressure Measured?

03.02.b2

High Atmospheric Pressure — Low Atmospheric Pressure

1013.25 mb

29.92 mb

Mercury

03.02.b1

◀ We can measure air pressure with an instrument called a *barometer*. The barometer shown to the left is a sealed glass tube fixed in liquid mercury. Changes in air pressure cause the liquid level in the tube to rise or fall, allowing the measurement of relative pressure. Such barometers have units of *inches (or centimeters) of mercury*. Pressure is also reported in units of a *bar*, with one bar being approximately equal to the average air pressure at sea level. Modern digital instruments record pressure in millibars.

▶ Meteorologists measure air pressure at vertical heights in the atmosphere using hydrogen- or helium-filled balloons, like this one. An instrument package called a *radiosonde* is suspended from the balloon (lower left of the photograph). Sensors measure pressure, temperature, humidity, and position (using a GPS) as the balloon ascends, and these measurements are transmitted via radio to a central computer. Wind speed and direction are inferred from successive positions of the radiosonde. The balloon eventually pops and the radiosonde parachutes to the ground.

C   How Does Air Pressure Vary Vertically?

1. Air pressure in the atmosphere is not constant. The largest variation is vertically, with an abrupt decrease in pressure upward from near the surface. The red curve on this figure shows how the air pressure, measured in millibars (mb), decreases from Earth's surface to the top of the atmosphere.

2. This diagram depicts the main layers of the atmosphere (troposphere, stratosphere, etc.) and highlights some of the features observed in each part, such as auroras in the thermosphere, shooting stars that mostly burn up in the mesosphere, and the restriction of most clouds and weather to the troposphere.

3. Colors along the left edge of the diagram convey temperature variations within and between the atmospheric layers. These vertical temperature variations affect the density of the air, impacting air motions caused by the Sun heating the Earth's surface.

03.02.c1

4. In the thermosphere and mesosphere, gas molecules are relatively sparse and temperatures are low (−90°C at the thermosphere-mesosphere boundary). As a result of the sparseness of molecules, air pressures are very low (less than one millibar).

5. The abundance of gas molecules increases down into the stratosphere, and this is accompanied by an increase in air pressure (the bending of the red curve to the right as it goes downward). Air pressures are slightly greater at the top of the stratosphere than in the overlying mesosphere, but at the base of the stratosphere (the tropopause) they have increased to about one-fifth of pressures measured at sea level.

6. The pull of Earth's gravity holds most gas molecules close to Earth's surface, in the troposphere. Air pressure increases downward in the troposphere because of a greater abundance of molecules downward and the larger total number of molecules pressing down from the layers above. The highest air pressures are close to the surface, and at the lowest elevations. Sea level is the reference level for air pressure, with an average pressure of 1,013 mb (a little over 1 bar).

D   How Does Air Pressure Vary Laterally?

1. Air pressure also varies laterally, from area to area, and from hour to hour, and these variations are typically represented on maps, like the one shown here. Such maps either show the pressure conditions at a specific date and time or show pressure values averaged over some time period, like a month or a year. To allow us to compare different regions and to see the larger patterns, the map uses pressure values that are corrected to sea level, or their sea-level equivalent. In this way, we eliminate the effects of differences in elevation from place to place.

03.02.d1

3. Most maps of air pressure feature the large capital letters H and L. An H represents an area of relatively higher pressure called a high-pressure area or simply a *high*. An L represents a low-pressure area, commonly called a *low*. An elongated area of high pressure can be called a *ridge* of high pressure and an elongated area of low pressure is a *trough*.

4. The map patterns change with time, corresponding to changes in air pressure that accompany changes in weather. Patterns typical for a region also change from season to season.

Before You Leave This Page Be Able To

☑ Describe the concept of pressure, what it represents, and how it changes when subjected to changes in temperature and volume.

☑ Explain how we measure pressure and what units we use.

☑ Sketch and explain how air pressure varies vertically in the atmosphere.

☑ Describe what a map of pressure shows, and explain the significance of an isobar, H, and L on such a map.

2. Such maps of air pressure contain numbered lines, called *isobars*, that connect locations with equal pressure. If you could follow an isobar across the countryside, you would follow a path along which the pressure values, once corrected to their sea-level equivalents, would be equal. Successive isobars are numbered to represent different values of air pressure, usually in millibars (e.g., 1,020), and there is generally a constant difference in pressure between two adjacent isobars (a 4 mb difference on this map). Note that isobars do not cross, but can completely encircle an area.

3.2

3.3 What Causes Pressure Variations and Winds?

THE MOVEMENT OF AIR IN THE ATMOSPHERE produces *wind*, or movement of air relative to Earth's surface. Circulation in the atmosphere is caused by pressure differences generated primarily by uneven insolation. Air flows from areas of higher pressure, where air sinks, to areas of lower pressure, where air rises.

A How Do We Measure the Strength and Direction of Wind?

Wind speed and direction are among the most important measurements in the study of weather and climate. On short time scales, wind can indicate which way a weather system is moving and the strength of a storm. When considered over longer time scales, winds indicate general atmospheric circulation patterns, a key aspect of climate.

03.03.a1

1. Wind directions can be assessed as easily as throwing something light into the air and tracking which way it goes, but it is best done with a specially designed measuring device (◄) that can measure the wind speed and direction. Wind speed is expressed in units of distance per time (km/hr) or as knots, which is a unit expressing nautical miles per hour. One knot is equal to 1.15 miles/hr or 1.85 km/hr.

2. Wind direction is conveyed as the direction *from which the wind is blowing*. Wind direction is commonly expressed with words (►), such as a northerly wind (blowing from the north). It can instead be described as an *azimuth* in degrees clockwise from north. In this scheme, north is 0°, east is 090°, south is 180°, and west is 270°.

3. The atmosphere also has vertical motion, such as convection due to heating of the surface by insolation. A local, upward flow is an *updraft* and a downward one is a *downdraft*.

03.03.a2

B What Causes Air to Move?

Air moves because there are variations in air pressures, in density of the air, or in both (recall that pressure and density are related via the Ideal Gas Law). Such pressure and density variations are mostly caused by differential heating of the air (due to differences in insolation) or by air currents that converge or diverge. The atmosphere is not a closed container, so changes in volume (i.e., air being compressed or expanded) come into play. These volume changes can make air pile up or spread out, resulting in variations in air pressure.

1. Movement of air occurs to equalize a difference in air pressure between two adjacent areas (►), that is, a *pressure gradient*. Air molecules in high-pressure zones are packed more closely together than in low-pressure zones, so gas molecules in high-pressure

03.03.b1

zones tend to spread out toward low-pressure zones. As a result, air moves from higher to lower pressure, in the simplest case (as shown here) perpendicular to isobars.

2. High-pressure zones and low-pressure zones can be formed by atmospheric currents that converge or diverge (▼). *Converging* air currents compress more air into a smaller space, increasing the air pressure. *Diverging* air currents move air away from an area, causing low pressure. Forces associated with converging and diverging air are called *dynamic forcing*.

3. Most variations in air pressure and most winds, however, are caused by thermal effects, specifically differences in insolation from place to place. This cross section (►) shows a high-pressure zone caused by the sinking of cold, high-altitude air toward the surface. In the adjacent low-pressure zone, warmer near-surface temperatures

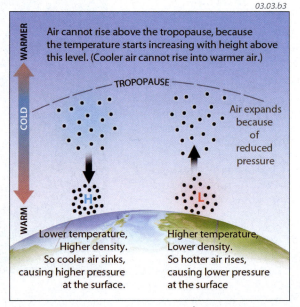

03.03.b3

Air cannot rise above the tropopause, because the temperature starts increasing with height above this level. (Cooler air cannot rise into warmer air.)

TROPOPAUSE

Air expands because of reduced pressure

H

L

Lower temperature, Higher density. So cooler air sinks, causing higher pressure at the surface.

Higher temperature, Lower density. So hotter air rises, causing lower pressure at the surface

have caused air to expand, become less dense, and rise, causing low pressure. Near the surface, air would flow away from the high pressure and toward the low pressure. Different air currents would form higher, in the upper troposphere, to accommodate the sinking and rising of the air. In an atmosphere that doesn't have strong updrafts or downdrafts, at any altitude the upward pressure gradient will approximately equal the downward force of gravity.

Converging Air

Diverging Air

03.03.b2

C What Forces Result from Differences in Air Pressure?

Differences in air pressure, whether caused by thermal effects or dynamic forcing, result in a *pressure gradient* between adjacent areas of high and low pressure. Associated with this pressure gradient are forces that cause air to flow. Pressure gradients can exist vertically in the atmosphere or laterally from one region to another.

03.03.c1

3. Lateral variations in air pressure also set up horizontal pressure gradients, and a pressure-gradient force directed from zones of higher pressure to zones of lower pressure. On the map below, the pressure-gradient force acts to cause air to flow from high pressure toward lower pressures, as illustrated by the blue arrows on the map.

03.03.c2

4. Places where isobars are close have a steep pressure gradient, and a strong pressure-gradient force, so movement of the atmosphere (i.e., winds) will generally be strong in these areas.

1. Elevation differences cause the largest differences in air pressure. At high elevations, there is less atmospheric mass overhead to exert a downward force on the atmosphere. As a result, density decreases with elevation, and air pressure does too.

2. These vertical variations in air pressure cause a pressure gradient in the atmosphere, with higher pressures at low elevations and lower pressures in the upper atmosphere. This pressure gradient can be thought of as a force directed from high pressures to lower ones. This *pressure-gradient force* is opposed by the downward-directed force of gravity, which is strongest closer to Earth's surface.

5. Places where isobars are farther apart have a more gentle pressure gradient, a weak pressure-gradient force, and generally lighter winds. Although winds tend to blow from high to low pressure, other factors, such as Earth's rotation, complicate this otherwise simple picture, causing winds patterns to be more complex and interesting.

D How Does Friction Disrupt Air Flow?

As is typical for nature, some forces act to cause movement and other forces act to resist movement. For air movements, the pressure-gradient force acts to cause air movement and friction acts to resist movement.

1. Friction occurs when flowing air interacts with Earth's surface.

2. As represented in this figure by the shorter blue arrows low in the atmosphere, wind is slowed near the surface, because of friction along the air-Earth interface. As the air slows, it loses momentum (which is mass times velocity). Some momentum from the moving air can be transferred to the land, such as when strong winds pick up and move dust or cause trees to sway in the wind. It is also transferred to surface waters, causing some currents in oceans and lakes and forming surface waves.

03.03.d1

3. Friction with Earth's surface, whether land or water, also causes the wind patterns near the surface to become more complicated. On land, air is forced to move over hills and mountains, through valleys, around trees and other plants, and over and around buildings and other constructed features. As a result, the flow patterns become more curved and complex, or *turbulent*, near the surface, with local flow paths that may double back against the regional flow, like an eddy in a flowing river. Friction from the surface is mostly restricted to the lower 1 km of the atmosphere, called the *friction layer*.

4. Stronger winds occur aloft, in part because these areas are further removed from the frictional effects of Earth's surface. Some friction occurs internally to the air, even at these heights, because adjacent masses of air can move at different rates or in different directions. Friction can also accompany vertical movements in updrafts and downdrafts.

Before You Leave This Page Be Able To

☑ Explain how we measure and describe wind velocity and direction.

☑ Sketch and explain how air motion and heating can form zones of low and high pressure.

☑ Describe the pressure-gradient force.

☑ Sketch and explain how friction affects wind.

3.3

3.4 How Do Variations in Temperature and Pressure Cause Local Atmospheric Circulation?

PRESSURE GRADIENTS INDUCE FLOW at all scales, including local and regional ones, arising from unequal heating by insolation and latent heat, by differing thermal responses of land versus sea, and even from the construction of large metropolitan areas. Such circulations contribute to the climate of a place, particularly in the absence of, or interaction with, more powerful circulations.

A What Causes Breezes along Coasts to Reverse Direction Between Day and Night?

People who live along coasts know that gentle winds sometimes blow in from the sea and at other times blow toward the sea. The gentle winds are called a *sea breeze* if it blows from sea in toward the land and a *land breeze* if it blows from the land out to sea. Such breezes are mostly due to differences in the way that land and sea warm up during the day and cool down at night.

The Sea Breeze (Daytime)

1. During daytime hours, land heats up significantly, particularly in summer. Land, which has a lower specific heat capacity than water, heats up more overall and more rapidly than water does. The hot air over the land surface rises, inducing a local low-pressure area over the land.

03.04.a2 Baja California, Mexico

2. At the same time, air over the water body is cooled by the relatively cool water temperatures and by cooling associated with evaporation. The relatively cool air over the water sinks, inducing a local high-pressure area over the water.

03.04.a1

3. The difference between the low pressure over land and the high pressure over the water represents a pressure gradient. The associated pressure-gradient force pushes air near the surface, from higher pressure to lower pressure. This flow is an *onshore breeze* or *sea breeze* that feels cool to people on the beach.

5. In this photograph (◄), heating of the hot, desert land causes air over the land to rise, drawing in moist air from the adjacent water. Rising of the moist air along the coast forms thin, scattered clouds, but can draw in much thicker masses of clouds, forming coastal fog and overcast skies.

4. Air aloft moves in the opposite direction, from land to sea. This is a response to an upper-level pressure gradient caused by the "extra" air rising over land and less upper-level air over the sea, due to air sinking.

The Land Breeze (Nighttime)

1. At night, land cools significantly, particularly if there isn't much cloud cover. The relatively cool air over the land surface sinks, inducing a surface high-pressure area. Sometime in the evening hours, the surface low over the land weakens and becomes a high.

03.04.a4 Philippines

2. The water body doesn't cool as much at night, partly because of water's higher specific heat capacity and because of mixing. So air over the water stays relatively warm compared to the air over the land. Therefore, this air rises, generating relatively low pressure over the water body.

03.04.a3

3. The difference between the high pressure over land and the low pressure over the water is a pressure gradient. The pressure-gradient force pushes near-surface air from higher to lower pressure, from the land to the water as an *offshore breeze* or *land breeze*.

6. In this photograph (◄), an early morning offshore flow, from the land to the sea, pushes moist air offshore, moving clouds out to sea and causing clear conditions along the shoreline.

4. Aloft, the pressure-gradient force causes the air that rose over the low to return to land to replace the air that sank to create the surface high.

5. The strength of sea and land breezes is proportional to the temperature gradient. The land breeze circulation strengthens through the night, begins to weaken at sunrise, stops sometime in the morning, and then reverses to a sea breeze, when it strengthens until peaking in the afternoon.

B How Does Topography Cause Pressure Gradients?

On land, topography can cause pressure gradients and winds that blow in different directions during the day versus at night. These are primarily related to differences in the way land and the atmosphere heat up and cool down, and to temperature differences as a function of elevation.

1. The Sun heats the surface more efficiently than it heats the atmosphere. During the day, the land heats up and in turn heats the adjacent air. Such heated air from lower elevations is especially warm and rises (▶), causing upslope winds, known as a *valley breeze*, or *anabatic wind*. Air that is farther from the surface, aloft in the center of the figure shown here, is less heated and so sinks to replace the rising air. Upslope winds are most obvious when winds from other types of circulation, like storms, are weak.

03.04.b1

03.04.b2

2. At night, the situation is reversed (◀). With no insolation to warm the surface, the surface will cool more rapidly than the air above it, as it emits longwave radiation to the atmosphere. Thus, the land and adjacent air at higher elevations cools significantly, and the resulting relatively cool surface air will move downslope, producing a *mountain breeze*, or *katabatic wind*. The cool air displaces warmer air in the valley upward, where it can flow to replenish the air lost at higher elevations. The end result is that valley bottoms can be substantially colder at night than areas at a slightly higher elevation.

3. Pollution in rugged terrain is mostly produced in the valleys where people are more likely to live, so the daytime valley breezes push polluted air upslope. But if the pollution is too intense or the slopes are too high and steep to push the air over the ridges and peaks before nightfall, the mountain breeze will send the polluted air back down into the valley to join the new polluted air produced during the next day. The cycle repeats day after day. Persistent high pressure (with its sinking air) over the area can make the situation even worse. Some of the poorest air quality exists in valleys surrounded by higher terrain, such as Mexico City.

C How Do Urban Areas Affect Pressure Features?

Human development of land can cause pressure gradients as natural ground cover is replaced by buildings, asphalt, and concrete. These human-produced materials have different properties from those of natural materials, and the resulting temperature differences can cause pressure differences.

1. Urban areas, such as cities and towns, are usually warmer than the surrounding rural environments, for several reasons. Urban areas often lack trees for shade and standing water for latent heating rather than sensible heating. They contain waste heat from engines, street lights, and fireplaces, along with urban building and road materials that absorb intense heat. The urban area causes the air to heat up more than air in the less developed surroundings. The warmer air over the city is less dense and rises, leaving behind a local low-pressure area.

2. The outlying areas are generally cooler because they are more likely to contain vegetation for shading and near-surface water, which absorbs insolation for latent heat rather than sensible heating. Outlying areas also have less human-generated waste heat.

3. This phenomenon is known as the *urban heat island* (UHI). Urban-rural temperature differences usually peak in the evening hours, after a day of surface heating and busy human activities that generate the urban heating. The UHI is responsible for some of the modern observed increase in surface temperatures.

03.04.c1

4. This urban-caused circulation of air can redistribute heat and pollution from the city center to outlying suburban and rural areas. Polluted air can be lifted out of the urban area, flow laterally, and descend into outlying areas. Higher pressure in the outlying areas may trap pollution near the ground. As with the sea breezes and valley breezes, the urban heat island can be diminished by storms and other strong air-circulation patterns.

Before You Leave This Page Be Able To

☑ Sketch, label, and explain the formation of sea and land breezes.

☑ Sketch, label, and explain the formation of valley and mountain breezes.

☑ Sketch, label, and explain what an urban heat island is and how it can cause the flow of air and pollution.

3.4

3.5 What Are Some Significant Regional Winds?

DIFFERENCES IN AIR PRESSURE cause a variety of regional to local wind conditions, such as those associated with storms, which are discussed in the chapter on weather. Some local winds are not so much related to weather systems as they are to differences in pressure that tend to occur at certain times of the year or after the establishment of an area of high pressure. These local to regional winds have interesting names, like Chinook winds or Santa Ana winds, and can have profound impacts on people.

A What Is a Chinook Wind and Where Do They Form?

The term *Chinook* originated in the Pacific Northwest and can refer to several types of winds. The most common usage is for a warm, dry wind that blows down the flanks of a mountain range. Chinooks are so warm and dry they are called "snow eaters," for the way in which they can cause a sudden melting of snow and ice on the ground. The onset of a Chinook wind can cause a sudden rise in temperatures, especially during the winter. A Chinook in Loma, Montana, caused temperatures to rise from −48°C (−59°F) to 9°C (49°F) within a 24-hour period, the most change recorded for a single day in the U.S. In Spearfish, South Dakota, a Chinook off the adjacent Black Hills caused temperatures to rise 27 C° (49 F°) in two minutes, the world's fastest rise in temperature ever recorded!

1. This figure depicts the formation of a *Chinook wind*. The process begins when winds push moist air against the *windward* side of a mountain, where windward refers to the side toward which wind is blowing (the left of the mountain in this figure). As the moist air rises up the mountain, it cools, causing the formation of clouds, a process that also releases latent heat. The heat warms the air, which continues rising toward the mountain peaks.

03.05.a1

03.05.a2

2. Once the air reaches the peak, it begins flowing down the other side—the *leeward* side, of the mountain (the side opposite the windward side). As the air descends, it continues drying out and is compressed and heated. It is also warm from the release of latent heat on the windward side. The warm, dry air descends from the mountain and spreads across the adjacent lowland, forming a Chinook wind.

3. This map shows the locations where Chinook winds are relatively common. As expected, Chinooks occur on the leeward side of mountain ranges (prevailing winds are from the west to east in this region), in Alaska, and elsewhere.

B How Do Katabatic Winds Affect Polar Regions?

1. On the previous two pages, we introduced the term *katabatic wind* for a wind that blows downslope, forming a cool *mountain breeze*. More regional and pronounced katabatic winds affect Antarctica and Greenland, both of which have a high central landmass surrounded by ocean. Air over the middle of the landmasses is very cold and so very dense, flowing off the central topographic highs and down the icy slopes.

03.05.b1

03.05.b2

03.05.b3 Antarctica

2. These two perspective views show that Antarctica (on the left) and Greenland (on the right) both have a broad, high area centered in the middle of the ice. Katabatic winds blow down off these high areas in all directions. These winds are especially strong where they are channeled down valleys, such as the famous Dry Valleys of Antarctica, so named because strong katabatic winds have stripped most ice off the land surface.

3. Katabatic winds in Antarctica generally involve cold but dry air, but they can interact with clouds along the coast, sometimes creating stunning effects as the cold air and clouds spill off the highlands, similar to the scene in the photograph above.

C How Do Santa Ana Winds Affect Southern California?

Winds in Southern California typically blow from west to east, bringing relatively cool and moist air from the Pacific Ocean eastward onto land, especially in areas right along the coast, like Los Angeles and San Diego. These coastal cities also often have onshore sea breezes during the day and offshore land breezes at night. At other times, however, regional winds, called *Santa Ana winds*, blow from the northeast and bring dry, hot air toward the coast, causing hot, uncomfortable weather and setting the stage for horrendous wildfires.

Setting of Santa Ana Winds

1. *Santa Ana winds* are regional winds that blow from the northeast (▶), typically developing during spring and fall, when high pressure forms over the deserts of eastern California and Nevada. Circulation of air associated with the area of high pressure pushes winds south and westward, toward the coast, in marked contrast to the normal onshore flow. This air is coming from the Mojave Desert and other desert areas to the north and east, so is very dry. The winds bring this dry air southwestward across the very dry Mojave Desert and toward the coast.

03.05.c1

2. Santa Ana winds from the Mojave Desert are partially blocked by the mountains on the northern and eastern sides of Los Angeles. The winds spill through mountain passes, such as Cajon Pass northeast of Los Angeles, and are funneled down the canyons and into the Los Angeles basin. The funneling effect causes winds to be especially strong within the canyons. As the air moves from higher deserts to the northeast and down toward Los Angeles, the air compresses and heats up. As a result, during an episode of Santa Ana winds, the coastal areas of Southern California experience much hotter and drier weather conditions than are normal. Due to this behavior of air flowing from high to low areas, Santa Ana winds are considered to be a type of katabatic wind.

Effect on Wildfires

03.05.c2 San Diego area, CA

1. The hills and mountains of coastal Southern California receive enough precipitation to be covered with thick brush (▲), such as oak, or by forests at higher elevations. During a Santa Ana wind, the fast winds dry out the brush, trees, and other vegetation, making it extremely prone to wildfires.

03.05.c4

3. This amazing image from NASA (◀) combines a satellite image of Southern California and adjacent states with the locations of fires (shown in red), as determined by processing a different kind of satellite data. The smoke produced by a number of fires trails off across the Pacific Ocean, clearly showing Santa Ana winds blowing from the northeast. These fires occurred in October of 2007 and killed 9 people and injured dozens of others, destroyed more than 1,500 homes, and burned more than 2,000 km² of forest, brush, and neighborhoods. The region was declared a federal and state emergency, as more than one million people were evacuated, the largest such evacuation in California history. The especially large amount of destruction from these fires was due to the combination of strong Santa Ana winds and a prolonged drought that had dried out the natural vegetation.

03.05.c3

2. The hills and mountains of Southern California experience some of the most spectacular but devastating wildfires of any place on the planet (◀). Santa Ana winds push these fires southwestward, toward the cities and through neighborhoods in the foothills. Wildfires associated with Santa Ana winds can burn thousands of homes, causing hundreds of millions of dollars in damage. Pushed by the strong winds down the canyons, the fast-moving fires can cause the deaths of firefighters and people who did not evacuate in time.

> ### Before You Leave This Page Be Able To
>
> ☑ Sketch and explain the origins of a Chinook and of a katabatic wind.
>
> ☑ Sketch and summarize the setting of Santa Ana winds and why they are associated with destructive wildfires.

3.5

How Do Variations in Insolation Cause Global Patterns of Air Pressure and Circulation?

SEASONAL AND LATITUDINAL VARIATIONS IN INSOLATION cause regional differences in air pressure, which in turn set up regional and global systems of air circulation. These circulation patterns account for many of the characteristics of a region's climate (hot, cold, wet, dry), prevailing wind directions, and typical weather during different times of the year. Here, we focus on vertical motions in the atmosphere resulting from global variations in insolation and air pressure.

A What Pressure Variations and Air Motions Result from Differences in Insolation?

1. On this figure, the top graph plots the average amount of insolation striking the top of the atmosphere as a function of latitude. On the surface below, the large letters represent high- and low-pressure zones. Arrows show vertical and horizontal airflow and are color coded to convey the overall temperature of air.

2. The maximum amount of insolation striking the Earth is along the equator and the rest of the tropics. This heats up the air, causing the warm air to expand and rise. The expansion and rising results in a zone of surface low pressure in equatorial regions. The upward flow of air helps increase the height of the tropopause over the equator, as shown by the dashed line.

6. This global pattern of rising and sinking air and resulting low and high air pressures dominates the motion of Earth's atmosphere. The pattern results from variations in insolation and is compensated by horizontal flows, both near the surface and in the upper atmosphere.

5. The amount of insolation reaches a minimum near the poles. Air near the poles is very cold and dense, sinking to form a zone of high pressure near the surface. The descending air flows away from the pole, toward 60° latitude. In the upper atmosphere, air flows toward the poles to replenish the air that sank.

03.06.a1

3. Surface winds flow toward the low pressure to replace the rising tropical air. The rising tropical air cannot continue past the tropopause, so as it reaches these heights it flows away from the equator (EQ) to make room for more air rising from below. This upper-level air descends in the subtropics, near 30° latitude, where it forms a zone of high pressure.

4. The amount of insolation decreases away from the equator, with a relatively sharp drop-off across the mid-latitudes (30° to 60° latitude). Descending air and high pressure in the subtropics causes surface air to flow toward higher latitudes (to the right in this diagram).

Differences in Responses of Land and Oceans at Different Latitudes

The figures below correspond to the same four regions described above, but focus on differences in behavior between land and water.

7. In equatorial regions, which have an energy excess, the rising warm air produces low air pressure. The land warms faster than the sea, so the air is even warmer over the land-masses and the air pressure is correspondingly lower. Rising warm air and low air pressure also prevail over the equatorial oceans, although temperatures tend to be cooler and pressures are not quite so low.

8. In the subtropics, at approximately 30°, cool air descends from the upper troposphere and causes higher-than-normal air pressures. In the subtropics, oceans are usually cooler than the surrounding continents. Air descending over the oceans will warm up less than air descending over the warmer continents, and so air pressures are particularly high over the oceans.

9. Air in the mid-latitudes is forced to ascend by surface air converging from the poles and subtropics. This zone has an energy deficit, and oceans retain their energy and are generally warmer than continents. Air forced to rise above the oceans is therefore warmer than that rising over the continents. As a result, the air over the oceans is more likely to rise, and the associated low-pressure zone tends to be stronger over the oceans than over the land.

10. The cold, dense, descending air over the poles produces a zone of high air pressure. This zone has a large energy deficit. In such zones, continents lose their energy faster, and are therefore cooler, than the adjacent polar oceans, so air pressures are extremely high over the cold continents. The slightly warmer temperatures of the adjacent oceans diminishes somewhat in these high air pressures.

B How Does Sea-Level Air Pressure Vary Globally?

Average Annual Air Pressure

1. This map shows sea-level equivalent air pressure averaged for 1981 to 2010. Observe the main pattern and compare these patterns with the figures on the previous page. Can you explain the larger patterns on this map?

2. Two belts of high pressure (shown in light gray) encircle the globe at about 30° N and 30° S (the subtropics). Between these two is a belt of lower pressure (shown in medium gray) straddling the equator, in the tropics. The equatorial low pressure and flanking high-pressure zones are due to the large air current that rises in the tropics and descends in the subtropics.

JANUARY TO DECEMBER: 1981 TO 2010
SEA LEVEL PRESSURE (MB) COMPOSITE MEAN

995   1012.5   1027.5   1045

3. A set of low-pressure areas occurs near 60° N. Note that the lows are best developed in the oceans, and are poorly developed on land.

4. A prominent air-pressure feature on this map is a belt of extremely low pressure (shown in dark gray) in the ocean just off Antarctica. This belt is so well developed in the Southern Hemisphere because of the abundant ocean surface, uninterrupted by continental landmasses at this latitude (60° S). An intense high-pressure belt (very light gray) occurs over continental Antarctica, in contrast to the oceanic Arctic.

January Air Pressure

5. Patterns of air pressure change with the seasons, following changes in insolation patterns. In January, one of the most prominent air-pressure features is a high-pressure area over Siberia, Russia, called the *Siberian High*. As discussed later, this high pressure helps drive the monsoon that affects much of southern Asia.

6. The broad belt of tropical low pressure moves toward the Southern Hemisphere, following the direct rays of the Sun (remember that January is in the southern summer). The migration is particularly noticeable over the hot land surfaces, like Australia.

JANUARY: 1981 TO 2010
SEA LEVEL PRESSURE (MB) COMPOSITE MEAN

995   1012.5   1027.5   1045

7. A large area of low pressure, called the *Icelandic Low*, develops over the northern Atlantic Ocean, wrapping around Greenland. As ocean waters retain their heat better at this cold time of year, a similar low, the *Aleutian Low*, develops in the northern Pacific Ocean west of Canada.

8. Elongated highs (shown in light gray) occur over the oceans in the Southern Hemisphere subtropics, but not over the adjacent continents. The highs are enhanced by the relatively cool oceans in this region. Farther south is the pronounced belt of low pressure in the cold oceans that encircle Antarctica.

July Air Pressure

9. Air-pressure patterns change markedly by July, the northern summer. In the Northern Hemisphere, the Siberian High has dissipated as warm air over interior Asia rises. Pressure gradients across the Northern Hemisphere are weaker in July than in January, because the equator-to-pole energy gradient isn't as steep in the summer as it is in the winter.

10. Typical patterns remain in the Southern Hemisphere, with belts of high pressure in the subtropics, flanked to the south by a continuous belt of very low pressure across the southern oceans. High pressure strengthens over the main landmass of Antarctica during the prolonged darkness of the southern winter.

JULY: 1981 TO 2010
SEA LEVEL PRESSURE (MB) COMPOSITE MEAN

995   1012.5   1027.5   1045

11. Bullseye-shaped high-pressure areas develop over the oceans, one over the central Atlantic (the *Bermuda-Azores High*) and another in the Pacific near Hawaii (*Hawaiian High*).

Before You Leave This Page Be Able To

☑ Sketch and explain the global air-circulation patterns, summarizing how they differ over land versus oceans.

☑ Summarize the main patterns of air pressure, identifying key features.

3.6

3.7 What Is the Coriolis Effect?

THE PRESSURE-GRADIENT FORCE drives airflow in the atmosphere, but winds do not blow in exactly the direction we would predict if we only consider pressure gradients. All objects — whether air masses, ocean waters, or airplanes — moving across the surface of the Earth display an apparent deflection from the objects' intended path. The cause of this deflection is the *Coriolis effect*. Why does this apparent deflection occur?

A What Is the Coriolis Effect?

1. The *Coriolis effect* refers to the apparent deflection in the path of a moving object in response to rotation of the Earth. The easiest way to envision this is by considering air that is moving from north to south or south to north. Earth's atmosphere, including any moving air, is being carried around the Earth by rotation.

2. The blue arrows show how much distance the surface rotates in an hour. The arrows are longer near the equator, indicating a relatively long distance that these areas have to travel, and therefore faster velocities. The distances traveled and the linear velocities gradually decrease toward the poles.

3. At the poles, the distance traveled and velocity are both zero — the surface has no sideways velocity due to the rotation. In 24 hours, an area directly at the pole would simply spin 360°, whereas an area at the equator would have moved approximately 40,000 km (the circumference of the Earth).

03.07.a1

4. As air moves toward the poles, it possesses the eastward momentum that it had when it was closer to the equator. So, it appears, from the perspective on Earth's surface, to be deflected to the right (to the east).

5. The opposite occurs as air moves toward the equator and encounters areas with a faster surface velocity. The air appears to lag behind, deflecting to the west as if it were being left behind by Earth's rotation. Note that in the Northern Hemisphere, air deflects to the right of the flow (not necessarily to the right as you look at it on a map), irrespective of which way it is moving (toward the pole, away from the pole, or in some other direction).

6. In the Southern Hemisphere, air moving toward the pole travels from faster rotating areas to slower ones, so it appears to be rotating faster than the surface — it deflects to the left. Moving air in the Southern Hemisphere deflects to the left irrespective of which way it is moving.

B Why Do Moving Objects Appear to Deflect Right or Left on a Rotating Planet?

To visualize why moving objects on a rotating planet appear to deflect left or right, examine these overhead views of a merry-go-round that is rotating counterclockwise (in the same way as Earth when viewed from above the North Pole). One person located at the center of the merry-go-round throws a ball to a second person standing near the outside edge of the merry-go-round. The path of the ball can be measured relative to two frames of reference: the two clumps of trees, which are fixed in our perspective, or from the persons on the merry-go-round, which is moving.

1. The person at the center of the merry-go-round slowly tosses a purple ball toward an outer person, in the direction of the upper two trees. The intended path of the ball is shown by the yellow arrow. The outer part of the merry-go-round moves faster than the center.

03.07.b1

2. After a short time, the ball is heading toward the two trees, but, relative to the intended path (yellow arrow) or from the perspective of the thrower, the purple ball seems to be veering away to the right, because the thrower rotated.

03.07.b2

3. With each passing time period, the intended receiver moves farther away from the ball as the ball goes toward the two trees. As viewed from the thrower, the ball deflected to the right relative to the intended path.

03.0

4. In the last figure, the ball's path traced upon the moving framework of the merry-go-round (open purple circles) reveals an apparent deflection to the right (shown with a dashed red line) of the intended path. However, relative to the fixed reference of the upper two trees, the ball has actually followed a straight line. This view is similar to one of the rotating Earth viewed from above the North Pole. The thrower and receiver are two locations at different latitudes, and the ball represents an air mass moving from the slow-moving pole toward the faster-moving equator.

C   Will Deflection Occur if Objects Move Between Points on the Same Latitude?

A similar deflection occurs if an object moves parallel to latitude on a rotating planet. To visualize why this is so, we return to the merry-go-round, which is still rotating counterclockwise, like Earth viewed from above the North Pole. As before, it is key to consider movements in terms of a fixed reference frame and a reference frame that is moving.

1. The person throwing the ball is on the outside of the merry-go-round along with the receiver. The intended path of the ball is shown by the yellow arrow. Since the players are the same distance out from the center, they are moving at the same rate.

03.07.c1

2. After a short time, the ball is heading along its original path (the purple path) relative to the upper two trees (the fixed reference frame). In the intervening time since the throw, however, the thrower and receiver have both moved (a moving reference frame).

03.07.c2

3. From the moving frame of reference of the thrower, the ball appears to be deflected to the right of the intended path, with the deflection shown by the orange dashed line.

03.07.c3

4. This example represents the apparent deflection of air (or any other object) moving *parallel to latitude.* So regardless of whether objects are moving in the north-south (meridional) or the east-west (zonal) directions, the objects appear to be deflected from their intended path. Moving objects have an apparent deflection to the right of their intended path in the Northern Hemisphere and to the left in the Southern Hemisphere. This left or right deflection due to the Coriolis effect accounts for the directions of prevailing winds, the paths of storms, and the internal rotation within hurricanes.

D   What Affects the Strength of the Coriolis Effect?

Since the Coriolis effect is related to the rate at which areas of the surface move during rotation of the Earth, we would suspect the strength of the effect may vary with latitude. It is also influenced by how fast objects are moving.

03.07.d1

1. When viewed from above the poles, the parallels of latitude constitute a series of concentric circles increasing in circumference from the poles to the equator. Moving from one latitude to another, like from the pole to 80° N, the percentage increase in circumference is much greater at high latitudes than nearer the equator. The difference in circumference for every 10° difference in latitude is plotted as orange triangles that represent the changing "slope" of the Earth's surface. Thus, the Coriolis effect is greatest at high latitudes, where the velocity of the moving reference frame changes most rapidly relative to the moving object.

03.07.d2 Hurricane Irene 2011

2. The Coriolis effect is expressed daily in many ways, including the shape of storms as viewed by satellites and featured on the daily weather report, the rotation of hurricanes (◄), and the changes in wind directions as a large storm approaches and then exits your town.

3. The Coriolis effect is stronger for an object with a large amount of mass, like a huge storm, or for an object that is accelerating (moving faster with time). In the case of a rotating storm, the acceleration can be related to movement of the entire storm across Earth's surface, rotations within the storm, and other motions. In any case, the Coriolis effect deflects faster moving objects more than it does slower moving objects.

SLOWER
Less Deflection

FASTER
More Deflection

03.07.d3

Before You Leave This Page Be Able To

☑ Explain what the Coriolis effect is and the direction of apparent deflection of moving objects in the Northern and Southern Hemispheres.

☑ Explain why a moving object on a rotating reference frame shows an apparent deflection.

☑ Explain what controls the strength of the Coriolis effect.

3.7

3.8 How Does the Coriolis Effect Influence Wind Direction at Different Heights?

PRESSURE GRADIENTS INITIATE MOTION in the atmosphere, but the actual direction in which the air moves is greatly influenced by the Coriolis effect. Close to the surface, where friction with the planetary surface is greatest and wind velocities are lowest, the pressure gradient dominates. Higher in the atmosphere, winds have higher velocities and the Coriolis effect dominates. The resulting patterns of airflow can dominate weather systems.

A Why Do Wind Speed and Direction Change with Height near the Surface?

The direction in which air moves is determined by three factors: 1) the pressure-gradient force (winds blow from higher pressure toward lower pressure); 2) the Coriolis effect, which appears to deflect objects moving across Earth's surface, and 3) friction with Earth's surface, which cannot change direction by itself but can interact with the other two forces to change the speed and direction of wind. Friction becomes less important upward.

1. This figure shows how wind direction can change upward, from the surface to well above the friction layer. Winds are fastest higher up and progressively slower down closer to the surface. This example is in the Northern Hemisphere.

2. The pressure-gradient force, caused by the pressure gradient between a high (H) and a low (L), causes winds right above the surface to blow from higher pressure to lower pressure, in an attempt to even out the pressure gradient. Even though friction is strong near the surface, it merely slows the wind; it does not by itself change the wind's direction.

03.08.a1

4. At some height, the deflection will be such that the winds flow parallel to the pressure gradient—no longer from higher to lower pressure. Further turning cannot happen because if it did, the flow would be going against the pressure gradient, from lower pressure toward higher pressure. The height at which this flow neither toward nor away from lower pressure (i.e., parallel to isobars) occurs is called the *geostrophic level*. It typically occurs when air pressure is about half that at the surface (i.e., the 500 mb level).

3. Higher up, friction is decreased, so winds are stronger. Since the Coriolis effect is proportional to wind speed, it begins deflecting air to the right (or to the left in the Southern Hemisphere). This causes successively more and more deflection with height.

-----

5. To illustrate how wind direction can change with altitude, examine what commonly happens to a balloon as it ascends from the surface. The red arrows indicate wind direction and the inset maps show isobars between high- and low-pressure areas. Follow the balloon's progress from the surface upward, starting with the text below.

6. Right after the balloon takes off, in Position 1 on this figure, the pressure-gradient force and friction are the primary factors that affect wind speed. The balloon will go from higher pressure to lower pressure, across the pressure gradient as it rises. Friction slows the circulation in this lowest part of the friction layer, which is called the *surface boundary layer*.

03.08.a2

9. Eventually, the balloon will ascend to a height (1–2 km above the surface) at which friction becomes negligible (above the *friction layer*). At that point, only the pressure-gradient force and the Coriolis effect are important in dictating flow. The wind speed is faster because of the reduced friction, so the Coriolis effect continues to pull the balloon to the right until it no longer flows toward lower pressure. Instead, the wind is perpendicular to the pressure-gradient force and parallel to isobars. This type of wind, flowing parallel to the isobars, is called a *geostrophic wind*.

8. As the balloon continues to rise, near Position 3, it experiences less and less friction, so the winds strengthen with height. This makes the Coriolis deflection even stronger with increasing height. The winds and the balloon move even less from higher pressure to lower pressure, instead turning at an angle to the isobars. This layer characterized by a turning of the winds with height (shown in Positions 2 and 3) is called the *Ekman layer*. It exists from about 100 m to about 1–2 km above the surface. The Ekman layer is a part of the friction layer.

7. In Position 2, the balloon has risen high enough from the surface that friction becomes weaker. This weakening of friction speeds up the balloon, and it also increases the impact of the Coriolis effect, because faster winds cause an increasing rightward deflection. This causes the balloon to move somewhat to the right of the direction that the pressure-gradient force would otherwise take it. The balloon therefore begins to travel more to the right as it rises.

03.08.a3 Namibia

B How Do We Depict Upper-Level Wind Patterns?

1. This map shows another way to represent pressure in the atmosphere, by contouring the height (in meters) at which 500 mb of pressure is reached. The contour lines represent lines of constant height and are called *isohypses*. Surface pressure is about 1000 mb, so the 500 mb level represents the height bounding the lower half of the atmospheric mass. This height is also generally considered to be the lowest level at which geostrophic winds occur. The higher the 500 mb level occurs, the higher the pressure. These "constant-pressure surface maps" are used by the National Weather Service and other organizations in making weather maps, because of the need for constant pressure surfaces in the aviation industry.

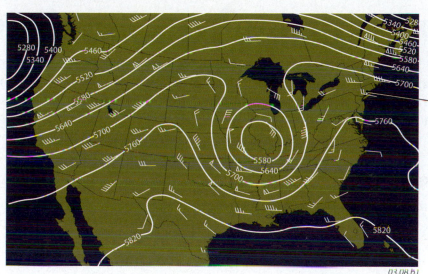

03.08.b1

2. This map also has symbols, called *wind barbs*, that show the direction and velocity from which the wind is blowing. The barb is parallel to wind direction, and each flag on the barb represents 50 knot (kt) winds (two flags indicate winds of 100 knots). Each full hash mark designates 10 kt, and each half-barb represents 5 kt. The wind barbs on this map show that winds generally flow parallel to the isohypses (they are geostrophic).

C Which Way Does Air Flow Around Enclosed High- and Low-Pressure Areas Aloft?

Aloft, in upper parts of the troposphere, well above the friction layer, wind directions are dominated by the Coriolis effect and so are typically geostrophic. As a result, air circulation can be nearly circular around upper-level lows and highs. In situations in which the flow is nearly circular, the low or high pressure typically lasts for a longer period of time than in cases where the flow is not enclosed as tightly, such as in an upper level *ridge* or *trough* of pressure. Closed circulations at the 500 mb level or above suggest a strong circulation that is not likely to dissipate soon. The figures below are map views.

Northern Hemisphere

03.08.c1

03.08.c2

Southern Hemisphere

03.08.c3

03.08.c4

For low pressure, the pressure-gradient force pushes air into the low from all directions, but at high altitudes the Coriolis effect deflects this air until it parallels the isohypses. In the Northern Hemisphere, the deflection into a low is to the right, so the upper-level flow goes *counterclockwise* around the enclosed low-pressure area. This type of curved flow is called *gradient flow*.

For a high-pressure area, the pressure-gradient force pushes air out of the high in all directions, but the Coriolis effect deflects winds to the right, parallel to the isohypses. Due to this deflection, the gradient flow goes *clockwise* around the high-pressure area.

Similar processes occur around lows in the Southern Hemisphere. Air pushed into the low from all directions is deflected to the left by the Coriolis effect, until it flows parallel to the isohypses. The gradient flow therefore goes *clockwise* around a low in the Southern Hemisphere.

In the Southern Hemisphere, deflection of air moving out from a high is to the left, causing the gradient flow to be counterclockwise around a high. In both hemispheres, pressure gradients tend to be weaker around highs, which are typically associated with fair weather, than around lows, which are commonly associated with stormy and windy weather.

Before You Leave This Page Be Able To

✓ Sketch and explain why wind directions change from the surface to higher levels in the troposphere.

✓ Explain what a geostrophic wind is and how you would recognize one on a map that represents air pressure.

✓ Sketch the upper-level flow around low-pressure and high-pressure areas in both the Northern and Southern Hemispheres.

3.8

3.9 How Do the Coriolis Effect and Friction Influence Atmospheric Circulation?

THE CORIOLIS EFFECT AND FRICTION affect the patterns of air movement set in motion by pressure gradients. These phenomena influence wind direction from local scales, affecting the rotation and wind patterns of individual storms, to global scales, affecting wind patterns of the entire planet. How do pressure gradients, the Coriolis effect, and friction explain wind patterns at such diverse scales? These two pages provide an overview of global and regional patterns of air circulation, with more in-depth coverage to follow in subsequent sections.

A What Influences the Patterns of Airflow Around High- and Low-Pressure Areas?

Once the pressure-gradient force starts air moving, the air is affected by the Coriolis effect and, if close to Earth's surface, is acted on by friction with the surface. The results of these interactions produce distinctive and familiar patterns of wind and circulation of air around areas of high and low pressure.

Low- and High-Pressure Areas

1. In an area of *low pressure*, the pressure-gradient force pushes air laterally into the low from all directions. If the Earth were not rotating and therefore had no Coriolis effect, this

03.09.a1

simple inward flow pattern would remain intact. Friction would likewise not perturb this pattern because friction does not change the direction of airflow, only the speed.

2. In an area of *high pressure*, the pressure-gradient force pushes air laterally out in all directions. As with a low pressure, this outward flow of air is influenced by the Coriolis effect

03.09.a2

and by friction near Earth's surface. As air flows outward from the high, it is replaced by air flowing down from higher in the atmosphere.

Northern Hemisphere

3. Earth does rotate, so the Coriolis effect, represented in this figure by green arrows, deflects the inward-flowing air associated with a low-pressure area. This deflection causes an inward-

03.09.a3

spiraling rotation pattern called a *cyclone*. In the Northern Hemisphere, the Coriolis deflection is to the right of its intended path, causing a *counterclockwise* rotation of air around the low-pressure zone. This counterclockwise rotation is evident in Northern Hemisphere storms.

4. In a Northern Hemisphere high-pressure area, the Coriolis effect deflects winds to the right, but friction again slows the winds. This results in an outward-spiraling rotation pattern called an

03.09.a4

anticyclone. In the Northern Hemisphere, circulation around an anticyclone is *clockwise* around the high-pressure zone.

Southern Hemisphere

5. For a low-pressure area in the Southern Hemisphere, the air flowing inward toward a low is deflected to the left by the Coriolis effect. This causes air within a cyclone in the Southern

03.09.a5

Hemisphere to rotate *clockwise*, opposite to what is observed in the Northern Hemisphere. In either hemisphere, friction causes wind patterns within cyclones to be intermediate between straight-inward winds and circular gradient winds.

6. For a high in the Southern Hemisphere, the air flowing outward from the high is deflected to the left by the Coriolis effect. This causes the air within an anticyclone in the Southern Hemisphere to rotate

03.09.a6

counterclockwise, opposite to the pattern in the Northern Hemisphere. Therefore, cyclones and anticyclones can rotate either clockwise or counterclockwise, depending on hemisphere.

03.09.a7

7. This satellite image (◄) shows a cyclone near Iceland (Northern Hemisphere). It is characterized by a circular rotation of strong winds around the innermost part of the storm. It displays a distinctive inward-spiraling flow of the clouds and air around an area of low air pressure. Within a cyclone, air moves rapidly down the pressure gradient toward the very low pressure at the center of the storm. In the Northern Hemisphere, the air is deflected to the right of its intended path, resulting in an overall counterclockwise rotation of the storm. The distinctive spiral pattern results from a combination of the pressure-gradient force (which started the air moving), the Coriolis effect, and friction with Earth's surface. A hurricane, which is a huge tropical cyclone, shows a similar pattern of rotation around the center of the storm—the eye of the hurricane.

B What Factors Influence Global Wind Patterns?

Insolation warms equatorial regions more than the poles, setting up pressure differences that drive large-scale atmospheric motion. Once set in motion, the moving air is acted upon by the Coriolis effect and by friction with Earth's surface. These factors combine to produce curving patterns of circulating wind that dominate wind directions, climate, and weather.

Rotation and Deflection

1. Earth is a spinning globe, with the equatorial region having a higher velocity than polar regions. As a result, air moving north or south is deflected from its intended path by the Coriolis effect.

2. Equatorial regions receive more sunlight than do areas closer to the poles and so preferentially heat up. As Earth rotates, the Sun's heat forms a band of warm air that encircles the globe and is re-energized by sunlight each day.

3. Warmed equatorial air rises and flows north and south, away from the equator. Air at the surface flows toward the equator to replace the air that rises. The Coriolis effect deflects this equatorial-flowing surface wind toward the west (to the left in the Southern Hemisphere and to the right in the Northern Hemisphere).

4. These flows of air combine into huge, tube-shaped cells of circulating winds, called *circulation cells*. Very fast flows of air along the boundaries of some circulation cells are *jet streams*.

Polar Jet
Polar Easterlies
Subtropical Jet
Westerlies
Northeast Trade Winds
Equator
Southeast Trade Winds
Westerlies
Sunlight

03.09.b1

Prevailing Winds

5. Wind direction is referenced by the direction from which it is coming. A wind coming from the west is said to be a "west wind." A wind that generally blows from the west is a *westerly*.

6. Polar regions receive the least solar heating and are very cold. Surface winds move away from the poles, carrying cold air with them. *Polar easterlies* blow away from the North Pole and have a large apparent deflection toward the west because the Coriolis effect is strong at high latitudes.

7. *Westerlies* dominate a central belt across the U.S. and Europe, so weather in these areas generally moves from west to east.

8. *Northeast trade winds* blow from the northeast and were named by sailors, who took advantage of the winds to sail from the so-called Old World to the New World.

9. *Southeast trade winds* blow from the southeast toward the equator. Near the equator, they meet the northeast trade winds in a stormy boundary called the *Intertropical Convergence Zone (ITCZ)*.

10. *Westerlies* also occur in the Southern Hemisphere and are locally very strong because this belt is mostly over the oceans and has few continents to generate additional friction to disrupt the winds.

11. *Polar easterlies* flow away from the South Pole and deflect toward the west but are mostly on the back side of the globe in this view. The Coriolis effect causes this large apparent deflection due to the very high latitudes.

03.09.b2

≥250 km/hr
210-249 km/hr
178-209 km/hr
154-177 km/hr
118-153 km/hr
63-117 km/hr
0-62 km/hr

12. The origins and paths of hurricanes in the North Atlantic and Caribbean basins in 2011 are shown here (◄). Note that few storms originate in the far southern part of this map, partly because the Coriolis effect is so small at these latitudes that it fails to impart the necessary rotation to the storms. Once formed, the paths of hurricanes are steered by the global wind patterns, which generally move Atlantic hurricanes west (guided by the trade winds) and then north and east once they enter the latitudes of the westerlies.

Before You Leave This Page Be Able To

☑ Sketch, label, and explain the combined effect of the pressure-gradient force, Coriolis effect, and friction, and how this is expressed in flow around anticyclones and cyclones in both the Northern and Southern Hemispheres.

☑ Locate and name on a map the main belts of global winds.

3.13 How Does Air Circulate Aloft over the Mid-Latitudes?

SURFACE WINDS IN THE MID-LATITUDES are generally from west to east in both hemispheres, but the pattern of air movement is less well developed higher in the troposphere. The main features are two currents of fast-moving air—jet streams—that encircle the globe near the boundaries of the mid-latitudes. What factors determine the direction and speed of airflow aloft in the mid-latitudes, such as in the jet streams?

A What General Circulation Occurs Aloft over the Westerlies?

The main direction of surface air in the mid-latitudes is as curving arcs that move poleward away from the subtropics and then bend increasingly to the east, becoming westerlies. What happens aloft? The figure below shows the setting of upper-level airflows in the mid-latitudes, between the polar cell and the Hadley cell.

1. Circulation of air is essentially restricted to the troposphere, so the geometry of the tropopause (the top of the troposphere) influences the flow of air aloft. The height of the tropopause is largely controlled by temperatures in the underlying troposphere and whether the underlying air rises or sinks. As a result, the tropopause is shallower (8 km or less) over the poles, where cold, dense air sinks, than over the equator (16 km or more), where air is warmer and rises; it increases in height from the poles across the mid-latitudes.

2. The upper part of the polar cell has west-to-east (westerly) winds aloft in both hemispheres, in response to the strong Coriolis effect at high latitudes. Surface winds are easterlies.

3. In the mid-latitudes, surface winds are from west to east (westerlies) in both hemispheres, as poleward-moving winds are turned east by the Coriolis effect.

4. In the upper segment of the Hadley cell, over the tropics, winds aloft move southwest to northeast in the Northern Hemisphere, but northwest to southeast in the Southern Hemisphere; upper-level flows in both areas are therefore westerlies.

5. If surface winds in the mid-latitudes and the winds on both sides are moving overall from west to east (in both hemispheres), it is logical that the winds aloft over mid-latitudes will also be westerly. Friction is less aloft than at the surface, so the upper-level westerlies over the mid-latitudes are stronger than westerlies at the surface.

03.13.a1

Tropopause — Mid-Latitudes — H — Hadley Cell — 30N — L — Polar Cell — 45-60N — H — 90N — Equator

B Is There a Circulation Cell in the Mid-latitudes?

1. If there were a cell of circulating air in the mid-latitudes, as there is in the tropical and polar latitudes, then the sinking from the poleward edge of the Hadley cell would cause shear that would also induce sinking at the equatorward edge of the mid-latitudes.

2. Likewise, the rising motion at the equatorward edge of the polar cell (the subpolar lows) would induce rising next to the poleward edge of the mid-latitudes. This rising motion, coupled with the sinking next to the Hadley cell, would logically set up upper-level circulation over the mid-latitudes, with air moving from the pole toward the tropics. This hypothetical mid-latitude circulation system is known as the *Ferrel cell*.

3. Descending air near the subtropical highs flows poleward along the surface and is turned toward the east by the Coriolis effect, contributing to the mid-latitude westerlies. Aloft, some rising air at the subpolar lows moves equatorward.

4. However, other factors cause the Ferrel cell to be developed very weakly or not at all. Thermal properties, which result in differences in air density, also dictate whether air will rise. Air on the subtropical side of the Ferrel cell (near 30° latitude) is relatively warm, and so it should be rising, not sinking. Air on the polar side of the cell is colder, and so it should be sinking, not rising. In other words, the thermal effects of the Ferrel cell's warmer air rising and cooler air sinking somewhat offset the dynamic effect of the shear from the adjacent Hadley and polar cells. The result is the near absence of a circulation cell in the mid-latitudes.

Subtropical Jet Stream — Polar Front Jet Stream — Tropopause — COLD AIR SINKING — WARMER AIR RISING — COLDER AIR DOES NOT RISE — SHEAR — Ferrel "Cell" — WARMER AIR DOES NOT SINK — SHEAR — COOLER AIR RISING — Hadley Cell — WARMER AIR RISING — Polar Cell — 90N — 45-60N — 30N — EQ — COLD AIR — WARM AIR — *03.13.b1*

C What Are Jet Streams and How Do They Form?

Fast-moving, relatively narrow currents of wind, called *jet streams*, flow aloft along the boundaries of the mid-latitude air currents. One jet stream is located along the edge of the polar cell and another is along the edge of the tropical Hadley cell. How do such jet streams form and what controls their locations?

1. Jet streams are strong air currents that are long and relatively narrow, resembling a ribbon or tube wrapping around the planet. This figure shows two jet streams: the *polar front jet stream* and the *subtropical jet stream*. They are shown for the Northern Hemisphere, but equivalents are also present in corresponding positions in the Southern Hemisphere. The jet streams are typically a few hundred kilometers wide and about 5 km thick, and are located high in the troposphere, near the tropopause. They generally follow paths that curve or meander and can locally split into separate strands that eventually rejoin. The four jet streams all blow from west to east as a result of Earth's rotation.

2. The subtropical jet stream circles around the globe at about 30° latitude (in both hemispheres), near the boundary between the Hadley cell and westerlies aloft in the mid-latitudes.

3. In both the north and south hemispheres, the polar front jet stream encircles the globe near the edge of the polar cell. It shifts position north and south from time to time, but typically resides between 45° and 60° latitude. As it shifts farther away from the pole, it brings cold air into the mid-latitudes, like the center of North America.

4. In addition to influencing weather, jet streams affect air travel in good and bad ways. A plane flying in the same direction as the jet stream (to the east) moves faster than normal, but a plane flying against the jet stream (west) is slowed. Jet streams also cause clear-sky turbulence on flights.

03.13.c1

D What Are Rossby Waves and How Do They Relate to Jet Streams?

1. Jet streams do not track around the planet along perfectly circular routes, but instead typically follow more irregular curved paths, like along the thick white line in the polar projection below. These meandering paths, when viewed in three dimensions, resemble curving waves, and are called *Rossby waves* after the scientist who discovered them. The lines on this map, which is viewed directly down on the North Pole, are isohypses with dark colors showing local areas of low heights (low pressure). Notice that these are concentrated in the subpolar areas.

2. Where the polar jet stream curves toward the pole, this part of the jet stream is called a *ridge*. Ridges bend poleward.

3. Where the polar jet stream curves away from the pole, we call that type of curve a *trough*. Troughs bend equatorward.

4. The size and position of ridges and troughs is not constant, but changes over days, weeks, and months. Troughs can get "deeper" or "shallower," while ridges get "stronger" or "weaker."

03.13.d1

5. Bends in the polar jet stream (Rossby waves) have a great influence on weather in nearby parts of the mid-latitudes. When a trough becomes more accentuated and shifts farther away from the pole (i.e., it deepens), it allows cold Arctic air to extend farther away from the pole. Also, as discussed in the weather chapter, Rossby waves influence surface high and low pressures near the polar jet and help strengthen, weaken, and guide mid-latitude cyclones as they migrate across the surface. Accordingly, we return to Rossby waves later.

Before You Leave This Page Be Able To

☑ Describe the flow of air aloft over the mid-latitudes.

☑ Sketch, label, and explain the two types of jet streams.

☑ Sketch and explain Rossby waves, including a trough and ridge.

3.13

3.14 **What Causes Monsoons?**

A COMMON MISCONCEPTION is that the word "monsoon" refers to a type of rainfall, but the word actually refers to winds that reverse directions depending on the season. One of these seasonal wind directions typically brings dry conditions and the other brings wet conditions. Monsoons impact a majority of the world's population.

A What Are the Features of the Asian Monsoon?

One way to characterize a monsoon is to compare maps showing wind directions for different times of the year. Such maps can then be compared to rainfall records to determine which seasonal wind directions bring dry conditions and which ones bring wet conditions. The maps below show climatological wind conditions for the years 1981 through 2010 for two months — January and July. Arrows show wind directions, and shading represents pressure at sea level, with light gray being high and dark gray being low. Examine the patterns of circulation for each month and then compare the patterns between the months.

January — This map shows typical wind conditions for Asia during January. In the center of the map, winds define a region where flow is clockwise and outward, centered on the light-colored area of high pressure. This high-pressure area (anticyclone) forms from cold, sinking air over Siberia. This circulation brings very dry air (from the cold interior of the continent) from the north over southern Asia and from the northwest across eastern Asia. We would predict from these wind patterns that little precipitation would occur in much of Asia at this time.

July — This map shows that wind conditions for the same region during July are totally different than they are for January. Circulation that marked the high pressure is gone, replaced by an area of inward and counterclockwise flow over Tibet (north of Kolkata). In the Northern Hemisphere, this pattern of circulation is diagnostic of a low-pressure area, which in this case is caused by warm, rising air that accompanies warming of the Asian landmass. This circulation brings very humid air from the southwest over southern and southeastern Asia. How do you think this circulation affects rainfall?

Seasonal Variation in Precipitation

Observe these graphs showing average monthly precipitation amounts for two very different parts of Asia: Kolkata, one of the largest metropolitan areas of India, and Seoul, the capital of South Korea. For both cities, notice prominent precipitation peaks that occur during the summer — the wet season. The increase in precipitation during the wet season results from the flow of moist air from oceans onto land, toward the *Tibetan Low*. The dry season, during the winter months, reflects the flow of dry air from the land, flowing outward from the *Siberian High*.

Effect of the Monsoon on Vegetation

These satellite images show increased vegetation due to monsoon-related rains along the western coast of India. The left image is during the dry season, when wind patterns bring in dry air. The right image is from the end of the monsoon. Note the increase in plant cover (green areas) during the monsoon-caused rainy season.

03.14.a5 India

03.14.a6 India

B What Other Regions Experience Monsoon Circulations?

West Africa

January—In January, near-surface winds in West Africa largely flow from the northeast bringing in dry air from inland areas, including the Sahara Desert, and carrying it southwest to coastal areas and farther offshore. Such *offshore flows* generally result in dry weather.

03.14.b1   03.14.b2

Winter   Summer

July—A shift in wind direction in July brings moist ocean air from several directions onto the very hot land where air has risen This change in wind direction causes enormous differences in precipitation, as shown by the graph below for Dakar, Senegal. Along with the increase in precipitation comes an increase in the amount of vegetation. In Dakar and much of the region, precipitation is nearly nonexistent in January and adjacent months.

03.14.b3

Northern Australia

January—In January (the southern summer), winds over northern Australia bring moist air from the ocean onto the heated land surface.

03.14.b4   03.14.b5

July—The wind shifts by July (winter) as the land surface cools, creating higher pressure over the land. This causes a large drop in precipitation, as shown by the graph below for Katherine, Australia. The monsoon flow in January results in plentiful rain.

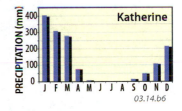

03.14.b6

Southwestern U.S.

January—Southwestern North America, most of which is desert, has a less dramatic, but still important monsoon effect. In the winter months, winds blow from various directions, and winter precipitation in this region is from brief incursions of cold, wet air (i.e., cold fronts) from the northwest.

03.14.b7   03.14.b8

July—During the late summer months, heating of the land surface and the resulting low pressure causes a shift in winds. Winds from the south bring moist air northward from the Gulf of Mexico and Gulf of California, and summer thunderstorms from when this air interacts with the heated land. These summer thunderstorms cause precipitation to peak in August, as shown by the graph below for Tempe, Arizona. First, note the different scale needed to show the relatively small amounts of precipitation. Also note that nearly as much precipitation falls in the winter from the cold fronts.

03.14.b9

The Effect of Monsoons on Cultures

Monsoons greatly influence the lives of people living in regions with seasonal shifts in wind. The main effects of a monsoon are seasonal variations in precipitation, which in turn affect water supplies, amount of vegetation, and overall livability for some normally dry landscapes. Many cultures plan their activities around these seasonal changes, conserving water during the dry season and taking advantage of the plentiful water during the wet season.

The monsoon pervades the psyches of people in southern and southeastern Asia, especially the region from India to Vietnam,

in ways not fathomable to most North Americans or Europeans. The influence on agriculture, including the cultivation of rice, and on flooding and other natural hazards is obvious, but the monsoon also appears in literature, art, music, architecture, and nearly every other aspect of culture. Ceremonies commonly mark the anticipated start of the monsoon. In years when the monsoon rains arrive later than usual, people become very concerned that harvests will suffer. The date of the onset of the monsoon rains varies by location, but generally proceeds from south to north with the onset in April and May.

Before You Leave This Page Be Able To

☑ Explain what causes a monsoon, using examples from Asia, West Africa, northern Australia, and southwestern North America.

☑ Describe some of the effects of shifting monsoonal winds.

3.14

How Have Global Pressures and Winds Affected History in the North Atlantic?

INTERCONTINENTAL TRAVEL AND TRADE have relied upon moving currents in the air and oceans. Before the 20th century, when transoceanic travel and shipping relied on wind power, global winds, such as the trade winds and westerlies, dictated which directions of travel were possible at different latitudes. The directions of global winds therefore greatly influenced the exploration and colonization of the Americas, and traces of that influence can be detected in past and present cultures.

This small globe (▶) shows the main belt of global winds, controlled by the pressure-gradient force and Coriolis effect. Can you locate and name the main wind belts? In the Northern Hemisphere, trade winds (shown in red) blow from northeast to southwest, allowing wind-powered sailing ships to travel from western Africa to Central America and the northern part of South America.

03.15.a2

The polar easterlies (shown in blue) likewise allowed westward travel, but under cold and stormy conditions.

In contrast, westerlies (shown in purple) allowed a return trip from North America to Europe, but only in the mid-latitudes. Ships commonly had to travel north or south along coasts to catch a prevailing wind going in the direction of intended travel across the Atlantic Ocean.

03.15.a3 Cotton plantation along the Mississippi River

◀ Plantations in the southeastern U.S. depended upon African slave labor to cultivate and harvest cotton, tobacco, and tropical agricultural products. In contrast, the economy of the northeastern U.S. was based more upon manufacturing and the refining of products grown on the southern plantations, with the final refined product being exported to western Europe. The traditional cultural divide between the northern states and southern states is the Mason-Dixon line (near 40° N), which corresponds roughly to the global climatological divide between the zones of excess climatic energy to the south and an energy deficit to the north. The excess and deficit energy settings controlled which crops could be grown in the two regions and allowed different kinds of activities by impacting the climate. This climate-energy boundary contributed to the political and social attitudes that existed then and today.

03.15.a4 Sugar plantation

Slaves and various provisions were brought from Africa on ships utilizing the trade winds on either side of the equator, winds that provided a direct conduit of trade from western Africa to equatorial parts of the Americas. Most slaves were put to work on labor-intensive tasks, such as agriculture, in and around the tropical regions of the Caribbean (◀). Their descendants still exert a great influence over the distinct culture of the islands and coastal areas in the region.

03.15.a5 Sir Francis Drake

After Portugal and Spain agreed to divide the "new" world into two separate spheres of influence, the Spanish encountered difficulty in transporting products and plunder from the Pacific empire across Portuguese territories, especially Brazil. Spanish goods were brought to Caribbean ports or were transported across the isthmus of Central America, before being stockpiled temporarily in Spanish ports in the tropics, including Havana, Cuba. Treasure-bearing sailing ships departing tropical ports had to sail north up the coast of Anglo-America (the yellow paths on the large globe) in order to reach the mid-latitude westerlies, which then carried the ships east to Spain. This coast provided an ideal safe haven for English privateers, like Sir Francis Drake (◀), who captured many ships and their content, enriching the privateers and England in the process, and laying the financial foundation for the British Empire.

03.15.a6 Remains of Viking settlement, Iceland

03.15.a7 Titanic

◄ The Viking explorer Leif Erickson probably visited North America about 500 years before Christopher Columbus. Moving westward from Scandinavia, the Vikings progressively colonized Iceland and Greenland before journeying across the Davis Straits to Baffin Island and the coast of Labrador (the blue path on the globe). The remains of a Viking settlement have been found on the most northerly tip of Newfoundland at L'Anse Aux Meadows. Viking ships would have employed the polar easterlies and associated ocean currents to aid their westward migration. Ironically, a shift to a much colder climate in the late 15th century caused the extinction of the settlements in Greenland at exactly the same time as Columbus was embarking on his explorations.

▲ When the steam-powered Titanic left Southampton, England (51° N), for New York City (41° N) in April 1912, global winds and currents were no longer such a strong determinant of oceanic travel routes. The planned northerly route (red path on the globe), however, crosses a treacherous part of the North Atlantic Ocean—the region with stormy weather near the subpolar lows. The Titanic sank at 42° N off the east coast of Canada, striking an iceberg brought south by global winds and ocean currents.

The prosperity of many trading ports in western Europe and the Americas (▼) during the early 19th century was based on various types of interconnected trade that followed a triangular path (green path on the globe). Manufactured goods from the emerging European industrial economies were loaded for trade with West Africa. There, African slaves were taken on board and carried west by the trade winds to the Americas. Once in the Americas, the slaves were traded for the products of the tropical plantation system. Using the westerlies, the raw goods were returned to western Europe for consumption or industry. The triangular path was designed to maximize favorable winds and to avoid the subtropical high pressures and calm winds of the horse latitudes.

03.15.a8 New Orleans, LA

03.15.a1

Departing from Cadiz, Spain (36° N), Christopher Columbus wanted to sail west, but first had to sail south down the northwestern coast of Africa to avoid the westerlies and instead pick up the northeast trade winds (the purple path on the globe). Portuguese navigators had known this strategy for many years, and Columbus probably learned of it when he attended the school of navigation in Sagres, Portugal. Global wind patterns therefore controlled Columbus's path and explain why in all likelihood, he first stepped ashore in the Bahamas (24° N), much farther south than Cadiz, where he started.

Before You Leave This Page Be Able To

☑ Sketch and describe the main features of atmospheric circulation in the northern Atlantic Ocean and how they influenced early travel.

3.15

What Occurs During Seasonal Circulation Shifts?

GLOBAL ATMOSPHERIC CIRCULATION responds directly to insolation. As the Sun's direct rays migrate seasonally, belts of winds, such as the westerlies, migrate too. In this investigation, you will examine the general circulation of the atmosphere, as expressed by data on air pressure, wind velocity, and cloud cover for two months with very different seasons—January and July.

Goals of This Exercise:

- Identify major patterns in air pressure, wind velocity, and cloud cover for each season.
- From these data, identify the major features of the global atmospheric circulation in each season.
- Assess and explain the degree of seasonal movement of these circulation features.

When examining large-scale patterns of the Earth, such as global circulation patterns, a useful strategy is to focus on one part of the system at a time. Another often-recommended strategy is to begin with relatively simple parts of a system before moving to more complex ones. For this investigation, you will infer global patterns of air circulation by focusing on the Atlantic Ocean and adjacent lands (▶).

This globe is centered on the central Atlantic, and its top is slightly tilted toward you to better show the Northern Hemisphere. As a result of this tilt, Antarctica is barely visible at the bottom of the globe. The colors on land, derived from satellite data, depict rocks and sand in tan and brown. Vegetation is in various shades of green, with the darkest green indicating the thickest vegetation (usually forests). Shallow waters in the Caribbean region (on the left side of the globe) are light blue.

Observe the entire scene, noting which areas on land have the most vegetation and which ones have the least. Compare these vegetation patterns with large-scale patterns of atmospheric circulation and air pressure, like subtropical highs and the Intertropical Convergence Zone ITCZ.

Consider what directions of prevailing winds would occur in different belts of latitude. For example, where in this globe are the two belts of trade winds (one north and one south of the equator)? How about the mid-latitude belts of westerlies in each hemisphere? Consider how these winds might blow moisture-rich air from the ocean onto land. After you have thought about these aspects, read the procedures below and examine the globes and text on the next page, which highlight average air pressure, wind velocity, and cloud cover for two months—January and July.

03.16.a1

Procedures

Complete the following steps on a worksheet provided by your instructor or as an online activity.

1. Study the two globes showing air pressure (on the next page), and note areas with high and low pressure. Locate the Icelandic Low and Bermuda-Azores High, and determine for which season each is strongest or if there is not much difference between the seasons. Then, locate a belt of low pressure near the equator and the adjacent belts of subtropical highs on either side. Mark and label the approximate locations of these features on the globe on the worksheet.

2. Next, examine the two globes that show wind velocity. In the appropriate place on the worksheet, draw a few arrows to represent the main wind patterns for different regions in each month. Label the two belts of westerlies and the two belts of trade winds. If the horse latitudes are visible for any hemisphere and season, label them as well. Mark any somewhat circular patterns of regional winds and indicate what pressure feature is associated with each.

3. Examine the two globes that show the average cloud cover for each month. From these patterns, label areas that you interpret to have high rainfall in the tropics due to proximity to the ITCZ or low rainfall due to position in a subtropical high. Examine how the cloud patterns correspond to the amount of vegetation, pressure, and winds.

4. Sketch and explain how the different features of circulation and air pressure change between the two months. Answer all the questions on the worksheet or online. OPTIONAL EXERCISE: Your instructor may have you write a short report (accompanied by a sketch) summarizing your conclusions and predicting how seasonal shifts would affect people.

Air Pressure

January

03.16.a2

July

03.16.a3

These two globes show average air pressure over the Atlantic and adjacent land areas during the months of January and July. Lighter gray indicates relatively high pressures, whereas darker gray indicates low pressures. Intermediate grays indicate intermediate pressures. The lines encircling the globe are the equator and 30° and 60° (N and S) latitudes.

Observe the main patterns on these two globes, noting the positions of high pressure and low pressure and how the positions, shapes, and strengths change between the two seasons. Then, complete the steps described in the procedures section.

Wind Velocity

January

03.16.a4

July

03.16.a5

These globes show average wind velocities for January and July. The arrows show the directions, while the shading represents the speed, with darker being faster. In this exercise, the directions are more important than the speeds, but both tell part of the story.

Observe the large-scale patterns, identifying those patterns that are related to global circulation (i.e., westerlies) versus those that are related to more regional features, such as the Bermuda-Azores High and the Icelandic Low. Note also the position of where winds converge along the equator and how this position changes between the two months.

Cloud Cover

January

03.16.a6

July

03.16.a7

The clouds that form, move above Earth's surface, and disappear can be detected and tracked with satellites, shown here for January and July 2012. On these globes, light colors that obscure the land and ocean indicate more abundant clouds (and often precipitation), whereas the land shows through in areas that average fewer clouds.

Observe the large-scale patterns, noting which areas are cloudiest and which ones generally have clear skies. Relate these patterns of clouds to the following: amount of vegetation on the land, air pressure for that month, and average wind directions. Answer all questions on the worksheet or online.

Atmospheric Moisture

MOISTURE IN THE ATMOSPHERE, in the form of water vapor, liquid water, and ice, controls most aspects of our weather and climate. Moisture moves back and forth from Earth's surface to the atmosphere and, once in the atmosphere, is transferred vertically and laterally by moving air. Atmospheric moisture is expressed as clouds, precipitation, storms, weather fronts, and other phenomena. In this chapter, we explore how and where moisture occurs in the atmosphere, what happens when moisture moves with air currents, and how such motions are expressed as clouds, precipitation, and other aspects of weather.

Examine this figure and observe the various features. Try to identify which features involve water in one form or another. Next, consider how moisture can move from one place to another, and from one state of matter (gas, liquid, solid) to another. After you have done this for each main feature shown, read the text surrounding the figure.

04.00.a2 Grand Cayman Islands

04.00.a1

Our world has various types of clouds, some of which are shown here in the large figure and the photograph above. Some clouds are thin and wispy, whereas others are tall and puffy. Some are high in the troposphere, while others are close to Earth's surface. Some clouds are associated with precipitation (rain, snow, hail, or sleet), but most are not.

What is in a cloud, how do we classify and name different types of clouds, and what do different types of clouds tell us about what is going on in the atmosphere?

04.00.a3 Plush, OR

Most clouds and precipitation are caused by cooling of moist air. Such cooling generally occurs when air rises into cooler parts of the atmosphere. Cooling of moist air can also occur when warmer air interacts with cooler land or water. As moist air cools, water vapor can change into a liquid (water drops) through the process of *condensation*. If cold enough, water vapor can instead form ice crystals through the process of *deposition*. Also, water drops and ice crystals can form from one another. In any case, the water drops and ice crystals are what forms clouds and, under the right conditions, cause various types of precipitation (◄).

For each place where a cloud is shown in the large figure, what are some possible reasons why the air might be cooling at that location?

TOPICS IN THIS CHAPTER

The amount of moisture in the atmosphere varies laterally from place to place and vertically between different heights in the atmosphere. We commonly use the term *humidity* to express how much moisture is in the atmosphere.

In this figure, predict which areas would have more humidity compared to other areas. How would this extra humidity be observed in the atmosphere, how would the air feel to you, and how would increased humidity affect your weather?

Insolation

Evaporation

Evaporation

Ocean Current

04.00.a4 Costa Rica

Water moves from the surface to the atmosphere through the process of *evaporation*, where liquid water becomes water vapor. Most evaporation occurs over the oceans, but significant amounts also occur over lakes, streams, and other surface water. Water also gets into the atmosphere from moisture released by moist soil and by plants through their leaves to the atmosphere, a process called *transpiration*. The combination of plant transpiration and evaporation from Earth's surface is *evapotranspiration*. The released water vapor is invisible but can condense into visible mists (◄).

On the large figure, examine the surface, both the water and land, and identify places where moisture could move from the surface to the atmosphere.

Expressions of Moisture

The large figure on these two pages shows a wide variety of ways that moisture is expressed on the surface and in the atmosphere. Water is easily identifiable in the large body of water (ocean) on the right, in the two lakes on land, and in the streams. This water is shown in liquid form, but under cold-enough circumstances there could also be ice, such as if the lakes were at least partially frozen or if the ocean had some sea ice (frozen seawater) or icebergs (large floating chunks of ice). We can easily recognize water in its solid form (ice) in the snow on the high mountains. There would also be a zone of especially moist air (more water vapor) over and adjacent to the ocean, lakes, streams, and wet parts of the land, but we cannot see it because water vapor is not visible.

Even less obvious is the water contained in the green plants and the underlying soil. This moisture moves from these sites to and from the atmosphere, such as rain that falls from the clouds, soaks into the soil, and is taken up by plants, which can release this moisture back into the atmosphere via transpiration.

Moisture in the atmosphere is expressed in the various types of clouds. Most clouds are visible to us because they contain tiny drops of liquid water — even if there is no rain associated with the cloud. The wispy, higher-level clouds consist of ice crystals, which are easily blown and streaked out by the strong upper-level winds. The higher parts of some of the taller lower-level clouds would also contain ice, which can remain aloft or eventually accumulate into such large particles that it falls to the surface as snowflakes or hail. In all, many aspects of our planet owe their existence to moisture in the atmosphere and to the exchange of moisture between the atmosphere and the surface.

4.0

4.4 What Is the Dew Point?

ANOTHER USEFUL MEASURE OF HUMIDITY is the *dew point*, the temperature to which a volume of air must be cooled to become saturated with water vapor. If the air temperature is at the dew point, the air is so saturated with water vapor that vapor begins to condense as drops of liquid water, such as drops in clouds or rain, or as water drops (dew) on solid surfaces. When you wake in the morning and everything outside is covered with moisture (dew), it signifies very humid air and that night-time temperatures cooled to the dew point.

A How Is the Dew Point Expressed?

04.04.a1 Colorado

◄ *Dew* is expressed as drops of liquid water that condense out of the atmosphere and onto plants, rocks, walls, or any solid surface. It typically forms at night, in response to the air cooling. If temperatures are at or below freezing, earlier formed drops of dew can freeze or ice can form directly from the air (deposition). In any case, drops of dew or crystals of ice disappear rapidly once the Sun rises and air temperatures begin to climb.

► To explore the formation of dew, examine the sealed glass container on the left. It contains liquid water and water vapor (represented as blue dots). The container is warm enough that all the available water vapor can exist in the air without condensing—that is, the air is unsaturated.

04.04.a2

If we cool that same container with ice cubes, the temperature of the water and air will decrease. Recall that cold air has lower water-vapor capacity than warm air, so as the air cools it moves closer to saturation. At some temperature, the cooled air reaches saturation and can no longer allow so much vapor to exist in it. As a result, the water begins to condense as drops on the inside of the container, forming dew. As we know from our experience with cans or glasses of an icy beverage, drops of dew can also form on the outside of the container, especially if the surrounding air has high humidity.

B What Is the Dew-Point Temperature?

1. To better visualize why cooling can cause the formation of dew, we return to a familiar graph, one that plots vapor pressure (*e*) versus temperature (▼). The curved blue line marks conditions where air is saturated in water vapor; it is labeled as e_s, where the *s* indicates saturation.

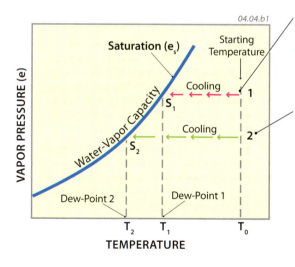

04.04.b1

2. We begin with an air mass at a certain temperature (T_0) and vapor pressure, marked as position 1, well within the unsaturated field of the graph. As this air cools, its position on the graph moves horizontally to the left (temperature is changing but vapor pressure is not), as indicated by the reddish arrows. It cools until it reaches saturation at position S_1, at which point the air has a relative humidity of 100%. The temperature at which saturation occurs is the *dew-point temperature* and can be read from the graph by drawing a vertical line down from S_1 to the temperature axis (T_1).

3. A second air mass, at position 2, starts at the same temperature (T_0) but has less water vapor in it (lower vapor pressure) than does air mass 1. If the air cools, as indicated by the green arrows pointing to the left, it also approaches the saturation curve. In order to reach saturation (at point S_2), however, it has to cool more than air mass 1. Air mass 2 has less water vapor than air mass 1, and as a result has a lower dew-point temperature (T_2).

4. To generalize, humid air at a given temperature has a higher dew-point temperature than does less humid air at the same temperature. If air is very humid, for example with a relative humidity of 80% to 90%, it may need to cool only a few degrees to reach its dew point. In contrast, air that is very dry (like the 5% to 10% relative humidities common in the Mojave Desert in California) has to cool significantly to reach its dew point, and on most nights does not.

5. The difference between current air temperature and the dew-point temperature is known as the *dew-point depression*. A large dew-point depression indicates that a large temperature drop is required to reach saturation. Thus, large dew-point depressions suggest dry air and small dew-point depressions suggest humid air so that only a minor drop in temperature will cause saturation. Dew-point depression is used in some agricultural and forestry applications and tells us how much temperatures might drop overnight, since air temperatures at night generally do not cool much below the afternoon's dew point.

6. Using dew-point temperature as an index of moisture overcomes most of the difficulties of using relative humidity, because dew-point temperature does not change as much through the course of a "normal" day as relative humidity does. In the Arizona desert, weather forecasters use dew-point readings to determine the official start of the summer monsoon. If the dew-point temperature is 55°F or higher for three straight days, summer rains and thunderstorms associated with the monsoon are likely.

Which Parts of the U.S. Have the Highest and Lowest Dew Points?

Dew-point temperature is an important measure of the humidity of air, as well as a predictor of dew and precipitation. The maps below show average monthly dew-point temperatures and air temperatures across the conterminous U.S. for January and July. Blue colors represent low (cold) dew-point temperatures, whereas orange and red indicate high dew-point temperatures. While observing the maps, identify areas that are unusually high or low and consider possible explanations for the patterns you observe. Find the values for the area where you live or would like to visit and think about how this relates to the climate there and how humid it feels at different times of the year.

January

1. Extremely low dew-point temperatures, as characterize the northern states in January, generally mean that the air is so dry that clouds do not form readily. Without cloud cover, night-time radiational cooling is intense, making January temperatures even more frigid.

2. Topography, such as mountain ranges in western North America, block moisture from penetrating to some regions, leading to low dew-point temperatures in the desert Southwest.

3. This map (▶) shows normal January air temperatures, averaged over the entire month. The warmest areas are in the Southwest and South Florida.

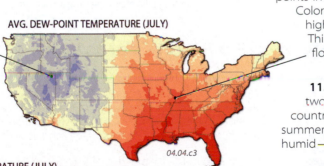

4. January dew-point temperatures are extremely low in northern areas. Little insolation is available to drive the evaporation process under such conditions, so there is limited moisture in the air. When dew-point temperatures are below 0°C, they are called "frost-point" temperatures, and ice forms instead of dew.

5. Dew-point temperatures in eastern North America generally are not far below the air temperature, so air there does not have to cool very much to become saturated. Thus, clouds form even if there are no weather fronts. This is why many eastern states have overcast days in the winter.

6. January temperatures in the Southeast, especially southern Florida, are more moderate, so there is enough energy available to evaporate some water. This results in relatively high dew-point temperatures compared to the rest of the country.

July

7. Summer dew-point temperatures are much higher than in the winter. Increased insolation allows more energy to drive up temperatures and cause more evaporation, which increases humidity.

8. Areas in rain shadows of the moisture sources, such as the Intermontane West, have dew points that remain far below their air temperatures shown on the lower map. A slightly larger amount of moisture flows into southern Arizona from the Gulf of California in the summer, as represented by the small orange spots on the southwestern edge of this map.

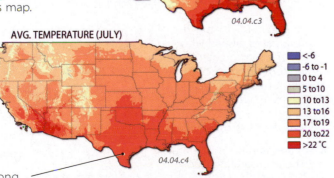

9. Summer dew-point temperatures are high in the Southeast, as depicted by the orange colors that fringe the Gulf Coast in the dew-point map for July. This high humidity, combined with high summer air temperatures, creates an oppressive summer climate. Strong clockwise flow around the high-pressure system near Bermuda (the Bermuda-Azores High) pumps moisture from the Gulf of Mexico, Caribbean Sea, and Atlantic Ocean deep into the heart of North America. There are no topographic barriers that would interfere with the movement of this moist air inland, so moisture from the ocean flows onto land, raising dew-point temperatures.

10. Notice the steep gradient in dew-point temperature across the Great Plains, from moderately low dew points in the plains of eastern parts of New Mexico, Colorado, Wyoming, and Montana, to much higher dew points toward the Mississippi River. This gradient mostly shows the northward flow of moisture from the Gulf of Mexico.

11. We can use the combination of these two maps to identify which parts of the country are hot and humid during the summer and those that are cool and less humid—useful in planning a summer vacation!

Before You Leave This Page Be Able To

☑ Describe what the dew point represents, using a graph to show how cooling can cause air to reach its dew point (saturation).

☑ Describe the general patterns of dew point in the U.S., identifying some of the factors that contribute to high and low dew-point variations from place to place and month to month.

4.4

4.5 What Happens When Air Rises or Sinks?

THE ATMOSPHERE IS A DYNAMIC ENVIRONMENT, with upward, downward, and sideways motion. What happens to air that rises and encounters different temperatures and pressures? What happens when air sinks? The vertical motions of air cause clouds, precipitation, and many other weather phenomena.

A At What Rate Does an Air Parcel Cool As It Rises?

To explore the vertical motions of air, it is useful to track what happens to a discrete mass of air, called a *parcel* (used in the same way we refer to a lot of subdivided land as a "land parcel"). As we might expect, air changes temperature as it rises or sinks, largely in response to changes in air pressure and volume that accompany vertical motions through the air. Let's examine what happens as we follow a rising parcel of air, beginning with text at the bottom left (number 1).

04.05.a1

3. When there is no energy exchanged between the parcel and its surroundings, as in the case shown here, the process is said to be *adiabatic*. When such an exchange does occur, the process is *diabatic*. Increasing insolation of the surface as noon approaches is an example of diabatic heating.

2. As the air rises higher, the pressure continues to decrease. If the parcel does not acquire or lose additional energy with its surroundings, then the decrease in pressure is accompanied by (1) an increase in volume (as shown by the increasing size of the air parcel), and (2) a decrease in temperature as molecules become more widely spaced.

1. If the parcel of air starts near the ground, its volume is confined by the weight of the overlying atmosphere, which exerts significant pressure. If the air begins rising, there is less air on top, which results in a decrease in air pressure.

4. If the rising motion can be assumed to be adiabatic, we can calculate how much a rising air parcel cools for a given rise in elevation, a rate called the *lapse rate*. Lapse rate is in units of degrees per vertical distance, usually in C°/km.

5. This graph (▶) plots the lapse rate for air that is rising adiabatically and remains unsaturated. The lapse rate is calculated simply as the change in temperature for the change in height, and reported as a ratio (C°/km). In the case shown here, the air cools 10 C° in one kilometer, or 10 C°/km.

04.05.a2

$$\text{UNSATURATED LAPSE RATE} = \frac{5\,C°}{0.5\,km} = 10\,C°/km$$

6. If any parcel of air *rises* adiabatically and does not become saturated during its ascent, it always cools at 10 C°/km. This constant lapse rate, called the *unsaturated adiabatic lapse rate*, applies everywhere adiabatically rising air remains unsaturated. By convention, positive lapse rates are for temperatures that cool with height, as shown by the orange line on the graph above.

7. If a parcel *descends* adiabatically, it warms at the same rate that it cooled on ascent—10 C°/km. An adiabatically descending parcel of air will remain unsaturated, because the slightest bit of warming will increase the water-vapor capacity and therefore decrease the relative humidity. If the relative humidity is below 100%, the air is unsaturated.

B How Much Does a Saturated Air Parcel Cool as It Rises?

1. A different situation occurs if the atmosphere is saturated. As a saturated parcel of air rises, the cooling decreases the water-vapor capacity of the air and, since the air is already saturated, the moisture must come out of the vapor via condensation, freezing, or deposition, as illustrated in the diagram below.

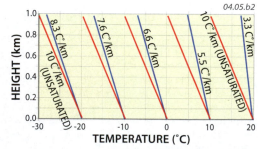

04.05.b1

2. As water vapor molecules condense, they release latent heat that warms the local environment. This warming counteracts some of the cooling caused by the expansion of the parcel during adiabatic ascent. As a result, a rising parcel of saturated air cannot cool as much with increasing height as does unsaturated air. The lapse rate for saturated air will be less than the unsaturated adiabatic lapse rate.

3. The rate at which saturated air cools with adiabatic ascent is called the *saturated adiabatic lapse rate*. This rate depends on temperature because at higher temperatures more water vapor is available for condensation, freezing, or deposition. The figure below plots unsaturated adiabatic lapse rate in red and the saturated adiabatic lapse rate in blue, for different starting temperatures on the surface (on the bottom of the graph).

4. The slope of the lapse rate for unsaturated air (red lines) remains the same, regardless of starting temperature. The lapse rates for saturated air (blue lines), however, vary for different temperatures, as reflected by different slopes of the blue lines.

5. For any starting temperature, the blue line (saturated lapse rates) consistently plots to the right of the red line (unsaturated rate). This means that rising unsaturated air cools at a faster rate than saturated air (which is warmed by latent heat).

04.05.b2

6. Note that the slopes for the unsaturated and saturated air are only slightly different for low temperatures (not much water vapor), but become more different at higher temperatures (more water vapor). This means that with increasing starting temperatures (farther right on the graph), warmer saturated air parcels cool less with height.

C   What Is the Lapse Rate of Air Adjacent to a Rising Air Parcel?

Air adjacent to a rising parcel of air can have a totally different lapse rate than the rising air, and these rates can vary significantly from day to day and place to place, depending on the local weather, climatic setting, and other factors.

1. The temperature change with height for air *surrounding* the adiabatically moving air parcel is known as the *environmental lapse rate*. The environmental lapse rate can be quite variable. On this graph, each green line represents the environmental lapse rate under certain atmospheric conditions. The different lapse rates are expressed as different slopes (inclination of the lines). When dealing with lapse rates, physical geographers call a lapse rate "steep" if the parcel cools a lot with height (as with the left line), even though that line does not look steep on this kind of graph. An environmental lapse rate is for air that is not rising or sinking, but instead represents the ambient conditions.

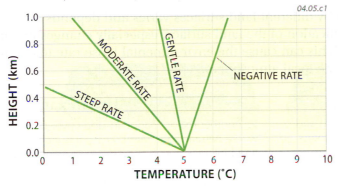

04.05.c1

5. The rightmost line has a negative lapse rate. It depicts a situation where temperature *increases* with increasing height. In such situations the environmental lapse rate is said to be *negative* because the temperature cools negatively (i.e., warms) with increasing height. A situation with a negative environmental lapse rate is called a *temperature inversion*.

4. The third line (gentle rate) represents an even more gentle lapse rate. This air starts at 5°C at the surface, but the temperature only decreases to 4°C one kilometer higher in the air. This equates to an environmental lapse rate of 1 C°/km. The gradual nature of this change is why this type of rate is called a gentle lapse rate (even though the line looks steep on this graph).

3. The second line from the left has a moderate (less steep) lapse rate. For this lapse rate, temperature decreases more slowly upward in the atmosphere. The air is 5°C at the surface, but temperatures cool to 1°C by one kilometer higher. The change in temperature is therefore 4 C° between the surface and one kilometer in altitude, yielding an environmental lapse rate of 4 C°/km.

2. The left line has the steepest environmental lapse rate. The air temperature starts at 5°C at the surface but decreases 5 degrees (Celsius) by a height of 0.5 km. This yields an environmental lapse rate of 5 C°/0.5 km, or 10 C°/km. For this graph, the left line has the steepest lapse rate—its temperature cools the most rapidly with height.

D   What Do the Different Types of Lapse Rates Indicate About the Stability of Air?

The three types of lapse rates—unsaturated adiabatic lapse rate, saturated adiabatic lapse rate, and environmental lapse rate—each indicate something about conditions in the atmosphere. By comparing the three types of lapse rates, we can predict whether air will rise, sink, or not move vertically. That is, we can tell how stable the air mass is.

1. This graph compares different types of lapse rates for an adiabatically moving air parcel that starts from the surface at 10°C. The unsaturated adiabatic lapse rate is shown by the red line, and the saturated adiabatic lapse rate is depicted by the blue line. In green are three different environmental lapse rates with varying steepness.

04.05.d1

4. The rightmost green line represents a gentle environmental lapse rate, where the air cools relatively slowly with increasing height. As a result, an adiabatically rising air parcel (shown by the red or blue lines) would be cooler than the surrounding environment. In this case, the air parcel would cease rising because it is cooler and has a higher density than the surrounding air. At any given height, it will more likely sink, a condition known as a *stable atmosphere*. This will happen any time the environmental lapse rate is more gentle than the saturated rate (the green-shaded area). If the environmental lapse rate is negative (sloping up to the right), it forms a temperature inversion—a very stable atmospheric setting.

2. The green line in the lower left part of the graph has a steep environmental lapse rate, steeper than either the unsaturated or saturated lapse rates. If an air parcel starts to rise, it changes temperature according to one of the adiabatic lapse rates (red line or blue line, depending on whether it is unsaturated or saturated). Once either type of air has risen even a small amount, it will be warmer and less dense than the surrounding environment at a given height, so it will tend to continue rising. This will occur for either unsaturated or saturated adiabatically rising air if the environmental lapse rate falls within the orange region on this graph. In such situations, any rising air parcel will remain warmer than its surroundings and so will tend to continue rising, producing conditions that are called *unstable* (where air is rising).

3. A different situation occurs if the environmental lapse rate is moderate (the middle green line), plotting between the unsaturated and saturated lapse rates. In this case, if the adiabatically rising air is saturated (blue line), it will remain warmer than its surroundings (green line) and will tend to continue rising. If it is unsaturated, however, it will become cooler than its surroundings (green line is to the right of the red line in this case) and tend to stop rising. For this reason, the situation represented by the yellow area is termed *conditionally unstable*. In this situation, the addition of moisture to the air increases the likelihood for rising motions and other unstable conditions.

Before You Leave This Page Be Able To

✓ Sketch, describe, and contrast the unsaturated and saturated adiabatic lapse rates, explaining why the saturated adiabatic lapse rate is less than the unsaturated adiabatic lapse rate.

✓ Describe the relationship between the three lapse rates and atmospheric stability.

4.5

4.7 What Mechanisms Can Force Air to Rise?

AIR RISES FOR VARIOUS REASONS, some caused by differences in density between the air and its surroundings, and others a result of externally imposed factors, like a mountain. If an air parcel rises because atmospheric conditions are unstable, *free convection* results. If air is forced to rise due to external factors, it is called *forced convection*. Forced convection can occur in a number of settings, even if the atmospheric conditions are stable.

A How Do Convergence and Divergence of Air Cause Rising?

As air moves laterally across the surface of the Earth and at higher levels, it can encounter obstacles that impede continued motion. Such obstacles can be mountains or other features on the land surface, or they can be other air masses moving in an opposing direction.

Convergence of Low-Level Air Masses

04.07.a1

04.07.a2

Low-level convergence (▲) occurs when winds from two systems are moving on a collision course. The easiest way for the colliding air to move is upward. In such cases, the air parcel must rise whether it is stable or unstable. If the air reaches its dew point as it ascends and cools, water vapor condenses into tiny drops that form clouds and perhaps precipitation.

The best example of a location where low-level convergence is important at continental scales is the *Intertropical Convergence Zone* (ITCZ), which encircles the Earth, more or less centered on the equator. In the satellite image above (▲), the ITCZ appears as the belt of moist air and clouds shown in red for its position during the northern summer and blue for its position during the southern summer. Along the ITCZ, the northeast trade winds, coming from north of the equator, converge with the southeast trade winds coming from south of the equator. The two sets of trade winds converge at the ITCZ, causing cloud cover as moist air is uplifted to a height at which it cools to its dew-point temperature.

Interaction of Moving Air with the Land Surface

04.07.a3

04.07.a4

04.07.a5

Low-level convergence also occurs by a wind that is slowed down, such as by friction with the surface. The slowed air causes faster moving air behind it to begin to pile up, forcing some air to rise. A common scenario for this involves a sea breeze, where winds blowing across a large water body slow down as they encounter the more irregular land surface.

Mountainous areas represent obstructions to low-level winds. As air encounters topography, it slows down, piles up, and rises, called an *orographic effect*. As the air rises, it cools by adiabatic expansion, forming a cloud if it cools to the dew-point temperature. This orographic effect is why thunderstorms are usually more common over mountain peaks than over the adjacent valleys.

During daytime, the Sun heats the surface more efficiently than the air above it. As the surface of a mountain warms, the adjacent air warms, which then flows upslope. When winds flowing up opposite sides of the mountain converge at the summit, they rise more. This is yet another reason why cloud formation seems to occur preferentially over mountains.

Convergence and Divergence Caused by Rossby Waves

1. Convergence can also occur from the way that upper-level air circulates around bends in the polar front jet stream, as illustrated in the figure to the right. These bends in the jet stream are called *Rossby waves*.

2. In upper levels of the troposphere, air that turns around ridges of high pressure will tend to move slightly faster than air around low-pressure troughs. This is because air turning around ridges moves to the right in the Northern Hemisphere (as shown in this example), in the same rotation direction as the Coriolis effect. The two forces work together on this type of bend.

3. Around troughs, air turns to the left (in the Northern Hemisphere), in opposition to the Coriolis effect. Along these bends, the Coriolis effect works against the flow of air, and the air moves slightly more slowly.

04.07.a6

4. The slowing down of air as it moves from the upper-level ridge of high pressure to the upper-level trough of low pressure causes air to *converge* aloft on the ridge-to-trough side of the Rossby wave. This piled-up air then sinks and warms adiabatically, causing high pressure near the surface.

5. On the opposite side of the trough, the air speeds up as it turns into the next ridge. This speeding up causes the air to *diverge* and spread apart. Air from below flows into this zone of divergence, causing low pressures near the surface. This promotes rising of the near-surface air. All of the directions of rotation shown here are opposite in the Southern Hemisphere.

B How Do Mountains Affect Adiabatic Temperature Changes?

Mountains interact with moving air masses in interesting ways. In the example below, moist air encounters the side of a mountain and is forced to move upslope. Once it reaches the top, the air flows down the other side. Examine this setting and think about what happens to the air on its way up and on the way down.

1. As the air encounters the base of the mountain and begins moving up the slope, it is initially unsaturated and cools according to the unsaturated adiabatic lapse rate, which is 10 C°/km everywhere.

2. As the air flows upslope, at some elevation it has cooled enough that it becomes saturated (reaches the dew point), and at this level clouds form. This height is called the *lifting condensation level*. As this air continues rising, it now cools at the saturated adiabatic lapse rate, which is somewhat less than the unsaturated adiabatic rate.

04.07.b1

3. By the time the air reaches the top, it has been warmed by latent heat. As it moves down the other side it will compress and warm, and the air will become unsaturated (warm air has higher water-vapor capacity than cold air). It will remain unsaturated, thereby causing a drying effect on this side of the mountain. It warms according to the unsaturated adiabatic rate of 10 C°/km.

4. As a result of this process, the air at the leeward side will be warmer than the air at the same elevation on the windward side.

C How Do Fronts Induce Rising Motion?

Whenever two air masses with significantly different temperatures meet (▶), the warmer air will be forced up and over the colder air because the warm air is less dense. This process is called *frontal lifting*. As this warm air rises, it will cool adiabatically. If it has abundant moisture, it will not need to rise very high before it cools to the dew-point temperature. At that level, active condensation or deposition begins and a cloud forms. This can lead to precipitation, as shown by colors on this weather map (▶) near a weather front marked by the blue line. The teeth on the line indicate which way the front is moving. The weather associated with fronts is presented in the next chapter.

04.07.c1

04.07.c2

Before You Leave This Page Be Able To

✓ Sketch and explain four settings in which convergence of air masses induces rising atmospheric motion.

✓ Describe how mountains influence the humidity, lapse rate, and temperature as air moves up one side and then down the other side of a mountain.

✓ Explain how fronts can force air to rise.

4.7

4.8 What Do Clouds Tell Us About Weather?

CLOUDS ARE ACCUMULATIONS of liquid water and ice suspended in the air. The types and amounts of clouds vary from place to place, from time to time, and from season to season. What are the different types of clouds, and how does each type form? Clouds provide clues not only about the amount and distribution of moisture at a given place and time, but also about conditions in the atmosphere, specifically the stability of the atmosphere and the type of convection that is occurring. This, in turn, informs us about important aspects of our weather.

We classify clouds based on three main factors: their form, their altitude, and whether they are associated with precipitation. Each of these aspects is summarized below, and examples of the main types of clouds are pictured in the large central figure.

Cloud Form

The three main forms of clouds are cumuliform, stratiform, and cirriform. *Cumuliform* clouds generally are taller than they are wide, or at least have a lumpy appearance. "Cumulus" in Latin means "a heap", and forms the root of words like accumulation. *Stratiform* clouds have the form of one or more layers (stratus means "to stretch" in Latin). *Cirriform* are feathery, wispy clouds, named after cirrus, the Latin word for a "lock of hair."

Cloud Altitude

We also subdivide clouds as to whether they are relatively low in the troposphere (low clouds), are near the top of the troposphere (high clouds), or are at an intermediate elevation (called mid-level clouds or designated with the prefix "alto"). For example, a mid-level cumuliform cloud is referred to as an altocumulus cloud. Some clouds extend from low levels up to high ones.

Precipitation

Clouds are composed of drops of liquid water, crystals of ice, or commonly some of each. Most drops and crystals are held aloft within the cloud because the buoyancy forces from unstable rising air exceed the gravitational force that pulls them toward Earth's surface. When the drops or crystals become too large to stay aloft, or the rising force weakens, the drops and crystals fall to the surface as rain, snow, hail, or some other type of precipitation. Clouds experiencing precipitation are designated with the prefix "nimbo" or the suffix "nimbus". A stratiform cloud that is precipitating is a nimbostratus cloud, and a cumulus cloud that is precipitating is a cumulonimbus. Even with all the possible combinations of words for form, height, and precipitation, there are only about ten main types of clouds, which are illustrated here.

Cumuliform Clouds

Cumulus clouds all have a somewhat puffy aspect but have different appearances, depending on their altitude and how unstable the air is (i.e., how strongly the air is rising). Cumulus clouds are produced by air rising under its own impetus, or free convection, and indicate that air is unstable to some degree.

04.08.a2

▲ Mid-level clouds contain both liquid water and ice particles, with the level at which ice can be present being lower in winter. Cumuliform clouds at these levels are termed *altocumulus*. They are puffy and only somewhat wispy.

04.08.a3

▲ Typical clouds on many days are low-level clouds with a puffy (cumuliform) appearance, or cumulus clouds, which can be widespread across the sky or localized over mountain peaks. When the atmosphere is only slightly unstable and water vapor is not too abundant, cumulus clouds are not very tall. Such clouds are typical during morning hours before surface heating causes air to rise more rapidly, and at times when conditions are only slightly unstable.

04.08.a4

► High in the atmosphere, clouds are composed almost entirely of ice. When such high-level clouds are partly cumuliform (lumpy) and partly cirriform (wispy), they are *cirrocumulus clouds*.

► Cumulus clouds can occur in discrete layers, called *stratocumulus clouds*. Like normal cumulus clouds, stratocumulus clouds are typically low clouds that are composed entirely of tiny drops of liquid water.

04.08.a5

04.08.a6

▲ If moisture is abundant and the atmosphere is very unstable, strong lifting in the atmosphere can form cumulus clouds with great vertical extent. Tall cumulus clouds associated with precipitation are *cumulonimbus clouds*, and commonly are accompanied by lightning. A cumulonimbus cloud contains liquid water near its base but ice particles near its top, permitting hail to form. If well developed, such clouds develop an anvil shape at the top that points downwind, as the stronger upper-level winds push the moisture and storm cloud sideways.

Cirriform Clouds

▶ *Cirrus clouds* are high in the troposphere and have a distinctive wispy appearance, as if smeared on a photograph. At these levels, air temperatures are well below freezing, so these clouds are composed of ice crystals, which are easily blown and "smeared" about by the strong, upper-level winds. The ice refracts (bends) the light, enhancing the wispy, shimmery appearance of most cirrus clouds. Cirrus clouds often indicate that stormier weather is approaching, usually within several days.

A high-flying jet can leave a linear streak of clouds behind it, called a *contrail* (short for condensation trail). These form when water vapor produced by the engines encounters the much colder outside air, causing condensation of water drops or the formation of ice crystals. Once formed, contrails can be dispersed and sheared by the wind. They also contribute to the formation of cirrus clouds.

04.08.a7

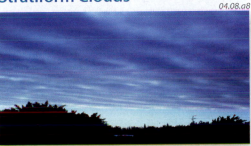

04.08.a8

Stratiform Clouds

▲ Stratiform clouds have a layered or spread-out aspect. They are the clouds we observe when skies are overcast, but they may obscure just a part of the sky. Such layers of clouds form when air is forced upward (forced convection), such as by surface convergence or movement of a frontal boundary. Stratiform clouds are oriented laterally rather than vertically, implying that vertical motion is resisted by the atmospheric conditions.

Stratiform clouds in the upper troposphere are high enough to be composed solely of ice and so have a somewhat wispy (cirriform), sheetlike appearance (▶). Accordingly, they are called *cirrostratus*.

04.08.a9

Mid-level stratiform clouds contain both ice and liquid water and can be expressed as high overcast. These mid-level clouds are termed *altostratus*. Stratus clouds can form at several levels (▶).

04.08.a10

04.08.a1

04.08.a11

A stratiform cloud that is precipitating is a *nimbostratus* cloud, as shown here (◀). This is a typical cloud associated with regional storms. A low-level stratiform cloud that is not precipitating is simply called a *stratus* cloud. Stratus can form a low blanket of clouds that causes overcast conditions.

Fog is a cloud that hugs the ground surface. It is described in detail in the next two spreads.

Before You Leave This Page Be Able To

☑ Sketch each of the main types of clouds, explaining for each type the reasons for its appearance and how it forms.

4.8

4.9 What Conditions Produce Fog?

FOG IS SIMPLY A CLOUD at ground level, so the same conditions that create a cloud also produce fog. Specifically, fog is produced by cooling of the air, by increasing the humidity of the air, or some combination, with the end result that water vapor content in the air reaches saturation. Fog forms in several distinct settings, each of which causes air at or near ground level to become saturated.

A What Is Fog?

Fog is a ground-level cloud that obscures visibility, such as with this lighthouse (◄). It is most common along coastlines, over large water bodies, and in certain low areas that have the right setting to produce fog, but fog can form nearly anywhere, under the right conditions.

04.09.a1 Marshall Point, ME

04.09.a2

Fog indicates that the relative humidity has reached 100% and the air temperature has reached the dew-point (or frost-point) temperature, causing saturation. One way in which air can reach saturation is by increasing the moisture content until vapor pressure reaches the water-vapor capacity, moving up on this graph (◄).

Another way to reach saturation is by cooling the air until it reaches the dew-point (or frost-point) temperature, represented on this graph as moving to the left.

A combination of cooling and increasing humidity can also cause saturation.

B What Conditions Can Add Enough Humidity to the Air to Form Fog?

Fog can form where enough moisture is added to the air to cause saturation (100% relative humidity), but more commonly fog forms where the addition of moisture occurs simultaneously with the chilling of the air. Addition of moisture to the ambient air may happen when a nearby source of water for evaporation exists, or when precipitation evaporates as it falls through drier air.

Evaporation Fog

04.09.b1 Central Idaho

1. When a warm water body underlies colder air (◄), the environmental lapse rate is steep and the atmosphere is unstable. As the unstable air rises, it can incorporate water molecules that evaporate from the water body, increasing the humidity of the air. The increase in humidity and cooling causes the rising air to reach saturation, forming a type of fog called an *evaporation fog*. The instability causes an evaporation fog to rise in vertical columns, as shown in the photograph to the left. This phenomenon is common in fall, when water bodies are likely to remain warm from the long summer, but air above them begins to cool.

Valley Fog Produced by Increased Humidity

2. In rugged terrain, cold nocturnal (nighttime) downslope winds can trap atmospheric moisture in valleys. If moisture is added to the valley air, such as from wetlands, irrigated fields, canals, and lakes, the humidified and cooled air can reach the dew point, forming a *valley fog*. Perhaps the most famous valley fog in the U.S. is the Tule fog of the Central Valley of California, the white area in the center of this regional satellite image (►).

04.09.b2

Precipitation Fog

3. Fog can also form in association with precipitation, such as occurs along a warm front. A warm front is a mass of warm air that moves laterally, trying to displace cold air in its path. Along a warm front, warmer, less dense air is forced to rise at a gentle angle over more dense, colder air.

4. As the moist, warm air rises over the cold air, it is cooled and can reach saturation, forming clouds and rain. The released rain then begins to fall through the underlying cold air and toward the cold surface.

04.09.b3

5. If the underlying colder air is dry, it can cause some or all of the falling raindrops to evaporate, increasing the humidity of the cold air. Since cold air has a lower water-vapor capacity than warm air, the increase in humidity can cause the cold air to reach its dew point, forming fog and other low-level clouds. Rain that reaches the ground can also evaporate, increasing the humidity of the near-surface air, potentially forming more fog. In both cases, evaporation of precipitation is causing the *precipitation fog*.

C In What Settings Does Enough Near-Surface Cooling Occur to Form Fog?

Fog can also form where moist air is cooled enough to reach its dew-point temperature. This can occur where moist air moves over a colder part of Earth's surface, such as over snow or over a cold ocean current. It also occurs when moist air cools if it moves vertically. Several types of fog form primarily from cooling of moist air.

Radiation Fog

Although the term "radiation fog" sounds like something in the latest zombie movie, it is actually related to cooling of the ground surface by radiant heat loss. If skies at night are clear (i.e., cloud cover is sparse), and winds are light or calm, then the longwave radiation loss from the surface to the atmosphere will be relatively efficient, causing the surface to cool significantly.

Under these conditions, especially if near-surface air is humid, the surface temperature can drop to the dew point (or frost point) and cool the immediately overlying air, causing it to become saturated in water vapor. The condensation of water forms fog, in this case called a *radiation fog*. If the surface cools substantially, a temperature inversion will occur (colder air below warmer air), stabilizing the atmosphere and keeping the fog in place. Sometimes such fog can become concentrated in valleys and other topographically low places, therefore representing another way to form a valley fog (described below).

04.09.c1

Advection Fog

When warm air flows over a colder surface, the air loses some of its energy to the surface and therefore cools. If it cools to the dew point or frost point, the moisture in the air will condense and produce fog, as occurred when moist air moved over this glacier (▶). The movement of air laterally is called advection, so a fog produced in this manner is an *advection fog*.

04.09.c2 Athabasca Glacier, Alberta, Canada

A similar process produces fog along many coastlines. If moist air over an ocean or large lake blows onshore, it can encounter ground that is colder than the body of water.

04.09.c3

The cold ground lowers the temperature of the moist air, often enough to cause saturation and an advection fog. Advection fogs also form where warm, moist air blows over a cold ocean current (not shown). Advection fog is common along coastal California, which has a cold ocean current, called the California Current, offshore, bringing cold waters south along the coast.

Upslope Fog

As air is forced uphill by the prevailing local wind, it will cool because of the decreasing atmospheric pressure above it. At some level above the surface, the air may cool to the dew-point (or frost-point) temperature, and a cloud will form. But on the mountain slope, the cloud forms on the ground, so it is a fog (▼). This type of fog is commonly called an *upslope fog* to indicate that the fog is caused by the upslope motion of air.

04.09.c4

04.09.c5 Cumberland Gap, KY

Valley Fog Caused by Descending, Cold Air

Valley fog can form in yet one more way. In mountainous terrain, air near the surface cools at night because of longwave radiative loss. This air will flow downhill because it is more dense than warmer air. It may remain colder than the surrounding air, even as it warms adiabatically on descent. If so, the descending air may chill the ambient air in the valley. If the ambient air is cooled to the dew-point or frost-point temperature, it will become saturated and form a valley fog.

04.09.c6

Before You Leave This Page Be Able To

☑ Sketch and describe three ways that increasing humidity, perhaps accompanied by some cooling, can form fog.

☑ Sketch and describe four settings in which near-surface air can be cooled enough to form fog.

4.9

4.10 Where and When Is Fog Most Likely?

CERTAIN SETTINGS AND CONDITIONS are conducive to the formation of fog, so we might anticipate that some regions will have more fog than others. Also, as temperatures and humidity change with the seasons, some times of the year are likely to be foggier than other times. Considering all the places you know, which ones have more fog, and why do you think this is so?

A In Which Settings Is Fog Likely?

Formation of fog occurs when the combination of moisture and temperature causes the air to reach saturation (100% relative humidity). What settings would most likely have fog?

1. Examine this diagram and try to identify any factors that you think could influence the amount of fog. Then read the different blocks of text, which will describe some of these factors—you may identify more than are described.

2. An important factor is the climate of a place, especially the temperatures and humidity levels that are typical. Regions at lower latitudes will generally receive more direct insolation than regions at higher latitudes, and this variation greatly affects the temperature, which in turn controls how much moisture the air can evaporate.

04.10.a1

6. Topography and elevation are clearly key factors, since they control temperatures and influence local wind directions. Mountain slopes have upslope winds at some times and downslope winds at others. They also interact with the prevailing winds.

5. Prevailing wind directions are important since they can bring moist or dry air into an area. Especially important is whether breezes along a shoreline blow onshore or offshore (toward the water).

3. Another important control on the frequency of fog will be the typical humidity of a place. Is it a humid place, commonly covered with clouds, or a drier, inland desert that commonly has clear skies?

4. Proximity to the ocean or other large bodies of water will clearly be a major factor in fog formation, since such water bodies can contribute abundant moisture to the air, if they are warm enough. Water also warms up and cools down more slowly than land, so it tends to help moderate any temperature swings of the adjacent land (cooler days and warmer nights than land farther away from water).

B What Regional Settings Are Favorable for the Formation of Fog?

This map shows major ocean currents and land cover. Warm ocean currents are in red and cold ocean currents are blue. On the land, green colors show abundant vegetation, tan colors show areas with less vegetation, and white areas are covered by ice and snow. Examine this map and think about how these ocean currents and regional patterns could influence the frequency of fog.

Ice and snow on the land and on frozen oceans greatly cools the adjacent air, causing intense temperature inversions over the polar ice caps. These inversions lead to widespread radiation fog, especially beneath the polar highs.

Fog is frequent along the windward sides of mountain ranges, such as those along the western side of North and South America. Near these mountains, fog is more common where moisture is abundant, and this is controlled by regional patterns in winds and ocean currents.

Radiation fogs are common in inland areas, which often experience clear skies and a significant drop in temperature at night. Radiation fog can persist when there is a temperature inversion to stabilize the atmosphere and hold the fog in place, low to the ground.

Latitude plays a major role in whether fog forms because it influences regional temperatures, humidity, wind directions, ocean currents, and the types of storm systems.

04.10.b1

Great Britain is very well known for its cold and damp advection fogs.

Over the open oceans, evaporation fogs are common as water evaporates and rises into overlying cooler air.

Ocean currents greatly influence fog development. Advection fogs are common where cold ocean currents extend into otherwise warm areas near the coast, such as along the western coasts of South America and southern Africa. Sometimes these fogs provide the major source of moisture in desert areas, such as the Atacama Desert of Chile and the Namib Desert of Namibia.

C Which Parts of the U.S. Have the Most and Least Fog Days?

Mean Number of Fog Days Per Month (average for entire year)

This map shows the average number of days of fog experienced by areas in the conterminous 48 states, with darker colors indicating more fog. Observe the patterns on this map, noting how common fog is in the area where you live or visit.

The Pacific Northwest frequently has fog. As saturated marine air is pushed by the mid-latitude westerlies toward the Pacific Northwest, it gets chilled from beneath by a cold ocean current, causing an advection fog. Other fog is caused by air moving upslope on the western sides of the Cascades and other mountains. San Francisco experiences advection fog for the same reason, and the mountains surrounding the bay cause the fog to be funneled and focused in the Bay Area.

The Great Lakes and other large bodies of water support evaporation fog, particularly in fall.

In New England and parts of the Appalachian Mountains, fog often occurs as cyclones and anticyclones over the Atlantic push moist and relatively warm maritime air over the colder continent. This creates advection and precipitation fog. Cyclones and anticyclones also produce upslope fog farther south, along the flanks of the Appalachian Mountains. Valley fogs are also common in this region.

The continental interior often experiences clear nights, which may cool the air to the dew-point temperature and create radiation fog. Fog is less common in the Desert Southwest, which often lacks abundant moisture.

The Gulf Coast experiences advection fog as warm, moist air from the Gulf of Mexico is advected onshore due to clockwise circulation around the Bermuda-Azores high-pressure area.

04.10.c1

Legend: 1 2 3 4 5 6 7 8 9

Seasonal Variations in the Number of Fog Days Per Month

Winter

► In January, advection fogs occur frequently in the Pacific Northwest and other coastal areas. These are caused by relatively warm air over the oceans moving over much colder land, which cools the air and leads to the formation of fog.

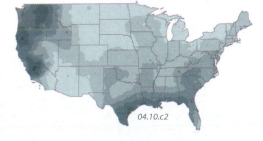
04.10.c2

Spring

By March, fogs become less frequent across many parts of the country. Still, radiation fogs tend to be fairly common across the middle of the country under clear nighttime skies. The Gulf Coast experiences frequent fog caused by moist air from the Gulf of Mexico (▼).

04.10.c3

04.10.c5

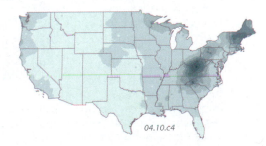
04.10.c4

Fall

▲ Fall, like summer, is a time of less frequent fog, except perhaps for upslope fogs in the Appalachians. In addition, downslope winds at night can trap moisture in the adjacent valleys. In subsequent months, the patterns characteristic of winter begin to become reestablished.

Summer

▲ Fog is least common in most of the U.S. in summer. Even though the vapor pressure is high, higher temperatures mean that it is difficult for the air to reach saturation. Fog is most common in New England and in mountainous areas, such as West Virginia, where fogs are largely attributable to upslope, precipitation, and radiation fogs.

NUMBER OF FOG DAYS PER MONTH
1 2 3 4 5 6 7 8 9

Before You Leave This Page Be Able To

✓ Sketch and describe some factors that influence the frequency of fog.

✓ Summarize how these factors affect the global, regional (U.S.), and seasonal distributions of fog.

4.10

4.11 How Does Precipitation Form?

THE PROCESS OF PRECIPITATION is vital to life on Earth, helping to redistribute water from the oceans to the atmosphere to the land. Precipitation is the ultimate source of all the fresh water on the planet, which we depend on in our daily lives. How does precipitation occur? What is going on inside clouds that forms raindrops, snowflakes, and hail? It turns out there are two important mechanisms by which precipitation droplets form, one of which is somewhat surprising.

A What Is Precipitation?

1. Precipitation is the process whereby liquid droplets of water (raindrops), solid bits of ice (snowflakes and hail), or some combination of these fall from the sky. Examine this figure and consider all the processes that have to occur to cause rain, snow, or hail.

2. The cycle begins with evaporation of water in the oceans and other parts of Earth's surface, a process that puts water vapor into the atmosphere. Next, the water vapor forms the various types of clouds, which can contain tiny drops of water, ice crystals, or some of each.

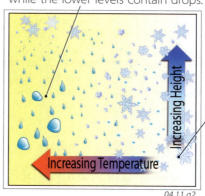

04.11.a1

3. For precipitation, some processes are occurring within the cloud that make the water droplets or bits of ice heavy enough that the pull of gravity can overwhelm the buoyancy forces (atmospheric instability) that uplift air within the cloud (the rising air is how most clouds form).

4. Whether a cloud contains drops of liquid, ice crystals, or some combination depends primarily on the temperature of the cloud. If ambient temperatures near the cloud are warm, the cloud will contain mostly drops of liquid. Temperatures decrease upward within a cloud, however, so the upper levels can contain ice, while the lower levels contain drops.

Increasing Height

Increasing Temperature

04.11.a2

5. Under cold ambient conditions, a cloud contains mostly ice throughout its vertical extent. Clouds at intermediate ambient temperature, or intermediate altitudes, will contain a mix of drops and ice.

B How Do Water Droplets Form and Grow?

1. An important factor in how water droplets in a cloud grow to become raindrops is the immense differences in size between the various players. Raindrops are huge compared to the size of water droplets that form a cloud, as shown by their relative sizes in the properly scaled diagram below.

CLOUD DROPLETS
(0.02 mm)

TYPICAL RAINDROP
(2 mm)

CONDENSATION NUCLEI
(0.002 mm)

04.11.b1

2. It is energetically difficult for the tiny water droplets in clouds to just form by themselves, but it is easier for them to form if they condense around even tinier particles, such as dust, salt, and smoke. Due to this role, such particles are called *condensation nuclei*.

3. The figure to the right illustrates what can happen to a moving water drop that interacts with smaller cloud droplets around it.

4. Some larger drops form when liquid water droplets in clouds merge and grow to a size that can be pulled down by gravity. A water droplet begins to fall as soon as the downward-directed gravitational force exceeds the upward-directed buoyancy force caused by instability. This occurs sooner for larger droplets than for smaller droplets, and the larger drop overtakes the smaller droplets on their descent, making the falling drop even larger. Eventually it falls as a raindrop. The growth of a drop in this manner is called *collision-coalescence*.

FALLING RAINDROPS

CLOUD DROPLETS

Growing

No Change In Size

Splitting

CONDENSATION NUCLEI

04.11.b2

5. In other cases, the smaller droplets simply slide past the falling drop, because the collision is not enough to break the surface tension that tends to keep each drop intact. In this case, the falling drop will not increase in size as it falls. This situation is actually very likely under many atmospheric conditions.

6. Alternatively, wind resistance can reshape the falling drop until it becomes easier to break the drop into separate drops than to retain its contorted shape. In this way, a falling raindrop becomes smaller, which can also occur if water molecules on the outside of the drop simply evaporate into the surrounding air.

How Do Ice, Liquid, and Vapor Interact Within a Cloud?

Many clouds are in environments in which the temperature is below freezing, even in summer. Exceptions are clouds in the tropics and low in the atmosphere, or a cloud being warmed by latent heat during precipitation. Even when temperatures are below freezing, some liquid water droplets exist. This sets up a situation in which water exists in all three phases — vapor, liquid, and ice (solid). The interaction between these three coexisting phases allows precipitation to form much more easily than would otherwise be possible if only vapor and liquid were present.

1. The diagram to the right shows the familiar saturation curve for water vapor, but focuses on conditions slightly above and slightly below freezing. At temperatures below freezing, the saturation curve for water vapor near ice (the dashed line) diverges from that for water vapor and liquid (the solid line). The divergence of the two curves begins at the freezing point because ice can't easily exist above freezing.

2. The area between the two curves, shaded orange, depicts temperatures at which air is unsaturated with water vapor relative to a liquid but above saturation for air next to ice in the cloud. If a parcel of air has the water-vapor capacity and temperature to plot in this orange area of the plot, vapor will become deposited on ice crystals while liquid water drops are evaporating at the same time.

3. Another way to state this is that saturation is reached with slightly less water vapor when air is in contact with ice in the cloud than when it is in contact with liquid water, even at the same temperature. That is, less water vapor can exist in air that is in contact with ice than in air that is in contact with liquid water, if both are at the same temperature.

4. This difference in saturation levels aids in the formation of precipitation. If ice, water vapor, and drops of liquid water coexist at the conditions specified by the orange field in the previous figure, then the air will be unsaturated next to the liquid drop. As a result, water molecules will evaporate from the drops, increasing the relative humidity of the surrounding air.

5. As the liberated water vapor molecules diffuse into the air, they cause an increase in water vapor adjacent to a nearby snowflake. The air near the snowflake therefore becomes saturated or supersaturated with respect to the ice. This in turn causes water vapor molecules to be deposited on the snowflake, enlarging it.

6. Depending on the air temperature, the ice can remain as a solid or if temperature increases slightly, the ice can melt, going directly from ice to liquid. In this way, liquid water becomes water vapor, which then becomes ice, which then becomes drops, an easier path (under many conditions) than trying to overcome surface tension to grow drops or the difficult energetics of nucleation required to make new drops. This process of precipitation is often called the *Bergeron process* after one of the scientists who discovered it.

How Cloud Seeding Works

Technology exists to "help" precipitation occur more efficiently in drought-stricken areas. Strategies generally involve injecting particles into clouds to enhance droplet or ice crystal growth, either via a ground-based delivery system or more commonly airplanes that fly through the clouds, as shown here. One strategy uses dry ice to cool the cloud to temperatures so low that vapor deposits onto ice more readily. A second strategy is to inject into the cloud microscopic solid particles that

act as condensation nuclei. The idea is that water vapor can condense or deposit much more easily when it has something to "hold onto" during the phase change. Silver iodide has often been used for this purpose because its crystalline structure maximizes opportunities for vapor to attach to it. These cloud seeding efforts have achieved mixed results. In some cases, precipitation has been enhanced or shifted to other areas. Interventions into natural processes, however, often have unintended con-

sequences, such as, in this case, silver-iodide pollution from the chemicals used to seed the cloud.

Before You Leave This Page Be Able To

☑ Explain what precipitation is and how the size of a drop can change.

☑ Sketch the water-vapor capacity curve as a function of temperature, with respect to liquid water and ice, and describe how this leads to a Bergeron process for precipitation.

☑ Sketch and explain cloud seeding.

4.11

4.12 How Do Sleet and Freezing Rain Form?

RAIN AND SNOW ARE THE MOST COMMON forms of precipitation, but other types of precipitation are also important. The term *sleet* is used for a mixture of snow and rain or for precipitation as small partially frozen pellets that are too small to be called hail. *Freezing rain* is precipitation that reaches the ground as raindrops, but that immediately freezes upon contact with cold objects on the surface, such as the ground, roads, plants, and power lines. Can you envision the situations that could cause sleet or freezing rain?

A Under What Conditions Do Sleet and Freezing Rain Form?

Sleet and freezing rain are often confused, and they have several similarities. Both result from snow and ice pellets that melt on their way down to the surface as they pass through a layer that is warmer than freezing.

Sleet

This graph shows a hypothetical temperature profile (the green line) that would allow sleet to form. Note that this temperature profile shows a subfreezing layer close to the surface, with temperatures that are colder than those slightly higher in the atmosphere—a temperature inversion. As snow falls, it crosses from freezing conditions through the zone of above-freezing temperatures and then back into freezing ones some distance above the surface. As a result, the snow will at least partially melt on the way down, forming a drop of water that then partially refreezes before it reaches the surface. This creates a tiny pellet that bounces on impact with the surface. This type of precipitation is called *sleet*.

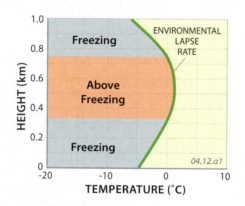

Freezing Rain

This similar graph shows the conditions needed for freezing rain. As for sleet, the temperature profile displays a temperature inversion, but this time with a thinner zone of subfreezing near the surface. As snow falls through these conditions, it will at least partially melt when it encounters the mid-level interval of warmer-than-freezing air. As the drop continues falling, it again encounters freezing temperatures, but not until it is at or right above the surface. As a result, the drop does not have enough time to refreeze, and so hits the surface as rain that sticks to the ground instead of bouncing like sleet. Upon encountering the cold ground and objects on the surface, the drop immediately freezes, forming coatings of ice on the ground and objects.

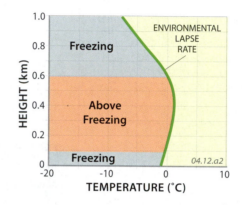

B Where Do Sleet and Freezing Rain Form?

The scenario that provides the best opportunity for the temperature inversion required for the formation of sleet and freezing rain is ahead of the warm front, as shown in the cross section below.

1. A warm front is a mass of relatively warm air that moves across the surface, displacing cold air in its path. In the cross section shown here, the warm air is on the left side (on top) of the frontal boundary and cold air is on the right side of (below) the boundary.

2. The warm air is less dense than the cold air, so it is wedged up and over the colder air mass. If warm air resides above colder air, it creates a temperature inversion. As the warm air is forced to rise, it cools, forming clouds and rain above and along the boundary.

3. Close to the surface position of the front, rain produced along the front only falls through a short vertical interval of cold air. If raindrops pass through this zone without refreezing, they will reach the ground as normal rain. This is the first type of precipitation that announces the arrival of the warm front.

6. Snow can be associated with a warm front if the warm air rises so high off the ground that it is below freezing. In this case, the entire profile remains below freezing, allowing snow to form aloft and reach the ground.

5. Even farther back from the surface warm front, rain produced aloft must pass through a thicker column of cold air. Thus, raindrops have enough time to freeze on the way down, forming sleet.

4. Farther back from the surface location of the front, the ground may still be at freezing temperatures. In this case, the rain freezes upon encountering the ground and other cold objects on the surface, producing freezing rain. As the warm front passes over an area hovering near freezing temperatures, therefore, normal rain follows freezing rain.

C What Regions Have a Higher Frequency of Freezing Rain?

04.12.c1

1. This map of central North America (▶) shows the average annual number of days with freezing rain. Examine this map for regional patterns and try to identify if the place where you live or visit has frequent freezing rain—you probably already know this answer, because freezing rain is not easily forgotten. A map of sleet would show a similar distribution. Since latitude and elevation largely control temperature, they also influence where freezing rain occurs.

3. Freezing rain is also frequent in the highland areas of the Northeast, including New England and some parts of the Appalachian Mountains. The freezing rain and sleet are the result of the same type of cyclonic storm system as in the Midwest. Sleet and freezing rain are also common in other humid uplands in mid-latitudes around the world, such as the Alps, Japanese highlands, and parts of the Andes of South America.

2. On this map, freezing rain is frequent in the Midwest, exhibiting a pattern that follows the typical paths of storms as they move eastward across the country. Such storms have a northward (counterclockwise) flow along their eastern, leading edge, bringing relatively warm air from the south that in many cases forms a warm front.

D How Does Terrain Contribute to the Distribution of Sleet and Freezing Rain?

Movement of air against mountains can also cause freezing rain and sleet. When a cold air mass is next to a mountain range, the circulation associated with it will cause air to move upslope as it encounters higher terrain, as shown here (▶). The movement of air can be in response to larger scale clockwise flow (in the Northern Hemisphere) around an anticyclone, or from counterclockwise flow (in the Northern Hemisphere) around a cyclone.

04.12.d1

A cold air mass beside the mountain is dense, while the air wedged above it may be warmer, leading to a temperature inversion. While the inversion would also stabilize the local atmosphere above the cold surface air mass, the movement upslope by the circulation around the pressure system may result in forced convection that produces rain at the lowest elevations, freezing rain at higher elevations, sleet farther upslope, and snow at the highest elevations. Mountains that block flow in this way are sometimes referred to as "cold air dams," as shown in the photograph below. Mountain-caused freezing rain, sleet, and snow are frequent next to the Rocky and Appalachian Mountains in winter.

04.12.d2

The Impacts of Freezing Rain

Freezing rain can cause extensive damage and poses severe hazards to animals, plants, and any object on the surface. Freezing rain can coat sidewalks and roadways with a layer of dangerously slippery ice, causing pedestrians to fall and automobiles to skid out of control. Freezing rain resembles rain as it falls, so people may believe that they are only driving in rain when in fact they are driving on an icy road. Even the rain that falls on windshields gives no clue that it will freeze on impact with the surface, because automobile defrosters and friction created by windshield wipers

prevent freezing that might alert the driver. The fact that sleet bounces informs motorists to its presence as ice, making it less dangerous.

Freezing rain adds considerably to the weight of trees, especially in fall or early winter, when leaves are still attached to tree limbs and branches. This may cause limbs to fall, perhaps onto power lines, which are themselves weighed down considerably by freezing rain. On average, freezing rain causes more deaths and property damage per hour than any other type of storm.

04.12.t1

4.12

4.13 What Is the Distribution of Precipitation?

THE AMOUNT OF PRECIPITATION varies from region to region, season to season, and day to day. The daily variations are related to short-term changes in weather, but variations between regions and seasons reflect differences in the overall climatic setting, such as differences in latitude, prevailing wind directions, proximity to large water bodies, any nearby ocean currents, elevation of the land, and countless other factors. We explore some of these by examining regional variations in precipitation over North America and the world.

A How Does Precipitation Vary Across North America?

January

1. Examine this map, which shows average amounts of precipitation for January, averaged over several decades. In northern and high-elevation regions, much of this precipitation is as snow. Note any regional patterns and the precipitation that is typical for where you live.

2. Heavy rainfall along the coast of Canada and the Pacific Northwest is due to proximity to the ocean combined with orographic effects that cause moisture-laden Pacific air to move upslope under the influence of the mid-latitude westerlies.

3. Inland from the western mountains, precipitation totals decline dramatically. These areas are blocked from oceanic moisture sources by mountains both to the west and to the east.

4. Precipitation decreases inland away from moisture sources and northward with lower temperatures. Cold air has lower water-vapor capacity and therefore will not produce clouds with a sufficient amount of moisture to allow for much precipitation.

JANUARY AVERAGE PRECIPITATION (cm)

0 3 6 9 12 15 18 21 24 27 30 33 36 39 42

5. The Great Lakes are much warmer than the overlying air in winter. This generates a steep environmental lapse rate, which makes an adiabatically moving air parcel likely to be warmer than its surrounding environment. Thus, the lakes destabilize the atmosphere, increasing local "lake effect snow" totals downwind.

6. Precipitation in both the Atlantic and Pacific oceans at this time of year is focused near the boundary between cold air and warm air, in part because these regions experience frequent passage of fronts.

July

7. This map shows precipitation totals for July, the middle of summer. Note the regional patterns and then compare this map with the one for January.

8. July totals are higher than in January in most of North America, except along the West Coast. In most areas, local surface heating destabilizes the atmosphere, but the cold California Current off the West Coast negates this effect.

9. Summer precipitation totals are low in western North America and over the adjacent Pacific Ocean, even in places that are relatively wet in winter. This is because of the drying effects of sinking air on the eastern side of the Hawaiian high-pressure feature, which makes its closest approach to North America in summer.

10. July totals are much higher than those in January in the far north. The warmer air has more energy to evaporate water, producing water vapor that can then precipitate. Also, the region is much closer to the cold-warm boundary in summer than in winter, facilitating frontal lifting.

JULY AVERAGE PRECIPITATION (cm)

0 3 6 9 12 15 18 21 24 27 30 33 36 39 42

11. The Great Lakes are cool relative to the overlying air in July. This weak environmental lapse rate (or even an inversion) stabilizes the atmosphere, the opposite of winter conditions, limiting the amount of precipitation.

12. In summer the Bermuda-Azores high-pressure area reaches its northernmost extent. The clockwise circulation around this feature supplies warm, moist air into interior North America.

13. The prong of higher precipitation that occurs in western Mexico is related to a northward flow of moisture as part of the "Arizona monsoon." Near the mountains, intense surface heating and orographic lifting cause summer thunderstorms.

B How Do Annual Precipitation Totals Vary Globally?

The map below shows average annual precipitation at a global scale. Examine the regional patterns and consider what factors might be causing them. You may want to refer back to the globes showing specific humidity earlier in this chapter. An area that most people consider to be "wet" generally has high precipitation and high humidity. How dry an area seems is a function of low precipitation, low humidity, and evaporation rates that exceed precipitation rates.

1. Globally, precipitation is concentrated in equatorial oceanic areas, expressed as a belt of blue and green colors near the equator. Precipitation generally decreases poleward, inland, and downwind from major mountain ranges.

8. The idea that Siberia receives extremely large amounts of snowfall is a myth. The intense Siberian High is driven by cold, dense air, and it produces a sinking motion. The cold surface also acts to decrease the environmental lapse rate, making it more likely that an adiabatically moving air parcel will remain colder than its surrounding environment, enhancing atmospheric stability.

2. The tropical rain forests of Indonesia, along with adjacent parts of the Indian and Pacific oceans, are among the wettest places on Earth. They are part of an east-west belt of high precipitation that roughly follows the intertropical convergence zone (ITCZ). Along the ITCZ, convergence of moisture-laden air carried by the northeast and southeast trade winds causes lifting and precipitation. The Hadley cell migrates with the seasons, so the position and influence of the ITCZ do, too.

04.13.b1

ANNUAL AVERAGE PRECIPITATION (cm)

0   36   73   109   146   183   219   255   292   328   365   402   438   474   511

7. The east coasts of mid-latitude continents are generally moderately wet, as cold and warm air meet frequently, producing frontal precipitation. In summer, convective precipitation supplements the totals.

6. The Amazon basin, with extensive tropical rain forests, is another region of high precipitation. This is due to the ITCZ and to orographic lifting of the moisture-laden trade winds over the eastern flanks of the Andes Mountains.

3. Locations in the subtropics (30° N and S) tend to be very dry. This belt of latitudes contains many of the world's deserts, such as the Sahara and those of the Arabian Peninsula, southwest Africa, central Australia, and the American Southwest. Even though warm air would ordinarily tend to rise in these warm areas, this effect is counteracted by descending air in the downward-directed part of the Hadley cells. Some of these areas are especially dry because they are in the rain shadow of major mountain ranges. Exceptions to the dry subtropics occur along the east coast of continents (for example China), where there is an abundance of warm, moist air in association with warm surface ocean currents that flows around subtropical anticyclones.

4. Polar areas are generally too cold to have much water in the atmosphere. Limited moisture, combined with the sinking of the intense cold air, opposes vertical cloud development. Furthermore, these areas are typically far from boundaries between warm and cold air, which can produce warm and cold fronts and their attendant storms. As a result, places near the poles receive very little precipitation.

5. Relatively little was known about precipitation over the oceans until recent years when satellite data became available. Oceanic patterns are similar to continental ones, except for the absence of complications caused by topography.

How We Measure Precipitation

Rainfall is measured using a standard cylinder with a circumference that is 1/10 of the opening. The amount of rainfall can be read from graduated markings on the cylinder or from a measuring stick with hashmarks exaggerated 10 times. This allows for the observation of rainfall totals to the hundredths of an inch. Some rain gauges, like the one shown here, contain two cylinders, one for measuring typical rainfall amounts and a second larger cylinder used when extreme amounts of precipitation cause the first cylinder to overflow. When it

04.13.t1

snows, the snow is allowed to melt and the amount of liquid can be measured as described above. Snow typically has an average density of 8% that of water, so the equivalent of one centimeter of rain will produce, on average, more than 12 cm of snow (an inch of rainfall will produce a foot of snow), but this ratio can vary widely, depending on local conditions.

4.13

How Can Moisture Extremes Be Characterized?

THERE ARE GREAT EXTREMES in the distribution of moisture, with some regions being very dry for an extended period of time, as during a drought, whereas other regions receive huge amounts of precipitation in a short period of time. The cause of these extremes in moisture can be broad regional shifts in the climate or more localized shifts, such as a slight alteration in wind direction that changes local patterns of orographic lifting. Extremes of precipitation, whether too much or too little, are of great concern to physical geographers as they have the greatest impacts on people. Here, we describe several types of extreme events, how we document such events, and how we assess the risks for either too little or too much moisture.

A What Causes a Drought and How Can We Quantify It?

Drought is an unusually extended period when losses of surface water to evaporation and transpiration from vegetation in the area are uncharacteristically higher than the addition of water to the surface by precipitation. Drought places stress on plants and animals and can cause normally green and fertile land to dry up into eroded, dusty plains. Drought can cause water shortages by drying up streams and lakes, and by decreasing flow into groundwater. Three causes of drought are described below, but others are also common.

Drought can be caused by weather patterns that are atypical for an area. Persistent high pressure, with its sinking air (◄), can cause a region to be drier than normal. The map on the right (▶) shows a large area of high pressure over the Desert Southwest. If such high pressure is persistent, the region can shift into drought conditions.

An unusual change in wind direction (◄), such as one that brings dry air from the land over an area that normally experiences an onshore flow of moist air from the ocean, can cause a drought.

A temporary change in the direction or strength of an ocean current (◄) can last multiple years, causing a temporary change in climate, possibly including drought if a cold current results in less moisture in the air.

500 mb Geopotential Height (m)

04.14.a1

04.14.a2

B How Can We Quantify the Risk of Drought?

NDVI DEPARTURE FROM AVERAGE

04.14.b1

% of Average			
■ 1-54	■ 65-74	95-115	■ 136-150
■ 55-64	■ 76-84	85-105	■ 151-249
	■ 85-105	116-125	■ 126-135	■ 252-253

Satellite data, such as this eMODIS image (◄), are used to create drought indices based on vegetation health and other biophysical properties.

The numerically based *Palmer Drought Index* (PDI) uses climatic and soil moisture data to quantify the severity of the drought. Status is expressed relative to normal at that location and time of year, such as the values shown on the map here for August 2012 (▶).

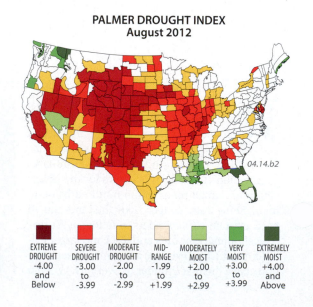

PALMER DROUGHT INDEX
August 2012

04.14.b2

EXTREME DROUGHT -4.00 and Below	SEVERE DROUGHT -3.00 to -3.99	MODERATE DROUGHT -2.00 to -2.99	MID-RANGE -1.99 to +1.99	MODERATELY MOIST +2.00 to +2.99	VERY MOIST +3.00 to +3.99	EXTREMELY MOIST +4.00 and Above

How Can We Quantify and Depict the Risk of an Extreme Precipitation Event?

An important aspect to consider in planning land use or other activities is how often events of extremely high or low precipitation are likely to occur. We can calculate the average length of time that we would expect to see between events that equal or exceed a given magnitude, the *return period*. This, also called a *recurrence interval*, allows geographers, land-use planners, farmers, and others to assess the risk of extreme events and to plan accordingly.

1. To calculate a return period, the first step is to specify the magnitude and duration of a hypothetical extreme event of interest. The magnitude is the amount of precipitation that fell. The duration is the length of time that was required for that amount of precipitation to fall. We then use historical data to find where this event ranks with other events of similar duration. The results are entered into the equation below.

$$\text{Return Period (RP)} = (\text{number of year of record} + 1) / (\text{rank of event})$$

2. To illustrate this concept, we present a simple example. If 8.4 cm of precipitation fell over a 12-hour period at a given place, we could estimate its return period by finding its rank among the largest 12-hour rainfall totals in the 50 years of historic records at that location. If our 12-hour total ranks 30th among all 12-hour totals at the location (in the 50 years during which measurements were taken at that site), the return period would be 1.7 years. This means that we would expect 8.4 cm of precipitation to fall within a 12-hour period at that place once every 1.7 years, on average. Enter these numbers in the equation above and see what you get.

3. There are two inherent dangers with such approaches. First, it is extremely dangerous to extrapolate far beyond the observed historic data. How can we know what a 100-year event is if we only have 50 years of data? And second, people often forget that these numbers are just average return periods. If a 5-year event happened last year, that does not mean that another 5-year event will not occur again this year; it simply means that the magnitude would be *expected* to occur once every 5 years, on average.

Maps of Return Period

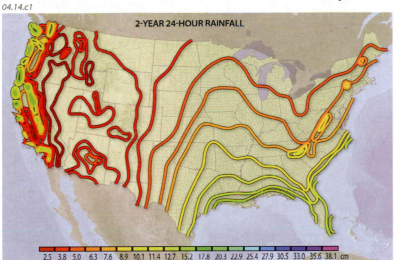

04.14.c1

2-YEAR 24-HOUR RAINFALL

2.5 3.8 5.0 6.3 7.6 8.9 10.1 11.4 12.7 15.2 17.8 20.3 22.9 25.4 27.9 30.5 33.0 35.6 38.1 cm

04.14.c2

10-YEAR 24-HOUR RAINFALL

2.5 3.8 5.0 6.3 7.6 8.9 10.1 11.4 12.7 15.2 17.8 20.3 22.9 25.4 27.9 30.5 33.0 35.6 38.1 cm

04.14.c3

100-YEAR 24-HOUR RAINFALL

2.5 3.8 5.0 6.3 7.6 8.9 10.1 11.4 12.7 15.2 17.8 20.3 22.9 25.4 27.9 30.5 33.0 35.6 38.1 cm

4. Maps of the magnitude of rainfall events for various return periods and durations have been produced. Three are shown here for a 24-hour duration, including one for a 100-year return period. The "100-year storm" is often of interest because many structures are designed to withstand the effects of a 100-year storm. Examine these maps for the area in which you live or have some interest. Consider how such maps might be used, such as when planning land-use near a river or how high to build a new bridge.

Before You Leave This Page Be Able To

☑ Describe the concept of drought, and sketch and explain three possible causes.

☑ Explain the concept of a return period, and detail how one is calculated.

4.14

4.15 What Caused the Recent Great Plains Drought?

PARTS OF THE GREAT PLAINS, stretching from Texas to Montana, experienced severe drought in 2012, resulting in huge crop losses and other problems. These losses occurred in spite of recent technological advances, such as improved soil and water management practices and the use of drought-resistant varieties of crops. Some effects, such as crop loss, resulted directly from the event, but others, such as less food for livestock, were indirect results of the drought. What caused this drought, and what were some primary and secondary impacts?

A What Was the Meteorological Setting of the Drought?

Examine each pair of maps. The top maps show averages of the daily pressure (as represented by the height of the 500 mb pressure) for May and July of 2012. The bottom pair of maps shows specific humidity for the same two months. Compare the air pressure maps to the specific humidity maps and consider the relationship between the patterns of air pressure and humidity, if any exist.

Air Pressure

In May (▶), the air pressure patterns and upper atmosphere wind directions over the lower 48 states were essentially west-to-east, a condition meteorologists call a *zonal flow*. Pressures over the Great Plains were neither much higher nor lower than areas to the east and west. For most places, a zonal upper-level pattern results in typical weather, because it neither supports nor suppresses storms. For Texas, a west-to-east zonal flow brings in dry air from northwestern Mexico and the Desert Southwest (e.g., Arizona).

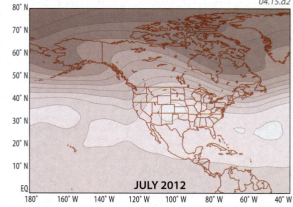

500mb GEOPOTENTIAL HEIGHT (m)
5440   5600   5720   5880

By July (◀), a strong ridge of high pressure developed over the Great Plains states, with the entire southern U.S. under high pressure (high 500 mb heights). The sinking air associated with this high pressure caused abundant adiabatic warming and compression, and little cloudiness and precipitation. High pressure also blocked or deflected storm tracks away from the region. As a result of these factors, drought extended across most of the central U.S., including much of the Great Plains.

Specific Humidity

These two maps show specific humidity (the mass of moisture in a mass of air) for the same two months as shown above. Darker shades of green represent more moisture, whereas lighter shades indicate drier air. During May (▶), dry air in the Desert Southwest (e.g., Arizona) and in the deserts of adjacent northern Mexico spread into west Texas, but moderately humid conditions existed in much of the Great Plains, such as in Kansas and Nebraska.

1000mb SPECIFIC HUMIDITY (g/kg)
5   10   14   19

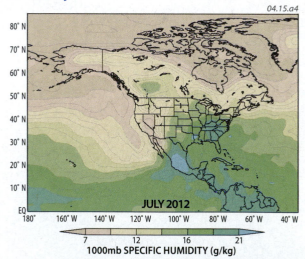

1000mb SPECIFIC HUMIDITY (g/kg)
7   12   16   21

The specific-humidity map for July (◀) is not too different from the one for May. The clockwise flow around the Bermuda-Azores High, which makes its northern-most approach in July, has advected moisture into the interior of North America. A conclusion you can reach from comparing the four maps on this page is that the upper-level conditions were very different, but the amount of moisture in the air was similar, although the patterns differ in detail.

B What Were the Patterns of Drought in the Middle of 2012?

The two maps below show the Palmer Drought Index for the same two months as depicted on the previous page. Examine the patterns of drought on these maps and compare them to the maps of air pressure and specific humidity. From comparing each set of maps, decide if the drought in the Great Plains was mostly caused by a lack of humidity or by upper-level air pressure and patterns of flow.

On these maps, areas experiencing severe and extreme drought are shown in orange and brown, respectively. Areas in light yellow are experiencing mid-range conditions, meaning not overly dry or overly moist. In May (▶), areas of the Great Plains with extreme drought are mostly restricted to isolated areas in northern Texas, eastern New Mexico, and especially southeastern Colorado. In the central and northern parts of the Great Plains, from Oklahoma to North Dakota (a north-south strip through the center of the map), conditions were in the mid-range.

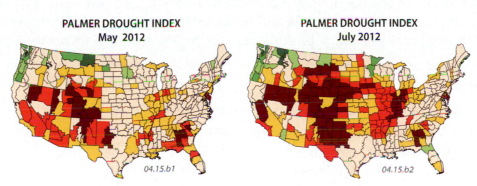

PALMER DROUGHT INDEX
May 2012

04.15.b1

PALMER DROUGHT INDEX
July 2012

04.15.b2

EXTREME DROUGHT	SEVERE DROUGHT	MODERATE DROUGHT	MID-RANGE	MODERATELY MOIST	VERY MOIST	EXTREMELY MOIST
-4.00 and Below	-3.00 to -3.99	-2.00 to -2.99	-1.99 to +1.99	+2.00 to +2.99	+3.00 to +3.99	+4.00 and Above

By July (◀), extreme and severe drought had spread across more of northern Texas, eastern New Mexico, and into parts of Kansas, Nebraska, Wyoming, and the Dakotas. Nearly all of the Great Plains was experiencing at least moderate drought (dark yellow, orange, or brown). From your comparisons of these maps with those of air pressure and specific humidity (on the previous page), what do you conclude about the main cause of the drought inflicted on the Great Plains in the summer of 2012?

Primary and Secondary Impacts of Drought

In considering the effects of drought, some effects, such as drying out of agricultural fields and dying of crops, are due directly to the drought; these are *primary impacts*. Other effects are related to drought, but in a more indirect way, like an increase in the price of seeds that resulted from the initial dying of crops; such effects are *secondary impacts*. Here, we examine the primary and secondary impacts of the 2012 drought, focusing on Texas, but similar consequences of the drought were felt in other states.

An obvious primary effect of the drought was damage to crops due to the extreme dryness of the soil and lack of rainfall. In Texas, half of the cotton crop was lost, and the corn harvest decreased by 40%. In a similar manner, it is estimated that up to 10% of the trees in Texas were damaged, but it is still too early to be certain about the actual amount of loss. Wildlife populations dwindled as their habitats were degraded. Ranching likewise suffered the effects of the drought, with a sharp decrease in the number of cows. Such losses were largely caused by a diminished supply of drinking water or an increase in water contamination as ponds and other sources of surface water dried up. Throughout the drought-stricken region, the supplies of sur-face water and groundwater decreased, with reservoirs estimated as 40% below normal.

There were also secondary impacts. The dry conditions caused grass and brush to dry out (a direct impact), causing a series of wildfires that burned over 3 million acres in Texas, causing more than $750 million in damage to homes (a secondary impact). Trees and plants stressed by the drought became more vulnerable to later disease and pest infestations. High temperatures and low water levels (primary impacts) led to low amounts of dissolved oxygen in some lakes, ponds, and streams, resulting in fish dying (a secondary impact). Reduced seed production altered the food chain, even for species that weren't affected by the drought directly. Shrinking supplies of agricultural products resulted in higher prices.

Other secondary impacts are even less obvious. Transportation was delayed or halted along highways because of wildfires. Tourism decreased as fewer people visited lakes because the low water levels left an unpleasant and potentially dangerous situation for boaters. Also, the amount of recreational fishing declined for the same reason and because of the previously mentioned fish kills. Reduced water supplies in cooling res-ervoirs used by power-generation plants caused a reduction in power output, at the same time that energy demand peaked due to the higher temperatures (more air conditioning needed). This resulted in rolling blackouts, providing an excellent example of a secondary impact. Geographers also use the concepts of primary and secondary impacts for other types of meteorological events, such as hurricanes and flooding. These terms are also considered in the context of global and regional changes that result from climate change and other events.

Before You Leave This Page Be Able To

✓ Summarize the patterns of upper-level air pressure and specific humidity that accompanied the 2012 drought in the Great Plains.

✓ Distinguish between primary and secondary impacts of an event, like drought, providing examples of each type of effect from the 2012 drought in Texas.

4.16

What Do Smoke Plumes Tell Us About Atmospheric Conditions?

THE INTERACTION OF MOISTURE and varying degrees of instability in the atmosphere produce clouds, fog, and precipitation. Clues about local moisture and stability characteristics can be gained simply by looking at evidence in the everyday world. Plumes of smoke or steam coming from the top of smokestacks provide such evidence. In this investigation you will observe smoke plumes and make inferences about the stability of the local atmosphere. Through this process, by simply observing a plume, you can be a reasonably effective weather forecaster.

Goals of This Exercise:

1. To recognize the variety of forms displayed by plumes of smoke and steam.
2. To correlate the shape and intensity of these forms to atmospheric moisture and stability conditions.
3. To use the evidence from plumes to predict the local weather conditions.

A What Are Some Possible Shapes of Plumes of Smoke and Steam?

When smoke or steam exits from the top of a smokestack, several things can happen. If there is no wind, the smoke or steam will likely simply rise and spread out, diffusing into the local atmosphere. If the wind is blowing, however, the plume will deflect sideways in a downwind direction. The path of the plume depends on how rapidly the plume is rising relative to how fast the wind is blowing. The degree to which the plume moves vertically depends on the stability conditions of the atmosphere, so the shape of a plume can be used to predict the weather. Five types of plumes can be used to assist in making these predictions. Observe each of the five types and then read the text associated with each.

04.16.a1 04.16.a2 04.16.a3

Fanning Plume—This type of plume spreads out horizontally, but not as much vertically. This form of plume suggests that the local atmosphere resists upward or downward motion.

Coning Plume—Some plumes spread out evenly up, down, and sideways, resulting in the shape of a sideways-oriented cone. In such plumes, the smoke shows no real preference for moving up or down.

Looping Plume—The plume shown above has the form of up-and-down curves or "loops." A "looping" form indicates that turbulence and vertical movement are occurring. The plume alternately encounters upward and downward rushes of air, with lulls in between.

04.16.a4 04.16.a5

Fumigating Plume—This form of plume expands downward, causing the area downwind to become fumigated with smoke or steam. A fumigating plume indicates that the air is preventing the plume from rising but allowing it to expand downward.

Lofting Plume—A lofting plume is the opposite of the fumigating plume, in that it expands upward but not downward. The atmospheric conditions are allowing the plume to rise and expand upward, but not to expand downward.

Note: Keep in mind that wind speed affects the plume's appearance. A strong wind would favor a fanning plume, as wind pushes the plume downwind faster than it can spread out. In contrast, a light wind would accentuate any looping characteristics of a plume. Also, the dispersion characteristics shown here are only representative of the atmosphere at the top of the smokestack. Conditions higher in the atmosphere might be different, and the plume could have a different shape if it were higher (or lower).

B What Do Plume Shapes Indicate About Stability of the Atmosphere?

The form of a plume provides clues about whether the atmospheric conditions favor rising of the air (an unstable atmosphere), sinking of the air (very stable atmosphere), or something in between. The relative stability or instability of the atmosphere in turn provides some constraints on how the saturated and unsaturated adiabatic lapse rates compare to the environmental lapse rate, as summarized below. After reviewing these considerations, match each of the plumes shown on the facing page with the most likely configuration of the environmental and adiabatic lapse rates, from the choices presented in the five graphs near the bottom of this page.

1. The graph, similar to ones presented earlier in the chapter, plots different possible environmental lapse rates associated with an adiabatically moving parcel of air. The unsaturated adiabatic lapse rate is shown by the red line—we are assuming an unsaturated atmosphere in this example. In green are four different environmental lapse rates with varying steepness. In the case of smokestacks, we are interested in comparing the unsaturated and environmental lapse rates.

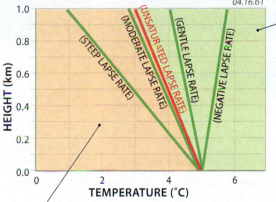

04.16.b1

2. If the environmental lapse rate plots to the left of the unsaturated adiabatic lapse rate (as in the orange area on this graph), the adiabatically rising air parcel is warmer than the surrounding environment at a given height, so the air continues rising and the atmosphere is *unstable*.

3. If the environmental lapse rate plots to the right of the unsaturated adiabatic lapse rate (the green area on this graph), the adiabatically rising air parcel is cooler than the surrounding environment at a given height. The air parcel will cease rising because it has a higher density than the surrounding air, and will tend to sink, so the atmosphere is *stable*. This is typical, for example, when there is a temperature inversion—warm air above colder air, a negative lapse rate and a very stable situation.

4. If the environmental lapse rate and unsaturated adiabatic lapse rate are about the same (i.e., the two lines plot close to one another as in the center of the graph), the atmosphere can be thought of as having a relatively neutral effect on rising or sinking air.

04.16.b2

The graphs below represent five different atmospheric conditions by comparing the environmental lapse rate (the green lines) with the unsaturated adiabatic lapse rate (the red lines). Examine each graph and consider what that graph indicates about atmospheric conditions—whether the air is unstable, stable, neutral, or features a temperature inversion. Then, determine the best match between a graph and the expected form of a plume under those conditions. Enter all your answers in the worksheet or online.

| Graph 1 | Graph 2 | Graph 3 | Graph 4 | Graph 5 |

04.16.b3-7

1. Graph 1 has an environmental lapse rate that plots to the left of the unsaturated adiabatic lapse rate. At all levels, adiabatically rising air will be warmer than the air around it.

2. Graph 2 shows an environmental lapse rate that is to the right of the unsaturated adiabatic lapse rate below the height of the smokestack, but plots to the left for levels above the height of the smokestack.

3. Graph 3 has the environmental lapse rate slightly to the left of the unsaturated adiabatic lapse rate below the height of the smokestack, but to the right for levels higher than the smokestack.

4. Graph 4 has the environmental lapse rate and the unsaturated adiabatic lapse rate nearly parallel and plotting very close to one another.

5. Graph 5 has an environmental lapse rate well to the right of the unsaturated adiabatic lapse rate. At all levels, adiabatically rising air will be cooler than the air around it.

4.16

Weather Systems and Severe Weather

OUR ATMOSPHERE IS DYNAMIC, featuring various types of weather systems, some relatively benign and beneficial, while others, like hurricanes and tornadoes, are dangerous and destructive. What types of weather do we experience, what causes weather, and what determines if a weather system is relatively gentle or severe? In this chapter, we explore the main types of weather systems on Earth.

05.00.a1 Eastern Colorado

05.00.a2 Pickens County, AL

05.00.a3 Tuscaloosa, AL

Powerful tornadoes are one of nature's most spectacular but frightening phenomena. A tornado forms beneath a thunderstorm as a swirling vortex of winds that can reach speeds of several hundred kilometers per hour. They can have a tapered, almost delicate shape, or can be thick, dark, and ominous columns filled with debris.

How do tornadoes form, how wide do they get, and how are they related to thunderstorms?

Tornadoes rampaged through the southeastern U.S. in April 2011, as part of the largest outbreak of tornadoes in history. Between April 25 and April 28, 358 tornadoes tore across Alabama and nearby states, completely destroying some houses, like the ones above, and killing more than 300 people.

How strong do tornadoes get, where do most occur, and what times of the day and year are they mostly likely to develop?

The tornadoes carved swaths of destruction, totally leveling parts of neighborhoods (▲), while leaving nearby houses essentially undamaged, except for uprooted and toppled trees, downed power lines, and broken windows. In addition to the destruction caused by the tornadoes, deaths and damage resulted from hail, lightning, and flooding.

How do hail and lightning form, and are they related to one another?

05.00.a4

05.00.a5

The tornado outbreak of late April 2011 was part of a large weather system, shown in the satellite image above. A swirling storm centered over the upper Midwest (in the upper part of the image) has a distinctive comma shape, with a long tail curving to the south and southwest. The tornadoes formed in the cluster of bright clouds to the southeast—out in front—of the main system.

What is a weather system, how does one form, and why do some acquire a curved shape, like a comma, or a coil?

Weather radar systems tracked each tornado as it moved across the region, as shown by this map depicting colorful lines (red signifies a very strong tornado) of tracks for each tornado. Some tornadoes stayed on the ground for hours and crossed multiple states. Radar images are a mainstay of daily weather reports.

How do radar systems work, what do they measure, and how do such data help us anticipate a storm's path and severity? Once a tornado or other severe storm is identified, how is the public notified?

TOPICS IN THIS CHAPTER

05.00.a6

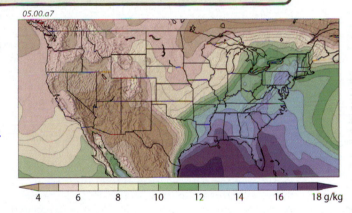
05.00.a7

A map of temperatures for the day of peak tornado activity (◄) shows a boundary between warm air (shown in green and yellow) in the southeast against cold air (shown in purple and blue) in the upper Midwest. A map of specific humidity (►) for that day indicates that the warm air was also humid, but the cold air was relatively dry.

Why are some masses of air warm, while others are cold? How are boundaries between warm and cold air related to thunderstorms and tornadoes?

05.00.a8

Regional wind directions during this time (◄) show distinct patterns, with winds blowing from the south to the north in the area affected by the tornadoes (such as Alabama and Georgia). This wind direction brought warm, moist air northward. The colors on this map indicate wind speeds, with yellow and green representing the fastest winds.

Do regional wind patterns help us predict which way a storm will move?

05.00.a9

Heavy precipitation was concentrated in the same region as the tornadoes (◄). In this map, purple and blue indicate the highest amounts of precipitation, which for these storms was as much as 40–50 cm (15 to 20 in.). The heavy rains, falling in a relatively short time period, produced severe flooding that destroyed roads and buildings and caused additional deaths.

Does precipitation influence the strength of tornadoes?

What Is Weather?

When we use the term *weather*, we are referring to *conditions in the atmosphere at some specific time and place*, whether it is right now, an hour ago, or sometime next week. Weather refers to the temperature, humidity, and windiness, and if it is fair or there is precipitation. It also refers to whether an area is experiencing severe storms, such as a thunderstorm, tornado, hurricane, or snow storm, or is in a time of exceptional dryness. Weather refers to the various aspects featured on these two pages, and much more! Examine the four maps on this page and figure out what the weather was like on April 27, 2011 where you live.

Weather affects all of our lives, from deciding what clothes to wear, what week to plant crops, or how soon to take shelter from some type of severe weather. In this chapter, we explore the fundamental processes that control weather and then examine the main types of weather phenomena, including rain and snowstorms, freezing rain, thunderstorms, hail, lightning, hurricanes, and, of course, tornadoes. Along the way, we explain features you can observe, like the types of clouds, to understand and predict the weather around you.

5.0

5.1 Why Does Weather Change?

THE ATMOSPHERE IS SO THIN that a single regional mass of cold or warm air can dominate the entire vertical extent of the troposphere. With time, such cold or warm air masses can expand or shrink, become colder or warmer, and can remain stationary or move great distances. In the narrow zone where the two air masses meet, weather conditions can change drastically across short distances and times.

A What Are Some Fundamental Controls of Weather?

Weather is controlled by the processes described in previous parts of this book, and in fact those processes were described to get to this point—an understanding of weather and climate. Weather is a response to variations of insolation and resulting energy imbalances, to large-scale motion in the atmosphere, and to the important role of atmospheric moisture, including latent heat. Here, we briefly review the large-scale circulation patterns. Only the Northern Hemisphere is shown below, but similar patterns are also present in the Southern Hemisphere.

The most important features that control weather are the huge circulation cells that essentially divide each hemisphere into thirds (by latitude, not by area). Areas at high latitudes, poleward of approximately 60°, are under the effect of the *polar cell* and *polar easterlies*. High pressure dominates the pole.

The edge of the polar cell is marked by the *polar front jet stream*, a belt of high-speed winds that encircle the globe from west to east, high above the surface. The surface in this area has changeable and stormy weather due to rising air. Bends in the polar front jet stream are called *Rossby waves*, which exert a strong control on many weather systems.

A second jet stream, called the *subtropical jet stream*, encircles the globe (also west to east) high above the surface near 30° latitude. Between this jet stream and the equator, near-surface winds blow from the northeast, forming the *northeast trade winds*. Along the *equator*, high amounts of insolation-related heating cause rising air and associated low pressure. Winds converge from both sides of the equator, forming the *Intertropical Convergence Zone* (ITCZ).

Between the two jet streams, in the mid-latitudes and part of the subtropics, the general circulation is from the west—the *westerlies*. There is not a well-established, upper-level circulation cell over this region.

Between the subtropical jet stream and ITCZ is the *Hadley cell*, fed by rising air in the ITCZ. The cell descends in the subtropics, causing the *subtropical highs* (high pressure), with their associated normally dry conditions.

05.01.a1

B What Is an Air Mass?

The pattern of global circulation helps form large bodies of air, called *air masses*, that have relatively uniform temperatures and moisture over large horizontal distances. We identify different types of air masses according to their temperature, moisture, and geographic characteristics.

The air-temperature map to the right depicts a cold air mass, represented by purple and blue, over the northern part of North America, including the northern Great Plains. A warmer air mass exists farther south, covering most of the southern and eastern U.S. The boundary between the two air masses is along the steep color gradient (a large change in color over a short distance) between the purple and blue-green areas.

	Dry	Humid
Cold	**A, cP:** Stable, often with surface inversions; fair weather	**mP:** Unstable, often with cold air overlying warmer air adjacent to oceans
Warm	**cT:** Unstable, with hot surface and cooler air aloft, but lack of moisture	**mT, E:** Often unstable, with abundant moisture, especially E

The classification system for air masses uses a two-letter abbreviation. The first letter is lowercase and signifies the moisture characteristics, where "c" is for continental (dry) and "m" is for "maritime" (humid). The second letter is uppercase and refers to the temperature characteristics: "A" for Arctic (or Antarctic), "P" for polar (somewhat warmer than Arctic), "T" for tropical, and "E" for equatorial. An "A" air mass is inherently continental (dry), because water in such areas is often frozen. An "E" air mass is inherently maritime, because so much of the air at equatorial latitudes is over or near a source of water.

December 25, 2012

| -33 | -23 | -13 | 0 | 7 | 17 | 27 C |
| -27 | -10 | +8 | 32 | 44 | 62 | 80 F |

05.01.b1

Continental Air Masses

Hot, dry air masses, typically occurring over land, are continental tropical (cT) air masses. Note that cT air masses occur in the Desert Southwest of North America, southeastern South America, the Sahara and Arabian Deserts, Tibet, and central Australia. In areas under the influence of cT air masses, conditions can be oppressively hot and dry for many months during the year.

Arctic or Antarctic (A) air masses are formed near the North and South Poles. Not only are they extremely cold, but they are also very dry because the water-vapor capacity is low in such cold environments. Cold, dense air has a tendency to sink, and therefore, these air masses reinforce the region of polar high pressure (polar high) that resides over each pole. Surface winds near the pole move east-to-west, so the surface *A* air masses tend to be steered westward as they drift away from the poles. The paths of *A* air masses can also be affected by curves in the upper-level polar front jet stream (Rossby waves), which can steer an *A* air mass toward the northern U.S. or into central Eurasia. These are the "Arctic blasts" that bring very cold air into the northern U.S.

Continental polar (cP) air masses are located farther away from the poles than *A* air masses. They are cold and dry, but less so than *A* air masses. cP air masses form over cold, inland areas, like northern Canada, Siberia, Mongolia, and north-central Europe. The southern edge of cP air masses is the polar front. In the winter, cP air masses can move significant distances equatorward, bringing cold air into the U.S. and southern Eurasia. In the summer, cP air masses are not as cold and do not reach as far south.

05.01.b2

Maritime Air Masses

Due to their proximity to the equator, *equatorial (E) air masses* are very warm and moist. The high temperature and moisture support instability to great heights in the atmosphere, resulting in tropical clouds and storms. Storm development is aided by the general circulation of the atmosphere, which in equatorial regions promotes rising of the air, especially along the ITCZ.

Maritime tropical (mT) air masses form over warm, oceanic regions. mT air masses are warm and moist, and they have influence over much of the Earth because so much of the world (particularly in the low latitudes) is covered by ocean. For example, mT air tends to make most of the Atlantic Ocean and adjacent eastern U.S. warm and humid in summer. mT air masses are commonly near the subtropics and under the high pressure associated with this region.

Maritime polar (mP) air masses are cool and humid. They are not as cold as *cP* air masses because oceanic regions are not as cold as continental areas at the same latitude, particularly in winter. mP air masses are a cold, damp influence in places like Seattle and most of western Europe, as the westerlies push the cool, damp air eastward from their oceanic source region.

05.01.b3

Any air mass acquires characteristics of the region where it forms—its *source region*. Air masses that form inland or over frozen ocean areas (e.g., *cP*) will be dry, whereas air masses that form over oceans (e.g., *mT*) will be humid. Air masses that form in the low latitudes will generally be warmer than those that form over high latitudes. When an air mass migrates from its source region across a different region (◄), its temperature and moisture will be modified, which can in turn affect the lapse rate, stability of the air, and how much mixing occurs within the air mass.

5.1

5.2 What Are Fronts?

THE NARROW ZONE separating two different air masses is called a *front* and is often the site of rising atmospheric motion. Whenever different air masses meet along a front, the less dense, warmer one will be pushed up over the more dense, colder one. If the rising air cools to its dew-point temperature, cloud formation will begin, perhaps followed by precipitation. There are several types of fronts, depending on the manner in which one air mass is displacing the other and whether the front is moving.

A What Processes Occur Along Cold and Warm Fronts?

1. The map below shows several weather fronts as decorated lines, among the areas of high pressure (H) and low pressure (L). The thin gray lines are isobars (lines of equal pressure), and the background of the map is a satellite image showing cloud cover. Examine each of the features on this map and note any correspondence between the fronts, high and low pressure, and cloud cover.

2. The blue lines represent *cold fronts*, the leading edges of cold air masses. Blue triangles along a cold front indicate the direction that the cold air mass is pushing. The east-facing triangles on the long cold front in the center of the U.S. indicate that the cold front is pushing eastward. Cold fronts are usually associated with *cP* but can also occur along the leading edge of an *A* or *mP* air mass.

3. The red lines represent *warm fronts*, the leading edges of warm air masses. The red semicircles along a warm front indicate the direction in which the warm air mass is pushing. The warm front over northwestern North America is moving east. Warm fronts are usually associated with *mT* or *cT* air masses.

05.02.a1 Nov. 11, 2011

6. To the north is a north-moving warm front. Air south of the warm front (like at "D") is warm and less dense, so it gets wedged over the more dense, cooler air north of it (at "C"). The gentle slope of this rising motion causes a shield of stratiform clouds over eastern Canada.

5. Warmer air ahead of the cold front (such as at B) moves up and over the colder air, creating cumuliform cloud cover. The air west of the cold front (at "A" for example) is cold and dense, so it tends to sink, associated with increasing surface pressure (H). The pressure gradient is very steep, as reflected by closely spaced isobars near the cold front.

4. Let's examine the weather in the eastern half of the U.S., beginning with the long cold front. As the front and associated mass of cold, dense air move east, it displaces and mechanically forces up the less dense warmer air that is east of the front.

Cold Fronts

The figure below shows a cross section through a cold front along line A-B on the map above.

The cold air mass displaces the warm air mass. Cold air is more dense and forces the warm air upward. If the rising warm air cools to its dew-point temperature, cloud cover and perhaps precipitation occur along the front.

05.02.a2

Warm Fronts

The figure below shows a cross section through a warm front along line C-D on the map above.

As the warm air mass follows and displaces the cool air mass, it slides up over the colder air. If the rising air cools to the dew-point temperature, stratiform clouds and perhaps precipitation occur along the warm front, but spread out over a significant width.

05.02.a3

Comparing Warm and Cold Fronts

Compare the cold-front and warm-front figures, observing similarities and differences, then read the text below.

The slope at which a warmer air mass is pushed upward along a cold front is much steeper than the slope at which it is forced upward along a warm front. This tends to support cumuliform cloud formation along cold fronts, but stratiform cloud cover along warm fronts. As a result, precipitation along cold fronts tends to be intense, localized, and short-lived, whereas precipitation along warm fronts tends to be light, widespread, and long-lived. Use these tips the next time a weather front crosses your hometown.

B What Is a Stationary Front?

Sometimes, for a period of several hours or days, neither the cold air mass nor the warm air mass is displacing the other. There is a temporary stall in motion of the front, and so such a stalled front is a *stationary front*. On a weather map, a stationary front, such as occurs here along the Atlantic Coast of North America, is represented by alternating blue and red line segments with alternating blue triangles and red semicircles. The blue triangles point away from the cold air, and the red semicircles point away from the warm air, as shown in the cross section here.

05.02.b1 Nov. 14, 2012

05.02.b2

On this stationary front, the blue triangles point east, so the cold air is trying to push east and must therefore be on the opposite side of the front (on the west side). The warm air is therefore on the east side. A band of clouds and precipitation is along the front, mostly on the cold air side (west of the front). Since a stationary front is not moving, prolonged precipitation can occur as the front sits over the same place for an extended time.

C What Types of Weather Are Characteristic of Cold, Warm, and Stationary Fronts?

1. On this map, examine the weather fronts, locations of high and low pressure, and the underlying satellite image showing the distribution of clouds. Find a cold front, warm front, and stationary front and for each one sketch a simple cross section across the front, showing which way it is moving (or if it is not moving). After you have done this, read the rest of the text around the figure.

2. For the storm system in the southwest, the blue line marks a cold front that is moving to the south. The red line is a warm front moving to the northeast, but a warm front slopes opposite to the way it is moving (see the cross section on the previous page), to the southwest in this case.

05.02.c1 Dec. 26, 2012

4. In the east is an east-west oriented red line that marks a warm front. The semicircles point north, so this is the way the front is moving.

5. In the southeast, the blue line is a cold front. The cold air is on the northwest side, and the front is moving to the southeast. The warm front and cold front join at an area of low pressure, marked by the red L. Symbols for the four types of fronts are shown below. We discuss occluded fronts later.

05.02.c2

COLD FRONT   WARM FRONT

STATIONARY FRONT   OCCLUDED FRONT

3. The red and blue line along the mountains in the west-central part of the country is a stationary front. The cold air is trying to move to the west, so it must be on the opposite side (east). Based on the cross section above, the front slopes away from the cold air.

6. This figure (▶) focuses on the low-pressure area in the east where the cold front (blue) meets the warm front (red). In this situation, the two fronts divide the region into three fairly distinct areas, called *sectors*. We reference each sector by the temperature of the air in that region relative to the two other sectors. The *cold sector* is behind the cold front and naturally has the coldest air.

05.02.c3 Dec. 26, 2012

7. The air in front of the cold front is the warmest and so is called the *warm sector*. The counterclockwise rotation around the low is bringing air north from lower latitudes and from oceans in the warm sector. The third region is intermediate in temperature and is referred to as the *cool sector*.

8. Clouds along the cold front are clumpy and probably cumuliform, but clouds in other areas are less clumpy and so are probably mostly stratiform.

Before You Leave This Page Be Able To

✓ Identify cold, warm, and stationary fronts on a weather map and suggest what types of air masses could be on each side of the front.

✓ Sketch a cold, warm, and stationary front in cross section, including typical cloud types and the weather likely to be associated with each front.

5.2

Where Do Mid-Latitude Cyclones Form and Cross North America?

WEATHER FRONTS generally do not exist in isolation, but typically form, migrate, and fade away as part of a larger system called a *mid-latitude cyclone*. Cold and warm fronts usually trail from a central core of low pressure — a *cyclone*. While popular culture uses the term "cyclone" to refer to a tornado or other form of wind storm, a cyclone is really any enclosed area of low pressure. Mid-latitude cyclones migrate across Earth's surface guided by large-scale atmospheric circulation, like the polar front jet stream and the westerlies.

A What Surface and Upper-Level Conditions Form Mid-Latitude Cyclones?

Cyclones originate in various places as long as conditions in the lower and upper atmosphere are favorable for cyclogenesis. Commonly, however, they develop downwind of mountain ranges or just offshore of cold land. In addition to forming near certain surface conditions, cyclogenesis is favored by flow in the upper atmosphere, especially along bends in the polar front jet stream — Rossby waves.

Leeward Side of Mountains

As air moves downslope on the lee side, it becomes stretched vertically as the depth of the atmosphere increases. In doing so, it acquires an increasing tendency for counterclockwise spin, or *vorticity*, wherever cold and warm air masses meet. In the U.S., westerlies cause cyclones to form east of large mountain ranges.

05.03.a1

Offshore of Cold Land

Cyclones also form over water, offshore of cold land. This can occur during the colder times of the year, when the coastal land has cooled more rapidly than the adjacent water. Rising air over the ocean forms an area of low pressure just offshore. This, plus winds blowing toward the pole, cause the cold-warm interface along the coast to initiate potentially violent storms because of the abundance of moisture.

05.03.a2

Polar Front Jet Stream and Rossby Waves

1. A jet stream can develop bends, much like a meander in a river. In the Northern Hemisphere, a trough in the polar front jet stream is a bend to the south and a ridge is a bend back to the north. As the fast-moving air passes from a trough to a ridge it usually accelerates.

05.03.a3

2. As the air spreads out laterally, the decrease in pressure sets up an upward flow of air from lower in the atmosphere to the part of the jet stream that is diverging. As a result, an area of low pressure forms at the surface and begins rotating, forming a cyclone. Times when the Rossby waves are most accentuated (curved) are favorable for the formation of mid-latitude cyclones.

3. The map on the right shows how Rossby waves appear on a map showing upper-level air pressures. To show upper-level pressures, we commonly use a *constant-pressure map*, which plots lines, called *isohypses*, to represent the height (in meters) that you would have to go up in the atmosphere in order to reach a certain pressure. Higher heights correspond to higher pressure aloft. In this map, darker is low heights (pressure) and light shading is high heights and pressure. The dark, nearly circular feature is an area of low pressure, or a *trough*.

4. This map shows isohypses for 300 mb of pressure, high enough that air at this height only experiences approximately 30% of the atmospheric pressure compared to sea level (an average of 1,013 mb). A 300 mb pressure is up about 9,000 m (9 km), at the heights where the jet streams operate. The scale for the map is given here (▶), with heights in meters.

05.03.a4

Upper-Level Map

Sept 30

Geopotential Height at 300m

| 8950 | 9200 | 9400 | 9650m |

05.03.a5

5. Cyclone B is on the flank of the deep trough of low pressure, on the trough-to-ridge side of a Rossby wave. In this position, an upward flow of air into the upper atmosphere helps strengthen the cyclone below. The cyclone will move to the northeast, continuing to be strengthened.

6. Note that the pressure gradient is weaker near cyclone A. There is some lateral spreading (divergence) of the air, indicated by the shape of the isohypses, north and east of A, where the isohypses spread apart. Remember that air at these heights generally flows parallel to lines of equal pressure (equivalent to isohypses), so if isohypses diverge or converge, the airflow does too. The divergence at the 300 mb level helps draw air up from the surface, strengthening the cyclone. In the Northern Hemisphere (the case shown here), air overall flows from west to east (left to right).

B What Are Common Tracks for Mid-Latitude Cyclones Across North America?

There are several common origins and paths of cyclones, influenced by the position of mountains and coastlines. Once formed, cyclones are guided by large-scale patterns of atmospheric circulation and so follow similar paths—or *storm tracks*—across the surface. The combinations of specific sites of formation and common tracks produce storms with similar characters, paths, and associated weather, like those shown for North America below.

1. Low pressure off the west coast of North America, in the Gulf of Alaska, can spawn cyclones that bring cold, moist air from the ocean inland across Canada and the Pacific Northwest. These storms can deliver heavy rain along the coast and snow at higher elevations. Due to orographic effects, precipitation is especially heavy along west-facing slopes of coastal mountains.

2. Cyclones also form on the leeward sides of major mountain ranges, like the Canadian Rockies in British Columbia and Alberta. A storm formed here commonly incorporates cold Arctic air and is called an *Alberta Clipper*. Similar storms form east of the southern Rocky Mountains in or near Colorado, but are not as cold.

05.03.b1

3. Wintertime lows also form in the Gulf of Mexico, and are called *Gulf Lows*. They arise because of the contrast between the cold land and the waters of the Gulf, which remain fairly warm.

5. Note that although the storm tracks converge near New England, when the storm systems arrive, they are often older, having spent most of their latent energy before they track to the New England area. So, New England has many cloudy days but less severe weather than places experiencing less mature storms.

4. During winter, low pressure commonly occurs over the Atlantic Ocean, often near Cape Hatteras in the Outer Banks of North Carolina. Storms formed at this general area, called *Hatteras Lows*, move northward where they are also called *Nor'easters*. They are challenging to forecast because a slight variation in track can mean the difference between a foot or more of snow and none at all in the heavily populated areas of the Northeast.

6. The maps below show a surface cyclone, located at the low-pressure area labeled "L." The bottom map is two days later than the top map. The lines are 300 mb isohypses and are color coded to indicate their position with regard to the upper-level trough in the center of the top map. On these maps, note the correspondence of the surface cyclone and upper-level patterns.

7. The maps below show weather maps of the same cyclone (and an anticyclone to the west as the blue "H"), with fronts depicted with standard symbols for warm, cold, stationary, and occluded fronts. The lines represent surface air pressures.

8. The region shown is governed by west-to-east winds (westerlies) and movement of the Rossby waves, so the cyclone and anticyclone both generally move toward the east. However, the cyclone turns northeast and the anticyclone turns southeast, both moving parallel to the upper-level isohypses (left pair of maps). Mid-latitude cyclones will generally track northeastward, following the trough-to-ridge side of the Rossby wave. Accelerating and diverging upper-level winds on these sides will move the surface cyclone toward the northeast and strengthen it. Surface anticyclones often move southeastward.

Before You Leave This Page Be Able To

✓ Explain how surface features and upper-level flow contribute to mid-latitude cyclogenesis.

✓ Sketch the preferred cyclone tracks across North America, and explain why and when these cyclones occur.

✓ Briefly describe how upper-level air pressures and flow can guide and strengthen a cyclone.

5.3

5.4 How Do Mid-Latitude Cyclones Move and Evolve?

ONCE FORMED, MID-LATITUDE CYCLONES migrate across the surface and commonly evolve through a series of steps, due to the way that winds circulate around the area of low pressure. Circulation around the cyclone will steer the trailing fronts in a counterclockwise direction (in the Northern Hemisphere), while west-to-east motion of the westerlies and Rossby waves shift the entire storm system from west to east. Here, we examine the life stages of a mid-latitude cyclone as it moves west-to-east across North America.

A What Are the Three Stages in the Life Cycle of a Mid-Latitude Cyclone?

The map on the left shows several fronts as decorated lines, areas of high and low pressure, and isobars as thin lines. A mid-latitude cyclone begins when a relatively cold air mass and a warmer one meet along a frontal boundary. Here, cold and warmer air masses meet along a stationary front labeled "Cyclone A."

Formation (Cyclogenesis)

At some point along this boundary, a small amount of surface convergence occurs because of local circulation features that push one air mass into the other, because of topographic influences, or by some other local mechanism. This convergence causes the low pressure to intensify.

This satellite image shows clouds and precipitation at the same time as represented by the weather map on the left. Note that the clouds and precipitation are associated with both cyclones on this map (cyclone A over Texas, and cyclone B over New England).

Maturity

At a later time, the low pressure has moved east and intensified with a stronger pressure gradient, causing more surface air to move toward it. As this air begins to converge from all directions, the Coriolis effect causes the air to have an apparent deflection to the right. This results in counterclockwise flow that pushes cold air southward, turning one segment of the stationary front into an eastward-advancing cold front.

On the other side (east of the low), warm air pushes northward, forming a north-moving warm front. The movement of the two air masses causes a clear bend in the front, an indication that cyclone A has matured into a mid-latitude cyclone. Note that cyclone B, now east of the continent, has a more curved shape than cyclone A, an indication that it is even more mature.

Both cyclones are visible in the satellite image above. Although not obvious on this image, cyclone A now has distinct cold, warm, and cool sectors of air. The "cold" sector is behind (west of) the cold front, whereas the "warm" sector is behind (south of) the warm front. The cool sector is in front of (north of) the warm front—neither the warm front nor the cold front have passed through this sector. On the satellite image, dense cumuliform clouds (along the south edge of the image) form along the cold front, while indistinct stratiform clouds form along the warm front and in the cool sector. Advancing cold air can easily displace warmer air, but the warmer air has more difficulty moving the cool air; with time the cold front advances faster than the warm front, as in cyclone B.

Occlusion

05.04.a5

Cyclone B

Cyclone A

Sea Level Pressure
10/01/2012

05.04.a6

1. With time, the wester-lies push both weather systems toward the east or northeast. Eventually, the cold fronts will catch up with, and even overtake, the warm fronts as both flow counter-clockwise around the low-pressure areas. When this happens, the warmer air mass is lifted above the surface, because it is less dense than both the cold air and the "cool" air. This process is called *occlusion*, and a front that has experienced occlusion is an *occluded front*. In occlusion, the moisture is moved counter-clockwise around the low, so that the heaviest precipitation may be northwest of the low.

2. Occluded fronts, like that near cyclone B on the map above, are purple on weather maps with the triangles and semicircles pointing in the direction that the system is moving. Though many mid-latitude cyclones never occlude, occlusion often, but not always, signals the end of the mid-latitude cyclone's life, because by this time the cold air has warmed and the warm air has cooled. Some occluded fronts can continue to strengthen for some time.

3. Notice on the satellite image that the sky over the occluded front is filled with clouds. The cumuliform cloudiness associated with the cold front more or less merges with the stratiform clouds associated with the warm front. Occluded fronts, however, can have a wide variety of cloudy features.

4. This figure (▶) contains three cross sections that capture different stages in the process of occlusion. Occlusion begins nearest to the low-pressure core because there the cold front has the least distance to catch up to the warm front, and it proceeds away from the low over time. To visualize what occurs, move through the three cross sections, which represent different places, but also successive stages in the occlusion process.

F   D
COOL
E   C   B
WARM
COLD
A

05.04.a7

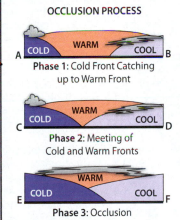

OCCLUSION PROCESS

A | COLD | WARM | COOL | B
Phase 1: Cold Front Catching up to Warm Front

C | COLD | WARM | COOL | D
Phase 2: Meeting of Cold and Warm Fronts

E | COLD | WARM | COOL | F
Phase 3: Occlusion

5. In this cross section, the cold front is moving to the east and the warm front is moving to the north, but the cold front is moving faster and so it is getting closer to the warm front.

6. Here, the cold front has caught up with the warm front, trapping the warm, moist air above the mass of cold air and the mass of cool air, which are in the process of joining together.

7. At segment EF, the two air masses are fully beneath the warm air, lifting it higher. As the warm air is forced up, it cools, forming clouds and precipitation. The air north of the warm front is cool but not cold because it is blocked from receiving a fresh supply of cold air from the air mass to the west.

B   How Do the Great Lakes Interact with Cyclones?

The Great Lakes in the north-central part of the U.S. are famous for interacting with winter storms, producing intense snowfall called *lake-effect snows*. The largest amounts of lake-effect snow occur after mid-latitude cyclones and associated cold fronts have moved eastward across the Great Lakes, and higher pressure behind them begins to approach the area. The circulation over the water causes moist air to flow onto land and dump huge amounts of snow on land south and east of the lakes, including the cities of Syracuse and Buffalo in upstate New York.

Moist Air

05.04.b1

As a cyclone moves across one of the lakes, it encounters water temperatures that are somewhat warmer than the air or the surrounding land. This is largely due to water changing temperature more slowly than air or land from fall to winter. In winter, especially early winter, the Great Lakes are much warmer than the air above them or the land beside them. This causes steep environmental lapse rates, destabilizing the atmosphere and strength-ening a cyclone passing over the Great Lakes.

Having flowed across the Great Lakes, the moist, destabilized air is pushed inland by an approach-ing anticyclone, depositing snow on adjacent downwind land, especially where topography, such as the mountains of upstate New York, cause orographic enhancement.

Before You Leave This Page Be Able To

✓ Sketch and describe the three stages in the life cycle of a mid-latitude cyclone (cyclogenesis, maturity, and occlusion).

✓ Sketch and explain how and why the process of occlusion occurs, using a map or cross section to illustrate the progression with time.

✓ Sketch and describe what causes lake-effect snows in the Great Lakes region.

5.4

How Do Migrating Anticyclones Form and Affect North America?

ANTICYCLONES ARE ROTATING ZONES of high pressure. Most migrate across Earth's surface, guided by large-scale atmospheric circulation, but some stay or reappear in the same general area year after year. As migrating anticyclones approach, pass over, or leave an area, they cause changes in wind direction, temperature, and moisture. They are typically associated with non-stormy weather, but they sometimes cause fog and an increase in smog. They can block the motion of cyclones or guide them one way or another, and they influence our weather in other ways.

A Where Do Anticyclones Occur?

There are two types of anticyclones: *semipermanent anticyclones* that remain in the same general area for entire seasons and year after year, and *migrating anticyclones*. Both types are zones of high pressure associated with clockwise rotation in the Northern Hemisphere and counterclockwise rotation in the Southern Hemisphere, opposite to the rotation of cyclones. Unlike cyclones, no fronts trail from anticyclones because the sinking air doesn't allow for cooling, cloud formation, and the release of latent energy needed for generating strong winds. So pressure gradients around anticyclones are generally weak, as weak pressure gradients are associated with light winds. Where cyclones tend to bring stormy, unsettled weather, anticyclones bring atmospheric stability and fair weather.

05.05.a1

This map shows two surface anticyclones, each shown as a blue "H," and two surface cyclones, each marked by a red "L." The polar front jet stream is represented by large blue arrows. The lines are isohypses at the 300 mb level, and so depict *upper-level* pressures. The gray shaded areas show *surface* air pressures, with darker gray indicating lower pressures. Each cyclone is associated with an area of low *surface* air pressure.

The cyclones are beneath the east sides of the upper-level troughs (on the trough-to-ridge side of a Rossby wave). Note the positions of the anticyclones. In both cases, the anticyclones are beneath the east side of an upper-level ridge as defined by where the 300 mb isohypses bend toward the pole. That is, the anticyclones are beneath the ridge-to-trough side of a Rossby wave.

Since the positions of the surface anticyclones are set to some degree by the position of the Rossby waves, they can remain relatively fixed as long as the ridges in Rossby waves remain in about the same location. In this way, the positions of some anticyclones remain broadly fixed year after year, but they may strengthen, weaken, or shift laterally from season to season. These are the semipermanent anticyclones, and they include some familiar high-pressure areas, like the Bermuda-Azores High, the Hawaiian High, and the Siberian High. Just as there are these semipermanent anticyclones, there are *semipermanent cyclones*, like the *Aleutian Low* and the *Icelandic Low*. Which pressure feature is the northernmost low-pressure zone shown on this map?

B Where Do Mid-Latitude Anticyclones Commonly Form?

High-pressure zones associated with anticyclones form in different settings than cyclones. Once formed, most mid-latitude anticyclones migrate in the direction of large-scale atmospheric circulation (west to east), all the while rotating clockwise (in the Northern Hemisphere). Many anticyclones result from convergence in Rossby waves.

Where upper-level air moves from the ridge to the trough of a Rossby wave, the air converges and slows down slightly. This in turn forces some of the air toward the surface, causing an area of surface high pressure below the convergence. The air around this high pressure rotates, clockwise in the Northern Hemisphere and counterclockwise in the Southern Hemisphere. In this manner, a surface anticyclone forms below the ridge-to-trough part of a Rossby wave. As it descends, the air heats up adiabatically.

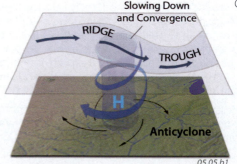

05.05.b1

Once formed, the mid-latitude anticyclone will migrate across the surface, guided by regional wind directions, such as the westerlies, and by the upper-level air circulation patterns. As we observed previously, anticyclones commonly move parallel to the isohypses, moving east and southeast across the U.S. Zones of high pressure can be formed in other ways, such as by descending air in the subtropical limb of the Hadley cell, forming subtropical high-pressure areas.

C How Do Anticyclones Typically Migrate Across North America?

Migrating mid-latitude anticyclones can cross the entire continent, driven east by the prevailing westerlies. They form and are likely to grow beneath the ridge-to-trough side of the Rossby wave, and they tend to migrate beneath this side of the wave as well, in a general east-southeast direction.

1. The maps below track the path of a surface anticyclone (shown with a blue "H" and clockwise rotation) during the course of several days. The shades of gray represent surface air pressure, with lighter gray indicating high pressure and darker gray indicating lower pressures. The yellow lines are isohypses at the 300 mb level, depicting upper-level conditions. The map also shows several surface cyclones (each as a red "L"), but our focus here is on the anticyclone. The legend is to the right (▶).

Geopotential Height at 300mb
(Rossby Waves)

05.05.c1

SEA LEVEL PRESSURE
992.5   1005   1015   1027.5 mb

2. In this first map, the anticyclone is located over western Wyoming, and a cyclone lies just to its southeast. There is a strong pressure gradient between the two, as implied by the rapidly changing shades of gray, resulting in strong 300 mb winds, implied by the tightly packed isohypses in the area.

3. On this second map, the anticyclone has moved south. The clockwise circulation around the high causes a different wind direction on each side of the high. In this relatively dry region of the continental interior, high pressure generally means clear skies.

4. In this map, the anticyclone has moved to south Texas as the ridge overlying it flattened out. On this, and the previous map, the anticyclone is on the east flank of an upper-level ridge—that is, the surface anticyclone is beneath the ridge-to-trough side of a Rossby wave.

D What Changes in Weather Occur When an Anticyclone Migrates Over an Area?

An anticyclone is characterized by descending air, so it tends to bring stability to the atmosphere, resulting in clear skies and fair weather, or fog under certain circumstances.

1. Air descending (▶) in the atmosphere is generally cooler, but warming. The clear skies, however, allow for more insolation during the day, warming the land and air. The same clear skies also allow much of this heat to radiate into space at night, resulting in cool nights and a large temperature contrast between day and night.

05.05.d1

2. The moisture in the surface air may be cooled and trapped by this descending motion, causing fog in certain settings. The downward flow can trap smog and other air pollution near the surface, especially if there is a temperature inversion.

3. The clockwise circulation can pin cold air against mountains when the cold air is too dense to move uphill. This often occurs on the western side of the Appalachian Mountains, and, if temperatures are cold enough, often creates temperature inversions with sleet, freezing rain, and snow.

4. As an anticyclone approaches an area (▶), winds will change direction depending upon the direction of approach (usually from the west or northwest). In front of an eastward progressing high (on the right in this view), cold air blows in from the north. After the high passes, warm winds blow from the south. North or south of the high, winds blow east or west. By employing this pattern, we can use wind directions to assess the larger weather picture. This approach works for cyclones, too, but the wind directions are opposite.

05.05.d2

Before You Leave This Page Be Able To

✓ Sketch and explain the relationship between mid-latitude anticyclones and upper-level features, including Rossby waves.

✓ Summarize how most anticyclones migrate across North America.

✓ Describe the weather associated with anticyclones and how wind directions change when an anticyclone migrates over an area.

5.5

5.6 What Conditions Produce Thunderstorms?

THUNDERSTORMS ARE COLUMNS of moist, turbulent air with variable amounts of rain, strong wind, lightning, and hail. They are perhaps the most fundamental of all organized weather systems. They can provide needed rainfall for crops, but they can be accompanied by severe weather. We begin by examining the formation, growth, and decay of an individual thunderstorm that is independent of larger storm systems—the *single-cell thunderstorm*—and then explore the formation of thunderstorms that occur in clusters or in large, consolidated masses.

A   What Stages of Development Characterize Single-Cell Thunderstorms?

Cumulus Stage

1. The initial *cumulus stage* of develop-ment of a single-cell thunderstorm occurs when the surface heats more rapidly than the atmo-sphere above it. This causes the heated near-surface air to

05.06.a1

rise relative to adjacent air, forming an *updraft*. Usually, such conditions begin to appear in mid-to-late morning, particularly in summertime. If the air is moist, one or more isolated *cumuliform* clouds will begin to form, starting out as a cumulus cloud. No rain is falling during this stage—the cloud is still growing, as indicated by the puffy, cauli-flower-shaped top. Most developing cumuli-form clouds never get past the first (cumulus) stage, either because the atmo-sphere is not sufficiently unstable, or is not moist enough, or both.

Mature Stage

2. The beginning of rainfall signals onset of the next stage—the *mature stage*. At this stage, the updrafts that allowed the cloud to grow in the cumulus stage are accompa-nied by downdrafts induced by falling

05.06.a2

precipitation. Most *lightning* occurs during this stage. If the thunderstorm is too tall, its top can flatten against the tropopause, where the very stable conditions in the temperature inversion of the stratosphere forbid higher vertical develop-ment. The strong upper-level winds produce an anvil-shaped cloud top. Evaporation of some cloud water drops cools the downdrafts, particularly on hot summer afternoons. Upon hitting the surface, the downdraft will move laterally in the direction of the winds that are pushing the storm, producing the cool wind gusts we feel before a storm.

Dissipating Stage

3. Eventually, enough rain falls to cool the surface, removing the unstable conditions that caused rising motions. This *dissipating stage* is characterized by continued rainfall

05.06.a3

until much of the moisture is rained out of the cloud. At this point, the cloud begins to disappear, taking on a ragged, feathery appearance, as it mixes with dry air from outside the cloud and evaporates into the surrounding air. Water in the cloud that does not precipi-tate may eventually evaporate into the surrounding drier air.

05.06.a4 Nova Scotia, Canada

4. The entire life cycle of an individual single-cell thunderstorm (▶) is on the order of 1–2 hours, but some last much longer, and several single-cell thunderstorms can occur in sequence. Unless they occur in association with larger systems like a mid-latitude cyclone, single-cell thunderstorms are far more likely to occur in summer than in winter. They are also far more likely to occur sometime between late morning and late afternoon than during any other time of day. Longer-lived and more severe thunderstorms can occur if conditions are especially favorable, such as abundant moisture or some other atmospheric phenomenon that enhances and strengthens the thunderstorm. Such thunderstorms can cause extensive damage, as described on subsequent pages.

Cloud Mergers

5. A single-cell thunderstorm can form or grow when two cumuliform clouds are forced together by winds. In this figure, adjacent fair-weather cumuliform clouds are supported by uplift associated with an

05.06.a5

unstable atmosphere, but no cloud has enough moisture and uplift to be dangerous by itself. When the clouds collide, the instability and associated uplift beneath the collision point dictate that the moisture from both clouds must move upward. This causes very rapid vertical development and a sudden storm that may be difficult to anticipate.

05.06.a6

Upper-Wind Effects

6. A thunderstorm that forms beneath strong winds aloft, such as the polar front jet stream, has an increased likelihood for development or intensification. Any increase in horizontal wind speed with height is called *vertical wind shear*. The fast winds in the top part of a tall thunder-storm, driven by the jet stream, may allow some of the rain from the storm to fall in areas separated from the updrafts. This reduces the possibility that the downdrafts associated with the falling precipitation will get in the way of, and cool, the updrafts that are sustaining the system. If vertical wind shear allows the precipitation to remain separated from the zone of uplift, the surface remains warm, allowing updrafts and overall atmospheric instability to continue for an extended time. This in turn allows the thunderstorm to persist and possibly grow stronger, increasing the possibility of more severe weather.

B What Conditions Form Large Thunderstorms or Clusters of Thunderstorms?

Thunderstorms can also occur in clusters, called a *multi-cell thunderstorm*, or can coalesce into a much larger single storm, called a *supercell thunderstorm*. Clusters of thunderstorms can form along a weather front or when the individual storms are sufficiently close to reinforce and strengthen one another. In either case, the thunderstorms can cause severe weather.

1. This figure (▶) shows two thunderstorms that are close enough to affect one another. In well-organized thunderstorms, downdrafts from one thunderstorm can contribute to updrafts in an adjacent thunderstorm. This happens because the downdraft of cold, dense air spreads out when it hits the surface, pushing warmer air upward, thereby strengthening a nearby thunderstorm or even triggering a new one. The downdrafts and resulting outflow can cause warm air ahead of the outflow (A) to be displaced upward to such an extent that it triggers another thunderstorm, or even a row of thunderstorms, as shown in the radar image to the left. These types of systems, where down-drafts coming from a line of storms produces another line of storms, proceed in a "conveyor belt" fashion and can affect huge regions. These types of storms are called *multi-cell thunderstorms*.

05.06.b1
Enhanced Uplift at "A" Causes Cloud Growth
A

05.06.b2 Binghamton, NY

05.06.b3
cP
cT
mT

2. Lines of thunderstorms commonly form along cold fronts, especially where the two air masses have very different moisture characteristics. This often happens as hot, dry (*cT*) air masses from the west interact with warm, moist (*mT*) air masses from the southeast (◀). This interaction can be driven by movement of a cold front and mid-latitude cyclone, but both of these air masses will be in the warm sector of the cyclone, as shown here. Such air masses meet at a typically north-south boundary called the *dry line*. Along the dry line, the more humid air mass will be lifted up and over the drier air mass, assuming that both are at the same temperature, because humid air is less dense than dry air. The resulting storms can have an ominous front (▶), a *squall line*, and so are called *squall-line thunderstorms*. They can be very severe due to the abundant moisture in the humid air mass, often even more severe than the storms generated by the subsequent passage of the cold front.

05.06.b4 Belleville, KS

3. Some single-cell thunderstorms develop into *supercell thunderstorms* (▶), tall and powerful storms. Like other large thunderstorms, supercell thunderstorms commonly have a flat-topped, anvil shape. The long point of the anvil generally indicates the direction the storm is moving (left to right in the figure below), because the strong upper-level winds push the entire surface storm in the same direction that they push its top. The anvil indicates the presence of vertical wind shear (increasing wind speed with height), which assists storm development by enhancing uplift and allowing precipitation to fall away from the updrafts.

05.06.b6 Yellowstone NP, WY

4. Intense rain and hail form within supercells and fall toward the ground, often in brief but heavy downpours. Note that hail can form in other types of storms, too. *Lightning* and *thunder* are also present in most supercells. Lightning may discharge in the air or reach the ground.

7. As in multi-cell storms, outflow from downdrafts can be very strong and can be the catalyst of other storms. In the case of a supercell, the system is so strong that much of the available latent energy released in condensation and deposition is consumed in the large and long-lasting supercells. Such storms can last several hours, be 20–50 km in diameter, and support more catastrophic severe weather than other systems.

05.06.b5

5. Supercells feature a rotating vortex called a *mesocyclone*. All supercells have at least one mesocyclone, and some supercells may support several. The rotating vortex is interpreted to have started out horizontal, where it was formed by horizontal shearing of the wind against the surface. Later, updrafts within the thunderstorm distort the vortex into a steep orientation.

6. Subsequent *updrafts* can spiral into the mesocyclone, strengthening it by tightening its rotation and ultimately leading to formation of a tornado. About one-third of mesocyclones form tornadoes.

Before You Leave This Page Be Able To

☑ Sketch and explain the processes that occur during the life cycle of a single-cell thunderstorm, and what can enhance them.

☑ Describe the formation of thunderstorms as part of multi-cell or supercell thunderstorms.

5.7 Where Are Thunderstorms Most Common?

THUNDERSTORMS CAN OCCUR as isolated single-cell thunderstorms, forming over mountains and other local features, or can be embedded in larger weather systems, like mid-latitude cyclones or multi-cell thunderstorms. By knowing how thunderstorms form, can you predict where they should be most common?

A How Does Thunderstorm Frequency Vary Globally?

Space-based detections of lightning can be used to determine the global distribution of thunderstorms. Optical sensors on satellites use high-speed cameras to look into the tops of clouds that the human eye cannot detect. Observe the map below, which shows the average yearly counts of lightning flashes per square kilometer based on data collected by NASA satellites.

1. These data were collected between 1995 and 2002. Places where fewer than one flash occurred (on average) each year are gray or purple. Places with moderate numbers of flashes are green and yellow, and places with the highest frequency of lightning strikes are red or black. Observe the main patterns before you read the associated text.

2. The globes show that lightning strikes are much more common on land than over the oceans. Large land-masses can heat up during the day and cause free convection, resulting in thunder-storms. Above the oceans, thunderstorms are more frequent near coastlines than in areas far from a coast; this reflects the sharp contrasts in air masses that occur along coastlines.

05.07.a1

4. North America exhibits frequent thunderstorms, reflecting a supply of warm, moist air from the Gulf of Mexico, abundant solar heating of the land, and the frequent passages of weather fronts. Europe commonly lacks much contrast between air masses, as the westerlies push a continual parade of *mP* air masses across the continent. Also, the mountains mostly trend east-west, parallel to the westerlies, minimizing the interaction between very hot and very cold air masses. As a result, Europe has fewer thunderstorms than might be expected, considering the frequency of frontal passages and rainy days.

3. Most of the 2,000 or so thunderstorms occurring at any time in the world are located within the Intertropical Convergence Zone (ITCZ), especially in the afternoon. Abundant tropical moisture, strong surface heating, and vigorous trade-wind convergence are responsible. Note the high density of thunderstorms in central Africa, attributable to the ITCZ and a large continent. ITCZ-related thunderstorms are also abundant in southern Central America, Southeast Asia, and Indonesia (northwest of Australia).

05.07.a2

05.07.a3

5. Polar regions, including Antarctica, are too cold to have moisture sufficient for the condensation and deposition required to support thunderstorms, even in the summer. Both poles are among the places least likely to experience thunder-storms and lightning.

Lightning Flashes/km²/Year
0.1   0.4   1.4   5   20   70

Seasonal Variations

6. These maps show the variations in lightning (thunderstorms) between summer and winter. The top map is for winter in the Northern Hemisphere and summer in the Southern Hemisphere. The bottom map shows the opposite (northern summer; southern winter). Examine these maps and compare the number of thunderstorms in summer versus winter for different regions.

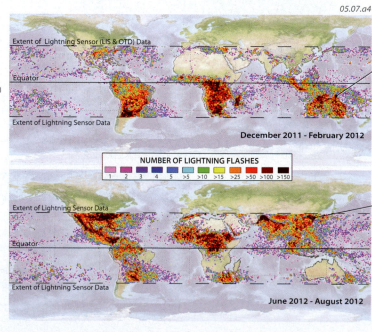

05.07.a4

Extent of Lightning Sensor (LIS & OTD) Data

Equator

Extent of Lightning Sensor Data

December 2011 - February 2012

NUMBER OF LIGHTNING FLASHES
1   2   3   4   5   >5   >10   >15   >25   >50   >100   >150

Extent of Lightning Sensor Data

Equator

Extent of Lightning Sensor Data

June 2012 - August 2012

7. From December through February, there are many more thunderstorms south of the equator than north of it. Note, for example, the difference in Australia and Southeast Asia between the two maps. During this time, thunderstorms are also abundant in southern Africa and in South America.

8. During June through August, the locus of thunderstorm activity shifts to the Northern Hemisphere, where it is summer. The difference in thunderstorm activity between the two maps is especially noticeable in North America and Central America. Thunderstorms in mainland Asia are tied closely to the summer monsoon, when surface heating combines with moist onshore flow to create strong convection. Central Africa, straddling the equator, reports thunderstorms during both seasons. Sinking air in the descending limb of the Hadley cell, combined with limited moisture, suppresses the development of thun-derstorms in the Sahara and Arabian Deserts.

B How Does Thunderstorm Frequency Vary Throughout the Continental U.S.?

1. This map shows the frequency of thunderstorms in different parts of the U.S., as expressed by the number of days per year that an area has thunderstorms.

2. Few thunderstorms occur on the Pacific Coast, even though there are many rainy days. Surface heating is seldom strong enough to create free convection. The difference between air masses across frontal boundaries is too slight to generate vigorous uplift. Cold air masses are not bitterly cold, and warm air masses are not too hot due to the moderating influence of the Pacific Ocean. Orographic uplift can generate an occasional thunderstorm, but even these are not too frequent as most of the coastal ranges are not very high and the highest ones are farther north, next to cold ocean waters.

3. Thunderstorms are somewhat more common elsewhere in the West, even though moisture is relatively scarce. More thunderstorms occur on the peaks of the highest mountains, where surface daytime heating moves air upslope from all sides, converging at the peaks and forcing additional uplift.

4. In the Midwest and Northeast, thunderstorms increase in frequency from north to south. Water-vapor capacities in the northern states are so low most of the year that they limit the supply of atmospheric moisture available for release as latent heat during condensation, which is essential to generate severe storms.

05.07.b1

ANNUAL THUNDER DAYS

	<5
	5–10
	11–20
	21–30
	31–40
	41–50
	51–60
	61–70
	>71

5. The Southeast has the highest frequency of thunderstorms in the U.S., with the Florida peninsula experiencing the most thunderstorms in the country. Winter and early spring thunderstorms mostly originate from cold fronts trailing from Gulf Lows and Colorado Lows. Spring, summer, and, to some extent, fall thunderstorms occur because of convective afternoon thunderstorms in an unstable atmosphere. When the surface heats up and air rises, a "double sea breeze" causes onshore flow from both the Gulf of Mexico and the Atlantic to fill in the low pressure over Florida. Late summer and fall also bring tropical cyclone-induced thunderstorms. The region has a combination of warm seawater, warm air, a sea-land contrast, and a position near storm tracks.

Seasonal Variations

These maps show the frequency of thunderstorms for different months. In March and April, thunderstorms occur inland along the Lower Mississippi Valley because cold and warm air masses come into contact with each other at strong cold fronts, accompanied by moisture from the Gulf of Mexico.

During May, the peak in Texas results from the combination of early summer convection on some days with late season cold-front activity on others.

In June, the location of thunderstorm activity shifts farther north and inland, onto the Great Plains. Land surfaces are starting to warm up, but cold air is nearby in Canada and sometimes flows into the plains. Cyclonic activity enhances the likelihood of thunderstorms in the central U.S. Thunderstorm activity begins to pick up in Florida, with the arrival of summer.

MAR

APR

MAY

JUN

SEP

AUG

JUL

05.07.b2

	<0.5		6.5–8.4
	0.5–2.4		8.5–12.4
	2.5–6.4		12.5–15.4
			>15.5

The September peak in Florida is the result of tropical cyclones enhancing the convectional thunderstorm activity. Hurricane season peaks in August and September. The rest of the southeastern U.S. has the same ingredients for thunderstorm formation as Florida, except for the "double sea breeze."

Most of the U.S. sees a summer peak in thunderstorm days, so activity picks up in July and August. By July, surface heating allows for destabilization of the atmosphere during afternoon hours. Summer cold fronts supplement these convectional thunderstorms in the Northeast and north-central states, while tropical cyclones (of strengths varying from tropical waves to hurricanes) can add to the frequency of thunderstorms in the Southeast. Increased thunderstorm activity in the Southwest and intermountain states of the West largely reflects the onset of the Arizona monsoon.

Before You Leave This Page Be Able To

☑ Summarize variations in thunderstorm frequency around the world, explaining changes with season and from land to oceans.

☑ Describe which parts of the U.S. experience frequent thunderstorms, explaining the factors influencing the frequency and seasonality.

5.8 What Causes Hail?

HAIL IS A BALL OF ICE that forms under freezing temperatures within a cumulonimbus cloud and that subsequently falls toward the surface. Large hail that reaches the surface can do so at high enough speeds and with enough force to smash windows, dent cars, and destroy entire fields of crops. How does hail form, what controls the size of individual chunks of hail, and where and when is hail most common?

A How Does Hail Form?

Most hail forms in association with a supercell thunderstorm, and much less commonly with other types of thunderstorms. The photograph to the right shows pea-sized pieces of hail, or *hailstones*, and the figure below illustrates the interesting way in which hail forms. Examine the figure and then begin reading clockwise from the lower left as we follow a single hailstone through the storm. Hail, lightning, and thunder are commonly associated with one another, so the thunderstorms that cause these phenomena are commonly called *thunderheads*.

05.08.a2 South Mtns., AZ

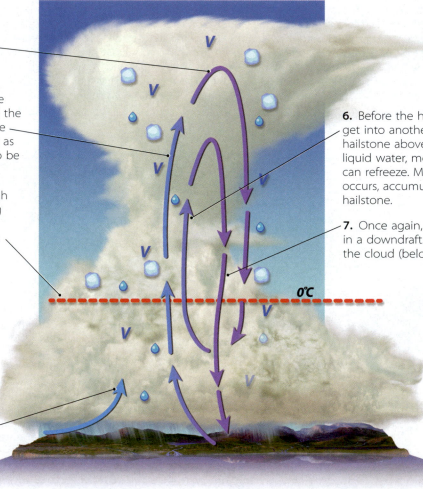
05.08.a1

4. Eventually the hailstone becomes influenced by a nearby downdraft, shown here as the beginning of a downward-arcing flow path.

3. As the central part of the storm continues to rise, more vapor deposition occurs and the hailstone becomes larger. The hailstone will continue rising as long as it is not too heavy to be lifted by the updraft.

2. Once air rises high enough that it enters the subfreezing part of the cumulonimbus cloud (above the red line on this figure), the Bergeron process causes water vapor (shown as a "v" on the diagram) to deposit onto an ice particle, causing it to grow. The resulting hail begins as a small more-or-less-spherical object.

1. Supercell thunderstorms and other strong cumulonimbus clouds have strong updrafts, capable of lifting raindrops, chunks of ice, and even planes unfortunate enough to fly into the storm. Such updrafts are associated with the mature stage of a thunderstorm, when air rises against gravity because of a very unstable atmosphere.

5. If the downdraft causes the hailstone to be carried below the freezing height (the red line), the outer surface of the ice can partially melt. At this stage, a core of ice is coated with a layer of liquid water. While temperatures are above freezing, more liquid water can coat the hailstone.

6. Before the hailstone can fall to the ground, it might get into another updraft. After the updraft lifts the hailstone above the freezing height, the coating of liquid water, most of which was probably melted ice, can refreeze. More deposition of vapor onto ice occurs, accumulating another ring of ice around the hailstone.

7. Once again, the developing hailstone gets caught in a downdraft. After falling to the warmer parts of the cloud (below the red line), more melting occurs.

8. This cycle can repeat many times—freezing and deposition of ice (via the Bergeron process) onto the hailstone above the freezing level, alternating with melting of an outside layer of ice when the hailstone is below the freezing level. Each trip up and down results in another layer of ice added to the hailstone. Eventually the developing hailstone becomes too heavy to be lifted by the updraft, and gravity pulls it to the surface. The more vigorous the updraft and the greater the instability, the larger the hailstone will be before it is finally pulled to the ground by gravity. The larger the hailstone and the faster it hits the ground, the more force it has and the more damage it can cause.

9. A number of factors influence the process of hail formation. For example, the freezing level is higher in summer than in winter. Hail formation is most likely at the time of year when the cloud extends substantially both above and below freezing. If a cloud is entirely warmer than freezing (which is unlikely), it will only produce rain. If the entire cloud is below freezing temperatures, it will produce snow. Other important factors include the amount of moisture, vertical extent of the cloud, the strength of the updrafts and downdrafts, and the temperature contrast between air masses across a front.

B Where and When Does Hail Tend to Occur?

Hail is associated with thunderstorms, so its distribution across the globe and in the U.S. is similar to the distribution of thunderstorms. As in the case of most other forms of severe mid-latitude weather, the U.S. experiences more than its share of hail compared to other locations on Earth. These maps show the geography and seasonality of hail in the contiguous U.S. Observe the general patterns shown on the large map below, and then examine the patterns for a certain month in the series of smaller maps. Try to explain the main patterns before reading the associated text.

This map represents the average number of days per year when a county experiences hail. From the legend, we can see that some areas average less than one day of hail per year, whereas others average more than 10.

Frequencies are low on the Pacific Coast, where contrasts between air masses are typically weak. Frequencies are also low in most of the interior of the West, which is typified by relatively dry air.

Frequencies are slightly higher in a north-south band through Arizona, which experiences thunderstorms during a summer monsoon.

Hail occurs most frequently over the Great Plains, in the center of the map, in a wide, north-south band stretching from central Texas to the Canadian border (and beyond). The average size of hailstones is larger over the Great Plains than anywhere else in the U.S., and perhaps anywhere else in the world. Most damage from hail is from larger hailstones (e.g., >5 cm), and the Great Plains often experiences such hail sizes.

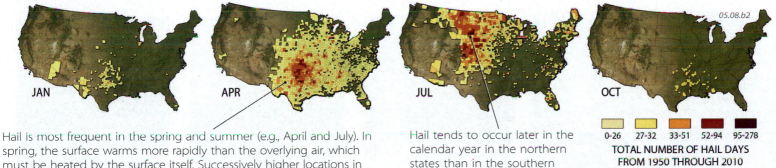

1 – 2
2 – 4
4 – 7
7 – 15

MEAN ANNUAL NUMBER OF HAIL DAYS

05.08.b1 Data: 1950-2010

Frequencies in the eastern U.S. are intermediate and spotty in distribution. The size of hailstones generally becomes smaller toward the coast.

Florida and areas along the Gulf Coast have frequent thunderstorms, but low to intermediate frequencies of hail. Here, thunderstorms are seldom tall enough and well-organized enough to form hail. Furthermore, temperature contrasts between air masses is usually not as great as over the Great Plains.

Seasonal Variations

The four maps below show hail frequencies for four different months, each representative of a season (winter, spring, summer, and fall, from left to right). How does hail vary from season to season?

JAN   APR   JUL   OCT

05.08.b2

0-26   27-32   33-51   52-94   95-278

TOTAL NUMBER OF HAIL DAYS FROM 1950 THROUGH 2010

Hail is most frequent in the spring and summer (e.g., April and July). In spring, the surface warms more rapidly than the overlying air, which must be heated by the surface itself. Successively higher locations in the atmosphere require longer to heat. This sets up a steep environmental lapse rate, in which the surface can be quite warm while the atmosphere is cold. This destabilizes the atmosphere, setting the stage for the cumulonimbus clouds that are required for hail to form. In fall, the pattern is reversed as the surface cools more rapidly than the air above it, and the atmosphere is seldom so unstable.

Hail tends to occur later in the calendar year in the northern states than in the southern states. This is no surprise because the boundary between the cold and warm air masses migrates northward as summer approaches. In fall (October) and winter (January), the surface is simply too cold in most of the U.S. to support the atmospheric instability needed for frequent cumulonimbus clouds and intense uplift.

Time-of-Day Variations

This graph plots average frequency of hail versus time of day for five cities. What do you observe? All of these cities show a strong tendency for most hail in the afternoon, when the ground has warmed enough to destabilize the atmosphere. These patterns are very similar to those for severe thunderstorms.

05.08.b3

St. Louis, MO
Indianapolis, IN
Columbus, OH
Birmingham, AL
Nashville, TN

HAIL FREQUENCY 1950-2010

1AM   6AM   12PM   6PM   12AM
TIME OF DAY

Before You Leave This Page Be Able To

✓ Sketch and explain the formation of hail within a cumulonimbus cloud, identifying all essential components.

✓ Explain the factors contributing to the geography, seasonality, and time-of-day variations for hailstorms in the continental U.S.

5.8

5.9 What Causes Lightning and Thunder?

LIGHTNING IS A DANGEROUS but intriguing weather phenomenon. It results from electrical currents within a storm cloud, releasing tremendous amounts of energy that can strike the ground. In the U.S., more than 80 deaths per year are due to lightning. Although this number is much less than those killed simply by extreme heat and extreme cold, lightning is responsible for more deaths than any other severe weather in the U.S.

A How Is Lightning Generated?

Lightning is a sudden and large discharge of electrical energy generated naturally within the atmosphere, typically expressed as a bolt of electricity that flashes briefly and then is gone. Most lightning occurs as *cloud-to-cloud* lightning or occurs within a single cloud, but about 25% of lightning strikes the ground (called *cloud-to-ground lightning*). Some lightning is less discrete, appearing to form diffuse sheets of lightning, called *sheet lightning*.

05.09.a1

1. Most lightning occurs as a single discrete bolt that can be relatively straight or extremely irregular with many obvious branches (◄). Lightning is associated with cumulonimbus clouds, the kind that form thunderstorms (►). Ice particles in the upper part of such clouds have a positive electrical charge, but the charge is mostly negative in the lower cloud where there is no ice. This charge separation is believed to happen because of the Bergeron process and different physical properties of ice particles and liquid water, but the exact process is unknown. The charge imbalance between the top and bottom of the cloud sets in motion a specific sequence of events, depicted in the series of figures below.

05.09.a2 West Central, IL

2. As a result of this charge imbalance, there is an excess of positive charges in the top of the cloud relative to the bottom of the cloud. Usually the Earth has a negative charge too, but as the cloud passes over a place, the negatively charged cloud base induces a positive charge directly below, and for several kilometers around, the cloud.

3. Recall that conduction requires energy transfer from one molecule to another, and therefore is most efficient in metals and other conductive materials. Air is a poor electrical conductor, so the flow of electricity through air cannot occur until the electrical potential becomes great enough to overcome the insulating effect of air.

05.09.a3

4. Lightning starts with negative charges near the base of the cloud. The negative charges begin to form an invisible channel of ionized molecules through the air, from the base of the cloud toward the positively charged ground. As the channel develops, it separates into a series of branches, all still invisible, and repeatedly starts and stops, advancing in a zigzag path, resembling a staircase. For this reason, it is called a *stepped leader*.

Stepped Leader

5. As the negatively charged stepped leader approaches the ground, positively charged streams of energy, called *streamers*, move upward to join the leader, commonly reaching up from the highest object on the ground, such as a tree, pole, building, or other tall object.

6. When the streamer makes contact with the stepped leader, it establishes a pathway of ionized molecules through which electricity can flow between the cloud and the ground.

Streamer

7. Once the streamer and stepped leader meet, an electrical connection is established and electrons flow from the base of the cloud to the ground, but this flow is still invisible.

8. As the negative charge approaches the ground, the positive charge begins to flow upward in what is called a return stroke. The return stroke begins near the surface.

9. The return stroke is nearly instantaneous, lighting up the sky with the discharge of energy. This is the main, bright flash we observe and recognize as lightning.

10. Once the electrical pathway is established, negative charges flow from the cloud to the ground, continuing to light up the sky with no obvious interruption after the return stroke.

11. In most cases, there is then a short pause, after which a new negatively charged leader, called a dart leader, starts down the established channel from the cloud to the ground.

12. The discharge of the dart leader causes another flash of light, distinctly after the initial flash. Lightning strikes can have several or several dozen flashes from dart leaders.

Thunder

◄ *Thunder* results from rapid heating and expansion of the air along the path of the lightning bolt. Around a lightning discharge, the air temperature approaches 15,000–30,000°C, which is much hotter than the surface of the Sun. The extreme heat causes this air to expand quickly away from the lightning stroke, as shown by the red arrows on this figure.

The expansion creates a momentary vacuum around the stroke. Almost instantly after the expansion, air rushes in from all sides toward the location where the lightning stroke passed. This is depicted by the green arrows. The sound of the matter in the various air streams colliding is what makes the sound of thunder.

05.09.a4

Using Thunder to Estimate Distances to Storms

Lightning travels at the speed of light. At about 300,000 km/s (186,000 mi/s), it can be considered instantaneous. But thunder only travels at the speed of sound (1,225 km/hr, or 760 mi/hr). Since there are 3,600 seconds in an hour, this means that the sound of thunder travels about 1.7 km or 1 mile in 5 seconds. So if you start counting when you see lightning, and can count to five before you hear the thunder, you are one mile away from the lightning. But beware: lightning does not necessarily flow straight down the middle of a storm. The strike could be farther ahead or behind the cloud from which it originated.

Elves, Sprites, and Jets

Intense lightning can cause a flat disk of dim reddish light, called an elve, to form about 96 km above the Earth. The light of an elve radiates outward in every direction, spreading over a huge area of sky.

Immediately after a very energetic bolt of lightning strikes the ground, ghostly red lights, called red sprites, may shoot straight up from the top of a thunderhead. Some red sprites soar up to 100 km into the atmosphere.

Blue jets are dim, blue streaks of light. They look like quick puffs of smoke that burst out of a thunderhead, arc upward, and then fade away. Blue jets can climb as high as 30 km into the atmosphere.

05.09.a5

Safety Tips Around Lightning

Lightning is extremely dangerous. The map below shows the number of lightning-related fatalities for each state from 1959 to 2011. How dangerous are states with which you are familiar? What factors contribute to this spatial pattern? How at risk are you for lightning?

Conduction of electricity is easiest in solids and most difficult in gases. Therefore, direct contact with any solid object, particularly a tall one that conducts electricity well, (such as a tall metal post) is extremely dangerous in a thunderstorm. Tall objects minimize the distance that the electrical energy needs to travel through a gas (air) during lightning.

A house or other enclosed building offers the best protection from lightning. Open shelters on athletic fields, golf courses, and picnic areas provide little or no protection from lightning. When inside a building, stay away from windows and doors, and avoid contact with plumbing, corded phones, and electrical equipment. Basements are generally safe havens, but avoid contact with concrete walls that may contain metal reinforcing bars. A car is a relatively safe place, because the metal body and frame carry the electrical current if struck.

If no safe shelter is available, stay away from trees and other tall objects, and crouch down with your weight on your toes and your feet close together. Lower your head and get as low as possible without touching your hands or knees to the ground. Do not lie down!

Consider the 30-30 rule: if the time between seeing a flash and hearing the thunder is 30 seconds or less, the lightning is close enough to hit you, so seek shelter immediately. After the last flash of lightning, wait 30 minutes before leaving a shelter.

05.09.t2

WHEN THUNDER ROARS **GO INDOORS!**

NUMBER OF LIGHTNING FATALITIES BY STATE, 1959–2011

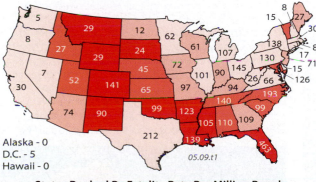

Alaska - 0
D.C. - 5
Hawaii - 0

05.09.t1

States Ranked By Fatality Rate Per Million People

| 1–10 | 11–20 | 21–30 | 31–52 |

Before You Leave This Page Be Able To

☑ Sketch and explain how and why lightning and thunder form, and describe the factors that set the stage for their formation.

☑ Summarize some strategies for remaining safe from lightning and the reason for using each strategy.

5.9

5.10 | What Is a Tornado?

A TORNADO IS THE MOST INTENSE type of storm known, expressed as a column of whirling winds racing around a vortex at speeds up to hundreds of kilometers per hour. The May 1999 tornadoes in Oklahoma City, Oklahoma, were estimated to have wind speeds over 500 km/hr. Fortunately, most tornadoes are less than several hundred meters wide, so they usually only damage narrow areas, but larger tornadoes can remain on the ground for hours and create a path of near-total destruction across one or more states.

A | What Is a Tornado and How Does One Form?

A *tornado* is a naturally occurring, rapidly spinning vortex of air and debris that extends from the base of a cumulonimbus cloud to the ground. Tornadoes typically have diameters of tens to hundreds of meters, but the largest ones can be 2–3 km wide. Most individual tornadoes remain on the ground for minutes to tens of minutes and have paths on the ground of less than a few kilometers. Some very rare, long-lived tornadoes track over 100 km, last a few hours, and destroy everything in their direct path.

05.10.a1

◄ This photograph shows the characteristics of a well-developed tornado, with the classic shape of a downward-tapering funnel. The tornado is visible because of the condensation of moisture, expressed as water drops, within the vortex. A brown cloud of debris picked up from the ground typically surrounds the base of the tornado, extending out farther than the vortex of strongest winds. As in the movies, if you are within the debris field, you are in trouble. Some tornadoes have wider and less elegant shapes, like a stubby, thick pillar.

05.10.a2

05.10.a3

Formation of a Tornado

1. The formation of a tornado is complicated and not fully understood. Different sets of circumstances are present in different tornadoes, but tornado formation generally begins with horizontal winds moving across the surface. Wind speeds are usually higher aloft and decrease downward due to friction with the surface. This difference in wind speeds causes a *vertical wind shear* within the wind column. Wind shear can result in a rotating, subhorizontal vortex (▲). Vertical wind shear generates a type of rotation known as *mechanical turbulence*, a normal occurrence, which by itself does not lead to severe weather.

2. In an unstable atmosphere, rising air (an updraft) can perturb the rotating tube of turbulent air (◄). With enough lifting, the tube will become more vertically oriented (►). As the tube is lifted, it generally splits into several short segments, some of which may even be rotating in opposite directions.

05.10.a4

3. Eventually a segment of mechanical turbulence may become vertical, at which point it merges with the original and unstable updraft. Once it spans the distance from the cloud to the ground, it is a tornado. If a funnel-shaped cloud does not reach the ground, it is a *funnel cloud*. In the Northern Hemisphere, the parent storm system often rotates counterclockwise, so the tornado also spins counterclockwise, except in a few rare cases. As a tornado tightens its rotation, its wind speeds increase, making it a stronger, more powerful, and potentially more dangerous storm.

4. This figure (►) shows the essential aspects needed to form a tornado. The main ingredient is a well-developed cumulonimbus storm cloud, especially one associated with a supercell thunderstorm. A favorable regional setting is another prerequisite, such as a mid-latitude cyclone that puts cold air adjacent to much warmer, moister air. The position and motion of the cyclone are in turn controlled by such factors as the position of the polar front jet stream.

5. Another necessary ingredient is vertical wind shear near the surface, caused by higher speed winds aloft and slower winds near the surface. Mountains and other topographic irregularities tend to disrupt a regular wind-shear pattern, which is one reason why tornadoes are more common in areas that have relatively low relief (little topography) than in areas with mountains.

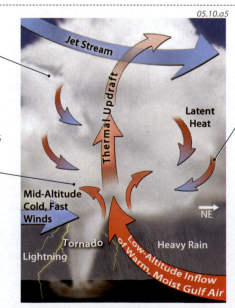
05.10.a5

Jet Stream

Thermal Updraft

Latent Heat

Mid-Altitude Cold, Fast Winds

NE

Tornado
Lightning

Heavy Rain

Low-Altitude Inflow of Warm, Moist Gulf Air

6. Formation of a tornado—and its host thunderstorm—requires an unstable atmosphere. Without it, there will not be enough lifting to form the cloud and to realign a vortex into a vertical orientation.

7. A storm must have strong updrafts in order to lift the horizontally spinning tube of air. The host storm must also possess strong internal rotation. Such rotation, which can be observed on weather radar, is one aspect weather forecasters use to gauge whether or not a severe thunderstorm is likely to spawn tornadoes. Storms that spawn tornadoes commonly have a *hook echo* pattern on radar (▼).

05.10.a6 Lacombe, '

B What Weather Systems Produce Tornadoes?

Tornadoes are related to strong thunderstorms, which form in a number of settings. Most tornadoes occur in association with multi-cell or large supercell thunderstorms, which are generally associated with cold fronts. Others occur embedded in tropical cyclones or with more isolated single-cell thunderstorms.

Associated with Cold Fronts

05.10.b1

Most catastrophic tornadoes are associated with multi-cell or supercell thunderstorms of the Great Plains and Southeast, usually in the warm sector of cold fronts. On the map above, a cold front (in blue) stretches south from a cyclone (marked by an "L") over the Great Lakes. In front of the cold front are yellow lines showing the tracks of tornadoes. These tornadoes formed in association with *squall-line thunderstorms* along the dry line. Such tornadoes would more likely occur in the afternoon and early evening when thunderstorms are most active.

Tropical Cyclones

05.10.b2 Hurricane Isaac, 2012

Tropical cyclones, like hurricanes, spawn tornadoes too, but a large percentage of these are not terribly strong because the upper-air pattern favored for tornado development is obliterated by tropical cyclone circulation. In the Northern Hemisphere, tornadoes are most common in the northeastern quadrant of a storm, such as embedded in the thicker swirls of clouds in the northeastern (upper right) part of Hurricane Isaac (▲). Tornadoes closer to the center of the tropical cyclone may occur at any time of day and are generally weaker than those farther from the center, which mostly occur in the afternoon.

Single-Cell Thunderstorms

05.10.b3

Single-cell thunderstorms can spawn tornadoes, but the tornadoes are generally small and weak, in part because the storm system is itself typically not very large or powerful. Tornadoes formed in this setting are generally restricted to the afternoon, when surface heating destabilizes the atmosphere and causes a peak in thunderstorm activity. Tornadoes formed in any setting can have a distinctive, sharp, low-forming cloud, called a *wall cloud* (▲), indicating that a tornado may be near. Storm chasers (people who purposely try to get close to a storm to observe and photograph it), are on the lookout for wall clouds.

C How Are Tornadoes Classified and What Type of Damage Do They Cause?

Tornadoes are classified using the Enhanced Fujita (EF) scale (shown in the table below), which is based on estimated wind speed as evidenced by damage. The amount of damage a tornado does is controlled by its strength (i.e., its EF rating), how long it stays on the ground, and what type of features are in the way (e.g., a city versus farmland).

This tornado is an awe-inspiring sight, but such thinly tapered tornadoes are typically not the strongest. The blue sky in the background suggests this may be related to a single-cell thunderstorm. It is passing over agricultural fields, so it may cause relatively limited damage.

05.10.c1 Colorado

This photograph illustrates the type of near-total destruction that tornadoes can cause. The wood-frame houses in the foreground were completely disassembled by the strong winds within a tornado, while the houses in the background are still standing (kind of) but suffered major damage.

05.10.c2

Enhanced Fujita Scale			
Strength	3-Second Wind Gust		Examples of Damage
	(km/hr)	(mi/hr)	
EF0	105–137	65–85	Branches/small trees fall
EF1	138–178	86–110	Mobile homes overturned
EF2	179–218	111–135	Roofs blown off houses
EF3	219–266	136–165	Walls blown off houses
EF4	267–322	166–200	Strong frame homes leveled
EF5	>322	200+	Everything obliterated

Before You Leave This Page Be Able To

✓ Sketch and explain what a tornado is and how one forms.

✓ Describe the settings of thunderstorms that produce tornadoes.

✓ Summarize the Enhanced Fujita classification of tornado damage.

5.10

5.11 Where and When Do Tornadoes Strike?

THE INTERIOR OF NORTH AMERICA has by far the most frequent tornado activity in the world. Tornadoes also occur, but far less frequently, in Europe, in the southeastern parts of South America and South Africa, and in a few other regions. Large parts of some continents, like all of North Africa and northern Asia, do not report any tornadoes. Where and when should we expect tornadoes in the U.S., and what causes the huge disparity between the central U.S. and most of the rest of the world?

A How Does Tornado Frequency Vary Across the Continental U.S.?

The continental U.S. is the site of numerous tornadoes each year, especially in a north-south belt through the middle of the country. Most of these tornadoes occur in the warm sector of a mid-latitude cyclone, ahead of an associated cold front that has great thermal contrasts between the warm and cold sectors. Such situations are most common in the mid-latitudes.

Examine this map, which shows the frequency of tornadoes in different counties across the continental U.S. From this map, three areas experience the highest incidence of tornadoes, one in the center of the country, one along the Gulf Coast, and one in Florida. The large-scale patterns reflect the typical types of fronts and other weather activity that occur in the different regions. The geography of North America also contributes to the high frequencies. The major mountain ranges are oriented from north-to-south rather than east-west, which allows warm air and cold air to meet in the center of the continent.

Florida experiences frequent tornadoes, many forming in association with single-cell thunderstorms or with tropical cyclones. These are mostly weak tornadoes and are overall not as deadly. For those related to tropical cyclones, people have already evacuated from less sturdy structures because of earlier warnings about the approaching larger cyclone.

05.11.a1

TOTAL NUMBER OF REPORTED TORNADOES FROM 1950-2012

0
1-30
31-60
61-120
121-180
181-220
221-252

The region of most tornadoes is along a north-south belt, running from north Texas northward to the Dakotas and into Canada—this is justly called *Tornado Alley*. Many of these tornadoes are strong, but some storms lack enough moisture to produce violent tornadoes.

Another concentration of tornadoes appears along the Gulf Coast and adjacent inland areas—a region called *Dixie Alley*. Dixie Alley tornadoes are often more deadly than their Tornado Alley cousins, in part because they are often hidden in thunderstorms amid abundant Gulf of Mexico moisture.

Seasonal Variations

These maps show frequencies of tornadoes for four different months, each representing a season, from 1950–2010. Compare the frequencies in different seasons and try to explain regional patterns.

1. In winter, tornadoes are rare except in Dixie Alley and Florida—areas near relatively warm waters of the Gulf of Mexico. Boundaries between warm and cold air masses are commonly located in this part of the Southeast, more so than anywhere else at this time of year.

2. In spring, the land starts to warm but warming of the atmosphere lags behind, resulting in an unstable atmosphere. The polar front jet stream tends to be strong and displaced southward at this time, which gives upper-level support for tornadoes in the form of wind shear. Tornadoes become more common.

3. Warming continues in the summer, and the boundary between warm and cold air migrates northward. Tornado activity follows into the northern Great Plains and areas to the east. In addition to increased surface heating and convection, moisture flows in from the south.

4. The frequency of tornadoes decreases in fall because lingering summer warmth in the atmosphere remains while the land begins to cool quickly. This stabilizes the atmosphere. Tropical cyclones as a source of tornadoes remain a threat along the Gulf Coast and Florida.

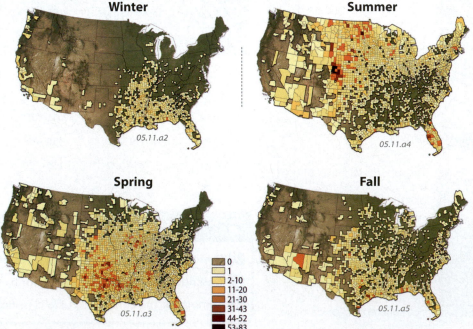

Winter — 05.11.a2
Summer — 05.11.a4
Spring — 05.11.a3
Fall — 05.11.a5

0
1
2-10
11-20
21-30
31-43
44-52
53-83

B How Does Tornado Frequency Vary Globally?

Tornadoes occur in other parts of the world, but are less frequent than in the tornado-prone parts of the U.S. Examine the map below, where each dot represents an area where tornadoes occur, and think how it compares to the previous one displaying the global distribution of thunderstorms, as expressed by lightning measurements.

The central U.S. clearly experiences the highest frequency of tornadoes. Tornadoes occur north into the Prairies of Canada, and others in Hawaii (on the western edge of the map).

Unlike North America, Europe has large, east-west-oriented mountain ranges that inhibit the meeting of very warm and very cold air. Therefore, tornado frequencies are lower in Europe than in North America. Nevertheless, a mid-latitude location, a high population density, and sophisticated detection abilities ensure that reported sightings of tornadoes are fairly high.

Central and northern Asia report few tornadoes, but part of this may be due to under-reporting in the vast, less populated areas.

05.11.b1

Few tornadoes are known to occur over open seas because surface heating is far less intense over oceans than over land. The apparent lack of tornadoes is partly due to a lack of radar coverage over the oceans to detect them. Regions covered by sophisticated detection systems, such as the U.S. and Canada, tend to report more tornadoes.

The lowlands of mid-latitude South America and Africa also have some tornadoes. Recall that the maps showing lightning indicated thunderstorms here.

The high frequency of tornadoes in south Asia and parts of Australia is largely attributable to their proximity to a supply of warm, moist air. These areas also showed a high density of activity on the lightning-strike map.

05.11.b2

168 Tornadoes in the Area of St. Louis, Missouri 1950-2010

(graph: NUMBER OF TORNADOES vs TIME OF DAY, 1AM, 6AM, 12PM, 6PM, 12AM)

05.11.b3 Salina, KS 2012 (EF4)

Time-of-Day Variations

This graph shows the average frequency of tornadoes versus time of day, using data for St. Louis, Missouri. Here, tornadoes typically occur in the afternoon but continue into the evening and night. These nocturnal tornadoes are often the most deadly, for two reasons. First, people are asleep or otherwise less likely to be alert for tornado watches and warnings. Second, nocturnal tornadoes are proportionately more likely to occur in association with storm systems that are well organized, long-lived, and strong, such as multi-cell storms. Other locations in Tornado Alley, Dixie Alley, and Florida display similar temporal patterns.

The tendency for tornadoes to strike in the afternoon and early evening partly accounts for the relatively dark and foreboding aspect (◄) of most photographs of tornadoes—the Sun is low in the sky or has started to set. The thick, dark clouds help too, especially when the tornado occurs in multi-cell or supercell thunderstorms.

Before You Leave This Page Be Able To

☑ Sketch or summarize the frequency and seasonality of tornadoes across the U.S., identifying the main factors that explain the observed patterns.

☑ Briefly describe the factors contributing to variations in tornado frequency around the world.

☑ Draw a simple graph of tornado frequency by time of day, and explain the reasons for the pattern.

5.15 How Are Weather Forecasts Made?

THREE INGREDIENTS ARE NECESSARY for weather—energy, motion (both horizontal and vertical), and atmospheric moisture. The physical processes, such as energy transformations during phase changes of water (between solid, liquid, and gaseous states) are well understood, but several problems limit accurate weather forecasting. First, the equations describing the various transformations across three-dimensional space are complicated, requiring the highest powered computers. Data that are input into the models are often inadequate in quantity and quality. Also, it is impossible to know all the initial boundary conditions or other complexities.

A What Are the Sources of Data for Weather Forecasting Models?

05.15.a1

05.15.a2 Underwood, SD

Airborne Instruments—Every 12 hours, weather balloons (◄) are launched synchronously at about 800 sites around the world. These balloons are equipped with computerized sensors that report the vertical profiles of temperature, pressure, humidity, wind speed and wind direction, and sometimes other data, like ozone concentrations. The balloon ascends until it eventually bursts above the 200 mb level, and the instrument floats down on a parachute, to be used again.

Satellite Data—provide estimates of temperature, humidity, wind speeds, and other variables. They may be the only data available for the oceans and other sparsely inhabited parts of the Earth. Some weather satellites (►) trace paths around the world over periods of several hours, while others remain over the same location, following the Earth as it rotates on its axis.

Radar—Meteorologists rely extensively on radar, some of which is collected from ground and some from satellites. Doppler radar stations (◄) emit microwaves that are reflected back to the receiver from water droplets. Differences in wave frequency reveal the direction of movement, both of the water droplets and of the circulation steering them.

Marine Data—Some marine areas have moored buoys (►) that measure air and sea temperature, pressure, wind speed and direction, and wave speed and height. Portable buoys and a variety of floating objects are used to track ocean currents, although similar data can be collected by satellites.

05.15.a3

05.15.a4

All of these data are reported telemetrically to a central receiving station in suburban Washington, D.C., for processing. The received data are first checked for obvious errors and to make sure the sensors are working correctly. After this first crucial step, data from the different sources are spatially averaged and interpolated to make the huge data sets a bit more manageable. For example, the atmosphere is commonly depicted as a series of points, each representing the center of a cube of air, with a three-dimensional array of virtual cubes encircling the Earth and stacked dozens high. The interpolated data are most accurate where derived from closely spaced, actual measurements, and much less accurate for regions or levels of the atmosphere with few actual measurements.

Upper-level data used to determine the value of the center point for cubes higher in the atmosphere are far less abundant than surface data, so the accuracy of the interpolation typically decreases upward. The smaller the cubes, the higher the spatial resolution, but smaller cubes also increase the computation time. Maps for various vertical levels, called *analysis products* (as opposed to *forecast products* used to predict weather) show the values of the interpolated data. These maps are produced for the U.S. by the National Weather Service and for the world by meteorologists affiliated with the World Meteorological Organization.

B What Equations Are Used in Weather Forecasting Models?

All weather forecasting models are based on only seven equations. Three of the equations describe atmospheric movement and are manifestations of Newton's Laws of Motion. These equations represent velocity or acceleration in three dimensions—west-east, north-south, and up-down in the atmosphere (►).

05.15.b1

A fourth equation accounts for density changes resulting from the piling up or spreading out of air in each cube. The fifth equation describes the rate of change of atmospheric moisture and accounts for changes of state of water in each cube. The sixth equation describes the rate of change of incoming energy (diabatic heating) from the Sun, the energy that powers all motion. Finally, the seventh equation is the Ideal Gas Law, which describes the relationship among temperature, pressure, and density in each cube, and how these intrinsic attributes change in response to all motions and all the other processes.

C How Are the Input Data Used in Computer Models?

1. Once the data have been collected, checked for accuracy, and interpolated into a three-dimensional grid, like the one in the figure below, the next step is to use the data as inputs into models of how our world works. To do this, the variables at the center point of each cube are input into the seven equations. Each equation (except for the Ideal Gas Law) calculates *a rate of change* of the variables in that equation. For example, an equation might calculate how fast the atmospheric velocity changes (i.e., accelerates or decelerates) in an west-to-east or south-to-north direction. Or, we could model how fast atmospheric density or moisture changes due to some change in the environment. Each step in the model forecasts these rates of change for a relatively short interval of time (usually minutes). This is called the *time step*.

2. In this way, answers to questions we have about weather and climate come from solving the seven basic equations simultaneously at each centerpoint. A model takes into account the values of each variable within a cube and in adjacent cubes (including the ones above and below it), for the thousands of cubes. Then, the model generates a forecast of the atmospheric conditions, starting at some specific time.

VARIABLES SIMULATED

OCEAN GRID
Current Vectors
Temperature
Salinity

SURFACE GRID
Ground Temperature
Water
Energy
CO_2 Flux

ATMOSPHERE GRID
Wind Vectors
Humidity
Clouds
Temperature
Chemical Species

SURFACE (LAND) GRID

Horizontal Exchange Between Cells

Vertical Exchange Between Cells

ATMOSPHERE GRID

OCEAN GRID

Vertical and Horizontal Circulation

05.15.c1

4. Any inaccuracies in input data cause errors in the forecast from the start. Also, with each time increment, the forecast will not be 100% accurate, because of an incomplete understanding of the physics involved in exchanges between the atmosphere and the land or ocean surface, or because processes smaller than the grid are at work. Given the complexity of the weather forecasting problem, especially the diverse set of spatially varying parameters, it is amazing that forecasts are as accurate as they are.

3. To forecast hours or days into the future, a model produces a series of forecasts, each representing one short increment of time into the future. Each incremental step uses the previous step's forecast (i.e., the "answers" for each cube) as a starting input for the next forecast. Many iterations of the model are run in order to eventually generate forecasts for hours to days into the future.

How Meteorologists Design, Use, and Intervene in Models

Even though computer models make various types of calculations and forecasts, humans are needed to determine which model output is most trusted in a given situation. For example, a meteorologist's experience and intuition may tell her that — at a particular time of year in a particular type of meteorological setting — one model is more successful at predicting precipitation than another model. Since weather forecasts can have profound implications for people's lives, such as whether to evacuate or take shelter, it is important to keep a human "in the loop."

Meteorologists determine what data and assumptions go into a model, and they design a model to answer a question at a specific temporal or spatial scale. Some models have shorter time steps (i.e., better temporal resolution) than others. This can be an advantage since a prediction is more likely to be accurate if it is made over a shorter time period. In addition, a poor temporal resolution could allow fast-forming weather systems to occur without being taken into account in subsequent calculations. Better temporal res-

olution, however, comes at the expense of computation time.

Likewise, models differ in their spatial resolution (including the number of vertical levels included), with finer resolution giving more precision, but requiring more computational power. The size of the area for which the forecast is generated is an important consideration. Modeling larger spatial domains gives more assurance that storm systems or other features will not creep into the area of interest undetected, but larger areas also require more computations.

Finally, some models express motion not in terms of what is happening at a series of three-dimensional gridpoints, but as a series of waves — these are called *spectral models*. They can be very accurate and may represent the "wave of the future" in weather forecasting models, particularly for longer forecasts.

In addition to collecting data, designing accurate data-collection strategies, and developing and interpreting models, atmospheric scientists research the physics and feedbacks involved in complex weather systems, with a goal of refining the models so

that forecasts and other analyses are increasingly accurate.

These complex systems are generally more important in models for predicting long-term climatic changes than for predicting weather.

Before You Leave This Page Be Able To

✓ Explain the origin of data for weather forecasting models.

✓ Describe the seven basic equations in all weather forecasting models and what each one represents.

✓ Summarize the various trade-offs between features offered by different weather forecasting models.

✓ Explain the role of meteorologists in weather forecasting, including the use of models.

5.15

6.4 What Causes Water to Rise or Sink?

MOVEMENT WITHIN THE OCEANS arises from other factors in addition to shearing induced by the wind. Such factors include variations in the density of water in response to changes in its temperature and salinity. Changes in temperature and salinity occur from flows of fresh water into and out of the oceans as a result of precipitation, evaporation, ice formation and melting, and inputs from rivers. The temperatures of ocean waters and the relative amounts of fresh water control the density of water, and these differences in density can alter circulation of ocean waters.

A How Does the Temperature of Water Control Its Density?

Like most substances, water changes its density when subjected to increases or decreases in temperature. Unlike other substances, however, water exhibits some peculiar density changes as it approaches freezing and then begins freezing.

Polarity of the Water Molecule

1. A key attribute of water that controls its behavior is its polarity. Recall that in a water molecule, the two hydrogen atoms are on one side of the oxygen, giving the molecule a polarity, which causes water molecules to be attracted to charged atoms (*ions*). This allows water to interact with other chemical substances, such as salt.

06.04.a1

06.04.a2

Hydrogen Bonding

2. Water molecules are attracted to each other as well as to ions. In water, a bond called a *hydrogen bond* forms between one molecule's hydrogen atom and another molecule's oxygen atom.

3. Hydrogen bonds form when the polarities of two adjacent water molecules attract one another, causing a bond to form between the molecules. The hydrogen bond is responsible for some of water's unique properties (e.g., viscosity and surface tension).

Density of Water as a Function of Temperature

4. This graph (▶) shows how the density of water changes with temperature. The curve on the right side shows how liquid water changes density in temperatures slightly above freezing. The vertical break marks 0°C, the temperature at which water freezes. The angled line on the left shows how ice changes density with temperature. Examine the main features of this graph before we explore it further. Read clockwise around the graph, beginning on the other side.

06.04.a3

5. At temperatures above 4°C, water, like most substances, becomes more dense as it cools, as shown by moving from right to left up the sloping line on the right side of the graph. With cooling temperature, individual molecules become less energetic and can occupy a smaller volume, thereby increasing the water's density.

6. At a temperature of 4°C, however, water begins to decrease in density, as is shown by the downward bend in the graph between 4°C and 0°C. As a result, water reaches its maximum density (1 g/cm³) at a temperature of approximately 4°C (39°F). This decrease in density occurs because, at temperatures below 4°C, the strength of the hydrogen bonds starts to exceed the disruptive random motions of the less energetic molecules, and the molecules start to organize themselves into a more regular arrangement that occupies more volume and decreases density.

7. As ice (the solid state of water) forms, the crystalline structure of the hydrogen bonds dominates, placing adjacent molecules about 10% more distant, on average, from their nearest neighbors than when in the liquid form at 4°C. This causes the density of ice to drop to about 0.917 g/cm³. This decrease in density when water freezes causes ice to float on water—less dense materials, like wood and ice, float on a more dense liquid, like water. Our world would be a very different place if ice were more dense than water. Ice would begin to grow upward from the ocean floor over time until it occupied a very large percentage of the oceans.

Sinking and Rising of Water as a Function of Temperature

8. These density changes of water as a function of temperature have huge implications for our natural world, many of which will be discussed later in this chapter and throughout the book.

9. At temperatures between freezing and 4°C, warmer waters are more dense than cooler waters (if they have the same salinity), and will therefore sink, helping to redistribute warmth downward in the ocean. Once ice forms, it is even less dense than warmer waters, so it floats. In most substances, the solid state is more dense than the liquid one, but not for water. Ice is approximately 9% less dense than liquid water.

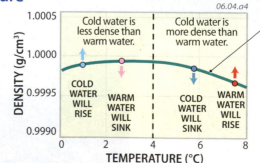

06.04.a4

10. At temperatures above 4°C, water behaves like most other substances, becoming less dense as it warms. At temperatures above 4°C, cooler water is more dense than warmer water and will therefore sink.

B How Does Change in Salinity Affect the Density of Water?

How Salt Occurs in Water and Changes Its Density

1. We all know that water can dissolve some solid materials, like salt, but how does water do this? When we add salt crystals, like the one in the center of this block, to a pot of water, each crystal is sur-rounded by water molecules. In the salt mineral halite, sodium atoms have loaned an electron and so have a positive charge (Na⁺). Such positive ions are called *cations*. Chlorine has gained an electron and so has a negative charge (Cl⁻). Such negative ions are *anions*.

2. The negatively charged chlorine anion is attracted to the positive (H) end of the water molecule. If this attraction is strong enough, it can pull the chlorine away from the halite crystal and into the water.

3. In a similar manner, sodium in halite is a positively charged cation (Na⁺), and thus is attracted to the negative side of any adjacent water molecule. This attraction can pull the sodium ion away from the halite crystal and into the water.

4. Once dissolved in water, the positively charged sodium cation (Na⁺) will be surrounded by the negative sides of water molecules, limiting its ability to rejoin the salt crystal. The negatively charged chlorine anion (Cl⁻) is likewise surrounded by the positive side of the encircling water molecules.

5. As the salinity of water increases so does its density, in a reasonably linear fashion. Elevated saline content in water increases its density, so it will sink. Less saline water will rise. The exact nature of the relation-ship between salinity and density also depends upon temperature.

Decreasing the Salinity of Water

6. One way to decrease the salinity of water is to dilute it with fresh water, such as by adding precipitation to the body of water.

7. Dilution of salt content also occurs from the introduction of fresh water, as from a stream. Most rivers contain dissolved salt, but much less than the sea.

8. Ice consists of fresh water because when it freezes it does not easily incorporate salt into its crystalline structure. When ice melts, it adds fresh water to the body of water, diluting the salt content and decreasing the salinity. Ice melting on land likewise provides abundant fresh water to streams, most of which drain into some body of water.

06.04.b6 Lemaire Channel, Antarctica

9. Several types of ice occur in water. The most familiar type is an *iceberg* (◄), which is a piece of ice broken off a glacier that flowed into a sea or lake. The second type is *sea ice* (not shown), which forms from freezing of seawater into a smoother layer of ice coating part of the sea.

Increasing the Salinity of Water

10. The most obvious way to increase the salinity of water is through evaporation, which removes water molecules but leaves the salt behind. The water vapor contains no salt and can be carried far away by motion in the atmosphere, eventually producing freshwater precipitation as rain, snow, sleet, or hail. This fresh water forms the streams and glaciers on land.

11. Salinity also increases when ice forms on the sea. Fresh water is removed during freezing, leaving behind the salt in the water. As a result, the salinity of seawater increases below and adjacent to *sea ice*. If the sea ice melts, as can occur during the summer, it introduces the fresh water back into the ocean.

Before You Leave This Page Be Able To

✓ Sketch a water molecule and explain why it has polarity.

✓ Sketch and explain the relationships between the temperature of water and its density, including changes at 4°C and 0°C.

✓ Describe how water dissolves minerals, and how this affects its density.

✓ Sketch and explain three ways to decrease the salinity of water and two ways to increase the salinity of water.

6.4

6.5 What Are the Global Patterns of Temperature and Salinity?

THE OCEANS VARY IN SALINITY from place to place and with depth. Spatial variations are due to differences in the amount of evaporation and formation of ice, which increase salinity, versus the amount of precipitation, input from streams, and melting of ice, all of which add fresh water that dilutes the salinity. Also, saline waters are more dense than fresh water, so there are differences in salinity with depth in the oceans.

A What Is the Observed Pattern of Sea-Surface Salinities?

The oceans vary in temperature and salinity, and therefore also in density, laterally across the surface and from shallower to greater depths. The maps below summarize the global variations in SST and surface salinity. Examine the patterns on each map and then compare the patterns between the maps, region by region.

1. This map (▶), now familiar to you, shows sea-surface temperatures, with red and orange being warmest and blue being coolest. Note where the warmest seawater is located, and think about what implication this might have for salinity. Do you think this warm water will be more or less saline than waters that are somewhat cooler, like those shown in yellow on this map?

2. The dominant feature of global variations in SST is the equator-to-pole contrast arising from the supply of insolation and the amount of energy available to heat the water.

3. Where are the coldest waters located? Naturally, they are near the poles. Note the cold end of the temperature scale below the map, which shows that some of the coldest areas are at or even below freezing. We would expect ice to be present in these areas. The ice could be forming and increasing the salinity of the seawater left behind, or it could be melting and introducing fresh water into the sea, decreasing the salinity. We might predict that ice would influence salinity.

4. This map (▶) shows salinities of the ocean surface, with red areas having the highest salinities and blue and purple having the lowest. First, observe the main patterns. Then, think about how you might explain these patterns by considering the processes described on the previous two pages, such as the amount of evaporation and cloud cover. After you have done these, read on.

6. High salinities occur in the subtropics, coinciding with locations of the subtropical high pressures. These areas have relatively clear skies and intense sunlight, which together cause high rates of evaporation, and the subtropics have low amounts of precipitation and dilution. The Mediterranean is very saline because it experiences high evaporation rates and has a limited connection to the Atlantic, with few high-discharge rivers flowing into it. Also note some moderate-salinity water in the northern Atlantic Ocean, near Greenland.

5. The lowest salinities, represented by the regions in blue and purple, are mostly at high latitudes, such as areas closer to the poles. We can largely explain these low salinities by high-latitude regions being cold, with low amounts of insolation, and therefore low amounts of evaporation. Note, however, that some areas of low salinities are well away from the poles. One low-salinity region is in the very warm water that is southeast of Asia; therefore, the warmest areas of the ocean are not necessarily those with the highest salinity.

7. Low salinities away from high latitudes occur (1) where large rivers, like the Mississippi, Amazon, and Ganges-Brahmaputra rivers, deliver fresh water to the sea, (2) where cold currents, like the California Current and Humboldt Current, bring less saline waters away from the pole, and (3) in areas of high precipitation, like west of Central America and in the previously mentioned warm-water area southeast of Asia. Find all of these areas on the map.

B How Do Temperature and Salinity Vary with Depth?

The oceans also vary in temperature and salinity as a function of depth. Seawater is warmest near the surface, where it is warmed by the Sun. The distribution of salinity is similar, with the most saline waters near the surface, where there is evaporation. Salinity has more complex patterns too, because saline waters are more dense than less saline water and so they can sink. We present the temperature and salinity data in two pole-to-pole cross sections.

1. This pole-to-pole cross section (taken at 20° W longitude) shows Atlantic Ocean temperatures from the surface to the seafloor. Warmer surface waters overlie colder, deep ones. When this occurs, less dense water overlies more dense water, a highly stable arrangement.

2. Only at polar latitudes, both in the far north (right) and far south (left), are surface and deep ocean water temperatures similar. When shallow and deep waters are similar in temperature and density, they can mix more easily, bringing deep water up toward the surface and causing shallow water to sink. Such mixing of waters from different depths is called *overturning* or *turnover*.

3. This cross section shows salinity (measured in PSU) along the same line, from the North Pole on the right to the South Pole on the left. Examine the patterns on this cross section and compare it with the temperature cross section above. Predict which way some of the water is moving at depth.

6. Last, but not least, a key feature is an extensive tongue of saline polar waters that descends near Iceland near 60° N to 70° N and then extends well south of the equator as the *North Atlantic Deep Waters* (NADW). As we shall explore shortly, this flow of saline waters down and to the south has a major role in moderating the climate of our planet.

4. In this cross section, a tongue of less saline waters from Antarctica, called the *Antarctic Intermediate Waters* (AAIW), forms a wedge below saline waters near the surface with somewhat more saline waters at depth. Along the bottom are some moderately saline, very cold and dense waters known as *Antarctic Bottom Waters* (AABW).

5. In the upper part of the middle of the cross section is a relatively thin veneer of extremely saline waters that formed from evaporation and limited rainfall in the subtropics. A thicker batch of salinity at 40° N marks saline waters exiting the Mediterranean Sea (MW), which explains the increased temperatures at 30° N to 40° N.

C What Are the Regional Relationships Among Temperature, Salinity, and Density?

1. To explore how temperature, salinity, and density vary from region to region, this graph plots temperature on the vertical axis and salinity on the horizontal axis. Contours crossing the graph represent differences in density in parts per thousand, as calculated from the temperature and salinity. The yellow star represents the mean temperature and salinity of the world's oceans. Examine this graph and consider why a region plots where it does.

2. Equatorial waters are warm due to high insolation and are less saline than average seawater due to heavy tropical rainfall. Warm temperatures and low salinity produce low densities (i.e., numbers on the contours).

3. Mid-latitude waters are moderate in temperature and salinity. Compared to the average ocean, they are cooler because they receive less insolation than average, but less saline because precipitation generally exceeds evaporation.

4. Ice-free polar regions, predictably, have relatively cold seawater. They are also slightly less saline due to lower amounts of insolation and associated evaporation at high latitudes.

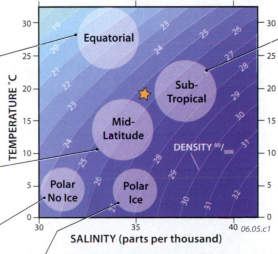

5. In polar zones where sea ice is actively forming, ice removes some fresh water from sea water, leading to greater salinities.

6. Subtropical waters are cooler than equatorial ones, but more saline, because surface evaporation exceeds precipitation. The salinity in particular leads to higher densities (shown by curved lines) of these waters. The most saline waters in the world are in the subtropics, especially where an area of the sea is mostly surrounded by continents and has somewhat restricted interchange with the open ocean. Such very saline waters are in the Red Sea and Mediterranean Sea.

Before You Leave This Page Be Able To

☑ Summarize the global patterns of ocean salinity and temperature, how different regions compare, and what factors cause the larger patterns.

☑ Summarize how salinity and temperature vary as a function of depth, identifying the major distinct types of water at depth.

6.6

What Processes Affect Ocean Temperature and Salinity in Tropical and Polar Regions?

THE EXTREMES IN SEAWATER TEMPERATURE AND SALINITY occur at high and low latitudes. The warmest and most saline waters occur in the tropics and subtropics, while very cold, less saline waters occur near the poles, away from where ice is actively forming. Here, we examine the processes occurring in some of the warmest and coldest parts of the ocean to illustrate how these important end members affect the planet's oceans and the planetary system.

A What Are Warm Pools and How Are They Formed?

1. Within Earth's oceans are regions where the water temperatures are higher than in adjacent regions, and these warm regions persist, or re-form, year after year. These regions of warm seawater are called *warm pools*. The figure here illustrates how they are formed and maintained.

2. Excess insolation (compared to outgoing longwave radiation) generates warm air and warm sea-surface temperatures (29–30°C). Abundant precipitation falls on the surface of the tropical and equatorial oceans where, despite the high temperatures, precipitation exceeds losses to evaporation.

3. Warm, rising air carries large quantities of water vapor into the troposphere. It cools adiabatically but more slowly than the air around it and therefore continues to rise due to instability, forming very deep convection.

06.06.a1

4. Deep convection over tropical and equatorial regions leads to the release of latent heat and an expansion of the troposphere, causing the tropopause to be higher (17–18 km) over the tropics than over the rest of the planet. At these heights, surrounding temperatures are as low as –85°C.

5. Air at the tropopause and in the lower stratosphere is very cold with extremely low moisture content (4 parts per million). It is no longer less dense than air around it since the stratosphere has a temperature inversion. Thus, air moves laterally toward the poles (the yellow arrow) as if along an invisible ceiling.

6. Air moves horizontally across the ocean surface to replace the unstable rising equatorial air, bringing water vapor evaporated from the subtropical regions, where evaporation exceeds precipitation. Once in the tropics, the air rises, condenses, and releases latent heat, which was derived from the subtropics. The tropics therefore accumulate excess energy, to be released in this restricted area of rising air and instability. The ocean warms, forming a warm pool.

Location and Influence of Warm Pools

7. These two globes show SST, but they focus on areas of warm pools, which occur in the warmest waters, those shaded red and orange.

8. Earth's largest warm pool is represented by the very warm waters near Southeast Asia, broadly between China and Australia. The warm pool is centered on the shallow seas that are partially enclosed by the islands of Indonesia; these are the world's warmest seas. Warm waters in this region also extend into the adjacent Indian Ocean, so this warm pool is commonly called the *Western Pacific Warm Pool*.

9. Another warm pool, often called the *Western Atlantic Warm Pool*, is centered on the Caribbean Sea, between North and South America. This warm pool becomes warmest and best developed in the summer and fall when the full impact of summer warming has occurred. The warm water occupies the Caribbean Sea, Gulf of Mexico, and adjacent parts of the western Atlantic Ocean. The shapes of North America and South America enclose the Caribbean part of this warm pool on three sides.

06.06.a2   06.06.a3

06.06.a4 Raja Ampat, Indonesia

10. The warm waters of the Western Pacific Warm Pool sustain rich and diverse life. These seas are famous for their coral reefs and clear, warm waters, stretching from Indonesia (◄) to east of the Philippines, in the island Republic of Palau (►).

06.06.a5 Palau

11. The Western Atlantic Warm Pool forms the warm waters of Caribbean vacation spots, like the Florida Keys, the Cayman Islands, and beaches near Cancun, Mexico (►).

06.06.a6 Akumal, Mexico

B   What Processes Occur Around the Poles?

1. In terms of temperatures and humidity, the poles are the exact opposites of a warm pool. Here, we explore the processes near the poles, using a simple example where the entire polar region is ocean (no land).

2. Cold air aloft moving poleward is extremely dry. Near the pole, it subsides and warms adiabatically. The combination of extremely low moisture contents and warming results in very few clouds.

3. The lack of water vapor and clouds encourages the loss of outgoing longwave radiation. When combined with the sinking and warming air, this creates a very stable atmosphere due to a near-surface temperature inversion (warmer air over cooler air), which inhibits cloud formation and precipitation.

4. Ice floats on the surface of the ocean, due to the unusual physical properties of water. The high albedo (reflectance) of snow and ice reflects much of the already limited insolation back toward space, while the absence of clouds and water vapor ensure considerable heat loss through longwave radiation back toward space. As a result, the surface remains cold.

06.06.b1

STRATOSPHERE
Dry
Stable
COLD POLAR AIR
High Surface Albedo
ICE
COOLING SALINE WATER

7. At the same time, cold polar air blows outward across the surface away from the pole, to ultimately replace the warm rising air at the equator. In this way, the oceans and atmosphere are both moving energy and material toward and away from the poles.

6. As the dense, polar waters descend, surface waters from lower latitudes flow toward the poles to replace them. This flow of dense saline waters downward, and their replacement by poleward surface flows of warmer water, sets up circulation involving both the shallow and deep waters.

5. The formation of ice on the surface of the polar oceans increases the salinity of the underlying waters, turning them into *brine*—very saline water. This and the very cold water temperatures increase the density of the surface waters to the point that they attain a higher density than that of the deep ocean waters below them, causing surface water to descend unstably.

06.06.b2 Antarctica

8. Ice shelves of Antarctica (◄) gain ice in three main ways, all of which result in ice that is fresh, not saline: (1) freezing of seawater below but excluding salt (called *brine exclusion*), (2) flow of glaciers from land to sea, and (3) snowfall and windblown snow on top. Brine exclusion makes adjacent water more saline, but melting of ice will release fresh water, decreasing the salinity. The resulting salinities indicate the relative contributions of freezing versus melting. The photograph to the right is of an ice shelf in Antarctica, but taken looking up at its underside, with a cooperative jellyfish for scale.

06.06.b3 McMurdo Sound, Antarctica

06.06.b4 Hekla Volcano, Iceland

9. Iceland is a large island in the north Atlantic Ocean. As are most regions at such high latitudes, Iceland is very cold. It is cloudier and stormier than most Arctic places, however, because of the effect of the Gulf Stream, which brings warmer water and moist air farther poleward than is typical.

10. Many scientists believe that this global-scale connection between the poles—along with feedbacks between the oceans, atmosphere, and cryosphere—is responsible for the abrupt changes (vertical gray lines) in climate recorded by ice cores from Greenland and Antarctica (▲). Rapid drops in Greenland temperatures are strongly associated with simultaneous Antarctic warming. These changes occur over short geologic time periods of one to two thousand years, but not as fast as Hollywood would have us believe.

06.06.b5

GREENLAND — Warmer / Cooler
ANTARCTICA — Warmer / Cooler
THOUSANDS OF YEARS BEFORE PRESENT

Before You Leave This Page Be Able To

☑ Sketch and explain processes that form low-latitude warm pools, and identify where they are located.

☑ Sketch and explain processes that occur in polar areas and how they affect the salinity of water.

6.6

6.8 What Connects Equatorial Atmospheric and Oceanic Circulation?

EQUATORIAL AND TROPICAL REGIONS are areas of greatest excess energy. The interaction of ocean and atmosphere systems there have great implications for climate and its variability in other parts of the world. In the same way that the Hadley cells establish connections between latitudes, there are large-scale, *east-west* connections within the equatorial atmosphere-ocean system. These connections are a major cause of global climate variability.

A How Do Continents Impact Ocean Circulation?

The equator crosses two broad areas of land—South America and Africa—and a large number of islands of the maritime southeast Pacific, including those in Indonesia. These various landmasses block east-west flow in the oceans, turning ocean currents to the north or south. They also cause deep ocean currents to upwell along their flanks.

1. If there were no equatorial or tropical continents, the world would be circumscribed by a fairly homogeneous band of warm ocean water. The trade winds, north and south of the equator, would exert a small stress on the surface of the ocean, causing a net westward movement of the waters.

06.08.a1

2. The convergence of winds produces a slight mound of water, which then spreads north and south of the equator. Once in motion away from the equator, the Coriolis effect causes the moving waters to deflect from their intended path, causing warmer surface waters to flow away from the equator in both directions—to the northwest, north of the equator, and to the southwest, south of the equator.

4. Continents and other landmasses are in the way of this westward flow of tropical ocean water, causing the warm water to accumulate, or pool, along the eastern side of the land, as expressed on this SST map (▼).

06.08.a2

5. Islands and peninsulas between Australia and Asia block westward movement of warm surface waters, causing a ponding, or a warm pool, in the western equatorial Pacific.

6. Cooler waters, an expression of equatorial upwelling, are visible in the eastern Pacific.

3. As surface waters move away from the equator, deeper and cooler water beneath the equator must rise, or "upwell" to replace the dispersing surface water. This results in slightly cooler water along the equator than to either side.

B How Do Warm Pools Influence Atmospheric Circulation Above Oceans?

1. Once formed along the eastern side of a continent, a warm pool can set up an east-west circuit of flowing air through the tropics.

2. Warm, moist air rises above the Western Pacific Warm Pool, where temperatures regularly exceed 28°C. As the air rises, it cools adiabatically, producing condensation, clouds, and intense rainfall, as well as releasing large quantities of latent heat into the troposphere immediately above the warm pool.

3. Unable to penetrate vertically beyond the temperature inversion at the tropopause, the air moves horizontally away from the warm pool. As it moves, it mixes with the surrounding cool air and cools.

4. Having cooled sufficiently, the air descends over the eastern Pacific, warming adiabatically.

5. This cooler, denser air from above turns west and passes back over the surface along with the trade winds, warming and evaporating moisture from the ocean surface as it proceeds westward.

6. This pattern of east-west equatorial atmospheric circulation, initiated by warm air rising above warm pools, is known as the *Walker cell circulation*. In 1923, Sir Gilbert Walker first reported "great east-west swayings" of the near-equatorial flow, in the manner represented on this figure.

06.08.b1

C Do Similar Circulation Patterns Exist over Other Oceans?

1. Warm pools are pronounced features of the global atmosphere-ocean system, as displayed on this SST map. Trade winds convey warm waters westward through the equatorial regions. In the Atlantic, this movement is blocked by South and Central America, ponding warm water off northern South America and in the Caribbean.

2. As shown in the wind-circulation patterns below the SST map, the Walker-like cell in the Atlantic begins with warm, rising air over the Amazon basin of South America — a hot, humid area of rain forest, and the Amazon River and its tributaries. To the west are the Andes along western South America, and the descending limb of the Pacific Walker cell, both of which restrict flow to the west from the Amazon.

5. The Indian Ocean differs as its warm pool is located at its eastern edge. This reflects the influence of the adjacent larger Pacific warm pool and the porous barrier between the two. The break between the Pacific and Indian Ocean basins at equatorial latitudes is discontinuous, marked by the islands and straits of Indonesia and Maritime Southeast Asia (collectively referred to as Australasia), through which warm waters may pass. Also, the Indian Ocean is unique in not extending beyond about 20° N. The large continental mass gives rise to the Asian monsoon system, whose summer winds are from the southwest, blowing waters back to the east. Low atmospheric pressure is experienced at the surface above the warm pools, and higher pressure is beneath areas where the air is descending.

06.08.c1

3. The Atlantic Ocean is narrower than the Pacific, so the pattern of circulation is not as wide. Air rises above the Caribbean warm pool and the extensive wetlands and forests of the Amazon lowlands, resulting in intense tropospheric convection, second only in its intensity to that of the Western Pacific Warm Pool. Air in the Atlantic cell descends in the eastern Atlantic and over the western side of North Africa, enhancing the regional dryness.

4. Heating and evapotranspiration from the moist forested surface of the Congo basin produces another zone of rising air and convection over this region, drawing moisture from the Atlantic deep into central Africa. The resulting Walker cell is narrow, being constrained on the west by the Atlantic cell and on the east by the Indian Ocean circulation and air motions in East Africa.

How These Circulation Systems Were Discovered

At the end of the 19th and into the early 20th century, India suffered several catastrophic droughts responsible for the deaths of millions. These droughts are displayed on the graph below, which shows departures of the total rainfall received annually in India from the summer monsoon compared to long-term averages (1870–2000).

Sir Gilbert Walker was head of the British Indian Meteorological Department during the droughts of 1918 and 1920. Walker was interested in understanding the causes of drought, which naturally led him to investigate regional wind patterns and ocean temperatures, both of which are critically important to the Indian monsoon that is so crucial to the lives of the people of India and adjacent countries. Walker began publishing his research just as severe droughts ceased. Thus, his contribution was undervalued in his lifetime, as reflected in his 1959 obituary which states sardonically, "Walker's hope was to unearth relations useful for ... a starting point for a theory of world weather. It hardly seems to have worked out like that."

Only in the late 1960s, as drought became a problem again, was the significance of Walker's discovery fully recognized, and this principal causal mechanism of equatorial circulation and climatic variability was named in his honor.

06.08.t1

Before You Leave This Page Be Able To

✓ Sketch and explain the concept of equatorial upwelling.

✓ Sketch the warm pools in each of the major ocean basins straddling the equator, and explain their distribution.

✓ Sketch and explain the Walker cell over the equatorial Pacific Ocean, and similar cells elsewhere.

6.8

6.14

What Influences Climates Near the Southern Isthmus of Central America?

THE SOUTHERN ISTHMUS OF CENTRAL AMERICA is geographically and climatically one of the most interesting places in the world. It is a narrow strip of land that connects Central America and South America and that separates very different waters of the Pacific and Atlantic Oceans. The climate of the region reflects intricately changing wind patterns and atmosphere-ocean interactions, including a migrating ITCZ, effects of the El Niño–Southern Oscillation (ENSO), and mountains with rain shadows on different sides at different times. The interactions between these various components lead to unusual climates and large variability within a single year and from year to year. The interactions have regional and global consequences.

1. The region around the Isthmus of Panamá includes the southward-tapering tip of Central America, including Costa Rica and Panamá, and the wider, northwest end of South America, including the country of Colombia. The spine of the isthmus is a volcanic mountain range in Costa Rica, but the Panamá part is lower (hence the location of the Panamá Canal). Colombia contains the north end of the Andes mountain range.

06.14.a1

2. West of the land is the Pacific Ocean, and to the east is the Caribbean Sea, a part of the Atlantic Ocean. The Caribbean Sea is bounded on three sides by land—North America to the north, South America to the south, and Central America, including the Isthmus of Panamá, to the west. On the east, the Caribbean is open to the main part of the Atlantic through gaps between individual, mostly small islands, of the Greater and Lesser Antilles. As a result of the Caribbean being partly enclosed and being within the tropics and subtropics, the waters of the Caribbean are generally warmer than those of the Pacific.

Regional Wind Patterns and Positions of the ITCZ

3. The isthmus and the adjacent parts of South America are north of the equator, but in tropical latitudes, centered around 10° N.

4. Crossing the region in an approximately east-west direction is the Intertropical Convergence Zone (ITCZ), which has a seasonal shift north and south. In September it is as far north as 13° N.

5. In January, the ITCZ shifts to the south to near 5° N, following the seasonal migration of direct-overhead insolation to the south.

06.14.a2

6. Much of the region is dominated by the northeast trade winds that blow from the Caribbean Sea, across the isthmus. Embedded in the trade winds is a low-altitude, fast-flowing current of air called the Caribbean Low-Level Jet (CLLJ).

7. In most tropical regions, the ITCZ is north of the equator during the northern summer (e.g., July) and south of the equator during the southern summer (e.g., January). In this region, however, the ITCZ stays in the Northern Hemisphere because of the cold Humboldt Current farther south. Related to this position are southwesterly winds that originated in the Southern Hemisphere as the southeast trades and are deflected back northeastward when they cross the equator as they flow toward the ITCZ. They reach farther north during times when the ITCZ is farther north.

Wind Patterns and Weather During the Winter

8. From November to April, while the ITCZ is displaced to the south, the northeast trade winds and the CLLJ flow southwest across the isthmus, carrying warm, moist air from the Caribbean Sea. The Caribbean is part of the Western Atlantic Warm Pool, similar in origin and setting to the warm pool of the Pacific. The Atlantic warm pool forms when trade winds blow water westward until it piles up against Central America. When the moist air is forced up over the mountains of the isthmus at Costa Rica, heavy orographic rains fall on the Caribbean flanks of the mountains, while the Pacific coast experiences a dry season from the rain-shadow effect.

06.14.a3

Wind Patterns and Weather During the Summer

06.14.a4

9. During the opposite time of year, from May to October, the Pacific coast of Costa Rica experiences its rainy season. This corresponds to the northernmost migration of the ITCZ across the region, which allows the southwesterly winds to bring in equatorial Pacific moisture. As the moist Pacific air is forced eastward over the mountains, orographic rain falls on the west side. At the same time, the rising air in the ITCZ is over the area, producing heavy rain on a nearly daily basis.

06.14.a5

Cold, Dry Outflow

Intense Storms to Tropopause

Descending Air

NICARAGUA

NE Trade Winds

Caribbean Low Level Jet

COLD

Caribbean Sea

COSTA RICA

Blown Offshore

WARM POOL

WARM

Pacific Ocean

COLD

Upwelling

Regional Wind Patterns and Positions of the ITCZ

10. In July and August, the rainy season of western Costa Rica is temporarily interrupted by a dry spell of several weeks called the "Veranillos" (little summer). This dry spell occurs when the Caribbean Low-Level Jet (CLLJ) reaches its maximum intensity. The jet accelerates through the topographic break in the mountain belt between Costa Rica and Nicaragua.

11. Recall that the Caribbean Sea is part of the Western Atlantic Warm Pool, so it has very warm water during July and August (during the northern summer). As a result, the trade winds and CLLJ bring moist air across the region. As this air climbs the eastern side of the mountains, much of the moisture is lost to precipitation, yielding a rainy season on this eastern (Atlantic) side of the country during these months.

12. During July and August, the vigor of the trade winds and CLLJ blow surface waters of the Pacific Ocean offshore, causing upwelling of cooler water in the Gulf of Papagayo at the Costa Rica/Nicaragua border and a temporary suppression of the convective activity in the ITCZ. This causes the short dry season (Veranillos) on the western (Pacific) side of the country.

13. Warm, moist air farther offshore produces deep tropical convection, and an outflow of cooler, drier air descends over northwestern Costa Rica, preventing convection over the upwelling.

ENSO Warm Phase   ## ENSO Cold Phase

14. The weather patterns are already very interesting and unusual, but this region also falls under the influence of ENSO. These two globes show SST during ENSO warm and cold phases, relative to the long-term averages. Red and orange show SST above their typical value for a region, whereas blue shows SST that are colder than normal for that area. What patterns do you observe near Central America?

15. During a warm phase of ENSO (El Niño), warmer waters appear in the eastern equatorial Pacific, as shown by the red colors along the equator. Note that Pacific waters west of the isthmus are warmer than usual, but waters of the Caribbean are colder than normal. The warmer than usual waters of the equatorial Pacific are a site of rising air, drawing the ITCZ south and west of its usual position. The displaced ITCZ reduces rainfall along the Pacific flank of Costa Rica and enhances the Caribbean trade winds. This increases rains along the Caribbean slope, while the Pacific slope suffers from an enhanced rain-shadow effect and an extended Veranillos.

06.14.a6-7

16. The opposite occurs during a cold phase of ENSO, as shown in this globe. Waters are colder than normal in the equatorial Pacific west of South America, but are more typical offshore of Central America. The waters of the Caribbean, however, are warmer than typical. The changes in water temperatures and air pressures between the warm and cold phases of ENSO cause changes in the directions and strengths of winds. For instance, the CLLJ in June through August is stronger in warm phases than in cold phases. Changes in position of the ITCZ and strengths of winds cause differences in the risk of flooding, as depicted in the graphs below.

Before You Leave This Page Be Able To

☑ Sketch and label the main wind features that affect this region, explaining how each one forms.

☑ Sketch and explain a cross section for each of the main settings for rainfall in the isthmus and in nearby South America.

☑ Summarize some changes that occur when the region is in a warm or cold phase of ENSO.

▶ **17.** These two graphs show the risk of flooding during warm and cold ENSO events. Flooding on the Pacific flank (left graph)

corresponds to the northward migration of the ITCZ during June through September. In warm phases (red bars), this migration is reduced, the CLLJ intensifies, and the Veranillos expands. In contrast, rivers on the Caribbean flank (right graph) generally flood when the CLLJ intensifies (JA and NDJ). Flooding intensifies in the Caribbean during warm phases (red bars) and in the Pacific during cold phases.

6.14

What Oceanic and Atmospheric Patterns Are Predicted for a Newly Discovered Planet?

PLANET W is a newly discovered planet that is similar to Earth. It has oceans, an atmosphere very similar to Earth's, and ice at both poles. Since no astronaut has yet ventured to the planet, we currently only know things we can observe from a distance, specifically the physical geography—the distribution of land, oceans, and sea ice, as well as the larger features on land. To guide future expeditions, you will use the known physical geography to predict the global patterns of wind, ocean currents, and atmosphere-ocean-cryosphere interactions.

Goals of This Exercise:

- Observe the global distribution of continents and oceans on Planet W.
- Create a map showing the predicted geometries of global winds, ocean currents, and major atmospheric features, like the ITCZ and polar front.
- Draw cross sections that portray the global patterns, predicting the locations of warm pools, zones of high and low precipitation, and possible sites for ENSO-type oscillations.
- Use your predictions to propose two habitable sites for the first two landing parties.

Main Procedures

Follow the steps below, entering your answers for each step in the appropriate place on the worksheet or online.

1. Observe the distribution of continents and oceans on the planet. Note any features that you think will have a major impact on the climate, weather patterns, and other aspects of atmosphere-ocean-cryosphere interactions. Descriptions of some features are on the next page.

2. Draw on the worksheet the predicted patterns of atmospheric wind circulation (trade winds, westerlies, etc.), assuming that the patterns are similar to those on Earth.

3. Draw on the worksheet the likely paths of ocean currents within the central ocean. Identify the segments of a current that will be a warm current, cold current, or neither.

4. Draw on the worksheet the likely locations of warm pools and shade in red any area on land whose climate will be heavily influenced by the presence of this warm pool. Shade in parts of the ocean that are likely to be more saline than others.

5. From the pattern of winds and ocean currents, identify which side of a mountain range would have relatively high precipitation and which side would have lower precipitation because it is in a rain shadow. Color in these zones on the worksheet, using blue for mountain flanks with high precipitation and yellow for areas of low precipitation. Note, however, that precipitation patterns along a mountain range can change as it passes from one zone of prevailing winds into another, or if wind directions change from season to season.

Optional Procedures

Your instructor may also have you complete the following steps. Complete your answers on a sheet of graph paper.

6. Draw four cross sections across the central ocean, extending from one side of the globe to the other. Of these cross sections, draw one in an east-west direction across each of the following zones: (1) between the equator and 30° (N or S); (2) within a zone of westerlies (N or S); and (3) within the south polar zone, south of 60° S. The fourth cross section can be drawn at any location and in any direction, but it should depict some aspect of the oceanic and atmospheric circulation that is not fully captured by the other cross sections. On each cross section, draw the surface wind patterns, zones of rising or sinking air, and possible geometries of a Walker cell or other type of circulation, where appropriate. Be prepared to discuss your observations and interpretations.

7. Draw a north-south cross section through the oceans, showing shallow and deep flows that could link up into a thermohaline conveyor. Describe how this might influence the climate of the entire planet and what might happen if this system stops working.

Some Important Observations About the Planet

1. The planet rotates around its axis once every 24 hours in the same direction as Earth. Therefore it has days and nights similar to Earth's. It orbits a sun similar to our own, at about the same distance. Its spin axis is slightly tilted relative to the orbital plane, so it has seasons.

2. The planet has well-developed atmospheric currents, with wind patterns similar to those on Earth.

3. The oceans are likewise similar to those on Earth. They are predicted to have ocean currents that circulate in patterns similar to those on Earth. The oceans are predicted to have variations in sea-surface temperature, salinity, and therefore density, but no data are available for these aspects.

4. The planet has ice and snow near both poles, some on land and some over the ocean.

1. A continent, simply called Polar Continent for now, is over the North Pole, but slightly off to one side. The western part of the continent extends south of 60°, but there are no data about whether this part has ice. There appear to be ice shelves and sea ice surrounding part of the continent.

2. There is a gap between Polar Continent and those continents to the south, providing a relatively land-free zone through which the ocean can circulate. Consider how this gap might influence speeds of winds and ocean currents here.

3. A nearly continuous mountain range runs along the coast of Eastern Continent.

4. There is extensive ice and snow over the South Pole, but the ice is floating on the ocean (there is no land at the pole). The sea ice extends to the Eastern and Western Continents, both of which also have glaciers that flow downhill toward the sea, where they melt.

5. Western Continent does not have significant mountains but, like Eastern Continent, has a curved coastline with large peninsulas that extend out into the ocean and embayments, where the ocean curves into the land.

06.15.a1

Climates Around the World

CLIMATE IS THE LONG-TERM PATTERN of weather at a place, including not just the average condition but also the variability and extremes of weather. It largely reflects an area's geographic setting, especially latitude, elevation, and position relative to the ocean. Climate is also influenced by atmospheric pressure and moisture, prevailing and seasonal wind directions, and the type of nearby ocean currents — in other words, everything we have explored so far in this book. Observed in this context, climate is the culmination of the various factors that move energy, air, and water (in all three states) in the Earth's climate system.

The climate of a region can be described, measured, and represented in many ways, including maps of annual average temperature and precipitation, like the two globes below (temperature is on the left). To examine long-term climate, as opposed to short-term weather, we consider data averaged over years or decades. Weather is what we get; climate is what we expect. Remember that climate includes not only average values but also extremes and variability about the averages.

What is the difference between weather and climate, and how are the two related?

Maps of climatic factors, like the ones below, visually represent large amounts of data for the land, sea, or atmosphere. For numerous measures, like temperature and precipitation, maps can depict annual averages, values for a particular month (perhaps the warmest month) as averaged over decades, or some measures of seasonal extremes or variability from one decade to another.

How can averages, seasonality, and variability in temperature and precipitation be combined to represent the climate of a region?

07.00.a1　　07.00.a2

Average annual temperatures vary greatly across the planet (▲). In the map above, high average temperatures are in red and orange, whereas cooler ones are in green and blue. What patterns do you observe? Much of Africa, but especially the central and northern parts, are very warm, much warmer than Europe. As anyone who has traveled to these places knows, they have very different climates. Africa straddles the equator, so some parts of it are likely to be tropical, whereas Europe is situated well north of the equator.

Do all areas with the same average temperature have the same climate? Are warm temperatures, by themselves, enough to define an area as tropical?

The average amount of precipitation (▲) is another factor most people consider when talking about the climate of an area. We might say that Scotland, the northern part of Great Britain, is a "cold, rainy place," but the Sahara of northern Africa is an "extremely hot and dry place." With such statements, we are expressing something about the average temperature or amount of precipitation but not what occurs on any given day (which is weather). Also, many regions have distinct wet and dry seasons, an important aspect of climate.

Can we combine measures of temperature and precipitation, and their seasonal variabilities, to represent the climate of an area?

TOPICS IN THIS CHAPTER

By considering temperature and precipitation, both in terms of their averages and in terms of average maxima, average minima, and seasonal variations, geographers developed several climate classification systems, including the one portrayed on this globe (▶). The colors and letters indicate what type of climate characterizes an area. Examine the color patterns and letters on this globe and compare these to the globes showing average annual temperature and precipitation (◀). What correlations do you note?

How do we classify climates, and what do the letters assigned to a region indicate about its climate?

The climate of a region is influenced by many factors discussed in this book, including latitude and its relationship to Sun angle, elevation, topography, the locations of semipermanent pressure features, prevailing wind directions, common storm tracks, sea-surface temperatures, ocean currents, humidity, atmospheric oscillations, and teleconnections. These and other factors act in concert to produce the characteristic climates around the world, such as these in Africa and Europe.

How do Sun angle, topography, pressure features, humidity, storm tracks, ocean currents, and other factors interact to produce the observed variations in climate?

Oceans have characteristic climates, just like land areas. Note on this map that the climate of a land area generally continues out across an adjacent body of water, whether it is the Atlantic and Indian oceans or the Mediterranean Sea. Evaporation of water in the oceans provides energy for weather systems and other dynamic processes (like flooding), so assessing the climate of the oceans is important in understanding Earth.

On land, we consider types of plants in an area when talking about its climate, but how do we assign a climate zone to part of an ocean?

Climate is not static. Variations happen from year to year, and longer-term changes also occur. Witness the rise of global temperatures in the last 150 years, since the end of a very cold period in the 1800s. Human activities introduce materials, such as pollution and CO_2, into the atmosphere, oceans, and land. Urbanization, as natural lands or less densely populated areas are converted into cities and towns, can affect the long-term averages of temperature, precipitation, cloud cover, and local winds—that is, urbanization changes the climate.

To what extent are the observed increases in global temperatures during the past century and a half attributable to human activities?

07.00.a3

7.0

7.1 How Do We Classify Climates?

CLASSIFICATION IS THE PROCESS of grouping similar items together and separating dissimilar items. Items that are similar in some ways can be different in others, and items that are in different groups can have certain similarities. Classifications allow us to examine general patterns, with the caveat that interpretations derived from classified groups are best done cautiously. Here, we examine a well-established method of classifying climates.

A What Is the Purpose of Classifying Climates?

Climates are classified to let us observe broad patterns and to simplify communication about the characteristics of a region. Historically there have been thousands of meteorological stations collecting information, and in the past few decades this wealth of knowledge has expanded with the addition of large quantities of remotely sensed information. Classification helps us to seek useful generalizations from this enormous amount of information. Climate can be classified in many ways, with the exact nature of the classification depending upon the research question that is of interest.

1. Colors on this map of an imaginary country represent mean annual temperatures (reds above 28°C, blue and purple below 0°C). The black contours represent mean annual precipitation in millimeters, and are called *isohyets*. Meteorological stations are shown as dots. Examine the patterns on this map and think about how you might classify different parts of the country (e.g., cold and wet).

2. The most southerly areas are hot, which is an important fact if we were investigating an outbreak of a heat-related sickness. But we might classify stations in the southeast differently than those in the southwest because of the differences in precipitation (and therefore in humidity and perhaps in mold).

3. A study of the timing and magnitude of floods might focus on (1) regions where winter temperatures reach below freezing, opening up the possibility of flooding due to spring snowmelt, and (2) the amount of winter precipitation, which controls how much snow accumulates.

4. Most crops have fairly well-defined climatological requirements for their growth. A climate classification could be developed based on "optimal," "fair," or "poor" for growing that crop. For example, the best conditions for coffee are average temperatures of about 20°C and 1,800–2,800 mm of precipitation. Areas for possible coffee cultivation in this country would be very limited.

B What Climate-Classification System Is Most Widely Used and Why Is It Useful?

Vladmir Köppen was a botanist interested in the global distribution of vegetation types. He surmised that the annual temperature and precipitation regimes determined the types of natural vegetation. Other influences on plant growth, such as which way a slope faces, the number of consecutive days without precipitation, and the type of soils, have more localized effects and so were not considered by Köppen.

Köppen first published his climate classification system in 1884. On his maps, the lines separating different types of climate also separate different realms of natural vegetation. For each climate type, Köppen carefully selected thresholds of temperature and precipitation that preserved the boundaries for that characteristic assemblage of vegetation. The original scheme underwent modification as more information on global climates and vegetation became available. The resulting *Köppen climate classification* is the most widely used system; an example is portrayed on this climate map of South America (▶). Notice how the patterns of colors correlate with latitude, elevation, proximity to the ocean, and other aspects of position within the continent. Some of the boundaries would be recognizable on a satellite image.

In reality, boundaries between climatic types are not static from year to year and not as clearly delineated as the map would suggest. Instead a boundary between vegetation and climate types usually occurs across a broad transition zone, known as an *ecotone*. In South America, an ecotone occurs between the Amazon rain forest and mountain climates of the Andes.

Köppen's system was designed to delineate vegetation realms, so it overlays well with worldwide natural vegetation zones—*biomes*. Both climate and vegetation are major controls on the types of soil that develop, leading to a notable similarity between the Köppen classification and maps of world soils. These linkages between climate, vegetation, and soils ensure that Köppen is still relevant in understanding the physical geography of our continents at planetary and regional scales. Here, because of data availability over oceans, we can compute the climate type over the ocean. But because Köppen was interested in vegetation realms, he never intended for the climate over an ocean surface to be classified.

How Is the Köppen Classification System Organized?

The Köppen climate classification system has five major categories, represented by the first (capital) letter of the labels on the South America map. These are the following: A – *tropical climates*, B – *arid climates*, C – *temperate mid-latitude climates*, D – *harsh mid-latitude climates*, and E – *polar climates*, each represented by a photograph below. The five categories are further subdivided by comparing the climate of an area to a succession of criteria of temperature and the availability of water, as illustrated by the flow chart below. In the A, C, and D climate types, the second letter is lowercase and indicates whether the climate is wet year-round (f), has a dry summer (s), has a dry winter (w), or experiences a monsoon (m). The second letter is uppercase for arid and polar climates. Subsequent pages list key characteristics of each type of climate, but follow the flow chart below first.

07.01.c1

B – Arid

07.01.c3 Namibia

A – Tropical

07.01.c2 Grand Cayman Islands

C – Temperate

07.01.c4 Germany Valley, WV

07.01.c5 Northern British Columbia, Canada

7.1

E – Polar

07.01.c6 Churchill, Canada

D – Harsh Mid-Latitude

Before You Leave This Page Be Able To

- ✓ Explain the purpose of climatic classification.
- ✓ Describe Köppen's intent in developing his climatic classification system.
- ✓ Discuss the five main categories of climates in the Köppen climatic classification system.

7.12 What Factors Influence Climate Change?

CLIMATE CHANGE IS ALWAYS OCCURRING, including global warming since the mid-1800s. There are many natural causes of climate change, including changes in Earth's orbit around the Sun and changes in sunspot activity. Many scientists propose that human activities, including the burning of fossil fuels and the clearing of forests, also contribute to climate change by releasing greenhouse gases to the atmosphere. Other factors may lead to *global cooling*, including ash from large volcanic eruptions and an increase in certain aerosols in the atmosphere. Here, we examine some of the factors that can influence our climate.

A What Processes Influence Atmospheric Temperature Change?

Earth's surface temperatures are dominated by energy from the Sun. Insolation heats the oceans, land, and atmosphere, but several factors influence how much of this energy reaches the surface and how much is retained.

Interaction of Insolation with Earth's Atmosphere, Oceans, and Land

1. Nearly all of Earth's heating at the surface comes from insolation, which heats the atmosphere, land, and oceans. Most of this energy escapes eventually back into space in the form of longwave infrared energy. The rest is delayed by interactions with Earth, keeping the planet warm by a process called the *greenhouse effect*. The amount of insolation hitting Earth varies regularly, by a small amount, due to orbital fluctuations and changes in the Sun's energy output, as expressed by changes in *sunspot* activity. Sunspots are the darker areas that, on average, appear and disappear on the surface of the Sun in an 11-year solar cycle.

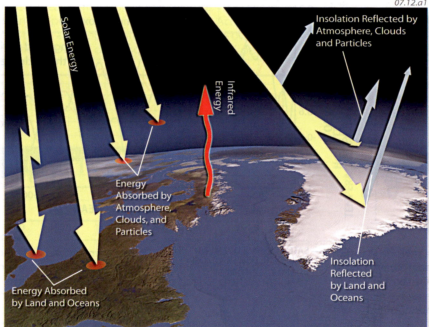

07.12.a1

2. Some insolation is *absorbed* by the atmosphere (shown as an orange disk in the figure), and some is *reflected* off the atmosphere. Much of the reflected insolation returns to space without heating Earth or its atmosphere, as depicted by the blue arrows.

3. Insolation is absorbed by clouds, by soot from burning, and by fine particles (aerosols), which are produced by volcanoes, industry, and automobiles. Some of this absorbed energy radiates back into space as *infrared* (longwave) *energy* (shown by the wavy red arrow). Clouds and particles also reflect some insolation.

4. Some insolation is reflected back to space from the land surface and oceans. Ice in continental glaciers is an effective reflector. As glaciers melt, darker land or ocean is uncovered. This increases the amount of energy that is absorbed by Earth and subsequently re-radiated back to the atmosphere.

5. Some insolation is absorbed by the land and the oceans, both of which then radiate infrared (longwave) energy back into the atmosphere. Some of this infrared energy is absorbed by atmospheric gases, such as water vapor (H_2O), carbon dioxide (CO_2), methane (CH_4), and nitrous oxide (N_2O), which are called *greenhouse gases*. Some portion of these gases is produced naturally and some is produced by human activities.

Orbital Variation

07.12.a2

Earth's orbit is affected almost exclusively by the Sun and Moon, causing it to cycle from more elliptical to more circular paths with one cycle completed over about 100,000 years. This influences Earth's climate, but the current global warming has occurred in less than 200 years. This is too short a time period for orbital changes to have caused the observed warming. The tilt and direction of Earth's axis of rotation also changes, which changes the contrast of the seasons, but these effects occur over tens of thousands of years.

Variations in Solar Radiation

07.12.a3

The Sun is a candidate for causing climate change. Solar activity, as presented by the number of sunspots, clearly correlates to earlier changes in temperature (◄), but less so to recent warming. Cosmogenic isotopes in ice cores and tree rings (proxies) suggest that the Sun's energy emission has varied by only 1% over the last 2,000 years. Both data sets are relevant when evaluating the Sun's role in recent warming.

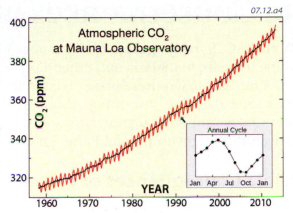

Greenhouse Gas Production

1. Several gases in Earth's atmosphere absorb infrared radiation emitted by Earth. This causes them to vibrate and heat up, and then to emit infrared radiation. This radiation can escape into space or be absorbed by other greenhouse gases, mostly lower in the atmosphere where the concentration of greenhouse gases is the highest. Since 1957, atmospheric scientists have collected air samples on the high peaks of Hawaii. The data for Mauna Loa are plotted on this graph (▶), which shows CO_2 content in the atmosphere as a function of time. The concentration of CO_2 in air, represented by the red line, has increased by 20% in the last 40 years. Most of this increase is attributed to humans, especially through the burning of fossil fuels.

2. Our oceans are a huge reservoir for dissolved gases and other chemical components, including CO_2. As the CO_2 of our atmosphere increases, a large amount of this goes into the oceans, where it affects seawater chemistry, especially the acidity. This *ocean acidification* and its impact on sea life lead to concerns about increases in atmospheric CO_2. Seawater can dissolve less CO_2 when it is warm, so it releases CO_2 back to the atmosphere as it warms, as is occurring now. So some increase in the atmospheric CO_2 could be a *result* of recent warming.

Greenhouse Gases and Temperature Change Records from Ice Cores

3. Long-term records of CO_2 and other greenhouse gases are contained within ice cores and can be traced back hundreds of thousands of years. Ice cores from Antarctica have been analyzed for CO_2, isotopic temperature, and isotopic age. The data, shown on this graph (▶), allow comparison between prior natural variations and changes in the last several hundred years.

4. How well do CO_2 concentrations in the atmosphere track temperature change? On this graph, the two trends are very similar. Graphs of other greenhouse gases, such as CH_4 versus temperature and N_2O versus temperature, show a similar correspondence. These large fluctuations in the past have natural causes.

5. Scientists have used data from ice cores and other proxies to interpret how atmospheric gases changed in the past few hundred years. These data indicate that levels of CO_2, CH_4, and N_2O have increased since the start of the Industrial Revolution of the last part of the 18th century. Many scientists infer that these increases in greenhouse gases are partly responsible for the recent increase in temperature, but there remains debate about how *much* our human activities have affected, or will affect, our climate.

Ocean Oscillations, Climate Change, and Climate Models

Recent discoveries have changed our view of the relationship between ocean currents and climate. Studies of the oceans have documented that separate ocean basins (e.g., the Pacific Ocean) display variations in temperature that occur over years or decades—*ocean oscillations*.

One long-recognized oscillation is the El Niño–La Niña effect and is called the El Niño–Southern Oscillation (ENSO for short). More recently, we have discovered other ocean oscillations, including the Pacific Decadal Oscillation (PDO) and the Atlantic Multidecadal Oscillation (AMO). As their names imply, these oscillations may last decades, whereas ENSO cycles last a

year or two. Variations in sea-surface temperatures (SST) and atmospheric pressure can be used to calculate a numerical *index* for each oscillation, as a way to represent the changes. The graph here shows values for the Atlantic Ocean (AMO), relative to its long-term average. The AMO represents SST in the Atlantic basin—high numbers indicate warm SST. This graph shows that Atlantic SST undergoes

cycles that last several decades. These cycles also relate to periods of abundant or rarer tropical cyclones, and they also show some similar patterns to global temperatures. Because of such correlations, climate scientists are increasingly examining whether oscillations are primarily a cause or effect of climate change.

Before You Leave This Page Be Able To

✓ Summarize the major factors that influence atmospheric temperatures.

✓ Summarize how atmospheric greenhouse gas concentrations can correspond to temperature changes.

7.12

7.13 What Are the Consequences of Climate Change?

CLIMATE CHANGE HAS MANY IMPACTS, ranging from obvious ones like an increase in global temperatures, to less obvious ones, such as a possible increase in malaria and other diseases. Some of these consequences are highly probable, whereas others are very speculative. On these pages, we briefly introduce some of the most likely results of climate change, specifically those related to global warming. Abundant information is available elsewhere, including reports by governmental and nongovernmental organizations. The most detailed and widely cited reports are those by the United Nations-sponsored Intergovernmental Panel on Climate Change (IPCC).

A What Are Some Possible Consequences of Climate Change to the Environment?

1. The figure below illustrates some consequences of global warming, but many more are possible. Examine the figure and think about what you know about the feature being depicted and how it might respond to an increase in average temperatures. Then read the text.

2. The most obvious result of global warming would be an *increase in global temperatures*. However, computer models of climate indicate that many areas, especially those near the poles, will indeed increase in temperature, but other regions might get colder.

3. Higher temperatures would be predicted to cause more *melting* of snow and ice, resulting in glaciers melting back and becoming less extensive, as has been observed.

4. Higher temperatures and altered ocean circulation could lead to more drought in some places and even the expansion of desert areas, the process of *desertification*.

5. Changes in global temperatures and accompanying changes in precipitation patterns can affect the *distribution of communities* of plants and animals. In response to warming, such communities may shift to higher elevations or higher latitudes to stay within an optimal temperature range. Climate change is predicted to decrease the geographic range of some communities, while increasing the range of others. Included in this consideration are croplands, some of which will benefit from global warming, while others will suffer.

6. An increase in global temperatures is predicted to increase *evaporation* from warmer surface waters. This would increase both humidity and water-vapor capacity of the air. Warmer temperatures and more humidity may increase *precipitation*. More precipitation, along with more melting of glaciers, would lead to more runoff from streams, with an associated increase in the amount of flooding. Increased runoff also means a larger influx of fresh water into the ocean.

7. Climate change has the potential to change the frequency and intensity of *severe weather*. For various reasons, global warming can either increase or decrease the frequency of certain types of severe storms. These are presented on the next page.

8. Global warming is predicted, and observed, to cause an increase in global sea-surface temperatures (SST) and a decrease in the amount of *sea ice*, at least in the Arctic.

9. The various consequences have complex interactions, as between SST, humidity, precipitation, runoff, and influx of fresh water into the ocean. Such interactions could result in other changes, like a change in ocean currents, including those involved in the *thermohaline conveyor*, an important moderator of global climate.

07.13.a1

10. The computer-generated images below show the decrease in the amount of sea ice in the Arctic from 1979 (left globe) to 2003 (right globe), the period of the most rapid global and regional warming. Additional loss of Arctic sea ice has occurred since 2003, with some year-to-year increases but an overall decrease.

11. Since the middle of the 1800s and before, sea level has been slowly increasing, rising 0.2 meters in 200 years. Currently, there is scientific debate about whether the rate of sea level rise is remaining the same or is accelerating with time.

1979

07.13.a2

2003

07.13.a3

B   How Might Climate Change Promote More Severe Weather?

In 2005, an unprecedented 28 named tropical storms formed in the Atlantic, and 19 formed in 2012. Patterns such as these have led people to wonder whether we are moving into an era of more numerous and destructive tropical cyclones, with the increase attributed to climate change. What does the science say about these important questions? Observed increases in the frequency of all forms of severe weather are at least partially the result of Doppler radar and other improved remote-sensing technology, which allow us to detect storms more easily. One approach is to compile data on severe storms over time, to examine whether the frequency and severity of storms correlates with observations of climate change. But remember that simple correlation does not demonstrate causation.

1. Observational and modeling studies suggest that climate change will increase SST, which in turn might increase the intensity and duration of the strongest storms. Warming might also lengthen the severe tropical cyclone season. In theory, increased SST would allow an increase in evaporated water, which increases the amount of latent heat in the atmosphere. Since latent-energy release—when the evaporated water condenses or deposits—is a key ingredient for any type of severe weather, increased evaporation can result in more energy for storms. The end result is larger, taller clouds fueling more powerful storms.

2. Warmer sea and land surfaces are also likely to affect atmospheric stability. Recall that with warmer air near the surface, the environmental lapse rate will steepen. A steep environmental lapse rate enhances the ability of air to rise, increasing the instability of the atmosphere. In other words, warmer air near the surface will contribute to taller clouds as long as moisture is present.

07.13.b1

3. Global warming could cause changes to upper-level circulation, which in turn could cause underlying areas to experience more frequent precipitation or drought. Changes in upper level circulation patterns might make the upper levels of the atmosphere more or less favorable for cyclone formation. Specifically, if climate change displaces or decreases the flow of upper-level westerlies, fewer developing tropical cyclones may have their tops sheared off as they move westward. Alternatively, global warming could increase vertical wind shear, which might cause mid-latitude storms to be more severe. What do the data say?

Frequency of Tropical Storms

4. The graph below plots the number of Atlantic storms for each year. The time periods represented on the graph correspond to the interval of time when global temperatures have demonstrably increased (although in detail, temperatures have gone up and down during several approximately 30-year cycles). Research using these and similar data suggests that short-lived Atlantic tropical cyclones may have become more frequent over the years, but moderate-duration storms have not.

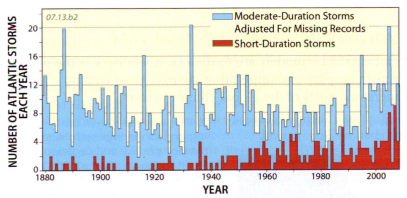

5. The research on this question is ongoing, with some studies predicting an increase in intensity, while others find a decrease in intensity. Other studies acknowledge that changes in tropical cyclones under climate change scenarios will be small in relation to other factors that create variability in tropical cyclones, such as ENSO.

Frequency of Tornadoes

6. Another question to evaluate is whether global warming has caused an increase in the number of mid-latitude cyclones or tornadoes. Some scientists have proposed that warming provides additional energy that might enhance the environment for such severe storms. Examine the graph below, which plots the number of tornadoes per year during the time when most warming occurred.

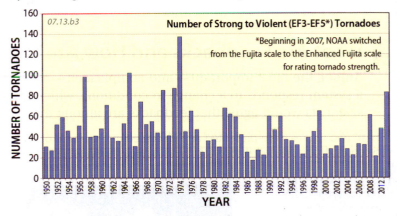

7. The data indicate that the number of strong-to-violent tornadoes in the U.S. has not increased over time. One possible explanation is that increasing surface temperatures would presumably be accompanied by increases in upper-level temperatures, leading to no net change in stability, as the adiabatically rising air and the air around it have both warmed.

Before You Leave This Page Be Able To

✓ Sketch and explain some possible consequences of climate change, specifically global warming, to the environment.

✓ Discuss some ways that climate change could increase the frequency and strength of severe weather and some relevant data.

7.15 How Does the Climate System Sustain Life?

GLOBAL ENERGY, HYDROLOGIC SYSTEMS, AND BIOLOGIC SYSTEMS are interdependent, linked by processes, like photosynthesis, evapotranspiration, and albedo, through a system of complicated feedbacks. How does the climate stay at conditions favorable for life, despite the fact that it is always changing? Alternatively, can the biosphere help sustain a favorable environment through such self-regulating processes? This possibility is demonstrated in an Earth-like planet populated only by black or white daisies—*Daisy World*. Daisy World illustrates the connections between various components of the climate, including life.

A How Are Energy, Water, and Plants Linked in Daisy World?

On this hypothetical planet, Daisy World (▶), there are two kinds of daisies—white daisies and black daisies. The black daisies have a low albedo, so they absorb more insolation than do white ones. In hot or bright conditions, however, the dark color is a disadvantage because the black daisies retain so much insolation that they overheat. The white daisies reflect more insolation, and so they do better in hot environments, but not as well in cold ones (they do not get enough energy). There are no "daisy-eaters," so the relative survival rates of the two daisies depend only on how they interact with the climate.

In too-cool times, black daisies out-compete white daisies.

Overheating

Overheating makes it too warm for black daisies.

Daisy World

Initial Condition

Dynamic Equilibrium

In too-warm times, white daisies outcompete black daisies.

Overcooling

Overcooling makes it too cold for white daisies.

07.15.a1

After observing the setup of Daisy World (◀), examine the figure below, which illustrates how energy, water, and the daisies interact to keep the climate within certain bounds of temperature and water vapor. The letter and number on each arrow references text adjacent to the figure. E arrows involve energy, W involve water, and B refers to the biosphere (daisies). Refer to chapters 2, 3, and 4 to review key topics discussed below.

Energy

E1. Insolation is identical in quantity and quality to that of Earth.

E2. Cloud albedo is similar to that on Earth, which ranges from 50% to 70%. The area of clouds varies, and is generated by the planetary water subsystem.

E3. The proportion of insolation reflected to space by clouds depends on the type and extent of clouds, which are controlled by the water subsystem.

E4. Insolation not reflected by clouds proceeds down toward the planetary surface.

E5. The surface is determined by planetary land cover (biosphere). The surface reflects a proportion of incident insolation back to space.

E6. Total energy absorbed by the biosphere is the sum of insolation (E5) and the re-radiated longwave radiation called *counterradiation* (E10).

E7. Insolation returned to space, or planetary albedo, comprises energy reflected from changing cloud and land surface covers.

E8. Energy that does not leave via planetary albedo must exit as longwave radiation, controlled by the temperature of the planet and the efficiency at which emission occurs for the gas.

E9. Water vapor (water subsystem) is the only greenhouse gas. Its quantity (clouds) therefore controls temperatures at the base of the clouds.

E10. Longwave radiation is radiated back to the surface by the atmosphere.

Water

W1. Like Earth, Daisy World is considered to be a closed system with respect to water, so no water leaves across the planetary boundary.

W2. Cloud cover is a function of the water vapor stored in the atmosphere (controlled by temperature and the energy subsystem), and the imbalance between evaporation inputs and precipitation outputs. When Precipitation (P) < Evaporation (E) cloud cover grows; when P > E cloud cover shrinks.

W3. Evaporation is driven by latent heat flux (governed by the energy and biosphere sub-systems), and by the quantity of water already in the atmosphere.

W4. Precipitation is positively related to cloud cover and based on the average precipitation on Earth of 1,000 mm.

W5. Temperature at the cloud base, which controls E10, is governed by surface temperature (biosphere subsystem) and the adiabatic lapse rate.

W6. In reality, the adiabatic lapse rate is controlled by moisture content, defined by the planetary cloud cover.

Shortwave Longwave

E7 E1 E8 W1

E3

E2 W2

ATMOSPHERE E9 W5

E6 E4 E P W4 W6

BIOSPHERE E5 E10 W3

B2 B1

Bare Ground

07.15.a2

Biosphere

B1. Black and white daisies share a constant mortality rate. Growth rates decline as temperatures above each daisy type diverge from a common optimum. In this way, the areas occupied by the two shades of daisy change with changes in the energy and water subsystems.

B2. Planetary surface temperature is a combination of temperatures over bare ground and black and white daisies, proportional to their planetary areas and albedos. Temperature rises as the number of black daisies increases, and vice versa.

B How Do Biophysical Features on Daisy World Interact?

Variables in Daisy World—and on Earth—all interact with, and adjust to, one other. Some adjustments are almost instantaneous on an annual scale, like changes in planetary albedo, while other linkages, such as the daisy growth rate, take time to fully impact the system. This diagram illustrates some of the ways in which Daisy World could evolve through time, through positive and negative feedbacks between energy, water, and biology. In many natural systems the variables are constantly adjusting to one another because of time lags between the input and response.

Constraints to the Model

1. The air above black daisies is always warmer than above white daisies because of the black daisies' lower albedo (25%) compared to that of white daisies (75%).

2. Planetary temperatures tend toward those associated with the prevailing daisy type (warmer over black daisies), although some of the planet is covered by bare ground.

3. As temperatures above each daisy type approach or diverge from the optimum (22.5°C here), growth rates increase or decrease.

4. Mortality rates of both daisies remain fixed, thus proportions of the planet covered by each daisy do not change instantaneously. However, there is an accelerated changeover from white to black daisies as higher temperature above black daisies approaches the optimum, while temperatures above white daisies are getting colder, diverging from the optimum.

5. Higher proportions of black daisies reduce planetary albedo and rapidly warm the planet.

Feedbacks Within Daisy World

6. Suppose that an area has a higher proportion of black daisies. What do you think will happen to temperature and atmospheric moisture in this scenario?

7. The black daisies have a lower albedo, so they reflect less insolation back to space and retain more of the energy, heating up the surface. A warmer planet leads to more evaporation, more water vapor in the air, more clouds, a more active hydrologic cycle, and a higher proportion of the planet covered in clouds.

8. Greater cloudiness increases planetary albedo and reduces the insolation reaching the daisies. However, water vapor is a greenhouse gas, thus the presence of additional cloud cover, while increasing global albedo, also restricts cooling from the loss of outgoing longwave radiation.

9. Eventually, the planet may get too hot for the heat-absorbing black daisies, so more black daisies die, causing the planet to cool, which then favors the white daisies.

Black Daisies

White Daisies

MOSTLY BLACK DAISIES

More Reflection
More Clouds

Less Reflection

More Heat

Less Reflection
Fewer Clouds

More Reflection

Less Heat

More Water Vapor

Less Water Vapor

WARMER   Too Warm for Black Daisies

Too Cold for Black Daisies   COOLER

System Stays Within Certain Bounds

07.15.b1

10. Suppose instead that an area has a higher proportion of white daisies. Predict what will happen to temperatures and atmospheric moisture in this scenario.

11. The white daisies have a higher albedo, so they reflect more insolation back to space, retaining less of the energy. This causes the surface to cool. A cooler planet has less evaporation, less water vapor, fewer clouds, and a less active hydrologic cycle. Less of the planet is covered in clouds, and there is less heat-trapping greenhouse gas. The proportion of white daisies continues to grow.

12. Following the peak abundance of white daisies, the high surface albedo (white daisies) and other factors cause a decrease in temperatures, decreasing the abundance of warm-adapted white daisies. At this point, the growth rate of white daisies declines at an ever-increasing rate, while that of the black daisies grows. A repetition of the cycle has then been initiated. Does Earth function in a similar way, with positive and negative feedbacks that keep the systems within certain bounds? We hope so!

Before You Leave This Page Be Able To

☑ Describe the link between characteristics of the two types of daisies in Daisy World and how this can alter temperatures, atmospheric water, and the proportion of white and black daisies, at both local and global scales.

☑ Explain what happens with the hydrologic cycle, particularly evaporation and cloud cover, in Daisy World as the planetary temperature increases, and describe the effects on planetary energy budget and planetary temperatures.

☑ Describe the concept of a planet as a self-regulating system, operating within certain limits because of positive and negative feedbacks. You can build your explanation around Daisy World or around Earth.

What Climates and Weather Would Occur Here?

PLANET S is a hypothetical replica of Earth. It has plate tectonics, oceans, and an atmosphere very similar to Earth's. You will map the climatic zones on this planet, identifying cold and warm ocean currents, prevailing winds, potential locations of rain forests and deserts, places where the climate and topographic setting would be suitable for agriculture, and sites at risk for hurricanes.

Goals of This Exercise:

- Observe the general map patterns and photographs from different parts of the planet. Note any clues about likely weather and climate.
- Create a map showing probable climatic conditions, based on ocean currents and prevailing winds.
- Locate likely sites for rain forests, deserts, agricultural areas, and places where storms will strike land.

Procedures

Follow the steps below, entering your answers for each step in the appropriate place on the worksheet or online.

1. Observe the distribution of continents and oceans on the map below and on the larger version on the worksheet. Examine the photographs of the various areas and infer the environments and climates that are represented. Each photograph has a letter that corresponds to a letter on the map. Some photographs are on the next page.

2. Carefully examine the various types of data on the next page. For each data set, examine the climatic and weather implications for different parts of the planet.

3. Draw on the worksheet your interpretations of whether each ocean current is warm, cold, or neither, and whether prevailing winds will bring warmth or coolness, and cause dryness or precipitation.

4. Map the main climatic zones. Include the likely locations of rain forests, deserts, areas suitable for croplands, and areas at risk for hurricanes and other major tropical cyclones. Be prepared to discuss your observations and interpretations.

5. Your instructor may provide you with climographs so that you can assign Köppen climate types to different regions.

07.16.a2

07.16.a3

07.16.a1

07.16.a4

07.16.a5

07.16.a6

07.16.a7

Some Important Observations About the Planet

1. The planet rotates around its axis once every 24 hours in the same direction Earth does. Therefore it has days and nights similar to Earth's. It orbits a sun similar to our own, at about the same distance. Its spin axis is slightly tilted relative to the orbital plane.

2. The oceans are likewise similar to those on Earth. They show similar variations in sea-surface temperature and have ocean currents that circulate around the edges of the main ocean basins and in gaps between the continents.

3. The planet has well-developed atmospheric currents, with wind patterns similar to those on Earth.

Preliminary Interpretation of Ocean Currents

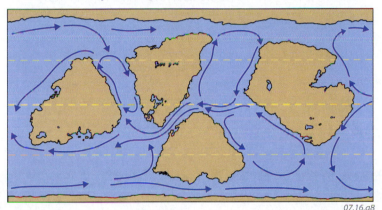

07.16.a8

This map shows the directions of ocean surface currents. From these data, identify which currents are probably cold, warm, or neither. Water temperatures will be key in inferring probable climates.

Preliminary Interpretation of Wind Data

07.16.a9

These data show the directions of prevailing surface winds, which are similar to the patterns on Earth. Different wind directions occupy bands encircling the planet, parallel to the equator.

Ocean Water Temperatures and Sea-Surface Pressures

07.16.a10

07.16.a12

07.16.a13

This map depicts satellite data for sea-surface temperatures (SST) during late summer. The numbers give the temperature values in degrees Celsius (°C).

31°

16°

2°

07.16.a11

Climate Analysis

On the version of this map on the worksheet, show the probable locations of the following features by writing the associated letters on the map: rain forests (R), deserts (D), potential agricultural areas (A), and lands at highest risk for hurricanes (H). You can use colored pencils to shade in the extent of land associated with each letter. Your instructor may also have you assign likely Köppen climate types to different sites or regions. Complete this step on the worksheet.

07.16.a14

7.16

Water Resources

WATER IS ARGUABLY THE MOST IMPORTANT RESOURCE provided by Earth—all life on Earth needs water to live and thrive. We are most familiar with *surface water,* water that occurs in streams, lakes, and oceans. Yet, the amount of fresh water in these settings is much less than the amount of fresh water that is frozen in ice and snow or that occurs in the subsurface as *groundwater.* This chapter is about surface water and groundwater and the important ways in which they interact.

08.00.a2 Shoshone, ID

The Snake River Plain, shown in this large satellite-based image, is a curved swath of low land that cuts through the mountains of southern Idaho. It contains a mixture of dry, sage-covered plains, water-filled reservoirs, green agricultural fields, and recent lava flows of dark-colored basalt, a volcanic rock. Most of Idaho's population lives on the Snake River Plain near the rivers and reservoirs.

The Big Lost River, Little Lost River, and adjacent streams that enter the plain from the north never reach the Snake River. Instead, the water from the rivers seeps into the ground between the grains in the sediment and through narrow fractures in the volcanic rocks. For this reason the rivers are called "lost."

Where does water that seeps into the subsurface go?

▲ **Within the Snake River Plain**, the Snake River has eroded a canyon down into layers of ledge-forming volcanic rocks and slope-forming sediment. The farmlands of the canyon bottom are on fertile sediment deposited by the river, and they receive water from rivers, springs, and wells drilled to extract groundwater.

Lost River Area

Craters of the Moon N.M

Snake River

Thousand Springs

Snake River

At Thousand Springs, large springs gush from the steep volcanic walls of the Snake River Canyon (◄). The canyon includes 15 of the 65 largest springs in the U.S., including those at Thousand Springs. The largest commercial trout farms in the U.S. use ponds fed by these springs.

What causes water from beneath the ground to flow to the surface as a spring, and where does the water in a spring come from?

08.00.a3 Thousand Springs, ID

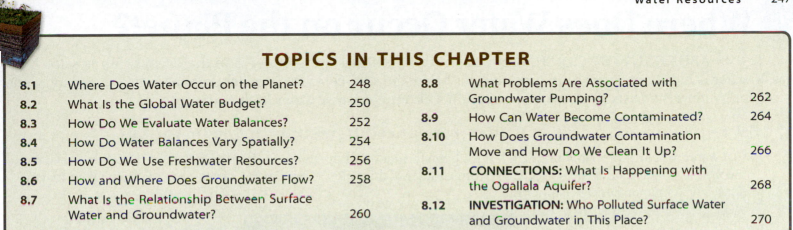

TOPICS IN THIS CHAPTER

The Snake River winds through mountains and then flows southwest and west across the Snake River Plain. *Where does this river, flowing across such a dry plain, receive its water?*

08.00.a4 Jackson Hole, WY

The river begins its journey in Jackson Hole, Wyoming (▲), from streams that drain the Grand Tetons and nearby Gros Ventre mountain range. The relatively higher rainfall and snowmelt in these highlands sustain the river as it flows westward across the dry plains. Farther downstream, rivers and streams entering the plain from the east and south flow directly into the Snake River, increasing its flow.

Where does the water in rivers come from, and do most rivers gain or lose water from groundwater?

08.00.a1

Many lakes and farms are situated next to the Snake River. Farmers irrigate millions of acres of agriculture with surface water derived from reservoirs, lakes, streams, and groundwater pumped to the surface. Chemicals used by some farms cause contamination of groundwater and surface water.

What happens if groundwater is pumped from the subsurface faster than it is replaced by precipitation and other sources? What do we do if water supplies are contaminated?

Disappearing Waters of the Northern Snake River Plain

Groundwater beneath the Snake River Plain is an essential resource for the region, providing most of the drinking water for cities and irrigation water for farms and ranches away from the actual river. Hydrologists and other scientists study where this water comes from, how it moves through the subsurface, and potential limits on using this resource.

Some water enters the subsurface from the Big and Little Lost Rivers, which flow into the basin from the north and then disappear as their water sinks into the porous ground. Other groundwater comes directly from the main Snake River and from tributaries that enter the basin from the south and east. The largest influx of water to the subsurface is seepage from irrigated fields and associated canals.

The surface of the Snake River Plain slopes from northeast to southwest. The flow of groundwater follows this same pattern, flowing southwest and west through sediment and rocks in the subsurface. Groundwater derived from the disappearing rivers flows southwest, along the north side and center of the basin. The groundwater does not flow like an underground river but as water between the sediment grains and within fractures in the rocks. Where the Snake River Canyon intersects the flow of groundwater, water reemerges on the surface, pouring out at Thousand Springs. This region illustrates a main theme of this chapter—surface water and groundwater are a related, interconnected, and valuable resource.

8.0

8.1 Where Does Water Occur on the Planet?

WATER IS ABUNDANT ON EARTH, occurring in many settings. Most water is in the oceans, but is salty. Most fresh water is in ice and snow or in groundwater below the surface, with a smaller amount in lakes, wetlands, and streams. Water also exists in plants, animals, and soils and as water vapor in the atmosphere.

A Where Did Earth's Water Come from and Where Does It Occur Today?

Most water on Earth probably originated during the formation of the planet or from comets and other icy celestial objects that collided onto the surface. Over time, much of this water moves to the surface, for example when magma releases water vapor during eruptions.

Oceans—Of Earth's total known inventory of water, an estimated 96.5% occurs in the oceans and seas as *saline* (salty) *water*. The remaining 3.5% is *fresh water* held in ice sheets and glaciers, groundwater, lakes, swamps, and other water features on the surface.

Streams—Streams are extremely important to us and are the main source of drinking water for many areas. They contain, however, only a very small amount of Earth's fresh water.

Lakes—Water occurs in lakes of various sizes. Many are freshwater lakes, but those in dry climates are saline or *brackish* (between fresh and saline). Lakes contain a majority of the liquid fresh water at Earth's surface, but most of this water lies in a few large lakes around the world.

Swamps and Other Wetlands—These wet places contain water lying on the surface and within the plants and shallow soil. They constitute about 11% of the liquid fresh water on the surface.

08.01.a1

Atmosphere—A small, but very important, amount of Earth's water is contained in the atmosphere (0.001%). It occurs as invisible water vapor, as water droplets in clouds, and as rain, falling snow, and other types of precipitation.

Glaciers—Nearly 69% of Earth's fresh water is tied up in ice caps, glaciers, and permanent snow. A small amount also exists in *permafrost* and ground ice.

Soil Water—Earth's soils contain about as much water as the atmosphere, but like water in the atmosphere, soil water is crucial to our existence.

Biological Water—Water is also within the cells and structures of plants and animals. It is clearly important to us but represents an exceptionally small percentage of Earth's total water (0.0001%).

Deep-Interior Waters—An unknown, but perhaps very large amount of water is chemically bound in minerals of the crust and mantle. Some scientists think Earth's interior may contain more water than the oceans.

Groundwater—About 30% of Earth's total fresh water occurs as groundwater (water below the reach of tree and plant roots). Groundwater is mostly in the open pores between sediment grains or within fractures that cut rocks. Most groundwater is fresh, but some is brackish or saline.

▶ These bar graphs show estimates of the distribution of water on Earth's surface and the uppermost levels of the crust. The left bar shows that the oceans contain 96.5% of Earth's total free water (water that is not bound up in minerals), but this water is saline. Only 3.5% of Earth's water is fresh. These graphs do not include water in Earth's atmosphere or deep interior.

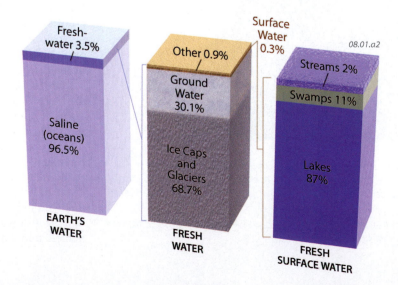

08.01.a2

◀ The middle bar shows that most of Earth's fresh water occurs in ice caps and glaciers. Almost all the rest is groundwater. Less than 1% occurs as liquid surface water in lakes, swamps, and streams.

◀ The right bar shows where Earth's small percentage of fresh, liquid, surface water resides. Most is in lakes, followed by swamps and streams.

B How Does Water Move from One Setting to Another?

Water is in constant circulation on Earth's surface, moving from ocean to atmosphere, from atmosphere back to the surface, and in and out of the subsurface. The circulation of water from one part of this water system to another, whether above or below Earth's surface, is called the *hydrologic cycle.* From the perspective of living things, the hydrologic cycle is the critical system on Earth. It involves a number of important and mostly familiar processes. It is driven by energy from the Sun.

Evaporation—As water is heated by the Sun, some of its molecules become energized enough to break free of the attractive forces binding them together. Once free, they rise into the atmosphere as water vapor.

Condensation—As water vapor cools, as when it rises, it condenses into a liquid or turns directly into a solid (ice, hail, or snow in deposition). These water drops and ice crystals then collect and form clouds.

Precipitation—When clouds cool, perhaps when they rise over a mountain range, the water molecules become less energetic and bond together, commonly falling as rain, snow, or hail, depending on the temperature of the air. Precipitation may reach the ground, evaporate as it falls, or be captured by leaves and other vegetation before reaching the ground.

Sublimation—Water molecules can go directly from a solid (ice) to vapor, a process called sublimation (not shown here). Sublimation is most common in cold, dry, and windy climates, such as some polar regions.

Infiltration—Some precipitation seeps into the ground, infiltrating through fractures and pores in soil and rocks. Some of this water remains within the soil, and some rises back up to the surface. Water can also infiltrate into the ground from lakes, streams, canals, or any body of water.

Groundwater Flow—Water that percolates far enough into the ground becomes groundwater. Groundwater can flow from one place to another in the subsurface, or it can flow back to the surface, where it emerges in springs, lakes, and other features. Such flow of groundwater may sustain these water bodies during dry times.

Ocean Gains and Losses—Most precipitation falls directly into the ocean, but oceans lose much more water to evaporation than they gain from precipitation. The difference is made up by runoff from land, keeping sea level close to constant.

08.01.b1

Surface Runoff—Rainfall or snowmelt can produce water that flows across the surface as runoff. Runoff from direct precipitation can be joined by runoff from melting snow and ice and by groundwater seeping onto the surface. The various types of runoff collect in streams and lakes. Most is eventually carried to the ocean by streams where it can be evaporated, completing the hydrologic cycle.

Transpiration—Some precipitation and soil water is taken up by root systems of plants. Through their leaves, plants emit water that evaporates into the atmosphere by the process of transpiration.

Before You Leave This Page Be Able To

☑ Summarize where most of Earth's total water resides.

☑ Describe the different settings where fresh water occurs, identifying which settings contain the most water.

☑ Describe or sketch the hydrologic cycle, summarizing the processes that shift water from one part to another.

8.1

8.2 What Is the Global Water Budget?

THE HYDROLOGIC CYCLE is key to many physical and biological processes on Earth as water moves in and out of different parts of the system. It also results in the conversion of solar energy to other forms of energy, such as when water that evaporated from the ocean later falls as rain, forms a river, and moves material in a streambed.

A How Much Water Moves Between the Global Stores of Water?

1. When considering motion of water in the Earth system, we want to know how *much* water a part of the system—called a *store*—holds, expressed as a volume (e.g., km³). We also measure how much water moves in and out of that store over some amount of time, which is a *flux*, in this case described with units of volume per time. The three main stores are the oceans, continents, and atmosphere (▶).

2. Annually 577,000 km³ of water evaporates and falls from the atmosphere. The daily quantity of energy involved in this process is approximately 2 million times the worldwide energy-generating capacity.

3. The ocean holds about 97% of Earth's known water, and evaporation over oceans accounts for 84% of all global evaporation. About 77% of Earth's precipitation falls on the oceans. These high percentages emphasize the significance of oceans as moderators of global temperatures.

4. In terms of fluxes between the atmosphere and the oceans, the ocean store would be reduced by 7% (84% minus 77%) annually.

5. Approximately 23% of global precipitation falls over continents. This is a little less than the percentage of the Earth's surface that continents occupy. However, only 16% of global evaporation comes from the continents—large areas of the continents are in arid (dry) climates, and much continental water is stored as groundwater or as ice and snow in regions of low insolation.

6. In terms of fluxes, input of precipitation to the continents exceeds evaporation output by 7% annually. That is, every year, the continents gain water from their exchange with the atmosphere. Without further outputs, the stores of fresh water on the continents should be increasing.

7. The imbalances between the atmosphere-ocean and atmosphere-continent exchanges (difference in inputs and outputs) is redressed by the flow of excess waters from the continents to the oceans by all the rivers in the world—*global runoff*. Note that approximately two-thirds of all water falling on the continents leaves as evaporation and one-third leaves by streams. Evaporation from the global oceans is 12 times larger than the combined flows of all the streams in the world.

ATMOSPHERE (12.9; ~0.001%)

08.02.a1

DIFFERENCE =7%

PRECIPITATION OVER OCEAN (77% OF TOTAL)

PRECIPITATION OVER LAND (23% OF TOTAL)

DIFFERENCE =7%

EVAPORATION OVER OCEAN (84% OF TOTAL)

EVAPORATION OVER LAND (16% OF TOTAL)

OCEANS (1,338,000; 97%)

GLOBAL RUNOFF (7% OF TOTAL FLUX)

CONTINENTS (47,660; 3%)

UNITS IN 1,000 KM³

B What Factors Control How Quickly Water Moves Through Each Store?

The amount of water in the stores is of practical interest, but says little about the dynamic nature of the planetary surface. Movement of water between global hydrologic stores represents energy available to do work in shaping the Earth's surface. A measure of the "dynamism" of each store is the average length of time that a drop of water remains in that store—its average *residence time*.

1. Consider a box (a store) with 10 spheres (think water molecule), including one of particular interest in red. In each time period, one sphere is removed (output) and a new sphere inserted (input). On average how many time periods will pass before the red sphere is selected (i.e., how long will the red sphere stay in the store)?

Volume of Store = 10 Balls

OUTPUT =1 Ball per Time Period

INPUT =1 Ball per Time Period

08.02.b1

2. Application of statistical theory indicates that the red sphere will most likely be removed (output) during the first sampling period (the time period over which, in this case, 10 spheres could have been output). It is less likely that the sphere would be around for successive samplings. The average length of time that the sphere stays in the store is 10 time periods.

Volume of Store = 10 Balls

OUTPUT =2 Balls per Time Period

INPUT =2 Balls per Time Period

08.02.b2

3. When the rates of inputs and outputs are increased (e.g., 2 spheres per time period), the chances of selecting the red sphere increase and the average time period declines to 5.

Volume of Store = 20 Balls

OUTPUT =1 Ball per Time Period

INPUT =1 Ball per Time Period

08.02.b3

4. If the size of the store (number of spheres) is increased to 20, then the chance of selecting the red sphere at any time period declines and the average time period increases to 20.

5. By making a few assumptions, like the size of the system remaining constant, we can calculate the average residence time for a molecule if we know the volume of the store and the average rate of flow into and out of the store (average rate of movement). Such calculations are useful in evaluating local and global water budgets.

C What Are Average Residence Times in the Global Stores of Water?

1. Water moves at different rates through each different subsystem (e.g., evaporation from the ocean) and its stores (e.g., ocean and atmosphere). This figure depicts the average rates at which water moves from one store to another. This rate, in turn, indicates how long, on average, that molecule will stay in the store—its *residence time*. To calculate the residence time, we simply divide the total volume of the store by the flux leaving it.

2. The volume of the atmospheric store is very small (12,900 km³), yet 577,000 km³ of water enters and leaves it annually, yielding average residence times of approximately 0.02 yr (i.e., 12,900/577,000), or 8 days. Thus, once evaporated into the atmosphere from the planetary surface, a water molecule stays there on average a very short time. The atmosphere is therefore very responsive to changes in energy conditions at Earth's surface.

4. Water stored in and on the continents (47,600,000 km³) is smaller than the amount of water in the oceans but much larger than the amount in the atmosphere. The input is precipitation and the outputs are evaporation and runoff. The relatively limited spatial extent of continents causes the average annual fluxes to be considerably smaller than between ocean and atmosphere. The combination of smaller fluxes and the intermediate size of the continental store results in an average residence time of approximately 350 years. Recall that the vast majority of fresh water on continents is stored in groundwater and ice, both of which may "lock up" water for centuries. Here, we have ignored water that percolates into the soil as groundwater.

PRECIPITATION ONTO OCEAN (444/YR)

PRECIPITATION ONTO LAND (133/YR)

EVAPORATION FROM OCEAN (485/YR)

EVAPORATION FROM LAND (92/YR)

SURFACE RUNOFF (40/YR)

Units in 1,000 km³   08.02.c1

3. Oceans are by far the largest store (1,338,000,000 km³). Inputs (precipitation and runoff) total 484,700 km³/yr, as does the output of evaporation. This yields an average residence time on the order of 3,000 years (1,338,000,000/484,700). The oceans are stratified vertically (warmer on top than on bottom) and evaporation varies geographically, but once water enters the ocean store it will likely stay there an average of several thousand years. Oceans therefore respond more slowly to changes, despite the fact that the atmosphere and oceans are strongly linked through evaporation and precipitation.

D How Are Fluxes of Water Linked to Shaping of Earth's Surface?

When we imagine the processes that reshape landscapes, we might think of events like powerful floods, scouring glaciers, turbulent mudflows, and blowing sand. All of these involve energy resulting from the movement of liquid water, ice, Earth materials, or the atmosphere. This type of energy is called *kinetic energy*—energy by virtue of motion. Also linked to the flux of water in some settings is *potential energy*—energy by virtue of being some height above a surface and capable of moving if allowed to.

1. Examine this photograph of a balanced rock, which obviously has the potential to fall sometime in the future—the rock has *potential energy* relative to the ground below (the datum). We can imagine how hard the rock would strike the ground when it fell, imparting a large force onto the ground it strikes. Clearly, this rock has a large amount of potential energy. ▶

2. With the help of gravity, the rock's potential energy is converted into *kinetic energy* (the energy of motion) when the rock begins to fall. The resulting motion of the rock can perform *work*, for instance the crushing of stones on which the rock falls.

3. After the rock has fallen and is in its new, lower position, it has lost its potential energy relative to the ground surface (the datum). We could increase its potential energy if we used a crane to lift the rock back onto its pedestal. This would require an expenditure of energy by the crane, and that energy expenditure is now stored in terms of the greater elevation and therefore greater potential energy. Note that once the rock is on the ground it could be moved by running water, such as that released by the thunderstorm in the background. The relatively short average residence times of water (calculated in the section above) show that water can quickly evaporate from the ocean, be lifted by storms, and be pulled by gravity back to the Earth as precipitation, renewing the water's ability to move across the surface.

08.02.d1 Arches NP, UT

08.02.d2

Condensation

Latent Heat Released

Evaporation

Energy Stored as Latent Heat

4. By means of latent heat transfer and the evaporation of water, the Sun's energy is able to renew the potential energy of water molecules by lifting them tens of thousands of meters above the datum (sea level). Following condensation and the formation of precipitation, the molecules can fall on high-elevation topography to begin the journey downslope, picking up kinetic energy. The short residence time within the atmospheric store ensures that this process is repeated over and over again with a very high frequency.

Before You Leave This Page Be Able To

✓ Sketch and explain the comparative magnitudes of the hydrologic fluxes between the global stores of water.

✓ Explain what is meant by the average residence time, and how the size of the store and magnitude of fluxes affect it.

✓ Explain potential and kinetic energy and how these could relate to the short average residence time of water in the atmosphere.

8.2

8.3 How Do We Evaluate Water Balances?

THE CONCEPT OF A WATER BALANCE is critical in understanding our world and concerns a broad range of topics, from evaluating the health of an ecosystem to gauging the sustainability of a city. Just as an accountant monitors the amount and flow of money in financial transactions, water balance is an accounting of the amount of water in each store and its rate of movement through the hydrologic cycle. The *climatic water balance* allows us to estimate water supplies and flows. Computer models based on some simple equations are used to assist us in calculating the climatic water balance at a local place.

A How Can the Gain or Loss of Water at the Surface Be Estimated?

To determine the water balance, we need to know how much is entering the system (the input) and how much is lost from the system (the output). On the previous two pages, we did this at a global scale, determining how continents were gaining water through precipitation and losing an equal amount through evaporation and runoff—they were in balance. Equally important is determining the water balance at regional and local scales, but first we need to know the inputs and outputs.

08.03.a1

Input (Gain)

1. Water is gained at the surface by precipitation (P) or by runoff entering the area, bringing in water that originated as precipitation from someplace else. Relatively simple instruments are usually used to measure P (in cm or inches), including a standard rain gauge (◄). Various factors cause measurements of P to be smaller than reality, including the effects of turbulence, water evaporating from inside the tube, and water drops that cling to the sides of the gauge and so are not measured.

2. When precipitation falls as snow, we can measure it with (1) a vertical rod marked with distances above the ground (►), (2) electronic height-measuring sensors, or (3) devices called snow pillows, which typically are several-meter-wide antifreeze-filled plastic sacks that measure the weight of snowpack and translate it into centimeters of water.

08.03.a2

3. More sophisticated tools, such as automated precipitation stations in the Automated Surface Observing System (ASOS) (◄), can measure and record P instantaneously. Radar is used to derive fast estimates of P, using a network of facilities. The resulting maps typically show highest rainfall amounts in orange and red. Satellites are also increasingly being used to measure P, especially in remote areas of the world.

4. There is a vast network of varied of weather stations across the U.S. and the more populated areas in the world (►).

08.03.a3

08.03.a4

08.03.a5

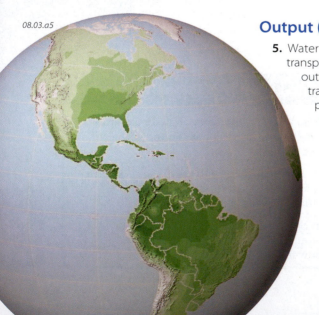

Output (Loss)

5. Water is lost from the surface to the atmosphere through direct evaporation or through transpiration—evaporation of water that has moved from a plant's roots, up through the stem, and out from a leaf's surface. Physical geographers usually don't distinguish between evaporated and transpired water, but instead combine the two as *evapotranspiration*. The theoretical maximum possible amount of water that can be evapotranspired from a wet surface, in centimeters or inches of precipitation over a given time period, is known as the *potential evapotranspiration* (PE). If P exceeds PE at a place for a certain period of time, there is "extra" water. This globe (◄) shows average annual potential evapotranspiration values across the Americas.

6. PE is estimated based on latitude and temperature. Latitude is important because it determines Sun angle (solar intensity) and length of daylight hours. Temperature is important because warmer air provides more energy to drive the evapotranspiration process. For more precise applications, other variables, such as humidity (drier air allows more PE), wind speed (faster winds cause more PE), and insolation are also used to estimate PE. But reliable measurements of these variables are only available at a few weather stations, so these factors are often absent from computer models.

B What Is the Influence of the Ratio of Precipitation to Potential Evapotranspiration?

When an area receives more precipitation (P) than it can lose through potential evapotranspiration (PE), the excess water must go somewhere. In contrast, there are many areas where PE routinely exceeds the amount of precipitation (P) and these are classified as arid and are commonly deserts.

P Greater than PE

1. When rain falls on the surface or snow melts, the resulting surface water can go several places. It can *evaporate* directly back into the air, as would be common in an area with a high PE. Or, it could instead be taken up by plants, commonly through the roots, to later be released back into the air during *transpiration*.

2. The water could also accumulate in the soil. After the soil becomes saturated, any extra water is called *surplus* and begins flowing downhill as *surface runoff* in streams and in less constrained sheets of water flowing across the surface (*sheetflow*). The running water can leave the area, carrying the surplus water away, or can accumulate locally in lakes and wetlands. Such *standing water* can later evaporate back into the atmosphere.

08.03.b1

3. Water that soaks into the soil is called *soil-water recharge*. This water exists in the part of the subsurface where much of the pore space is filled with air rather than water—the *unsaturated zone*. Soil water can be pulled back up to the surface by roots or by capillary action, or it can continue deeper into the subsurface until it eventually enters a zone where water fills nearly all the pore spaces and fractures—the *saturated zone*—becoming groundwater. This is where most water occurs in the subsurface. The top of the saturated zone is the *water table*, shown as a dashed red line. As with groundwater, the "excess" water can flow away.

PE Greater than P

4. During periods when PE exceeds P, in addition to evaporating the precipitated water directly, the atmosphere will evapotranspire water that is stored in soil in an attempt to meet its demand for water PE). This is called *soil-water use*. Soil can only hold a certain amount of water against gravity, termed its *field capacity*. A typical field capacity is about 15 cm (6 in) of equivalent precipitation. During times when PE exceeds P, the water stored in soil can be drawn down so much that the only soil water remaining strongly adheres to rock particles because of surface tension. Plants cannot easily get to this water, so this last amount of storage in the soil is termed the *wilting point*. Like many other water-related measurements, wilting point is measured in centimeters or inches of precipitation. Once soil water reaches the wilting point, any further unmet demand for water—the part of PE that is not provided by P or by soil-water use—is called *deficit*.

08.03.b2 Baja California Sur, Mexico

C What Are Some Applications of the Climatic Water Balance?

1. Weekly, the U.S. Climate Prediction Center runs a climatic water-balance program at hundreds of weather stations, using the most recent measurements of P and calculations of PE. The model calculates surplus or deficit for each station. The calculated values are then compared to normal surplus or deficits for that station and that time of year, and the results are mapped as the Palmer Drought Index (▶). Drought occurs when deficit exceeds "normal" deficit for that location at that time of year.

PALMER DROUGHT INDEX
May 2012

08.03.c1

| Extreme Drought −4.00 and Below | Severe Drought −3.00 to −3.99 | Moderate Drought −2.00 to −2.99 | Mid-Range −1.99 to +1.99 | Moderately Moist +2.00 to +2.99 | Very Moist +3.00 to +3.99 | Extremely Moist +4.00 and Above |

3. Places with abundant surplus may have other issues. Flooding may be a problem, and navigation on major rivers can be hazardous in such conditions. In extreme conditions, pressure on dams and levees can create major concerns. Places without large deficits or surpluses compared to "normal" are least likely to experience environmental stresses.

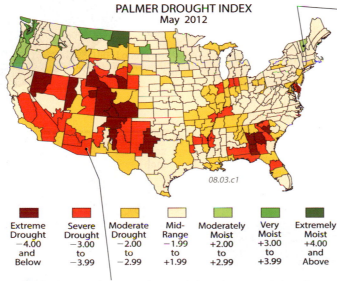
08.03.c2 Yuma, AZ

2. Places that are having a larger deficit than normal at that time of year for that location are considered to be in *drought*. In such places, forest fires may be a problem and planners may limit water use. Irrigation (◀) tries to supply enough water to keep agricultural soils above the wilting point, but this becomes more difficult and expensive during a drought.

Before You Leave This Page Be Able To

✓ Explain various ways in which precipitation is measured.

✓ Explain the concept of potential evapotranspiration and what happens when PE is greater than or less than P.

✓ Explain how we determine whether an area is experiencing drought or a water surplus.

8.3

How Do We Use Freshwater Resources?

WE USE LARGE QUANTITIES OF WATER each day, for a variety of purposes, especially power generation and irrigation of farms. How much water does each of our activities consume, and where does the water come from?

A What Are the Main Ways in Which We Use Fresh Water?

The U.S. Geological Survey (USGS) studied water use in the U.S. Its compilation shows that we use fresh water in six or seven main ways, depending on how we classify the data. For the year 2005, freshwater use in the U.S. was 349 billion gallons per day, most of which is from surface-water sources. Examine the graph below and think about the many ways you use water.

08.05.a1

◀ **1.** *Thermoelectric Power*— Electrical-generating power plants are the largest user of fresh water in the U.S., using 41% of the total amount of fresh water. Such power plants drive their turbines by converting water into steam and also use large amounts of water to cool hot components. Some of this water is from recycled sources. Power plants are also the largest users of saline (salty) water.

▶ **2.** *Irrigation*—Farms and ranches are the other large users of fresh water, using nearly 37% of fresh water. Farms use water from groundwater, streams, lakes, and reservoirs to irrigate grain, fruit, vegetables, cotton, animal feed, and other crops. Much of this water is lost through evaporation to the atmosphere before it can be used by plants. The sprinkler system shown here is designed to cut down on water use, as it delivers water more directly to crops than a central sprinkler.

08.05.a2 Ephrata, WA

▼ **3.** *Public and Domestic Water Uses*—The third largest use of fresh water is by public water suppliers and other domestic uses. We consume water by drinking, bathing, watering lawns, filling artificial lakes, and washing clothes, dishes, and cars. Much water from public water suppliers also goes to businesses. Most water for public and domestic use comes from streams and groundwater.

FRESH-WATER USAGE

Thermoelectric Power 41%

Irrigation 37%

Public and Domestic Use 12.6%

Industrial Uses 5.2%

Aquaculture 2.5%

Mining 1.1%

Livestock 0.6%

08.05.a4

08.05.a5 West Driefontane, South Africa

▲ **4.** *Industrial and Mining Uses*— Fresh water and saline water are extensively used by industries, including factories, mills, and refineries. Water is integral to many manufacturing operations, including making paper, steel, plastics, and concrete. Mining and related activities also use water in the extraction of metals and minerals from crushed rock.

08.05.a3 Phoenix, AZ

08.05.a6

◀ **5.** *Aquaculture*— According to the USGS study, we use approximately 2.5% of fresh water to raise fish and aquatic plants. Most of this use is in Idaho, near the Thousand Springs area. Such water is not totally consumed—much is released back into the Snake River.

▶ **6.** *Livestock*—Providing water for cows, sheep, horses, and other livestock accounts for only 0.6% of freshwater use, but much water is used for irrigation to raise hay, alfalfa, and other animal feed. Many ranches use small constructed reservoirs and ponds as the main water source for animals.

08.05.a7

B How Do We Use and Store Water?

08.05.b1

08.05.b2 Mississippi River, IA

08.05.b3 Charles River, MA

Electrical Generation—The movement of surface water can generate electricity. To do this, we build dams that channel water through turbines in a hydroelectric power plant.

Transportation—We use many large waterways, such as the Mississippi River, as energy-efficient transportation systems to move agricultural products, chemicals, and other industrial products.

Recreation—People use surface water in lakes and streams for many types of recreation, including swimming, tubing, rafting, boating, and fishing. We also use fresh water to fill ponds, fountains, and swimming pools.

08.05.b4 Folsom Dam, CA

◄ Surface waters are commonly stored in natural lakes and in constructed reservoirs behind concrete and earthen dams. Water for drinking and other municipal uses can be stored in underground or above-ground storage tanks.

► We construct canals, raised aqueducts, and large and small pipelines to move fresh water from one place to another, especially to irrigate farms or to bring water to large cities. Pumps in water wells move groundwater to the surface.

08.05.b5 Sunnyside, WA

C How Do We Refer to Volumes of Water?

Water-resource studies typically report volumes of water in one of three units: *gallons*, *liters*, or *acre-feet*. Gallons and liters are familiar terms, but the concept of an *acre-foot* of water requires some explanation.

How big is an acre? An acre covers an area of 4,047 m² (43,560 ft²). If a perfect square, an acre would be 64 m (210 ft) on a side. An acre is equivalent to 91 yards of an American football field. There are 640 acres in a square mile.

91 yards (273 feet; 83 m)
160 feet (49 m)
10 20 30 40 50 40 30 20 10
08.05.c1

An acre-foot of water is the volume of water required to cover an acre of land to a height of one foot. Imagine covering 91% of a football field (one acre) with a foot of water. An acre-foot is equivalent to about 326,000 gallons or more than 1.2 million liters of water.

Drinking-Water Standards in the U.S.

The U.S. Environmental Protection Agency (EPA) sets standards for safe drinking water. Nearly all *public* water supplies in the U.S. meet these standards, which can be found at *www.epa.gov*. These standards set a limit on the concentrations of selected contaminants in water. Small municipalities commonly have more trouble meeting these standards than large cities because of limited budgets for water analysis and for building and running facilities to remove contaminants. The EPA standards do not apply to private wells.

Many people prefer the taste of bottled water to public tap water, but there are generally no health reasons to buy bottled water so long as the public water provider meets all the federal, state, and local drinking water regulations. Commercially bottled water is monitored by the Food and Drug Administration (FDA) but is not monitored as closely as public water systems. The FDA requires a bottler to test its water source only once a year. Also, bottled water can cost as much as 1,000 times more than municipal drinking water.

8.5

8.6 How and Where Does Groundwater Flow?

GROUNDWATER FLOWS BENEATH THE SURFACE in ways that are controlled by several key principles. The direction and rate of groundwater flow are largely controlled by the porous nature of the materials, the slope of the water table, and the geometry and nature of the subsurface rock. Some rock types allow easy groundwater flow, whereas others prevent significant movement.

A What Is the Geometry of the Water Table?

The *water table* defines the boundary between unsaturated and saturated rock and sediment. It usually is not a horizontal surface but instead has a three-dimensional shape that mimics the shape of the overlying land surface. The shape of the water table commonly has slopes, ridges, hills, and valleys. These features control which way groundwater flows.

1. In most environments, the water table has the same general shape as the overlying land surface but is more subdued. Where the land surface is high, the water table is also high. The similarity in shape between topography and the water table is less straightforward in some arid environments and in places where humans have pumped out groundwater faster than it can be replenished by precipitation.

2. The water table generally slopes from higher to lower areas. It is generally deeper below the surface under mountains than under lowlands, so its slope is less steep than that of the land surface. The shape of the water table is largely independent of the geometry of rock units through which the water table passes.

3. Groundwater just below the water table flows down the slope of the water table. In this example, it flows from left to right, from areas with a higher water table to areas with a lower water table. The red arrows show flow directions of water right below the water table.

08.06.a1

4. Where the water table is horizontal, for example near this lake, groundwater may flow very slowly or not at all. Deeper water may flow in directions different from near-surface water.

6. Where the water table intersects the land surface, there may be lakes, wetlands, or a flowing stream. The stream in this figure occurs where the water table is at the surface. However, streams do not necessarily coincide with the water table, because some flowing streams are underlain by unsaturated materials.

5. The terminology used to describe features of a water table is derived from topography. A high part of the water table separating parts sloping in opposite directions is called a *groundwater divide*. Groundwater flows in opposite directions on either side of a groundwater divide into different *drainage basins*.

B What Controls the Rate of Groundwater Flow?

The rate of groundwater flow is typically measured in meters per day, but can be much slower. Rate is primarily controlled by *permeability*, a material's ability to transmit fluid, which can vary greatly from one material to the next. The rate is controlled to a lesser extent by the steepness of the water table because flow is driven by gravity. Other factors being equal, water flows faster down a steep water-table slope and slower down a more gentle one. The slope of the water table is called the *hydraulic gradient*.

08.06.b1

08.06.b2

The rate of groundwater flow is strongly controlled by the permeability of the rock type. In this diagram, flow is fastest in this highly permeable rock layer.

Flow is moderately fast in a somewhat less permeable layer.

Flow is slowest in layers that have a very low permeability.

C What Is an Aquifer?

An *aquifer* is a large body of permeable, saturated material through which groundwater can flow well enough to yield significant volumes of water to wells and springs. To be a good aquifer, a material must have high permeability.

1. The most common type of aquifer is an *unconfined aquifer* where the water-bearing unit is open (not restricted by impermeable rocks) to Earth's surface and atmosphere. Rainwater or surface water can seep unimpeded through the upper layers of rock and sediment into an unconfined aquifer.

2. A *confined aquifer* is separated from Earth's surface by rocks with low permeability. Here, a permeable aquifer is bounded above and below by layers of low-permeability rock.

08.06.c1

3. A low-permeability unit, such as the thin gray layer in the middle, can restrict flow. An *impermeable* unit blocks flow completely. Such units are the opposite of an aquifer and are referred to as an *aquiclude*.

D How Are Wells Related to the Water Table?

A well is a hole dug or drilled deep enough to intersect the water table. If the well is within an aquifer, water will fill the open space to the level of the water table. This free-standing water can be drawn out by buckets or pumps.

08.06.d1

This well has been drilled from the land surface downward past the water table. The aquifer is unconfined and has filled with water to the height of the water table.

08.06.d2

In dry seasons, or during periods of high groundwater use, some wells run dry. This occurs when the water table drops and the well was not drilled deep enough into the aquifer.

08.06.d3

A *perched water table* sits above the main water table and generally forms where a discontinuous layer or lens of impermeable rock blocks and collects water infiltrating into the ground. Perched water bodies can make it difficult to predict the best site to drill an adjacent well.

Artesian Wells and Water

We often hear the word *artesian* in the context of bottled water or certain beverages. What does this term imply? Does it mean that the water is better tasting, more natural, or more healthy? The short answer to these three questions is no, or at least not necessarily.

The term *artesian* means that groundwater is in a confined aquifer and is under enough water pressure that the water rises some amount within a well. The water does not have to reach the

08.06.t1

surface for the well to be called artesian, but many artesian systems have enough pressure to force the water all the way to the surface, creating a well or spring. Although it is a catchy advertising term, the term *artesian* is not indicative of how the water tastes, whether it is more natural than other types of groundwater, or whether it is healthy. It only means that the groundwater is confined and under pressure, and as a result rises some amount in the well.

Before You Leave This Page Be Able To

✓ Sketch and describe the typical geometry of the water table beneath a hill and a valley, showing the direction of groundwater flow.

✓ Summarize two factors that control the rate of groundwater flow.

✓ Sketch and describe the origins of perched water.

✓ Sketch and describe an unconfined, confined, and artesian aquifer.

8.6

8.7 What Is the Relationship Between Surface Water and Groundwater?

SURFACE WATER AND GROUNDWATER ARE NOT ISOLATED SYSTEMS. Rather, they are highly interconnected with water flowing from the surface to the subsurface and back again. Most groundwater forms from surface water that seeps into the ground, and some streams and lakes are fed by groundwater.

A How Does Water Move Between the Surface and Subsurface?

Surface water can soak into the subsurface and become groundwater if the surface material is permeable and the water table is deep enough so there is an unsaturated zone into which water can seep. *Percolation* of water into the groundwater system helps replenish or recharge water lost by wells, springs, or other parts of the system. Such replenishment is referred to as *groundwater recharge*.

As long as topography does not intersect the water table, the groundwater will remain at depth, generally flowing toward low elevations.

Where the water table intersects the surface, groundwater can flow out onto the land. Such flow forms many springs and can add water to lakes and streams, keeping them from drying up. Seen in this context, a spring represents the interaction between surface topography and the water table, and whether groundwater forms a spring depends on the geometry of both. The water table is the more subdued of the two, so a spring is usually along a topographic low spot or a steeper part of a slope—the spring shown in this figure is at both a low spot and on a steeper part of the slope.

08.07.a1

B What Causes Groundwater to Emerge as a Spring?

A *spring* is a place where groundwater flows out of the ground onto the surface. At most springs, the water table intersects the surface. This can occur in a variety of geologic settings, a few of which are summarized below. Some groundwater is heated by hot rocks, or perhaps magma, before coming to the surface, forming warm springs or hot springs. In rare cases, hot springs that are near the boiling temperatures of water can form a *geyser*, a kind of hot spring that intermittently erupts fountains or sprays of hot water and steam, like Old Faithful, a world-famous geyser in Yellowstone National Park of northwestern Wyoming and adjacent parts of Idaho and Montana.

08.07.b1

Many springs are related to limestone aquifers. In such rocks, water can flow easily along dissolved bedding planes or through caves or fractures widened by dissolution. Where the saturated zone in the aquifer intersects the surface, water can flow out in a spring. In the spring to the right, water rushes out of dissolved fractures and other passageways in limestone, producing a deafening roar and earning the spring its name—Thunder Springs. It is in the walls of a side canyon of the Grand Canyon, Arizona.

Many springs are related to boundaries between two different rock units. In the example shown here, a sequence of layered rocks sits on top of a granite (the lowest unit). The boundary at the top of the granite is a buried erosion surface, called an *unconformity*. Groundwater flows down through the layered rocks until it encounters the hard, less permeable granite. It then flows laterally until it encounters a low point in the topography, emerging as a spring. Some springs are fairly subtle, not thunderous, and are called *seeps* (▶). This seep has a yellowish color because the rocks are naturally sulfur-rich.

Unconformity

08.07.b2

08.07.b3 Grand Canyon, AZ

08.07.b4 Baja California Sur, Mexico

C How Are Lakes Related to Groundwater?

Lakes can have various relationships to groundwater. Most lakes occur where the water table intersects the ground surface, but some have a different setting. Most wetlands represent the interaction between rainfall, surface water, and groundwater and may be nourished by groundwater flow.

08.07.c1

1. Some lakes are perched above the water table. These lakes can be transient, lasting only a short time after precipitation. A perched lake can be permanent if the inflow of water into the lake is at least equal to the amount lost by outflow to the ground, by evaporation to the air, or by other means.

2. Most lakes mark where the water table intersects and rises above the land surface. A lake can be fed entirely or partially by inflow of groundwater.

3. Many lakes are along the bottoms of valleys where groundwater is commonly close to or at the surface. Such lakes may be nearly in equilibrium with the adjacent groundwater, neither gaining nor losing water.

4. Wetlands can form peripheral to lakes, commonly at the same level as the water table. Other lakes are perched on uplands that contain clay or other less permeable material close to the surface. The low permeability can trap precipitation and runoff, slowing the infiltration of water into the ground, forming a wetland from the ponded water.

D How Do Streams Interact with the Water Table?

Water in many small rivers and streams decreases to a trickle and entirely disappears farther down the drainage. In other cases, a stream will flow even though there has not been rain or snowmelt in a long time — what is the source of this water? These occurrences are a result of interactions with groundwater.

08.07.d1

1. Some streams and rivers are lower in elevation than the water table next to the stream, so groundwater flows into the stream or river. The red arrows show the direction of groundwater flow below the water table. A part of a stream that receives water from the inflow of groundwater is said to be *gaining* or to be a *gaining stream* or *gaining river*.

08.07.d3

3. Some losing streams and rivers disappear when they cross from hard, less permeable rocks to softer, more permeable materials. The water seeps into the ground, where it may continue to flow at a shallow depth in the loose sand and gravel in the basin.

08.07.d2

2. Other channels flow across an area where the water table is at some depth below the surface. The part of the stream that loses water from outflow to groundwater is said to be *losing* or to be a *losing stream* or *losing river*. The red arrows show that groundwater below the water table flows down and away from the channel.

Before You Leave This Page Be Able To

☑ Sketch and describe how the interaction of the water table with topography causes water to flow between the surface and subsurface.

☑ Sketch or describe what is required to form a spring and possible settings where this occurs.

☑ Sketch and describe ways that lakes and wetlands relate to groundwater.

☑ Describe gaining and losing streams and how a river or stream can lose its water entirely.

8.7

8.8 What Problems Are Associated with Groundwater Pumping?

THE SUPPLY OF GROUNDWATER IS FINITE, so pumping too much groundwater, a practice called *overpumping*, can result in serious problems. Overpumping can cause neighboring wells to dry up, land to subside, and gaping fissures to open across the land surface.

A What Happens to the Water Table if Groundwater Is Overpumped?

Demands on water resources increase if an area's population grows, the amount of land being cultivated increases, or open space is replaced by industry. Groundwater is viewed as a way to acquire additional supplies of fresh water, so new wells are drilled or larger wells replace smaller ones when more water is needed.

1. A simple case illustrates the problems with overpumping. The two figures below show what happens when an unconfined aquifer is pumped, first by a small-volume pump and later by a larger pump. The topography of this area is fairly flat, there are no bodies of surface water, and a single type of porous and permeable sediment composes the subsurface.

2. As people move into a nearby town, they drill a small well down to the water table to provide fresh water. The small well pulls out so little groundwater that the water table remains as it has for thousands of years, nearly flat and featureless. The well remains a dependable source of water because its bottom is below the water table.

Minor Groundwater Withdrawal

3. Across the entire area, groundwater flows from right to left, down the gentle slope of the water table. The red arrow shows the direction of flow for groundwater in the saturated part of the aquifer, right below the water table. This arrow is drawn above the water table so the arrow is visible, but water above the water table is not connected enough to flow in this way.

08.08.a1

Increased Groundwater Withdrawal

08.08.a2

4. As more people move into the surrounding area, they drill a larger well to extract larger volumes of water to satisfy the growing demand. The new, larger well pumps water so rapidly that groundwater around the well cannot flow in fast enough to replenish what is lost. This causes the local water table to drop, forming a funnel-shaped *cone of depression* around the well.

5. The direction of groundwater flow changes significantly across the entire area. Instead of flowing in one direction, groundwater now flows toward the larger well and into the cone of depression from all directions. The change in flow direction has unintended consequences. It may cause serious safety issues, since waste-disposal sites, such as landfills, are generally planned with the groundwater-flow direction in mind. The change in flow direction can bring contaminated water into previously fresh wells.

6. The original small well dries up because it no longer reaches the water table, which has been lowered by the larger well's cone of depression. A cone of depression is common around nearly all wells, but a large cone of depression, caused by overpumping of the aquifer, can have drastic consequences. It can dry up existing wells, change the direction of groundwater flow, and contaminate wells. In addition, overpumping can dry up streams and lakes, if they are fed by groundwater, or cause the roofs of caves to collapse. Effects similar to those described above can also occur in a confined aquifer.

Before You Leave These Pages Be Able To

☑ Sketch a cone of depression in cross section, describing how it forms and which way groundwater flows.

☑ Describe how a cone of depression can cause a well to become polluted.

☑ Sketch or describe some other problems associated with overpumping, including subsidence, fissures, and saltwater incursion.

B What Other Problems Are Caused by Excessive Groundwater Withdrawal?

Overpumping can cause the ground surface to subside if sediment within the underlying aquifer is dewatered and compacted. In certain settings, subsidence causes fissures to open on the surface.

Before Groundwater Pumping

1. Many areas have settings similar to this one: mountains composed of bedrock flank a valley or basin underlain by a thick sequence of sediment. Most water is pumped from beneath the sediment-filled basin and used by people in the valleys.

08.08.b1

2. Bedrock has interconnected fractures that give it some permeability, but it has a much lower overall porosity than sediment in the basin.

3. The water table slopes from the mountains toward the basin, across the boundary between bedrock in the mountains and sediment beneath the valley.

After Groundwater Pumping

08.08.b2 San Joaquin Valley, CA

4. If we overpump groundwater, the water table can drop over much of the area. In some real-world cases it has dropped more than 100 m (~330 ft).

08.08.b4 Chandler, AZ

5. As the water table drops, the upper part of the original aquifer is now above the water table and has been dewatered. Sediment within and below the dewatered zone compacts because water pressure no longer holds open the pore spaces.

08.08.b3

6. Compaction of the sediment causes the overlying land surface to subside by several meters. Once the sediment compacts, the subsidence and loss of porosity are both permanent and will not be undone if pumping stops and water levels rise again. On the left, a telephone pole in the San Joaquin Valley of California bears signs marking elevations of the land surface during subsidence—the land surface has dropped many meters. Along the coast, such subsidence could lower an area below sea level.

7. The granite cannot compact, so open fissures develop across the land surface along the boundary between land that subsided and land that did not (in the mountains). The earth fissure pictured here formed by this type of subsidence.

C How Can Groundwater Pumping Cause Saltwater Intrusion into Coastal Wells?

Some wells are by necessity near the coasts of oceans and seas. These wells have a special threat—overpumping can draw salt water into the well, a process referred to as *saltwater incursion* or *saltwater intrusion*.

08.08.c1

Along ocean coasts, fresh water commonly underlies the land, while groundwater beneath the seafloor is salty. Fresh water is less dense than salt water and forms a lens floating on top of salt water.

08.08.c2

When wells on land are overpumped, the interface between fresh water and salt water moves up and inland (saltwater incursion). Wells closest to the coast will begin to pump salt water and will have to be shut down.

8.8

9.4 What Are Some Common Sedimentary Rocks?

SEDIMENTARY ROCKS ARE THE MOST COMMON rock in many regions, forming many of the landscapes we encounter. They form when loose sediment is converted into a rock through the process of *lithification*, which involves *compaction* by overlying materials and *cementation* by calcium carbonate or other materials.

A What Are Some Common Clastic Sedimentary Rocks?

Some sedimentary rocks are composed of rock and mineral pieces, called *clasts*, and are called *clastic rocks*. We describe and classify clastic sedimentary rocks based primarily on the sizes of clasts, along with other aspects such as clast roundness. Common clastic sedimentary rocks are shown below and described using these criteria of classification.

09.04.a1

Gravel-Sized Clasts

◄ *Conglomerate* has large, rounded clasts (pebbles, cobbles, or boulders) with sand and other fine sediment between the large clasts. This conglomerate has well-rounded cobbles in a matrix of mostly quartz sand.

► *Breccia* is similar to conglomerate except that the clasts are angular. Breccia usually has a jumbled appearance because most has a range of clast sizes, such as boulders in a mud-rich matrix, as shown here.

09.04.a2

09.04.a3

Sand-Sized Clasts

◄ *Sandstone* consists of sand-sized grains. It can contain some larger and smaller clasts, but is mostly composed of sand. It generally has better-defined layers than conglomerate or breccia.

Mud-Sized Clasts

► *Shale* consists mostly of fine-grained clasts (i.e., mud), especially very fine-grained clay minerals. The minerals are aligned, so the rock breaks in sheetlike pieces or chips, as displayed here.

09.04.a4

B How Do Clastic Sediments Become Clastic Sedimentary Rock?

Compaction

As sediment is buried, increasing pressure pushes clasts together, a process called *compaction*. Compaction forces out excess water and causes sediments to lose as much as 40% of their volume. Originally loose sediment becomes more dense and more compact.

Sediments near Earth's surface, such as these sand grains, are a loose collection of clasts. The grains rest on one another but do not fit together tightly, so spaces, called *pore spaces*, exist between the grains. Pore spaces are generally filled with air and water.

As sand grains are buried, the weight of overlying sediment and other materials forces the grains closer together. The amount of pore space decreases as air or water is expelled, so the layer loses thickness. In this manner, sediment becomes compacted into sedimentary rock.

09.04.b1

Cementation

Even after sediment is compacted, adjacent clasts do not fit together perfectly. The remaining pore spaces are commonly filled with water that contains dissolved materials. The dissolved materials can *precipitate* (come out of solution) to form minerals that act as a natural *cement* that holds the sediment together.

When sand grains and other sediment are deposited, abundant pore spaces exist between the grains, even after compaction. If these spaces are interconnected, as they are with sand grains, water can flow slowly through the sediments carrying chemical components into or out of the sediment.

09.04.b2

As the sediment is buried, the water moving through the pore spaces can precipitate minerals in the pore spaces. Such minerals are called *natural cement*, and they help turn the sediment into hard sedimentary rock.

What Are Some Common Nonclastic Sedimentary Rocks?

Some sedimentary rocks are not composed of clasts and are therefore *nonclastic rocks*. Processes capable of producing nonclastic sedimentary rocks may be chemical, biological, or have both chemical and biological aspects. The photographs below show common nonclastic sedimentary rocks.

1. *Rock salt* refers to halite (NaCl) or to a rock mostly composed of halite. Halite commonly precipitates as the water that dissolved it evaporates. Most table salt forms in this way.

09.04.c1

3. *Limestone,* made mostly of calcium carbonate, forms when shells and coral skeletons are cemented together. Many limestones, like the one here, contain fossils, a record of past life.

09.04.c3

2. *Gypsum* refers to a type of mineral or a type of rock. Like halite, it mostly forms when seawater evaporates in tidal flats and narrow seas. It is the material in wallboard (sheetrock) in our houses.

09.04.c2

09.04.c4

4. *Chalk* is a soft, very fine-grained limestone that forms from the accumulation of the calcium carbonate remains of microscopic organisms that float in the sea. Chalk forms the famous White Cliffs of Dover, England.

09.04.c5

5. *Chert* is a silica-rich rock that forms in several ways. One way chert forms is in layers from the accumulation and compaction of tiny, silica-rich plankton shells that fall to the ocean bottom. It can also occur as layers and irregular masses within limestone that form from the mixing of water with different chemistries.

09.04.c6

6. *Iron formation* is a rock composed of centimeter-thick layers of iron oxide, iron carbonate, and iron silicate minerals, commonly intermixed with very fine-grained quartz. Most iron formations precipitated from seawater early in Earth's history. It is the main source of iron used in steel.

09.04.c7

7. *Coal* is formed from wood and other plant parts that have been buried, compacted, and heated enough to drive off most of its water and oxygen. Depending on the amount of heat and pressure, coal can be soft and dull or hard and shiny. There are different kinds of coal, including lignite, bituminous coal, and anthracite, in order of increasing quality.

Types of Natural Cement

There are four main types of natural cements that hold grains together: calcite, silica, clay minerals, and iron oxides. Other materials, such as gypsum, can function as cement but are less common.

Calcite ($CaCO_3$) is a common cement in sandstone and other sedimentary rocks. It holds grains together moderately well, but it is easily redissolved, so a calcite-cemented sandstone may become friable (crumbly).

Silica (SiO_2) acts as a cement in some sandstone and other sedimentary rocks. It

forms a strong cement that can tightly bind grains, forming a tough, resistant rock.

Clay minerals can cement together larger grains, including sand. They may have been deposited with the sediment or formed from the alteration of feldspar or volcanic ash.

Hematite and other iron oxide minerals precipitate from water as a natural cement between the grains. Iron oxide minerals commonly give sediment deposited on land a reddish color, as displayed in the spectacular red-rock landscapes of the Desert Southwest.

Before You Leave This Page Be Able To

✓ Describe the classification of common clastic sedimentary rocks.

✓ Describe what happens to clastic sediment as it becomes buried and converted into rock.

✓ Describe some common nonclastic sedimentary rocks.

✓ Describe the natural cements that are common in sedimentary rocks.

9.4

9.5 What Are Igneous Processes and Rocks?

IGNEOUS ROCKS FORM very differently from sedimentary rocks. They form by solidification of magma. Most igneous rocks have millimeter- to centimeter-sized crystals, but some have meter-long crystals, and others are composed not of minerals but of noncrystalline glass. Igneous rocks may contain holes, fragments, or ash that has been compacted. In what settings do different igneous rocks form?

A In What Settings Do the Different Igneous Textures Form?

Igneous rocks display various textures, which indicate the environment in which the magma solidified. Magma can solidify at depth, erupt onto the surface as molten *lava*, or be explosively erupted as volcanic ash. Examine the figure below and think about where each texture in the photographs on these two pages might form.

09.05.a2

◀ *Vesicles* are holes in volcanic rocks that form when gases dissolved in magma accumulate as bubbles at or near Earth's surface. Many lavas are vesicular, and much of the material in volcanic ash forms when the thin walls between vesicles burst, shattering partially solidified magma into the sharp particles that compose volcanic ash.

09.05.a3

Some volcanic rocks are composed of angular fragments, a texture called *volcanic breccia*. These form in many ways, including from explosive eruptions of ash and rock fragments, from a lava flow that breaks apart as it partially solidifies while flowing, or from volcano-triggered mud-flows and landslides.

09.05.a4

Volcanic glass forms when magma erupts on the surface and cools so quickly that crystals do not have time to form, as can occur in a lava flow or in volcanic ash.

Other igneous rocks cool slowly enough to grow crystals, with the size of the resulting crystals mostly revealing the time that was available for the magma to crystallize. A magma that is deeper needs more time to cool, and therefore solidify, than one nearer to the surface, so the size of crystals is mostly related to the depth, with larger crystals generally indicating greater depths. Examples of this variation in crystal size are shown below.

Some *volcanic ash* erupts vertically in a column and settles back to Earth. This ash cools significantly before accumulating on the surface, so it consists of ash and other particles that are not strongly compressed together, as in the rock called *tuff* (▶).

09.05.a5

Other volcanic ash erupts in thick clouds of hot gas and rock fragments, called *pyroclastic flows*, that flow rapidly downhill under the influence of gravity. The ash deposited by pyroclastic flows is very hot, and so most parts are compacted and welded together to some extent.

Some igneous rocks have a distinctive texture (called *porphyritic*) with larger crystals in a finer grained matrix (▼). Such rocks indicate that the magma cooled somewhat at depth to grow the larger crystals, but then rose and solidified on or closer to the surface, where it rapidly cooled to solidify the finer grained matrix.

09.05.a6

09.05.a1

09.05.a7

Fine-grained igneous rocks form if the magma cooled quickly and could only grow small crystals.

09.05.a8

Medium-grained rocks have larger crystals (mm but not cm) that are easily visible to the unaided eye.

09.05.a9

Coarse-grained igneous rocks form at greater depths, where slow cooling allows large crystals to grow.

09.05.a10

Some igneous rocks contain very large crystals (cm to m), and are called *pegmatite*.

B How Does the Composition of Igneous Rocks Vary?

Geoscientists organize igneous rocks according to the *size of crystals* and the *kinds of minerals* in a rock. Rocks with a light color and abundant quartz and feldspar are *felsic* rocks, whereas rocks that are dark and contain minerals rich in magnesium and iron are *mafic* or *ultramafic* rocks. *Intermediate* rocks are in between. The top two rows of photographs below show igneous rocks of different compositions and crystal size. The bottom two photographs show two other types of volcanic rocks.

	Felsic	**Intermediate**	**Mafic**

Coarsely Crystalline

09.05.b1

09.05.b2

09.05.b3

Granite is a coarsely crystalline, light-colored igneous rock. The light color is due to the light-colored, felsic minerals—quartz and feldspar. Most granites also contain some dark or light-colored mica, some oxide minerals, and certain other minerals.

Diorite, an intermediate rock, contains more mafic minerals than does granite. It is usually moderately dark from abundant dark silicate minerals, such as amphibole, and it also generally contains abundant plagioclase feldspar as the light-colored crystals.

Gabbro is a coarsely crystalline, mafic rock. It typically is dark and consists of dark-colored silicate minerals, like pyroxene. It can also contain light-gray feldspar. Some varieties are somewhat lighter colored, and some contain the greenish mineral olivine.

Finely Crystalline

09.05.b4

09.05.b5

09.05.b6

Rhyolite is the fine-grained equivalent of granite. It is mostly a finely crystalline rock, but it can contain glass, volcanic ash, pieces of pumice, and visible crystals.

Andesite, the fine-grained equivalent of diorite, is commonly gray or greenish. Andesite commonly has visible crystals of cream-colored feldspar or dark amphibole.

Basalt is a dark mafic lava rock, equivalent in composition to gabbro. Most basalt is dark gray to nearly black, and can have some crystals and vesicles, as shown here.

Other

09.05.b7

09.05.b8

Pumice is a volcanic rock containing many vesicles (holes). The holes are so numerous that most pumice floats on water. The solid material in pumice begins as volcanic glass, but over time it can convert into microscopic crystals.

Tuff is a volcanic rock composed of a mix of volcanic ash, pumice, crystals, and rock fragments. If the particles of ash and pumice remain hot, they become compacted by overlying materials, becoming a hard, dense rock with flattened pumice, as shown here.

> **Before You Leave This Page Be Able To**
>
> ✓ Sketch an igneous system and show where the main igneous textures form.
>
> ✓ Describe or diagram the classification of some common igneous rocks.

9.5

9.6 What Are Metamorphic Processes and Rocks?

METAMORPHISM INVOLVES CHANGING A ROCK that has become unstable. A rock may be unstable because the minerals it contains are unstable and as a result it develops new (metamorphic) minerals. It may be unstable because of the way the grains or layers are arranged, so it develops a new (metamorphic) texture. Depending on the conditions, a rock may develop new metamorphic minerals and a new metamorphic texture.

A What Causes Metamorphism?

For a rock to be metamorphosed, it must be subjected to conditions of temperature, pressure, and fluid chemistry that make it unstable. This generally occurs when temperature (T) and pressure (P) increase, possibly accompanied by forces that cause the rock to *deform* or by an influx of deep fluids. There are two main kinds of metamorphism: *contact metamorphism* is caused by local heating by magma, typically without deformation, whereas *regional metamorphism* involves deformation along with heating over a broader region.

09.06.a1

Pressure increases with depth in Earth because rocks are more deeply buried. Higher pressures compress the rocks and, in combination with high temperatures, may cause some minerals to be unstable. Forces deep in Earth can cause rocks to move and deform.

Temperature increases with depth and near magma. An increase in temperature usually causes new minerals to grow or existing minerals to grow larger. It can also weaken the rocks, allowing them to deform. The crust contains abundant water and other *fluids*, shown here as blue in water-filled fractures. Such fluids interact with minerals and carry material into, through, and out of rocks.

09.06.a2

Contact Metamorphism—Rising magma efficiently brings thermal energy higher into the crust, heating the wall rocks. This is called contact metamorphism because it occurs near contacts (boundaries) of magma. Heating causes new minerals to grow or existing minerals to increase in size. Heating may not be accompanied by deformation, so contact metamorphism commonly forms metamorphic rocks with little evidence of deformation.

09.06.a3

Regional Metamorphism—In regional metamorphism, heating is accompanied by enough force to cause deformation. The imposed forces can result from tectonics or burial. Regional metamorphism causes new minerals to grow and existing minerals to increase in size, while deformation during metamorphism generally results in the rock developing planar or linear metamorphic fabrics during metamorphism.

B What Processes Occur During Metamorphism and Deformation?

Many processes that operate during metamorphism change the minerals in a rock or change the arrangement of those minerals. Some metamorphic processes are related to heating or are chemical in character. Such processes cause grains to grow, recrystallize, redistribute themselves, or even dissolve in response to temperature, pressure, and any imposed forces. Some processes are physical and may deform or rotate individual crystals, grains, and layers, producing planar or linear metamorphic fabrics.

Chemical Processes

During metamorphism, movement of chemical elements can cause existing minerals to grow larger or can form new minerals. Minerals can grow in fairly random orientations, as occurs during contact metamorphism, which is due to an increase in temperature but not in forces. Or, minerals can grow with a preferred orientation, becoming aligned as in the example shown here. If minerals are arranged in lines, the rock has a *lineation*. If they are arranged in planes, the rock has a *foliation*. Many metamorphic rocks have a lineation and foliation.

09.06.b1

Physical Processes

Metamorphism is generally accompanied by deformation of any original constituents of the rocks, like crystals or pebbles. Deformation can flatten grains and clasts that were initially somewhat spherical into shapes like pancakes or the thin, long top of a skateboard. If a rock is flattened during metamorphism, the deformed objects will define a foliation. If a rock is stretched in one direction, it acquires a lineation, commonly along with a foliation. Physical and chemical processes both occur during most metamorphism.

09.06.b2

C What Rocks Form When Sedimentary Rocks Are Metamorphosed?

There are many types of metamorphic rocks, and these rocks indicate diverse starting rock types and various conditions under which rocks can be deformed and metamorphosed. Some metamorphic rocks form when sedimentary rocks are subjected to a change in temperature, pressure, or both. Photographs below show what happens to three common sedimentary rocks—shale, sandstone, and limestone. Igneous rocks can also be metamorphosed with similar results, and metamorphic rocks can be metamorphosed several times under varying conditions of temperature and pressure.

Increasing Temperatures

Shale

09.06.c1 Larder Lake, Ontario, Canada

Slate

When a shale is metamorphosed at low to moderate temperature, it can develop cleavage (a type of foliation) and become *slate* (◄). Slates are dull (not shiny) and commonly dark.

09.06.c2 Superior, AZ

Phyllite

At slightly higher temperatures, microscopic mica crystals give the rock a shiny aspect or *sheen* (◄). Such a rock is *phyllite*.

09.06.c3 Pioneer Mtns., ID

Schist

At higher temperatures, crystals of mica and other minerals become large enough to see. The resulting rock is a shiny rock called a *schist* (◄).

09.06.c4 San Gabriel Mtns., CA

Gneiss

At even higher temperatures, chemical constituents are mobilized and light- and dark-colored minerals separate, forming a foliation and banded rock called *gneiss* (◄).

Sandstone

09.06.c5 Baraboo, WI

Quartzite

Most sandstones are predominantly quartz (SiO_2), a mineral that is stable over a wide range of temperature and pressure conditions. During metamorphism, quartz grains grow together and become so tightly bonded that fractures break across the grains rather than around them. This type of rock is a *quartzite* (▲). Quartzite is made of quartz, just like the original sandstone, and can preserve beds and other original sedimentary features.

09.06.c6 Joshua Tree, CA

Coarse Quartzite

▲ With higher temperatures of metamorphism, the quartz in the rock begins to merge into larger crystals and can become a coarser grained quartzite, in some cases with no individual grains left. Quartz is soluble and mobile in metamorphic fluids and so at high temperatures can be redistributed into quartz veins, which are common in metamorphic rocks.

Limestone

09.06.c7 Ios, Greece

Finely Crystalline Marble

Limestone consists mostly of calcite, a chemically reactive mineral. Low-temperature (<250°C) metamorphism of limestone causes calcite to recrystallize slightly, forming a finely crystalline *marble* (▲).

09.06.c8 Franklin Mtns., El Paso, TX

Impure Marble

At moderate temperatures (~400°C), marble becomes medium grained (▲). When metamorphosed, impurities in the limestone, like clay and chert, may produce various new metamorphic minerals, like these red garnets.

09.06.c9 Naxos, Greece

Coarse Marble

▲ At higher temperatures, calcite crystals grow coarser to produce a coarsely crystalline marble. The one shown here is nearly 100% calcite, but coarse marbles commonly also contain other minerals.

Before You Leave These Pages Be Able To

☑ Summarize causes of metamorphism, and describe or sketch a chemical and physical process that can accompany metamorphism.

☑ Describe the changes different sedimentary rocks undergo as they metamorphose and the metamorphic rocks they become.

9.6

9.7 How Are Different Rock Types Expressed in Landscapes?

SEDIMENTARY, IGNEOUS, AND METAMORPHIC ROCKS can each have a distinctive appearance in the landscape, often allowing us to recognize these rocks from a distance. With some practice, we can drive down a highway, observe the characteristics of a hill or mountain, and make an educated guess about what type of rocks are exposed. Here, we provide a brief introduction into interpreting the rock types of landscapes from a distance.

A How Are Sedimentary Rocks Expressed in the Landscape?

There are diverse types of sedimentary rocks, displaying a wide variety in colors, thickness of layers, and resistance to erosion. The unifying feature of most sedimentary rocks is the presence of visible layers. Each main type of sedimentary rock has certain characteristics that allow us to identify it from a distance.

Sandstone

09.07.a1 Green River, UT

Some sandstone layers appear massive from a distance because they have little variation of grain size, as in the massive sandstone cliff in the center of this photograph. Thick layers of sandstone generally were deposited by wind as sand dunes.

09.07.a2 Ruby Canyon, CO

Most sandstone has layers that differ from one another in color, grain size, or composition of the grains. Such layers can be parallel beds, as shown here, or cross beds centimeters to tens of meters high. Numerous layers mean many changes in conditions during deposition.

Fine-Grained Clastic Rocks

09.07.a3 Grand Junction, CO

Shale and other fine-grained clastic sedimentary rocks are relatively easily eroded rocks. Where exposed, these rocks typically form soft slopes covered by small, loose chips derived from weathering of the thinly bedded rocks.

09.07.a4 Petrified Forest, AZ

Shale and associated fine-grained rocks form another distinctive type of landform—*badlands*. Badlands have a soft, rounded appearance that exhibits the softness of the rocks. Badlands also have an intricate network of small drainages and eroded ridges carved into the soft rocks.

Limestone

09.07.a5 Chamonix, French Alps

09.07.a6 Austrian Alps

09.07.a6 Guilin, China

In some relatively dry climates, limestone and other carbonate rocks are relatively erosion-resistant. The rocks can form light-gray to dark-gray cliffs and steep slopes composed of beds that may vary slightly in thickness and color. Most limestone has visible layers, on either a small or large scale.

Limestone is very soluble, and so in very wet climates it dissolves and weathers quickly, especially along fractures and layers, forming caves, sinkholes, pits, and depressions.

Weathered and dissolved limestone forms distinctive landforms with caves, sinkholes, and pillars. Such landscapes formed by dissolution of limestone are called *karst terrain*.

Before You Leave These Pages Be Able To

☑ Describe the characteristics of some common sedimentary rocks, including their expression in landscapes.

☑ Describe the appearance of some common igneous rocks in landscapes.

☑ Describe some characteristics displayed by metamorphic rocks, as exposed in landscapes.

B | How Are Igneous Rocks Expressed in the Landscape?

Igneous rocks form from magma, either from magma that solidifies below the surface, forming *intrusive* (plutonic) rocks, or magma that erupts onto the surface, forming *extrusive* (volcanic) rocks.

Granite makes up the bulk of the continental crust, and most granite is fractured. These fractures help speed up the weathering process, which wears away the edges and corners of fractures, resulting in rounded shapes in many granite exposures, a process called *spheroidal weathering*.

09.07.b1 Date Creek, AZ

09.07.b2 Grand Coulee, WA

Basaltic lava flows are the most common type of volcanic rock in many landscapes. Basalt flows are dark and spread out easily, forming distinct layers. Basaltic sequences generally have reddish zones from pyroclastic rocks that accompanied the eruption of lava, as shown in this photograph.

Volcanic rocks commonly have some sort of layers, and they also commonly have fractures perpendicular to the layers (▶). Depending on the rocks' compositions and the type of eruption that formed them, volcanic rocks can be various shades of gray, green, brown, tan, and cream colored.

09.07.b3 Copper Creek, AZ

09.07.b4 Grand Canyon, AZ

Many volcanic units, especially basaltic lava flows, have distinctive columnar fractures called *columnar joints*. Such joints form during the cooling and contraction of solidified igneous rocks. The size and orientation of the columns indicate how the rock cooled, but most columns are steep.

C | How Are Metamorphic Rocks Expressed in Landscapes?

Rocks that have been changed (or metamorphosed) by increased temperature and pressure are metamorphic rocks. Different types of metamorphic rocks result from different starting rock types and different conditions under which rocks can be metamorphosed and deformed. These differences, in turn, are expressed by a somewhat variable appearance of metamorphic rocks in the landscape.

09.07.c1 Pamour, Ontario, Canada

Many metamorphic rocks have layers that form platy, jagged outcrops and tabular slabs of rock, as in the left photograph below. They can include numerous bodies of granite and other igneous rocks, some of which help us observe how folded and deformed these metamorphic rocks really are (▼).

09.07.c2 Aurland Trail, Norway

09.07.c3 Grand Canyon, AZ

Metamorphic rocks can be shiny, even from a distance, if their mica minerals share a similar orientation and reflect light. These shiny metamorphic rocks preserve wavy ripple marks and a set of quartz-filled fractures that form small ridges. Such mineral-filled fractures are called *veins* and are common in most metamorphic rocks. Many such veins represent materials mobilized by elevated temperatures that accompany metamorphism.

9.8 What Controls the Appearance of Landscapes?

THE APPEARANCE OF A LANDSCAPE reflects the types of starting materials (rocks and loose sediment) and processes that act upon those materials, including precipitation, running water, wind, flowing ice, the pull of gravity, and the activities of plants, animals, and microbes. To understand the processes that shape landscapes, *geomorphologists*—physical geographers and geologists who study the spatial pattern, form, and evolution of landforms—often begin by observing a landscape to determine what is there. Most landscapes display a variety of features, such as different rock layers that we can distinguish by color, texture, and the way the rocks fracture.

A What Factors Influence the Development of Landscapes?

Examine the figure below and think about all the processes that might interact to produce a landscape.

1. A main control on the appearance of a landscape is the parent materials — what types of rocks, soils, and other materials are present. Some rocks are very hard and resistant to erosion, and these will tend to form peaks, cliffs, and steep slopes. Softer rocks will erode more easily, forming valleys or gentle slopes.

2. The orientation of a slope (i.e., which way the slope faces) has a strong control on the resulting landscape. In the Northern Hemisphere, south-facing slopes receive more sunlight than do north-facing ones, and so in some climates they will be drier and may have fewer plants. The relative lack of plants results in more soil erosion. This variation in slope appearance highlights the complex way all four spheres interact. If this figure shows an area in the Northern Hemisphere, are the north-facing slopes probably on the left or right side of the mountain range?

3. Climate has a great influence on the type of landscape that results. The amount and style of erosion are affected by the amount of rainfall and whether rainfall is spread out throughout the year or falls rapidly in a brief rainy season. Climate also influences the type and abundance of plants, which then determine whether soil develops and remains in place on the hillslope, held in place by roots. Another important climatic factor is whether the area experiences freezing temperatures, because ice freezing in fractures expands, further loosening rocks that can then move downhill.

09.08.a1

B What Features Do Landscapes Display?

Observe this photograph, trying to identify distinct parts of the scene and then focusing on one part at a time. After examining the photograph, read the accompanying text.

09.08.b1 Canyonlands NP, UT

Color commonly catches our attention. These rocks are various shades of red, tan, and gray. Close examination of these rocks by geomorphologists reveals that the rocks consist of consolidated sand and mud, and therefore are *sedimentary rocks*.

Another thing to notice is that this hill has different parts. A small knob of light-colored rocks sits on the very top.

Below the knob is a reddish and tan slope and a small reddish cliff.

There is a main, light-colored cliff, the upper part of which has a tan color and is fairly smooth and rounded.

Some parts of the cliff have horizontal lines that can be followed around corners of the cliff. These lines are the outward expression of *layers* within the rock. We call such layers in sedimentary rocks *beds* or *bedding*. These beds originally extended across the area prior to more recent erosion.

Lower parts of the cliff have a darker reddish-brown color and display many sharp angles and corners. Some of these corners coincide with vertical cracks, or *fractures*, that extend back into the rock. The reddish-brown color is a natural stain on the outside of the rocks.

Below the cliff is a slope that has pinkish-red areas locally covered by loose pieces of light-colored rock. A reasonable interpretation is that the loose pieces have fallen off the main cliff.

C What Are Some Strategies for Observing Landscapes?

Observe this photograph of Monument Valley, along the Arizona-Utah border, and try to recognize the types of features, such as layers and fractures, described on the previous page. After you have made your observations, read the text, which describes aspects to observe and some helpful strategies for looking at any landscape.

1. Most landscapes have a fairly complex appearance when viewed in their entirety, so a useful approach is to focus on one part of the landscape at a time. In the scene below, examine the left side of the image and compare it to the center. What similarities and differences do you observe?

2. Another approach is to let the landscape guide your observations from one part to another. In this scene, spend some time looking only at the cliff, and then focus on the reddish slope below the cliff. Next, pay attention to the piles of loose rocks that rest on the reddish slope.

3. Next, try focusing on one type of landscape feature at a time. In this photograph, start by concentrating on the fractures in the cliff. Are they steep, and are they evenly spaced? How do they affect the appearance of the cliff? Use the same approach to look at the ledges that cross the reddish slope.

09.08.c1 Monument Valley, AZ

4. Color is one of the first things we notice in any scene. Rocks, sediment, and soils have a range of colors depending on the composition of the materials and the environmental conditions imprinted on those materials. Some colors are integral to the rock, but others are a natural stain on the outside surfaces of the rock. The rocks of Monument Valley are reddish brown to tan inside, but are locally coated by a darker brown stain.

5. Some rock types are more resistant to erosion than others and have more dramatic expressions in the landscape. Cliffs and ledges generally represent rock types that are hard to erode, as shown by the cliffs of hard sandstone in this photograph. Slopes or soil-covered areas contain weaker materials, such as the loose wind-blown sand in the foreground of this image.

8. To visualize different *components* of a landscape, draw a sketch that captures the main features but leaves out less important details. Compare the sketch below with the photograph above. Note how the sketch changes the way you look at the photograph.

7. The *shapes* of eroded rocks depend on the hardness of the rock, thickness of layers, spacing of fractures, and many other factors. Landscapes change over time, so shapes seen today will evolve into different shapes on timescales of years to millions of years.

6. Obvious features in many landscapes are *layers* in the rocks. The cliff represents a thick layer of sandstone, whereas the underlying slopes and ledges are the expression of dozens of layers. In this location, the layers are nearly horizontal, but layers can be tilted or even folded. These differences in orientation have a great impact on the appearance of the resulting landscape. To understand how layers influence the landscape, you first observe the layers and recognize how they are oriented.

09.08.c2

Before You Leave This Page Be Able To

☑ Summarize the different aspects or features you can observe in a landscape.

☑ Draw a simple sketch of a landscape photograph, identifying the main components, like those shown on these pages.

9.8

How Are Landscapes Weathered and Eroded?

WEATHERING AND EROSION begin with starting materials and then sculpt them into the landforms we observe. *Weathering* involves the *in situ* disintegration of rocks and other materials, and we subdivide it into two varieties — physical weathering and chemical weathering. *Erosion* is the scouring and stripping away of materials from a landscape, usually those that have been loosened by weathering. Together, weathering and erosion reshape a landscape and leave behind many signs of their actions.

A How Do Physical and Chemical Weathering Affect Materials in Landscapes?

Weathering involves physical and chemical processes that attack rocks on or near Earth's surface, loosening pieces and dissolving some material. Weathering can loosen material to form sediment. It can even produce soil in situ. The type of weathering that occurs depends on the material that is weathered and the conditions during weathering. The processes of physical and chemical weathering are summarized below and are discussed in more detail in the chapter on soils.

Physical Weathering

Physical weathering is the physical breaking apart of rocks, soils, and other materials that are exposed to the environment. There are four major causes of physical weathering.

Near-Surface Fracturing — Many processes on or near the surface break material into smaller pieces. These include fracturing caused by rocks pulling away from a steep cliff. Fractures also occur when rocks expand as they are uplifted toward the surface and are progressively exposed to less pressure.

Frost and Mineral Wedging — Rocks, soils, and other materials can be broken as water freezes and expands in fractures. When the ice melts, the fractured pieces may become dislodged from the bedrock. Crystals of salt and other minerals that grow in thin fractures can also cause rocks to break apart.

Thermal Expansion — Rocks are heated by wildfires and by the Sun during the day. As rocks heat up, they expand, often irregularly, and may crack. This process probably plays a minor role in weathering, and scientists are currently debating its relative importance.

Biological Activity — Roots can grow downward into fractures and loose materials, prying them apart as the root diameter increases. Burrowing animals can transport and move rock and soil from depth to the surface where they are exposed to the elements, weathered, and eroded. Biological activity is especially efficient in soils and sediment, which are unconsolidated and therefore easy to move.

09.09.a1

Chemical Weathering

Chemical weathering includes several types of chemical reactions that affect rocks by breaking down minerals, causing new minerals to form, or by removing soluble material from the rock. Chemical weathering attacks both solid rock and loose rock fragments, and produces ions in solution, loose grains and other pieces, and a covering of soil.

Dissolution — Some minerals, like calcite, are soluble in water, especially weakly acidic waters that are common in nature. These minerals, along with the rocks, sediment, and soil that contain them, can dissolve. The dissolved material may be carried away in streams, or groundwater, or be used locally by plants.

Oxidation — Some minerals, especially those containing iron, are unstable when exposed to Earth's atmosphere. These minerals can combine with oxygen to form oxide minerals, such as hematite, which compose the reddish and yellowish material that forms when metal rusts.

Hydrolysis — When silicate minerals are exposed to water, especially water that is somewhat acidic, the water reacts chemically with the minerals. This process commonly converts the original materials to clay minerals and produces leftover dissolved material that is carried away by the water. Hydrolysis is responsible for the formation of many clay-rich soils.

Biological Reactions — Decaying plants produce acids that can attack rocks, and some bacteria consume certain parts of rocks. These biological processes are different from the "Biological Activity" discussed above because they cause minerals to break down into their constituent elements.

09.09.a2

B How Does Erosion Operate and What Are the Results?

During erosion, material is removed from a landscape and transported away, and such wearing away of the surface is called *denudation*. Erosion removes material loosened by weathering, but it can also scour and move unweathered rock.

Gravity—Rocks, soils, and other materials are under the constant pull of gravity. If they reside on a flat surface, they likely will remain in place or move very slowly, but material on slopes (▶) can fall, slide, or tumble down the slope and accumulate farther down the slope or near the base.

09.09.b1 Seton Lake, British Columbia, Canada

09.09.b4 Carmel, CA

Waves—Waves along shorelines, either along the ocean or the edge of a lake, constantly crash against the shore (◀). Wave action can carry away sand and other sediment and can erode into solid bedrock and undercut cliffs, causing even more material to fall onto the surf zone.

Running Water—Streams use the power of running water and the scouring ability of sediment carried by the water to cause erosion. Running water can remove soil, sediment, and other loose materials from a landscape, or carve into solid bedrock (▶).

09.09.b2 Cascade Creek, Cascade, CO

09.09.b5 Dry Valleys, Antarctica

Wind—Blowing wind can carry dust and fine sand grains, which act as sandpaper, scraping against and sanding away any materials exposed to the wind, including solid rock. The strange shapes here (◀) were scoured by blowing sand and ice (which also acts as a scouring agent).

Ice—Glaciers are large, moving masses of ice on land that persist for many years. Along their sides and bases (▶) they scour away any loose material and even solid bedrock, causing ice in those parts to become darker because of all the eroded material.

09.09.b3 Glacier Bay, AK

09.09.b6 Badlands NP, SD

Wind Erosion—Strong winds can remove so much material that the ground elevation is lowered (◀). This wind erosion can scour away soils, causing the land to be less productive. In this photograph, wind has sculpted soft sedimentary rocks into two pillars, remnants of the sedimentary layers that form the ridge in the background. This process is most efficient with soft and weak materials.

The Effect of Erosion on Topographic Relief

09.09.b7 Banff, Alberta, Canada

09.09.b8 Baja California Sur, Mexico

09.09.b9 Cima Dome, southern CA

When a mountain forms and for some time afterwards, it is relatively steep as streams and glaciers carve down into the landscape faster than gravity and other processes can lower the peaks—relief is high.

Over time, weathering and erosion lower the peaks, lessen the slopes, and produce an overall softer-appearing, less rugged landscape. The steepest topographic features are typically removed first because they are most unstable.

With even more time, erosion of bedrock can result in a gently sloping erosion surface. Such a surface is called a *pediment*, and can form a *pediment dome*, like the one above, or a gentle bedrock slope flanking a mountain.

Before You Leave This Page Be Able To

☑ Distinguish the main processes of physical and chemical weathering.

☑ Describe the agents that cause erosion and some of the results.

9.9

How Do Landscapes Record Transport and Deposition by Gravity, Streams, Ice, and Waves?

ONCE WEATHERED AND ERODED from a landscape, material can be transported farther away and later deposited. The main agents that transport earth materials are running water (streams), glaciers, waves in oceans and lakes, wind, and the ever-present force of gravity, which on its own moves materials down slopes and also drives or influences the other transport agents. Wind has some special characteristics and is discussed later.

A What Processes Transport and Deposit Sediment?

The same processes capable of erosion are also those that can simply continue transporting that material away. Gravity, streams, glaciers, waves, and wind are the main phenomena that transport material, and are often called *agents of transport*. Each works in a different way, but each affects the material being transported.

Gravity on Slopes

All materials on Earth are subject to the downward pull of gravity. Material on slopes, especially on slopes that are relatively steep, can succumb to this force by falling, sliding, tumbling, or slowly creeping down the slope (▶). Gravity can be aided by moving water, mud, and ice.

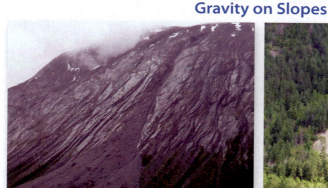
09.10.a1 Seton Lake, British Columbia, Canada

09.10.a2 British Columbia, Canada

Material that has moved downhill under the force of gravity, called *colluvium*, may not be obvious where mostly covered by soil or vegetation. Colluvium underlies the forested hillsides in this photograph (◀) but is only exposed where the slope was undercut by an eroding stream.

Water and Mud on Slopes

Water and mud can transport material, including huge boulders (▶), down slopes and into stream valleys and other areas below. Such slurries of mud and other debris are called *debris flows* and, as shown in this graph, can wreak havoc on anything in the way of the muddy torrent.

09.10.a3 Venezuela

09.10.a4 Death Valley, CA

Downhill, where the slope becomes less steep or adjoins a valley floor, the material being transported down the slope or in channels commonly accumulates at the bottom of a steep slope or at the base of the mountain, forming a cone-shaped feature known as an *alluvial fan* (◀).

Glaciers

Glaciers are typically loaded with rock debris collected along the way. Much of this material was eroded from bedrock and consists of angular clasts of various sizes. Glaciers also grind and pulverize material they transport, producing abundant particles of *silt* (between sand and clay in grain size).

09.10.a5 Mendenhall Glacier, AK

09.10.a6 Athabasca Glacier, Alberta, Canada

A glacier deposits material along its base, sides, or front, usually all at the same time. Glacially deposited materials can be blanket-like, covering the ground somewhat evenly, but generally are more irregular, forming curved ridges, hill-sized piles of sediment, or irregular terrain with pits and mounds (◀).

Running Water in Streams

09.10.a7 Rampart Creek, Alberta, Canada

09.10.a8 Icefields Parkway, Alberta, Canada

09.10.a9 Animas River, Durango, CO

Flowing streams are a powerful agent of transport, carrying huge volumes of sediment from higher elevations to lower ones. Some streams (▲) carry a wide range of clast sizes, including sand, pebbles, and cobbles. Clasts may be carried only a short distance or hundreds to thousands of kilometers.

In landscapes, sediment recently deposited by streams typically forms ribbons or blankets of debris near or within the stream channel (▲). These features are transient, being covered and rearranged by the next large flow of water down the stream. They generally include partially rounded cobbles, pebbles, and sand.

Even streams that seem to be moving in a tranquil manner (▲) are transporting sediment. A tan or brown color to such streams indicates that the stream is carrying abundant fine sediment, mostly mud within the visible water and sand along the streambed. These form muddy *floodplains* next to the stream.

Waves Along Shorelines

Waves along a shoreline can be very powerful as they crash onto bedrock or beach sediments along the shore. Through this process, they pick up sediment, transporting it back and forth within a relatively limited area, but over time they can transport material down the beach for long distances.

09.10.a10 Carmel, CA

09.10.a11 Naxos, Greece

Sediment moved by waves is mostly deposited along the beach. Depending on their setting and the types of materials that are available, beaches can be mostly covered by sand or by cobbles and other clasts, which became rounded when the waves pounded one clast against another.

B How Does Transport Affect the Size, Shape, and Sorting of Transported Material?

During transport by gravity, running water, ice, waves, or wind, clasts are subjected to tumbling, collisions with other clasts and with bedrock, and abrasion by fine sediment or ice. These actions may reduce the size of the clasts, increase their roundness, and sort the clasts by size. The distance that a clast has been transported is therefore a key factor.

Bedrock exposed in mountains or along cliffs breaks off to form blocks the size of boulders and cobbles. Clasts near their source are usually large, angular, and poorly sorted (i.e., they have a wide range of clast sizes).

As boulders and cobbles are transported by gravity, streams, ice, or waves, their sharp corners break off because they are the most exposed and weakest parts of a clast. The clasts become more rounded and smaller as pieces break off.

Far from their bedrock source, the clasts are worn into well-rounded pebbles and sand grains. River currents, beach waves, and wind separate clasts by size, eventually producing better-sorted sediments. Coarse materials are slowly left behind, and only the smaller clasts are carried far from the sediment source. Glaciers are an exception to this story because they continuously break rocks into angular clasts and can carry sediment long distances without doing any sorting.

09.10.b1

Before You Leave This Page Be Able To

✓ Summarize the main processes that transport and deposit sediment.

✓ Sketch and explain how transport affects clast size, shape, and sorting.

9.10

10.1 What Is Inside Earth?

UNDERSTANDING EARTH'S SURFACE requires knowing what is inside the Earth. You can directly observe the uppermost parts of Earth, but what else is down there? Earth consists of concentric layers that have different compositions. The outermost layer is the *crust*, which includes *continental crust* and *oceanic crust*. Beneath the crust is the *mantle*, Earth's most voluminous layer. The molten *outer core* and the solid *inner core* are at Earth's center.

A How Does Earth Change with Depth?

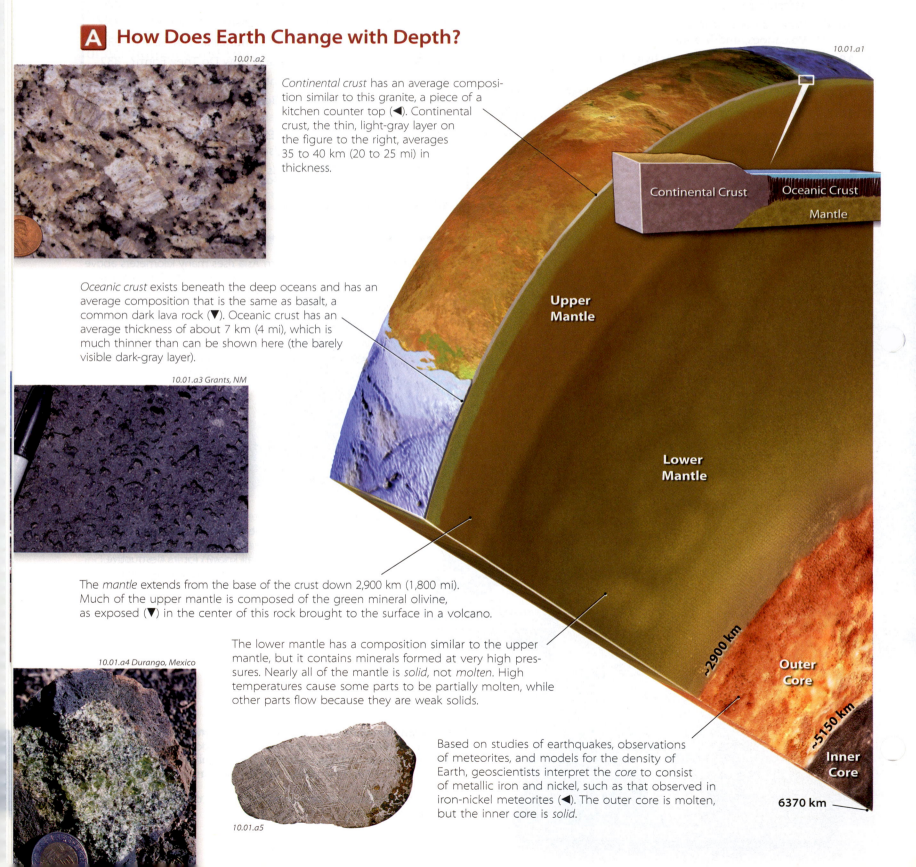

Continental crust has an average composition similar to this granite, a piece of a kitchen counter top (◀). Continental crust, the thin, light-gray layer on the figure to the right, averages 35 to 40 km (20 to 25 mi) in thickness.

Oceanic crust exists beneath the deep oceans and has an average composition that is the same as basalt, a common dark lava rock (▼). Oceanic crust has an average thickness of about 7 km (4 mi), which is much thinner than can be shown here (the barely visible dark-gray layer).

The *mantle* extends from the base of the crust down 2,900 km (1,800 mi). Much of the upper mantle is composed of the green mineral olivine, as exposed (▼) in the center of this rock brought to the surface in a volcano.

The lower mantle has a composition similar to the upper mantle, but it contains minerals formed at very high pressures. Nearly all of the mantle is *solid*, not *molten*. High temperatures cause some parts to be partially molten, while other parts flow because they are weak solids.

Based on studies of earthquakes, observations of meteorites, and models for the density of Earth, geoscientists interpret the *core* to consist of metallic iron and nickel, such as that observed in iron-nickel meteorites (◀). The outer core is molten, but the inner core is *solid*.

B Are Some Layers Stronger than Others?

In addition to layers with different compositions, Earth has layers that are defined by strength and by how easily the material in the layers fractures or flows when subjected to forces.

The uppermost mantle is relatively strong and solidly attached to the overlying crust. The crust and uppermost mantle together form an upper rigid layer called the *lithosphere* (*lithos* means "*stone*" in Greek), which averages about 100 km (about 60 mi) in thickness. The part of the uppermost mantle that is in the lithosphere is the *lithospheric mantle*.

The mantle directly beneath the lithosphere is mostly solid, but it is hotter than the rock above and can flow under pressure. This part of the upper mantle, called the *asthenosphere*, functions as a soft, weak zone under which the lithosphere moves. The word *asthenosphere* is from a Greek term for "*not strong*." In most regions, the asthenosphere is approximately 80 km to more than 150 km thick, so it can be deeper than 250 km.

C Why Do Some Regions Have High Elevations?

Why is the Gulf Coast of Texas near sea level, while the Colorado mountains are 3 to 5 km (2 to 3 mi) above sea level? Why are the continents mostly above sea level, but the ocean floor is below sea level? The primary factor controlling the elevation of a region is the thickness of the underlying crust.

The crust is less dense than the underlying mantle, and so rests, or floats, on top of the mantle. The underlying lithospheric mantle is mostly solid, not liquid.

The thickness of continental crust ranges from less than 25 km (16 mi) to more than 60 km (37 mi). Regions that have high elevation generally have thick crust. The crust beneath the Rocky Mountains of Colorado is commonly more than 45 km (28 mi) thick.

The crust beneath low-elevation regions like Texas is thinner. If the crust is thinner than 30 to 35 km (18 to 20 mi), the area will probably be below sea level, but it can still be part of the continent.

Most islands are volcanic mountains built on oceanic crust, but some are small pieces of continental crust.

Oceanic crust is thinner than continental crust and consists of denser rock than continental crust. As a result, regions underlain only by oceanic crust are well below sea level.

Density and Isostasy

The relationship between regional elevation and crustal thickness is similar to that of wooden blocks of different thicknesses floating in water (▼). Wood floats on water because it is less dense than water. Ice floats on water because it is less dense than water, even though ice and water have the same composition. Thicker blocks of wood, like thicker parts of the crust, rise to higher elevations than do thinner blocks of wood.

For Earth, we envision the crust being supported by mantle that is solid, unlike the liquid used in the wooden-block example. This concept of different thicknesses of crust riding on the mantle is called *isostasy*. Isostasy explains most of the variations in elevation from one region to another, and it is commonly

paraphrased by saying *mountain belts have thick crustal roots*. As in the case of the floating wooden blocks, most of the change in crustal thickness occurs at depth and less occurs near the surface. Smaller, individual mountains do not necessarily have thick crustal roots. They can be supported by the strength of the crust, like a small lump of clay riding on one of the wooden blocks.

The *density* of the rocks also influences regional elevations. The fourth block shown here has the same thickness as the third block, but it consists of a denser type of wood. It therefore floats lower in the water. Likewise, a region of Earth underlain by especially dense crust or mantle is lower in elevation than a region with less dense crust or mantle, even if the two regions have similar thicknesses of crust. Temperature also controls

the *thickness* of the *lithosphere*, and this affects a region's elevation. If the lithosphere in some region is heated, it expands, becoming less dense, and so the region rises in elevation. Thinner lithosphere also yields higher elevations because in this case less dense asthenosphere is replacing more dense lithosphere.

Before You Leave This Page Be Able To

✓ Sketch the major layers of Earth.

✓ Sketch and describe differences in thickness and composition between continental crust and oceanic crust, and contrast the lithosphere and asthenosphere.

✓ Sketch and discuss how the principle of isostasy can explain differences in regional elevation.

10.1

10.2 What Are the Major Features of Earth?

OCEANS COVER 71% OF EARTH'S SURFACE. Seven continents make up most of the rest of the surface, and islands account for less than 2%. We are all familiar with the continents and their remarkable diversity of landforms, from broad coastal plains to steep, snow-capped mountains. Features of the ocean floor, not generally seen by people, are just as diverse and include deep trenches and submarine mountain ranges. Islands exhibit great diversity, too. Some are large and isolated, but other islands form in arc shapes, ragged lines, or irregular clusters. What are the characteristics of each type of feature? In this chapter, we will explore how each of these features forms, so it is worth some time examining this map, which represents the world we call home.

This map shows large features on land and on the seafloor. The colors on land are from images taken by satellites orbiting Earth and show vegetated areas (green), rocky areas (brown), and sandy areas (tan). Greenland and Antarctica are white and light gray because they are mostly covered with ice and snow. Ocean colors show the depth of the seafloor and range from light blue where the seafloor is shallow to darker blue where it is deep. Examine this map and note the types of features you observe. Then, consider whether you have any ideas about how that type of feature forms.

Parts of the seafloor have mountains, the highest of which form islands such as Hawaii. Most mountains on the seafloor do not reach sea level, and are termed *seamounts*. Some islands, like Hawaii, and seamounts, are in long belts, which we refer to as *island and seamount chains*. Other islands and seamounts are isolated or form irregular clusters.

Some large islands, such as New Zealand, look like a small version of a continent.

Much of the ocean floor is moderately deep—3 to 5 km (9,800 to 16,000 ft)—and has a fairly smooth surface. Such a smooth, deep part of the seafloor is an *abyssal plain*.

10.02.a1

Arctic Ocean

Greenland

North America

Atlantic Ocean

Pacific Ocean

South America

Atlantic Ocean

Antarctica

A *mid-ocean ridge* is a broad, symmetrical ridge that crosses an ocean basin. Most ridges are 2 to 3 km (6,600–9,800 ft) higher than the average depth of the seafloor. One long ridge, named the *East Pacific Rise*, crosses the eastern Pacific and heads toward North America. Another occupies the middle of the Atlantic Ocean.

Cracks and steps cross the seafloor mostly at right angles to the mid-ocean ridges. Such a feature is an *oceanic fracture zone*.

Some continents continue outward from the shoreline under shallow seawater (light blue in this image) for hundreds of kilometers, forming submerged benches known as *continental shelves*. Which coastlines have broad continental shelves, like those surrounding Great Britain?

All continents contain large interior regions with gentle topography. Some continents have flat coastal plains, while others have mountains along their edges. Some mountains, like the Ural Mountains, are in the middle of continents.

Most continental areas have elevations of less than 1 to 2 km (3,300 to 6,600 ft). Broad, high regions, called *plateaus*, reach higher elevations, such as the Tibetan Plateau of southern Asia. Continents also contain mountain chains and individual mountains. Mount Everest, the highest point in the world, is almost 9 km (about 30,000 ft) in elevation.

Before You Leave This Page Be Able To

☑ Identify on a world map the named continents and oceans.

☑ Identify on a world map the main types of features on the continents and in the oceans.

☑ Describe the main characteristics for each type of feature, including whether it occurs in the oceans, on continents, or as islands.

Arctic Ocean

Asia

Europe

Africa

Indian Ocean

Atlantic Ocean

Australia

Antarctica

Ocean trenches make up the deepest parts of the ocean. Some ocean trenches follow the edges of continents, whereas others form isolated, curving troughs. Most ocean trenches are in the Pacific Ocean. Why are they here?

Crossing the seafloor are curving chains of islands, each known as an *island arc*. Most of the islands in an island arc are volcanoes, and many are active and dangerous. Most island arcs are flanked on one side by an ocean trench. Offshore of the Mariana island arc, located south of Japan, is the Mariana Trench, the deepest in the world at over 10.9 km (6.8 miles) deep—much deeper below sea level than Mount Everest is above sea level.

Some continents (such as South America) are flanked by an ocean trench, but other continents, such as Australia and Africa, have no nearby trenches.

Mid-ocean ridges and their associated fracture zones encircle much of the globe. In the Atlantic and Indian Oceans, they occupy a position halfway between the adjacent continents.

The oceans contain several broad, elevated regions, each of which is an *oceanic plateau*. The Kerguelen Plateau near Antarctica is one example, and another oceanic plateau lies northeast of Australia.

10.2

10.4 What Is the Distribution of Earthquakes, Volcanoes, and Mountain Belts?

EARTHQUAKES AND VOLCANOES are spectacular manifestations of our dynamic Earth. Many of these are in distant places, but some are close to human settlements. The distributions of earthquakes and volcanoes are not random, but instead define clear patterns and show a close association with mountain belts and other regional features. These patterns show important, large-scale Earth processes.

A Where Do Most Earthquakes Occur?

On this map, yellow circles show the locations of moderate to strong earthquakes that occurred between 1973 and 2000. Observe the distribution of earthquakes before reading on. What patterns do you notice? Which regions have many earthquakes and which have few? Are earthquakes associated with certain types of features?

Earthquakes are not distributed uniformly across the planet. Most are concentrated in discrete belts, such as one that runs along the western coasts of North and South America.

Most earthquakes in the oceans occur along the winding crests of mid-ocean ridges. Where the ridges curve or zigzag, so do the patterns of earthquakes.

Earthquakes are sparse in some continental interiors but are abundant in others, like the Middle East, China, and near the Himalaya.

10.04.a1

Large areas of the seafloor, especially the abyssal plains, have few earthquakes. Volcanically active islands, like Hawaii, in the middle of the Pacific Ocean, do have earthquakes.

Some continental edges experience many earthquakes, but other edges have few. Earthquakes are common along the western coasts of South America and North America, and these edges also have narrow continental shelves. There are few earthquakes along the eastern coasts of the Americas, where the continental shelves are wide.

Ocean trenches and associated island arcs have numerous earthquakes. In fact, many of the world's largest and most deadly earthquakes occur near ocean trenches. Recent examples were the large earthquakes that produced deadly ocean waves (tsunami) in the Indian Ocean in 2004 and in Japan in 2011.

Before You Leave These Pages Be Able To

☑ Show on a world relief map the major belts of earthquakes and volcanoes.

☑ Describe how the distribution of volcanoes corresponds to that of earthquakes.

☑ Compare the distributions of earthquakes, volcanoes, and high elevations.

B Which Areas Have Volcanoes?

On the map below, orange triangles show the locations of volcanoes that have been active in the last several million years. Observe the distribution of volcanoes and note which areas have volcanoes and which have none. How does this distribution compare with the distribution of earthquakes?

Volcanoes, like earthquakes, are widespread, but commonly occur in belts. One belt extends along the western coasts of North and South America.

Some volcanoes occur in the centers of oceans, such as the volcanoes near Iceland. Iceland is a large volcanic island along the mid-ocean ridge in the middle of the North Atlantic Ocean.

Volcanoes occur along the western edge of the Pacific Ocean, extending from north of Australia through the Philippines and Japan. Many are part of island arcs, associated with ocean trenches and earthquakes.

10.04.b1

Volcanic eruptions occur beneath the oceans, but this map shows only the largest submarine volcanic mountains. Volcanism is widespread along mid-ocean ridges, but it generally does not form mountains.

Some volcanoes form in the middle of continents, such as in the eastern part of Africa and China.

This map (▼) shows the topography of Earth's surface and seafloor, with high elevations in brown, low land elevations in green, shallow seafloor in light blue, and deep seafloor in dark blue. Using the three maps shown here, compare the distributions of earthquakes, volcanoes, and high elevations. Identify areas where there are (1) mountains but no earthquakes, (2) mountains but no volcanoes, and (3) earthquakes but no volcanoes. Make a list of these areas, or mark the areas on a map.

10.04.b2

10.4

10.5 What Causes Tectonic Activity to Occur in Belts?

MOST EARTHQUAKES AND VOLCANOES occur in belts around Earth's surface. Between these belts are vast regions that have comparatively little of this activity. Where are these belts of concentrated activity, and what explains this spatial distribution? What underlying processes cause these observed patterns? These and other questions helped lead to the *theory of plate tectonics*.

A What Do Earthquake and Volcanic Activity Tell Us About Earth's Lithosphere?

1. Examine the map below, which shows the locations of recent earthquakes (yellow circles) and volcanoes (orange triangles). After noting the patterns, compare this map with the lower map and then read the associated text.

2. On the upper map, there are large regions that have few earthquakes and volcanoes. These regions are relatively stable and intact pieces of Earth's outer layers. There are a dozen or so of these regions, each having edges defined by belts of earthquakes and volcanoes.

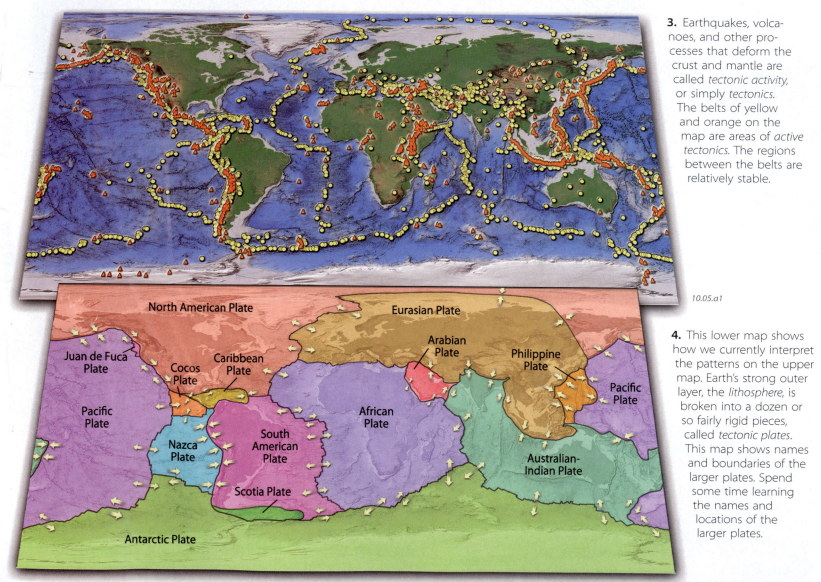

3. Earthquakes, volcanoes, and other processes that deform the crust and mantle are called *tectonic activity*, or simply *tectonics*. The belts of yellow and orange on the map are areas of *active tectonics*. The regions between the belts are relatively stable.

10.05.a1

4. This lower map shows how we currently interpret the patterns on the upper map. Earth's strong outer layer, the *lithosphere*, is broken into a dozen or so fairly rigid pieces, called *tectonic plates*. This map shows names and boundaries of the larger plates. Spend some time learning the names and locations of the larger plates.

North American Plate
Eurasian Plate
Juan de Fuca Plate
Cocos Plate
Caribbean Plate
Arabian Plate
Philippine Plate
Pacific Plate
Pacific Plate
African Plate
Nazca Plate
South American Plate
Australian-Indian Plate
Scotia Plate
Antarctic Plate

5. Compare the two maps and note how the distribution of tectonic activity, especially earthquakes, outlines the shapes of the plates. Earthquakes are a better guide to plate boundaries than are volcanoes. Most, but not all, volcanoes are near plate boundaries, but many plate boundaries have no volcanoes.

6. Some earthquakes occur in the middle of plates, indicating that the situation is more complicated than a simple plate-tectonic model, in part because some parts of a plate are weaker than others. Forces can be transmitted through the strong parts, causing weaker parts to break and slip, generating an earthquake within the plate. Generally, most tectonic activity occurs near plate boundaries.

B How Do Plates Move Relative to One Another?

Plate boundaries have tectonic activity because plates are moving *relative to one another*. For this reason, we talk about the *relative motion* of plates across a plate boundary. Two plates can move away, toward, or sideways relative to one another, resulting in three types of plate boundaries: *divergent*, *convergent*, and *transform*.

Divergent Boundary

10.05.b1

At a *divergent boundary*, two plates move apart relative to one another. In most cases, magma fills the space between the plates.

Convergent Boundary

10.05.b2

At a *convergent boundary*, two plates move toward one another. A typical result is that one plate slides under the other.

Transform Boundary

10.05.b3

At a *transform boundary*, two plates move horizontally past one another, as shown by the white arrows on the top surface.

C Where Are the Three Types of Plate Boundaries?

This map shows plate boundaries according to type. Compare this map with the maps in part A and with those shown earlier in the chapter. For each major plate, note the types of boundaries between this plate and other plates it contacts. Then use the various maps to determine whether each type of plate boundary has the following features:

- Earthquakes
- Volcanoes
- Mountain belts
- Mid-ocean ridges
- Ocean trenches

Divergent   Convergent   Transform

10.05.c1

Before You Leave This Page Be Able To

✔ Describe plate tectonics and how it explains the distribution of tectonic activity.

✔ Sketch and explain the three types of plate boundaries.

✔ Compare the three types of plate boundaries with the distributions of earthquakes, volcanoes, mountain belts, mid-ocean ridges, and ocean trenches.

10.5

10.6 What Happens at Divergent Boundaries?

AT MID-OCEAN RIDGES, Earth's tectonic plates diverge (move apart). Ridges are the sites of many small to moderate-sized earthquakes and much submarine volcanism. On the continents, divergent motion can split a continent into two pieces, forming a new ocean basin as the pieces move apart.

A What Happens at Mid-Ocean Ridges?

Mid-ocean ridges are divergent plate boundaries where new oceanic lithosphere forms as two oceanic plates move apart. These boundaries are also called *spreading centers* because of the way the plates spread apart.

1. A narrow trough, or *rift*, runs along the axis of most mid-ocean ridges. The rift forms because large blocks of crust slip down as spreading occurs. The movement causes fracturing, resulting in frequent small to moderate-sized earthquakes.

2. As the plates move apart, solid mantle in the asthenosphere rises toward the surface. It partially melts as it rises, because decreasing pressures can no longer confine the rock as a solid. The molten rock (magma) rises along narrow conduits, accumulates in magma chambers beneath the rift, and eventually becomes part of the oceanic lithosphere.

3. Much of the magma solidifies at depth, but some erupts onto the seafloor, forming submarine lava flows. These eruptions create new ocean crust that is incorporated into the oceanic plates as they move apart.

4. Mid-ocean ridges are elevated above the surrounding seafloor because they consist of hotter, less dense materials, including magma. Lower density materials and thin lithosphere mean that the plate "floats" higher above the underlying asthenosphere. The elevation of the seafloor decreases away from the ridge because the rock cools and contracts, and because the less dense asthenosphere cools enough to become part of the more dense lithosphere.

10.06.a1–2

Before You Leave These Pages Be Able To

☑ Sketch, label, and explain the features and processes of an oceanic divergent boundary.

☑ Sketch and label the characteristics of a continental rift (i.e., a divergent boundary within a continent).

☑ Sketch, label, and explain the stages of continental rifting, using East Africa and the Red Sea as examples of stages.

B What Happens When Divergence Splits a Continent Apart?

Most divergent plate boundaries are beneath oceans, but a divergent boundary may also form within a continent. This process creates a *continental rift,* such as the Great Rift Valley in East Africa. Rifting can lead to seafloor spreading and formation of a new ocean basin, following the progression shown here.

10.06.b1–4

1. The initial stage of continental rifting commonly includes broad uplift of the land surface as mantle-derived magma ascends into and pushes up the crust. The magma heats and can melt parts of the continental crust, producing additional magma. Heating of the crust causes it to expand, which results in further uplift.

2. Stretching of the crust causes large crustal blocks to drop down along faults, forming a *continental rift,* like in the Great Rift Valley. The down-dropped blocks may form *basins* that can trap sediment and water, resulting in lakes. Deep rifting causes solid mantle material in the asthenosphere to flow upward and partially melt. The resulting magma may solidify beneath the surface or may erupt from volcanoes and long fissures on the surface. The entire crust thins as it is pulled apart, so the central rift becomes lower in elevation over time. The Rio Grande Rift, which runs north-south through New Mexico, is another example of a continental rift in its early stages.

3. If rifting continues, the continent splits into two pieces, and a narrow ocean basin forms as seafloor spreading takes place. A modern example of this is the narrow Red Sea, which runs between Africa and the Arabian Peninsula. As the edges of the continents move away from the heat associated with active spreading, the thinned crust cools and drops in elevation, eventually dropping below sea level. The continental margin ceases to be a plate boundary. A continental edge that lacks tectonic activity is called a *passive margin.*

4. With continuing seafloor spreading, the ocean basin becomes progressively wider, eventually becoming a broad ocean like the modern-day Atlantic Ocean. The Atlantic Ocean basin formed when North and South America rifted away from Europe and Africa, following the sequence shown here. Continental edges on both sides of the Atlantic are currently passive margins. Seafloor spreading continues today along the ridge in the middle of the Atlantic Ocean, so the Americas continue to move away from Europe and Africa.

5. East Africa and adjacent seas illustrate the different stages of continental rifting. Here, a piece of continent has been rifted away from Africa, and another piece is in the early stages of possibly doing the same (▶).

6. Early stages of rifting occur along the East African Rift, a long continental rift that begins near the Red Sea and extends into central Africa. The rift is within an elevated (uplifted) region and has several different segments, each featuring a down-dropped rift. Some parts of the rift contain large lakes.

250 km

10.06.b5

7. The Red Sea represents the early stages of seafloor spreading. It began forming about 50 million years ago when the Arabian Peninsula rifted away from Africa. The Red Sea continues to spread and slowly grow wider.

8. The East African Rift is topographically lower than its flanks, but contains numerous active volcanoes (▶). Volcanoes are especially common in the Afar region of Ethiopia, located where the rift joins the Red Sea. The Afar region is one of the most volcanically active regions on the planet.

10.06.b6 *East African Rift, Kenya*

10.6

10.7 What Happens at Convergent Boundaries?

CONVERGENT BOUNDARIES FORM when two plates move toward each other. Convergence can involve two oceanic plates, an oceanic plate and a continental plate, or two continental plates. Oceanic trenches, island arcs, and Earth's highest mountain belts form at convergent boundaries. Many of Earth's most dangerous volcanoes and largest earthquakes also occur along these boundaries.

A What Happens When Two Oceanic Plates Converge?

1. Convergence of two oceanic plates forms an *ocean-ocean convergent boundary*. One plate bends and slides beneath the other plate along an inclined zone. The process of one plate sliding beneath another plate is *subduction*, and the zone around the downward-moving plate is a *subduction zone*. Many large earthquakes occur in subduction zones.

2. An *oceanic trench* forms as the subducting plate moves down. Sediment and slices of oceanic crust collect in the trench. This sheared, scraped-off material generally remains completely submerged, but is exposed in a few islands, like Barbados.

3. As the plate subducts, its temperature increases, releasing water from minerals in the downgoing plate. This water causes melting in the overlying asthenosphere, and the resulting magma is buoyant and rises into the overlying plate.

4. Some magma erupts, initially under the ocean and later as dangerous, explosive volcanoes that rise above the sea. With continued activity, the erupted lava and exploded volcanic fragments construct a curving belt of islands in an *island arc*. An example is the arc-shaped belt of the Aleutian Islands of Alaska. The area between the island arc and the ocean trench accumulates sediment, most of which comes from volcanic eruptions and from the erosion of volcanic materials in the arc.

5. Magma that solidifies at depth adds to the volume of the crust. Over time, the crust gets thicker and becomes transitional in character between oceanic and continental crust. Volcanic islands join to form more continuous strips of land, as occurred to form the island of Java in Indonesia.

10.07.a1

B What Happens When an Oceanic Plate and a Continental Plate Converge?

1. The convergence of an oceanic and a continental plate forms an *ocean-continent convergent boundary*. Along this boundary, the denser oceanic plate subducts beneath the more buoyant continental plate.

2. An oceanic trench marks the plate boundary and receives sediment from the adjacent continent. Again, sediment and other material are scraped off the oceanic plate, forming a wedge of highly sheared material near the trench.

3. Volcanoes form on the surface of the overriding continental plate in the same way the volcanoes form in an ocean-ocean convergent boundary. These volcanoes erupt, often violently, producing large amounts of volcanic ash, lava, and mudflows, which pose a hazard for people who live nearby. Examples include large volcanoes of the Andes of South America and the Cascade Range of Washington, Oregon, northern California, and southern British Columbia.

4. Compression associated with the convergent boundary squeezes the crust for hundreds of kilometers into the continent. The crust deforms and thickens, resulting in uplift of the region. Uplift and volcanism may produce a high mountain range, such as the Andes.

5. Magma forms by melting of the asthenosphere above the subduction zone. It can solidify at depth, rise into the overlying continental crust before solidifying, or reach the surface and cause a volcanic eruption.

10.07.b1

C What Causes the Pacific Ring of Fire?

Volcanoes surround the Pacific Ocean, forming the *Pacific Ring of Fire*, as shown in the map below. The volcanoes extend from the southwestern Pacific, through the Philippines, Japan, and Alaska, and then down the western coasts of North and South America. The Ring of Fire results from subduction on both sides of the Pacific Ocean.

10.07.c1

1. In the Pacific, new oceanic lithosphere forms along a mid-ocean ridge, the *East Pacific Rise*. The Pacific plate is west (left) of this boundary and several smaller plates are to the east (right). Once formed, new lithosphere moves away from the ridge as seafloor spreading continues.

2. Oceanic lithosphere of the Pacific plate and smaller places subducts beneath the Americas, forming oceanic trenches on the seafloor and volcanoes on the overriding, mostly continental, plates.

10.07.c2

3. Subduction of oceanic lithosphere also occurs to the west, beneath Japan and island arcs of the western Pacific.

4. More oceanic plate is subducted than is produced along the East Pacific Rise, so the width of the Pacific Ocean is shrinking with time.

D What Happens When Two Continents Collide?

Two continental masses may converge along a *continent-continent convergent boundary*. This type of boundary is commonly called a *continental collision*, and it produces huge mountain ranges, like the Himalaya and Mount Everest in the photograph below.

10.07.d1 Mount Everest

The large plate in the figure to the right is partly oceanic and partly continental. The oceanic part is being subducted to the right, under another continent at a convergent boundary.

As the oceanic part of the plate continues to subduct, the two continents approach each other. Magmatic activity occurs in the overriding plate above the subduction zone. The edge of the approaching continent has no such activity because it is not at a plate boundary yet.

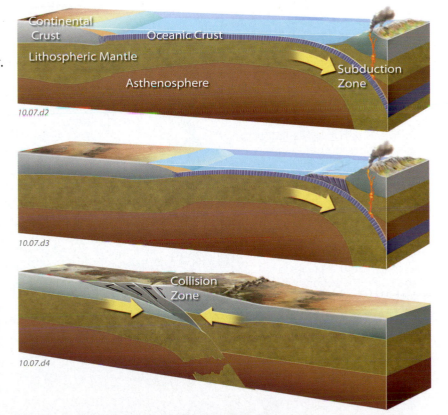

10.07.d2

10.07.d3

10.07.d4

When the converging continent arrives at the subduction zone, it may partially slide under the other continent or simply clog the subduction zone as the two continents collide. Because the two continents are thick and have the same density, neither can be easily subducted beneath the other and into the asthenosphere. Along the boundary, faults slice up the continental crust, stacking one slice on top of another. Continental collisions form enormous mountain belts and high plateaus, such as the Himalaya and Tibetan Plateau of southern Asia. The Himalaya and Tibetan Plateau are still forming today, rising at a rate of about 5 mm/year, as the continental crust of India collides with the southern edge of the Eurasian plate.

Before You Leave This Page Be Able To

☑ Sketch, label, and explain the features and processes associated with ocean-ocean and ocean-continent convergent boundaries.

☑ Sketch, label, and explain the steps leading to a continental collision (continent-continent convergent boundary).

10.7

10.8 What Happens Along Transform Boundaries?

AT TRANSFORM BOUNDARIES, PLATES SLIP HORIZONTALLY past each other along *transform faults*. In the oceans, transform faults are associated with mid-ocean ridges. Transform faults combine with spreading centers to form a zigzag pattern on the seafloor. A transform fault can link different types of plate boundaries, such as a mid-ocean ridge and an ocean trench. Some transform boundaries occur beside or within a continent, sliding one large crustal block past another, as occurs along the San Andreas fault in California.

A Why Do Mid-Ocean Ridges Have a Zigzag Pattern?

10.08.a1

To understand the zigzag character of mid-ocean ridges, examine how the two parts of this pizza have pulled apart, just like two *diverging* plates.

The break in the pizza did not follow a straight line. It took jogs to the left and the right, following cuts where the pizza was the weakest.

Openings created where the pizza pulled apart represent the segments of a mid-ocean ridge that are spreading apart. However, unlike a pizza, at a mid-ocean ridge, no open gaps exist because new material derived from the underlying mantle fills the space as fast as it opens, forming new oceanic crust.

The openings are linked by breaks, or *faults*, where the two parts of the pizza simply slide by one another. There are no gaps along these breaks, only *horizontal* movement of one plate sliding past the other. Arrows show the direction of relative motion. A fault that accommodates the horizontal movement of one tectonic plate past another is a *transform fault*. The spreading direction must be parallel to the transform faults and perpendicular to the spreading segments, so a zigzag pattern is required to allow a plate boundary to be curved.

Transform Faults Along the Mid-Ocean Ridge

1. Mid-ocean ridges, such as this one in the South Atlantic Ocean, have a zigzag pattern similar to the broken pizza.

2. In this region, spreading occurs along north-south ridges. The direction of spreading is east-west, perpendicular to the ridges.

3. East-west offsets are transform faults along which the two diverging plates simply slide past one another, like the breaks in the pizza. These transform faults link the spreading segments and have the relative motion shown by the white arrows.

4. Transform faults along mid-ocean ridges are generally perpendicular to the axis of the ridge. As in the pizza example, transform faults are parallel to the direction in which the two plates are spreading apart.

6. Continuing outward from most transform faults is an *oceanic fracture zone,* which is a step in the elevation of the seafloor. A fracture zone is a former transform fault that now has no relative motion across it. It no longer separates two plates and instead is within a single plate. Opposite sides of the fracture zone have different elevations because they formed by seafloor spreading at different times in the past, so they have had different amounts of time to cool and subside after forming at the spreading center. Younger parts of the plate are warmer and higher than older parts.

5. The zigzag pattern of mid-ocean ridges shows the alternation of spreading segments with transform faults. In this example, the overall shape of the ridge mimics the edges of Africa and South America and so was largely inherited from the shape of the original rift that split the two continents apart.

10.08.a2

B What Are Some Other Transform Boundaries?

The Pacific seafloor and western North America contain several different transform boundaries. The boundary between the Pacific plate and the North American plate is mostly a transform boundary, with the Pacific plate moving northwest relative to the main part of North America.

1. The Queen Charlotte transform fault, shown as a long green line, lies along the edge of the continent, from north of Vancouver Island to southeastern Alaska.

2. The zigzag boundary between the Pacific plate and the small Juan de Fuca plate has three transform faults, shown here as green lines. These transform faults link the ridge segments that are spreading (shown here as yellow lines).

10.08.b1

3. The Mendocino fracture zone originated as a transform fault, but it is now entirely within the Pacific plate and is no longer active. Oceanic crust to the north is higher because it is younger than oceanic crust to the south.

4. A transform fault links a spreading center (between the Pacific plate and the Juan de Fuca plate) with the Cascadia subduction zone and the San Andreas fault. The place where the three plate boundaries meet is a *triple junction*. The Mendocino triple junction is the meeting place of two different transform faults and a subduction zone.

5. The San Andreas transform fault extends from north of San Francisco to southeast of Los Angeles. The part of California west of the fault is on the Pacific plate and is moving approximately 5 cm/yr to the northwest relative to the rest of North America. South of this map area, the transform boundary continues across southern California and into the Gulf of California.

10.08.b2 Carrizo Plain, CA

6. Californians have a transform fault in their backyard. In central California, the San Andreas fault forms linear valleys, abrupt mountain fronts, and lines of lakes. In the Carrizo Plain (▶), the fault is a linear gash in the topography. Some streams follow the fault and others jog to the right as they cross the fault, recording relative movement of the two sides. In this view, the North American plate is to the left, and the Pacific plate is to the right and is being displaced toward the viewer at several centimeters per year.

Before You Leave This Page Be Able To

☑ Sketch, label, and explain an oceanic transform boundary related to seafloor spreading at a mid-ocean ridge.

☑ Sketch, label, and explain the motion of transform faults along the west coast of North America.

10.8

Why and How Do Plates Move?

THE PROCESS OF PLATE TECTONICS circulates material back and forth between the asthenosphere and the lithosphere. Some asthenosphere becomes lithosphere at mid-ocean spreading centers and then takes a slow trip across the ocean floor before going back down into the asthenosphere at a subduction zone. Besides creating and destroying lithosphere, this process is the major way that Earth transports heat to the surface.

A What Moves the Plates?

How exactly do plates move? To move, an object must be subjected to a *driving force* (a force that drives the motion). The driving force must exceed the *resisting forces*—those forces that resist the movement, such as friction and any resistance from other material that is in the way. What forces drive the plates?

Slab Pull—Subducting oceanic lithosphere is more dense than asthenosphere, so gravity pulls the plate downward into the asthenosphere. Slab pull is a significant force, and a plate being subducted generally moves faster than plates not being subducted. Subduction sets up other forces in the mantle that can work with or against slab pull.

Ridge Push—The mid-ocean ridge is higher than the ocean floor away from the ridge because lithosphere near the ridge is thinner and hotter. Gravity causes the plate to slide away from the topographically high ridge and push the plate outward.

Mantle Convection—The asthenosphere, although a solid, is capable of flow. It experiences *convection*, where hot material rises due to its lower density, while cold material sinks because it is more dense. Hot material rises at mid-ocean ridges, cools, and eventually sinks back into the asthenosphere at a subduction zone. Convection also occurs at centers of upwelling mantle material called *hot spots*, and it can help or hinder the motion of a plate. Another important source of forces is the motion of a plate with respect to the underlying mantle.

10.09.a1

B How Fast and in What Directions Do Plates Move Relative to One Another?

Plates move at 1 to 15 cm/yr, about as fast as your fingernails grow. This map shows velocities and relative motions along major plate boundaries, based on long-term rates. Arrows indicate whether the plate boundary has divergent (outward pointing), convergent (inward pointing), or transform (side by side) motion.

10.09.b1

Less than 3 cm/year   3 to 9 cm/year   More than 9 cm/year

C Is There a Way to Measure Plate Motions Directly?

Modern technology allows direct measurement of plate motions using satellites, lasers, and other tools. The measured directions and rates of plate motions are consistent with our current concept of lithospheric plates and with the theory of plate tectonics. Examine this map and determine which way your home is moving.

Global Positioning System (GPS) is an accurate location technique that uses small radio receivers to record signals from several dozen Earth-orbiting satellites. By attaching GPS receivers to sites on land and monitoring changes in position over time, Earth scientists produce maps showing motions for each plate. Arrows point in the direction of motion, and longer arrows indicate faster motion.

50 mm/yr

10.09.c1

Note the motions of different plates. Africa is moving to the northeast, away from South America. North America is moving westward and rotating counterclockwise in this view. The plate on which India rides is converging with Asia to form the Himalaya. In all, these motions match predictions from the theory of plate tectonics.

D Is the Age of the Seafloor Consistent with Plate Tectonics?

According to plate tectonics, oceanic crust forms from upwelling magma and spreading at a mid-ocean ridge. The oceanic crust then moves away from the ridge with further spreading. If so, the crust should be youngest near the ridge, where it was just formed, and should be progressively older away from the ridge. Also, oceanic crust near the ridge will not have had time to accumulate much sediment, but the sediment cover should thicken outward from the ridge.

Since 1968, ocean-drilling ships have drilled hundreds of deep holes into the seafloor. Geoscientists use drill cores and other drilling results to measure the thickness of sediment and examine the underlying volcanic rocks (basalt). Geoscientists analyze samples of sediment, rock, and fossils to determine the age, character, and origin of the materials.

Sediment

Basalt

10.09.d1

Drill core samples reveal that sediment is thin or absent on the ridge but becomes thicker away from the ridge. Age determinations on fossils in the sediment and from underlying volcanic rocks show that oceanic crust gets systematically older away from mid-ocean ridges. Such drilling results strongly support the theory of plate tectonics.

The map below shows the age of the seafloor, with letters marking the position of some mid ocean ridges (R) and trenches (T). Purple represents the oldest areas (about 180 million years), and the darkest orange represents very young oceanic crust.

The youngest oceanic crust is near mid-ocean-ridge spreading centers (R), as we expect from plate tectonics.

The oldest oceanic crust in any ocean is the most distant from mid-ocean ridges. None is older than about 180 million years, because all older oceanic crust has been subducted (destroyed).

10.09.d2

Age of the Seafloor

Before You Leave This Page Be Able To

☑ Sketch and explain the driving forces of plate tectonics.

☑ Describe the typical rates of relative motion between plates, describing one way to directly measure plate motion.

☑ Predict the relative ages of seafloor from place to place using a map of an ocean with a mid-ocean ridge.

10.9

10.10 How Is Paleomagnetism Used to Determine Rates of Seafloor Spreading?

PALEOMAGNETISM IS THE ROCK RECORD OF PAST CHANGES in Earth's magnetic field. The magnetic field is strong enough to orient magnetism in certain minerals, especially the iron-rich mineral *magnetite*, in the direction of the prevailing magnetic field. Magnetic directions preserved in volcanic rocks, intrusive rocks, and some sedimentary rocks provide an important way to determine the rates of seafloor spreading.

A What Causes Earth's Magnetic Field?

Earth has a metallic iron core, which is composed of a solid inner core surrounded by a liquid outer core. The liquid core flows and behaves like a *dynamo* (an electrical generator), creating a magnetic field around Earth.

1. The inner core transfers heat and less dense material to the liquid outer core. This transfer causes liquid in the outer core to rise, forming *convection currents*. These convection currents are limited to the outer core and are not the same as those in the upper mantle.

2. Movement of the molten iron is affected by forces associated with Earth's rotation. The resulting movement of liquid iron and electrical currents generates the magnetic field.

10.10.a1

▶ 3. Earth's magnetic field currently flows from south to north, causing the magnetic ends of a compass needle to point toward the north. This orientation is called a *normal polarity*.

10.10.a2

◀ 4. Many times in the geologic past, the magnetic field has had a *reversed polarity*, so that a compass needle would point south. The switch between normal polarity and reversed polarity is a *magnetic reversal*.

10.10.a3

B How Do Magnetic Reversals Help Us Infer the Age of Rocks?

The north and south magnetic poles have switched many times, typically remaining either normal or reversed anywhere from 100,000 years to a few million years. Scientists have constructed a magnetic timescale by isotopically dating sequences of rocks that contain magnetic reversals. This *geomagnetic polarity timescale* then serves as a reference to compare against other sequences of rocks.

10.10.b1

1. Scientists measure the direction and strength of the magnetism preserved in rocks with an instrument called a *magnetometer*. With this device, we can determine whether the magnetic field had a normal polarity or a reversed polarity when an igneous rock solidified and cooled or when a sedimentary layer accumulated.

▶ 2. This figure shows the series of magnetic reversals during the last 10 million years, the most recent part of the Cenozoic Era. The abbreviation "Ma" stands for millions of years before present.

3. The timescale shows periods of normal magnetization (N) in black and those of reversed magnetization (R) in white. Variability in the spacing and duration of magnetic reversals produced a unique pattern through time. Geoscientists can measure the pattern of reversals in a rock sequence and compare this pattern to the magnetic timescale to see where the patterns match. This allows an estimate of the age of the rock or sediment. Other age constraints, including isotopic ages or fossils, are used to further refine the age of the magnetized rocks. The magnetic timescale is best documented for the last 180 million years because seafloor of this age is widely preserved.

10.10.b2

How Are Magnetic Reversals Expressed at Mid-Ocean Ridges, and How Do Magnetic Patterns on the Seafloor Help Us Study Plate Tectonics?

In the 1950s, scientists discovered that the ocean floor displayed magnetic variations in the form of matching magnetic stripes on either side of the mid-ocean ridge. They interpreted the patterns to represent a magnetic field that had reversed its polarity, an idea that led to the theory of plate tectonics. Magnetic patterns allow us to estimate the ages of large areas of seafloor and to calculate the rates at which two diverging oceanic plates spread apart.

As the oceanic plates spread apart at a mid-ocean ridge, basaltic lava erupts onto the surface or solidifies at depth. As the rocks cool, the orientation of Earth's magnetic field is recorded by the iron-rich mineral magnetite. In this example, the magnetite records normal polarity (shown with a reddish color) at the time the rock forms.

If the magnetic field reverses, new rocks that form will acquire a reversed polarity (shown in white). Rocks forming all along the axis of the mid-ocean ridge will have the same magnetic direction, forming a stripe of similarly magnetized rocks parallel to the ridge. Once the rocks have cooled, they retain their original magnetic direction, preserving the magnetic polarity in the seafloor.

The magnetic poles have switched many times, and continued seafloor spreading produces a pattern of alternating magnetic stripes on the ocean floor. This pattern is strong enough to be detected by magnetic instruments towed behind a ship or a plane.

10.10.c1

10.10.c2

As magnetic instruments are towed behind a ship, the strength of the magnetic field is measured and plotted. Stronger measurements plot high on the graph and are called *positive magnetic anomalies*. The magnetic signal is weaker over crust that was formed under a reversed magnetic field because the magnetic direction in such rocks is opposite to and works to counteract the modern magnetic field. The reverse magnetization of the rocks slightly weakens the measured magnetic signal and will plot low on the graph, forming a *negative magnetic anomaly*.

The seafloor patterns are compared with the patterns on the geomagnetic polarity timescale to assign ages to each reversal. We can simplify and visualize these data as reversely and normally magnetized stripes on the seafloor, as shown in this cross section.

We can calculate rates of seafloor spreading by measuring the width of a specific magnetic stripe in map or cross-section view and then dividing that distance by the length of time the stripe represents:

rate of spreading for stripe = width of stripe / time duration

If a magnetic stripe is 60 km wide and formed over 2 million years, then the rate at which spreading formed the stripe was 30 km/m.y. This rate is equivalent to 3 cm/year. Spreading added an equal width of oceanic crust to a plate on the other side of the mid-ocean ridge, so the total rate of spreading across the ridge was 60 km/m.y. (6 cm/year), a typical rate of seafloor spreading.

The magnetic patterns on the seafloor, in addition to magnetic measurements on sequences of rocks and sediment on the seafloor and on land, demonstrate that Earth's magnetic field has reversed many times. Scientists are currently debating the possible causes of the magnetic reversals, with most explanations attributing reversals to chaotic flow in the molten outer core, which add to or subtract from the patterns caused by the dynamo, disrupting the prevailing magnetic field and causing a reversal.

Before You Leave This Page Be Able To

✓ Describe how Earth's magnetic field is generated.

✓ Describe how magnetic reversals help with determining the age of rocks and sediment.

✓ Describe or sketch how magnetic patterns develop on the seafloor.

✓ Calculate the rate of seafloor spreading if given the width and duration of a magnetic stripe.

10.10

10.11 What Features Form at Oceanic Hot Spots?

SUBMARINE MOUNTAINS, called seamounts, rise above the seafloor. In some places, they reach the surface and make islands. In many cases, islands and seamounts form linear chains, as in Hawaii. The seafloor also has relatively high and broad areas that are oceanic plateaus. How are seamounts and oceanic plateaus formed?

A How Do Island and Seamount Chains Form?

Linear chains of islands and seamounts, and most clusters of islands in the oceans, have two key things in common: they were formed by volcanism and they are near sites interpreted to be above unusually high-temperature regions in the deep crust and upper mantle—*hot spots*. Hot spots are interpreted to represent the surface manifestation of hot masses of material rising from the deep mantle.

1. This figure shows how linear island and seamount chains can be related to a plate moving over a hot spot. At a hot spot, hot mantle rises and melts, forming magma that ascends into the overlying plate. The magma generated by a hot spot may solidify at depth or form a volcanic mountain on the ocean floor. If the submarine volcano grows high enough above the seafloor, it becomes a volcanic island. Each of the Hawaiian Islands consists of volcanoes. Geomorphologists consider the hot spot to be currently below or near the eastern side of the Big Island, near Kilauea volcano.

2. If the plate above the hot spot is moving relative to the hot spot, the volcanic island can move off the hot spot. As it does, that part of the plate cools and subsides, so volcanoes that start out as islands may sink beneath the sea to become seamounts. As an island sinks, erosion can bevel off its top, forming a flat-topped seamount (▶), which are common on the seafloor.

10.11.a2

Hot Spot

10.11.a1

3. As a plate moves over a hot spot, volcanism constructs a chain of volcanic islands and seamounts, each created when it was over the hot spot. According to this model, volcanoes above the hot spot may be erupting today, those close to but not above the hot spot are relatively young, and those farthest from the hot spot are older. If a plate is not moving or is moving very slowly, the hot spot forms a cluster of volcanic islands and seamounts instead of a linear chain. The Galápagos, a cluster of volcanic islands in the eastern Pacific, are interpreted to be above such a hot spot.

B How Do Oceanic Plateaus Form?

Some large regions of the seafloor rise a kilometer or more above their surroundings, forming oceanic plateaus. These plateaus are largely composed of flood basalts, like the seafloor in general. How do these form?

1000 km
10.11.b1

10.11.b2

10.11.b3

10.11.b4

This perspective shows the Kerguelen oceanic plateau, which rises above the surrounding seafloor in the southern Indian Ocean. The plateau is several thousand kilometers long, but it only reaches sea level in a few small islands. The small sliver of land showing in the lower right corner is part of Antarctica.

Geologists interpret oceanic plateaus as forming at hot spots, above rising mantle plumes. The plumes travel through the mantle as solid masses, not liquids.

When the top of a plume encounters the base of the lithosphere, it causes widespread melting. Submarine flood basalts pour out onto the seafloor through fissures and central vents.

Immense volumes of basalt (as much as 50 million cubic kilometers) erupt onto the seafloor over millions of years. This volcanism creates a broad, high oceanic plateau.

C What Is the Distribution of Hot Spots, Linear Island Chains, and Oceanic Plateaus?

Hot spots have created many Pacific islands that we associate with tropical paradises and exotic destinations, like Tahiti and the Galápagos. Hawaii is the most famous island chain formed by movement of a plate over a hot spot, but several other linear island and seamount chains, in both the Atlantic and Pacific, formed in the same manner. We show oceanic and continental hot spots on the map below, but we discuss continental hot spots on the following two pages.

On this map, red dots show the locations of likely hot spots, many of which are located at the volcanically active ends of linear island chains. There is great debate, however, about which areas really are hot spots and how hot spots form.

The dark gray areas in the oceans represent linear island chains, clumps of islands, and oceanic plateaus, similar to this high area around Iceland, which is over a hot spot.

10.11.c1

The *Ontong Java Plateau* is the largest oceanic plateau on Earth, covering millions of square kilometers, nearly 1% of Earth's surface area. It formed in the middle of the Pacific Ocean 120 Ma and is no longer near the hot spot that produced it.

Volcanic islands near *Tahiti* define northwest-trending chains that are forming over several hot spots. In each chain, the islands to the northwest are older than those to the southeast, indicating that the Pacific plate is moving to the northwest relative to the underlying source of magma.

The *Galápagos* is a clump of volcanic islands west of South America. The western islands, shown in the satellite image to the lower left, are volcanically active and have erupted within the past several years. Eruptions build larger volcanoes and smaller volcanic cones, as shown in the photograph below.

Tristan da Cunha, a volcanic island in the South Atlantic Ocean, marks a hot spot just east of the Mid-Atlantic Ridge. Volcanism associated with the hot spot created a large submarine ridge (shown in gray) that tracks the motion of the African plate over the hot spot.

The *Kerguelen Plateau,* in the southern Indian Ocean, is the second largest oceanic plateau in the world. It mostly consists of basalt and was formed in several stages during the late Mesozoic (between 115 and 85 Ma).

10.11.c2

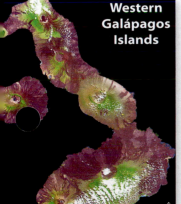

Western Galápagos Islands

10.11.c3 Galápagos Islands

Before You Leave This Page Be Able To

☑ Describe or sketch how a hot spot can form a linear chain of islands and seamounts, or a cluster of islands, providing several examples.

☑ Describe how oceanic plateaus are interpreted to have formed, and provide some examples.

10.11

10.12 What Features Form at Continental Hot Spots?

A HOT SPOT WITHIN A CONTINENTAL PLATE is marked by high elevations, abundant volcanism, and continental rifting. Hot spots can facilitate complete rifting and separation of a continent into two pieces and can help determine where the split occurs. Several continental hot spots are active today.

A What Features Are Typical of Continental Hot Spots?

Hot spots are volcanic areas interpreted to be above rising mantle plumes. Continental hot spots are associated with certain characteristics, including high elevations, volcanism, and the presence of rifts. Two examples are the Afar region of East Africa and the Yellowstone region of the western U.S.

Afar Region, East Africa

Continental hot spots have high elevations largely because of heating and thinning of the lithosphere by a rising plume of hot mantle. Many geoscientists interpret the Afar region of eastern Africa to be located above a hot spot that is currently active.

The East African Rift is within the African plate. It may or may not evolve into a full rift that fragments the continent into two parts and that leads to seafloor spreading.

10.12.a1

Near the hot spot, the Arabian Peninsula has pulled away from Africa along the Red Sea and the Gulf of Aden. Beneath these seas, seafloor spreading generates new oceanic crust. The divergent plate boundary is along the rift in the Red Sea and then turns eastward, following the zigzag oceanic ridge through the Gulf of Aden.

The Red Sea, Gulf of Aden, and East African Rift come together in the Afar region, branching off like three spokes on a wheel. The Afar region is among the most volcanically active areas on Earth and has experienced recent volcanic eruptions. Volcanism has been so prolific here that it has created a triangular area of new land in the corner of Africa from which the Arabian Peninsula pulled away.

Region Around Yellowstone National Park

Yellowstone is located in Wyoming, Idaho, and Montana and sits in a region that is higher in elevation than surrounding areas.

The Snake River Plain of southern Idaho is underlain by thick sequences of basalt and other volcanic rocks. It is the site of recent eruptions at Craters of the Moon National Monument.

10.12.a2

Three large volcanic centers, each issuing huge eruptions of volcanic ash, erupted in Yellowstone in the last 2.1 million years. Eruptions are not occurring today, but heat from hot volcanic rock and underlying magma drives the hot springs, geysers, and thermal pools for which Yellowstone is famous.

Yellowstone is interpreted to mark the present location of the hot spot, whereas the Snake River Plain records the track of North America as it moved southwest over the hot spot. Forming at the same time was the Basin and Range Province of Utah and Nevada, a broad continental rift adjacent to Yellowstone. The alternating valleys and mountains resulted from faults that uplifted some blocks of crust and downdropped others.

10.12.a3 Craters of the Moon, ID

Craters of the Moon National Monument (◄) in southern Idaho contains spectacular examples of recent, unweathered lava flows, whose patterns of molten flow are preserved in solid rock.

Yellowstone National Park, mostly in Wyoming, is named for the volcanic rocks turned yellowish (►) by hot waters that accompanied the eruptions and continue to form the famous hot pools.

10.12.a4 Yellowstone NP, WY

B How Do Continental Hot Spots Evolve?

Many continental hot spots underwent a similar sequence of events. They started with doming and ended with the formation of a new continental margin and a new ocean formed by seafloor spreading.

1. Hot spots mark where a mostly solid, hot mass rises, probably from the lower mantle, and encounters the base of the lithosphere. The rising material melts as a result of less pressure and also causes melting of nearby lithosphere.

2. As the upper mantle and crust heat up, a broad, domal uplift forms on the surface. Doming is accompanied by stretching of the crust, which commonly begins to break apart along three rifts that radiate out from the hot spot.

3. Some mantle-derived magma escapes to the surface and erupts as voluminous basalts called flood basalts. More felsic magma (more granitic in composition) forms where mantle-derived magma causes melting of the crust.

4. All three parts, or arms, of the rift are bordered by faults, which downdrop long fault blocks. The downdropped blocks form basins that contain lakes and are partially filled by sediment and rift-related volcanic rocks.

5. Complete rifting of the continent occurs along two arms of the rift. This results in a new continental margin and seafloor spreading in the new ocean basin. At the onset of spreading, the edge of the continent is uplifted because the lithosphere is heated and thinned due to the rifting.

6. The third arm of the rift begins to become less active and fails to break up the continent into more pieces. This *failed rift* is lower than the surrounding continent and commonly becomes the site of major rivers.

7. As seafloor spreading continues, the generation of new oceanic lithosphere causes the mid-ocean ridge to move farther out to sea. The continental margin cools and subsides and is covered by marine sediment on the newly formed continental shelf. This continental margin is no longer a plate boundary and is now a passive margin.

8. Sediment transported by streams down the failed rift will form a delta at the bend in the continent. This is currently occurring along the western coast of Equatorial Africa at the large inward bend in the coast (see the figure and text below).

Hot Spots and Continental Outlines

Geoscientists conclude that hot spots have helped define the outlines of the continents by shaping the boundary along which continents separate from one another. The best example of this is the inward bend of the western coast of Africa. This bend occurs at the intersection of three arms of a rift, two of which led to the opening of the South Atlantic Ocean. The third *failed rift* cuts northeastward into Africa and is the site of several major rivers. Large eruptions of basalt (flood basalt) occurred along the rifts, and active volcanism near the failed rift may mark the location of a hot spot. This figure shows what the area may have looked like 110 million years ago, after the continents started to rift apart.

Before You Leave This Page Be Able To

☑ Summarize the features that are typical of continental hot spots, providing an example of each type of feature.

☑ Summarize or sketch how continental hot spots evolve over time.

☑ Describe or sketch how hot spots influence continental outlines, providing an example.

10.12

10.13 What Are Continents and How Do They Form?

CONTINENTS ARE AMONG THE LARGEST FEATURES of our planet, but what are they, how did they form, and does a continent remain the same size over long time periods? Continents have interesting histories, which at times involved splitting apart or joining together. Through geologic time, continents overall have tended to grow larger as they collide with and incorporate island arcs, oceanic plateaus, seamounts, and other tectonic features.

A What Is a Continent?

When people think of a continent, most envision its shape as seen by its present-day coastline. This outline, however, changes as sea level goes up or down, as it has many times in the distant past. We should therefore consider a slightly different definition of a continent—one that indicates a less transient outline.

The continent of Australia (▶) nicely illustrates the issues involved with defining the edge of a continent. The familiar outline of the continent is the shoreline, the boundary between the browns and greens of land versus the blues of the adjacent ocean on this map.

10.13.a1

10.13.a2

One way to represent continents, and the rest of Earth's land and seafloor, is a graph that plots elevation versus the percentage of land or seafloor that is above that elevation (or depth). This graph, called a *hypsometric curve*, shows that somewhat less than 30% of Earth is continent at this time, with most of the rest being deep ocean. The major change between the continents and deep oceans occurs not at the shoreline, but across continental slopes. This graph illustrates that the continental shelves are more akin to the continents than to the deep seafloor.

Notice, however, that surrounding the land is an area of relatively shallow ocean, represented on this figure by light blue. This blue shelf surrounding the continent is similar to the rest of the continent, except that it is barely under water. Many geoscientists consider this shallow, continental shelf to be part of the Australian continent, putting the edge of the continent at the outer edge of the continental shelf.

B How Do Continents Form and Grow?

Examine this figure showing various tectonic features within and next to a central plate being consumed by subduction from both sides. If we predict where each feature will go as the plate motions continue, we expect that some of these features will collide with each other and with the continent to the right. When this occurs, they may be added to the front edge of the continent. A piece of exotic crust added to a continent is called a *tectonic terrane* (note the spelling).

Some terranes represent oceanic islands, oceanic plateaus, and other types of oceanic crust. Islands and plateaus have thicker-than-normal oceanic crust, so they are less likely to be subducted and more likely to be added to the edge of a continent. The oceanic plateau shown here is headed for the trench to the right and a collision with the edge of the adjacent continent.

Many terranes consist of volcanic and volcanic-related rocks that formed as island arcs. Island arcs often become terranes because they move across the ocean, potentially traveling long distances until they collide with, and become part of, another landmass, like a continent.

Some terranes have more continental characteristics, specifically thick granitic crust with continental sedimentary rocks. Such terranes generally represent pieces that were sliced or rifted off another continent and then tectonically transported until they collide with the edge of a continent.

For a terrane to become attached to a continent, it typically enters a subduction zone, where it is scraped off the sub-ducting plate and tectonically added to the continent. Through this process of adding terranes along subduction zones, a continent grows. In fact, incorporating terranes along their edges is probably the main way most continents grow, and this process was very important in constructing the continents we have today.

10.13.b1

C What Are Common Features of Continents?

Many continents display a similar pattern, with a central region of older igneous and metamorphic rocks surrounded by a relatively thin veneer of younger, nearly flat-lying sedimentary layers.

1. Many continents, including North America, have a central region called a *continental shield*. A shield consists of relatively old metamorphic and igneous rocks, often simply called *crystalline rocks*. The rocks exposed in the shield represent the kinds of rocks that underlie much of the continent. In North America, the continental shield is mostly in Canada and is called the *Canadian Shield*.

2. Surrounding the shield is a broad region called the *continental platform*. It is characterized by nearly horizontal sedimentary rocks that were deposited on top of the older, lower rock. The sedimentary layers commonly have been gently warped and then eroded, which results in higher and lower rocks being exposed at the surface from place to place. In North America, the entire center of the country, between the Rocky and Appalachian Mountains, is a continental platform.

3. The boundary between the flat-lying platform sedimentary rocks and the underlying crystalline rocks is an old, buried erosion surface (an *unconformity*), which separates rocks with very different ages and histories.

4. Sedimentary rocks in the interior generally have few major geologic structures, but some regions, such as the continental shield, have been gently warped upward, and others are warped downward, forming a low area called a *basin*. A large basin like this underlies nearly all of Michigan.

5. Most continents have at least one edge where elevation of the land slowly decreases toward the coast until it flattens out in a low plain barely above sea level. This low region is a *coastal plain* and is underlain by sedimentary units, many of which were deposited quite recently in geologic terms. In the U.S., the Coastal Plain warps around the southern and eastern edge of North America, from Texas to Florida to New Jersey.

Continental Shield · Basin · Continental Platform · Coastal Plain · Sedimentary Rocks · Unconformity · Crystalline Basement

10.13.c1

Michigan | Ohio · FINDLAY ARCH · Middle Paleozoic · Late Paleozoic · Ohio · Ohio River · West Virginia · 0 · −1,500 · Precambrian · −3,000 · Early Paleozoic · −4,500 · Northwest · 10.13.c2

◄ 6. This geologic cross section across the state of Ohio is typical of the geology of central North America and of continental interiors in general. In this area, sedimentary layers dip gently off the flanks of a dome, below which rocks of the crystalline basement are closer to the surface. The section is vertically exaggerated, so true dips are less than shown here, and the thicknesses of the layers are greatly exaggerated.

In Suspect Terrain

California is the area many geoscientists think about when they study how continents grow from the accretion of tectonic terranes. John McPhee's popular books, *In Suspect Terrain* and *Assembling California*, provide an accessible account of how terranes were recognized and how they added new real estate to North America.

This map shows the various types of terranes added to the continent. The terranes include slices of oceanic crust and sediment, parts of island arcs, and material scraped off oceanic plates that were subducted along a trench. The map does not show units that formed in place, such as granites in the Sierra Nevada.

Klamath Mountains · EASTERN AREA · Paleozoic Terranes · Part of N.A. · Sierra Nevada · Great Valley · San Andreas Fault · Mojave Desert · 100 km · N

WESTERN AREA
- Recent Sediments
- Accretionary Prism (Mz-Cz)
- Displaced Granitic Rocks (Mz)
- Oceanic Crust (Mz)
- Island Arcs (Mz)

10.13.t2

Before You Leave This Page Be Able To

✓ Sketch and describe a hypsometric curve and use it to explain how we define the edge of a continent.

✓ Describe what a tectonic terrane is, the main tectonic settings in which terranes originate, and how terranes are added to crust.

✓ Summarize some common features of continents and how these are expressed in North America.

10.13

10.14 How Did the Continents Join and Split Apart?

CONTINENTS SHIFT THEIR POSITIONS over time in response to plate tectonics. They have rifted apart and collided, only to rift apart again. Where were the continents located in the past, and which mountains resulted from their motions? The figures below illustrate which continents were joined and how they separated. We start with 600 million years ago (Ma) and work forward to the present, as if watching a movie of Earth's history.

600 Ma: The Supercontinent of Rodinia

The images on these pages show one interpretation of where the continents were located in the past. Geologist Ron Blakey created the artistic renderings of the continents, mountains, and oceans. For most time periods, he created two views, one focused on the Western Hemisphere (images on the left) and one on the Eastern Hemisphere (images on the right), generally with some overlap. We begin here with a single image, centered on the South Pole.

Before the Paleozoic, in the last part of the Precambrian, all of the major continents were joined. This supercontinent is called *Rodinia*. Nearly all of the other side of the globe is a huge ocean.

10.14.a1

North America was in the initial stages of rifting from Rodinia. This rifting outlined the western margin of North America, but geologists are not certain which continent was adjacent to North America. Options include Australia, Antarctica, and Asia.

Large parts of Rodinia were near the South Pole, and there is evidence of widespread glaciation in these polar regions. This far back in the past, there are large uncertainties about how the continents were arranged, but geoscientists are trying to match the ages and sequences of rocks between different continents.

500 Ma: Dispersal of the Continents

At 500 Ma, in the early part of the Paleozoic Era, North America and Europe were separate, moderate-sized continents that had not yet joined.

Antarctica, Australia, South America, and Africa were joined in the Southern Hemisphere, together forming the southern supercontinent of Gondwana, which was mostly located in the Southern Hemisphere (mostly out of view on these figures). Gondwana was separated from the northern continents by some width of ocean.

10.14.a2

With the continents still mostly clustered together, the other side of Earth was a single large ocean, much larger than the size of the present-day Pacific.

Island arcs surrounded Europe and parts of Asia. Some of the arcs would later collide with the continents, adding land to the continents.

370 Ma: Before Pangaea

In the middle of the Paleozoic, at 370 Ma, Europe and North America were joined but were not connected with Asia, which lay to the north. North America had collided with a microcontinent. This collision created mountains in what is now New England.

North America was approaching Africa and South America along a convergent margin. The continents were on a collision course as the intervening ocean became narrower over time.

10.14.a3

Gondwana remained mostly intact, except for some slices of continental crust that probably were rifted away from the larger supercontinent. The microcontinent that collided in New England was probably one of these rifted pieces, but it had broken away some time before 370 Ma, the time pictured here.

280 Ma: The Supercontinent of Pangaea

In the late Paleozoic, around 280 Ma, a continental collision between North America and the northern edge of Gondwana (South America and Africa) formed the Appalachian Mountains near the East Coast and the Ouachita Mountains (not labeled) in the southeastern U.S.

After this collision and a series of smaller collisions, all the continents were joined in a supercontinent called *Pangaea*.

A wedge-shaped ocean, the *Tethys Sea*, separated Asia from landmasses farther to the south. Southern Africa, Australia, and Antarctica were close to the South Pole and so at this time were partly covered with ice.

150 Ma: Gondwana and Laurasia

At 150 Ma, in the Mesozoic, North America had separated from Africa and South America. The Atlantic Ocean now existed, and continents on either side of the Atlantic were moving away from each other due to seafloor spreading. The left globe is rotated so that the central Atlantic Ocean is in the center of the image.

During this time, North America was still joined with Europe and Asia, forming the northern supercontinent of *Laurasia*. South America had not yet rifted away from Africa.

Antarctica was still attached to the southern tips of Africa and South America. India was attached to the northern edge of Antarctica. These continents were soon to be rifted apart, which would mark the end of *Gondwana*.

Present

This view shows the present-day configuration of continents and oceans. Examine these globes and think about the present-day plate boundaries, envisioning which way the continents are moving relative to one another. Use the relative motions to predict where the continents were likely to have been in the past. Check your predictions by examining each set of previous globes as you step backward in time.

When you are done working backward in time, start at 600 Ma and track the position of every continent you can and watch each ocean open. By viewing this sequence, you are observing how our present-day world came to be.

Finally, think about how the position of the continents will change if present-day plate motions continue into the future. What collisions are yet in store for North and South America?

Before You Leave This Page Be Able To

☑ Briefly summarize the general positions of the continents in the past, especially since 280 Ma.

☑ Identify times when the continents were joined in the supercontinents of Gondwana, Laurasia, Pangaea, and Rodinia.

10.14

10.15 Where Do Mountain Belts and High Regions Form?

MOUNTAIN BELTS AND OTHER HIGH REGIONS generally owe their high elevation to thick continental crust. Crust becomes thicker from deformation and additions from magmatism. In some cases, a region is higher than its surroundings due to processes originating in the mantle. Where are the world's main mountain belts and why did mountains form in these places?

A In Which Tectonic Settings Do Regional Mountain Belts Form?

Regional mountain ranges are hundreds or thousands of kilometers long. They are large enough that they can only be explained by major variations in the thickness and temperature of the crust and lithosphere. Most ranges occur near convergent plate boundaries or where there has been large-scale movement of material in the mantle.

10.15.a1

10.15.a2

10.15.a3

Subduction Zones—Convergent margins are high in elevation because the crust is thickened by magmatic additions from the subduction zone and by crustal shortening. Also, in these regions, lithosphere is heated and replaced by less dense asthenosphere.

Continental Collisions—Collision zones have high elevations due to an increase in crustal thickness as one continent is shoved over another. In these settings, crustal thickening occurs by thrusting, folding, and other forms of deformation.

Mantle Upwellings—Less dense asthenosphere can move upward into the lithosphere, causing regional uplift. This occurs near hot spots, plate boundaries, and in other settings; it is partly responsible for high elevations in parts of the western U.S.

B What Causes These Regions to Have High Elevation?

Western Canada has been a convergent margin for most of the last 100 million years. Its mountain ranges overlie crust thickened from major thrust faulting, from magmatic additions, and from collisions with island arcs and pieces of continental material.

The *Alps* mountain range of southern Europe is high because it has thick crust due to collisions between Europe and smaller continental blocks that came from the south.

The *Tibetan Plateau* and the *Himalaya* are extremely high because of very thick crust that resulted when the Indian continent collided with, and was partly shoved beneath, the continental crust of Asia.

10.15.b1

The *Andes* of South America are above a subduction zone. The underlying crust is hot and thick because of magmatic additions and crustal shortening.

The *East African Rift* is higher than most of Africa because of magmatic heating of the crust, thinning of the crust, and the presence of a hot spot related to mantle upwelling.

The *Great Divide Range* forms the eastern flank of Australia. There is currently no plate boundary here, and the age and cause of uplift are still being investigated.

C What Happens During the Erosion of Mountain Belts?

Mountains, once formed, are subjected to weathering and erosion. These processes wear mountains down but are countered by uplift related to *isostasy*. Uplift is driven by buoyancy due to the root of underlying thick crust.

Early Mountain Building

As a mountain belt forms, uplift is commonly faster than erosion, and the mountain becomes higher and more rugged over time. A high mountain belt results from uplift that is faster than erosion.

As soon as it starts forming, weathering and erosion begin to wear down a mountain, contributing sediment to streams. Sediment will be transported to adjacent low areas, perhaps in nearby oceans or other types of basins.

Erosion and Isostatic Rebound

As material erodes from a mountain belt, there is less weight holding down the thick crustal root. The buoyant crust can uplift, a process called *isostatic rebound*.

Sediment derived from the mountain is deposited in nearby basins, on both the sea and continental sides. The added weight of the sediment depresses the crust in these regional basins, making room for more sediment.

Late Stages of Evolution

Erosion and isostasy cause rocks deep in the crust to be uplifted and exposed at the surface. As a result, many mountain belts expose granite and other deep rocks.

Through simultaneous erosion and isostasy, the mountain is eroded down and the thick crustal root is gradually reduced in size. Material eroded from the mountain ends up in adjacent basins, increasing the crustal thickness beneath the basins.

D What Controls Regional Elevations in North America?

The vertically exaggerated topographic profile below illustrates how elevations vary from east to west across the U.S. It does not show the full thickness of crust, only the elevation of the land and depth of seafloor.

Western North America is high mostly because crust was thickened along a convergent margin.

The moderately high elevation of the Basin and Range is largely due to hot, thin lithosphere.

Compression and shortening within the North American plate thickened crust in the Rocky Mountains. Additional uplift is due to a hot, thin lithosphere and upwelling asthenosphere associated with rifting.

Elevation decreases from the Great Plains toward the Mississippi River because the lithosphere is cooler and thicker.

The Appalachian Mountains were once a region of thick crust, due to the collision between North America and Africa. Much of this thickness has been lost due to erosion, and so the range has lost elevation over time.

Rule of Thumb for Elevations

Regional elevations are relatively low for regions with thinner crust, and relatively high for regions with thicker crust, but by how much? A rule of thumb is that increasing the thickness of the crust by 6 km will result in an increase in elevation of 1 km (about 3,300 ft). Here is an example from Arizona.

Phoenix sits at an elevation of 300 m (1,000 ft), whereas Flagstaff is at more than 2,100 m (7,000 ft). This difference is about 2 km, so the crust beneath Flagstaff should be 12 km thicker than the crust beneath Phoenix ($2 \times 6 = 12$). Geophysical measurements show that the crust beneath Phoenix is about 28 km thick, whereas crust beneath Flagstaff is about 40 km thick. The difference is 12 km, the value we would predict from the rule of thumb.

ELEVATION ABOVE SEA LEVEL (m)

DEPTH TO BASE OF CRUST (km)

10.15.t1

Before You Leave This Page Be Able To

✓ Sketch and explain the main tectonic settings of high regions, providing an example for each setting.

✓ Summarize the settings of the world's high mountains and plateaus.

✓ Explain how erosion and isostasy help expose deeply formed rocks in eroded mountain belts.

✓ Summarize differences in regional elevation across North America.

10.15

10.16 How Do Internal and External Processes Interact to Form Landscapes?

EARTH'S LANDSCAPES REVEAL THE INTERPLAY between internal processes—those originating from within the Earth—and external processes—those imposed on the Earth by its envelope of moving fluids (atmosphere and hydrosphere). Internal processes tend to be constructive, such as when they construct a volcano or cause uplift of a mountain range. External processes, especially erosion, tend to wear away parts of the landscape, but the eroded materials are deposited in other areas. Internal and external processes interact in some surprising ways.

A How Do Internal and External Processes Interact?

Internal processes arise from within the Earth and are manifested in phenomena such as volcanism, earthquakes, deformation, and mountain building. They are largely driven by the planet's internal heat. External processes, like weathering and erosion, occur on or near the surface and are mostly driven by gravity and the energy from the Sun and the resulting movement of air and moisture.

Tectonics

10.16.a1 Baja California Sur, Mexico

Most internal processes are expressed as tectonics. Volcanism and mountain building can build new landscape features, such as a volcano or mountain, or can modify existing landscapes by faulting and uplift (▲).

Weathering

10.16.a2 Baja California Sur, Mexico

Once rocks and other materials are exposed at the surface, weathering (an external process) begins to loosen pieces and round off corners (▲), and in general acts to disintegrate the rocks. It creates clasts that can be moved.

Erosion, Transport, Deposition

10.16.a3 Baja California Sur, Mexico

Clasts, clay, and other materials produced by weathering are then eroded away and transported by streams, waves, ice, wind, and gravity. Erosion, transport, and eventual deposition are all external processes.

B How Do Plate Tectonics Affect External Processes?

External processes can disintegrate and transport materials in a landscape, but tectonics can influence the rates at which different external processes are operating. The current arrangement and topography of the continents and ocean basins help guide wind and ocean currents that redistribute heat and moisture around Earth. Tectonics can change topography, which changes the local or regional climate, the rate of weathering, and other factors.

Mountains and Uplift

10.16.b1 Last Hope Sound, Chile

Mountain ranges intercept wind and water vapor, causing orographic effects that concentrate rainfall and cause rain shadows and other climatic effects. Elevation and moisture in turn affect rates of weathering and erosion.

Seas and Subsidence

10.16.b2 Portage, AK

Tectonics can displace water from the ocean basins, flooding low parts of continents. Such regional flooding, or local tectonic subsidence (▲), changes local climates, which affects external processes.

Volcanic Gas and Dust

10.16.b3 Augustine volcano, AK

Volcanic activity releases CO_2 and water vapor, which cause atmospheric warming. Volcanic ash and SO_2 gas from volcanoes reflect solar radiation, which may cause regional or global cooling.

C How Do Plate Movements Affect Climate and the Rates of External Processes?

As a tectonic plate moves, it changes position relative to Earth's climatic zones. Plate movements open and close channels that connect oceans, thereby changing ocean currents. Tectonics can uplift mountain ranges, form new ocean basins, or bring continents together. These changes can alter a region's elevation, temperature, wind flow patterns, and the amount and frequency of precipitation, all of which influence the rate of external processes.

1. To illustrate how tectonics can influence external processes, we use the three maps below to show the geography of North America at different times: 260 Ma, 75 Ma, and 20 Ma, from left to right.

2. Examine each map and think about how the configuration of land and water might change regional climates. Also, consider how the mountains would influence wind patterns and the amount of precipitation. In the first map, North America was just north of the equator so would experience northeast trade winds. In the other two maps, it was farther north and so would be under the influence of westerlies, as it is today. Only the general patterns are important here.

3. About 260 Ma, North America was part of the supercontinent of Pangaea. The newly formed Appalachian Mountains were probably a huge, high mountain range, much like today's Himalaya. They blocked easterly trade winds, causing a rain shadow with deserts over the center of the continent, the same areas (e.g., upper Midwest) that receive abundant rainfall today.

4. At 75 Ma, shallow inland seas inundated the continent, bringing a warm, wet climate far inland, covering most of the continent. Tectonics along the west coast formed a large and fairly continuous belt of mountains inland from the coast. These changed the wind patterns and regional elevations, which affect the amounts, rates, and seasonality of rainfall, weathering, and erosion.

5. In the last 20 million years, mountains along the West Coast, such as the Sierra Nevada and Cascades, increasingly blocked prevailing westerly winds, causing increased precipitation along the coast and the western slopes of the mountains, but a rain shadow with deserts inland. The patterns of humidity, rainfall, weathering, and erosion probably started to resemble those of the present day.

D How Do Internal and External Processes Influence the Steepness of Slopes?

In actively uplifting mountain belts, hillslopes indicate the interplay between internal processes, which are uplifting the mountains and causing slopes to be steeper, and external processes that are eroding the rising mountains and causing slopes to be less steep. An example from the south-facing slopes of the Himalaya nicely illustrates this interplay.

As mountains are uplifted, rocks at the surface are removed by water, ice, landslides, and the slower movement of material downhill. In some mountain ranges, erosion is almost as rapid as uplift and so limits the heights the mountains can achieve. Also, uplift of the mountains leads to regional climate changes, which may act to lower the range through increased erosion. In other areas, uplift seems to outpace erosion, so the mountains get higher and generally steeper.

This graph is a topographic profile, showing the slope of the land. Mountain slopes of central Nepal have a curved topographic profile that may indicate a near balance between uplift and erosion.

Slopes in neighboring Bhutan have straighter topographic profiles, which may indicate that the mountains are uplifting faster than erosion can wear them down.

Before You Leave This Page Be Able To

✓ Describe how internal and external processes interact.

✓ Describe how plate tectonics affects external processes.

✓ Summarize ways that movement of plates affects climate.

✓ Summarize how the balance between external and internal processes influences slopes.

10.16

10.17 Why Is South America Lopsided?

THE TWO SIDES OF SOUTH AMERICA are very different. The western margin is mountainous, while the eastern side and center of the continent have much less relief. The differences are a result of the present plate boundaries and of the continent's geologic history during the last 200 million years. South America nicely illustrates many aspects of plate tectonics, including the connections between the tectonics of the land and seafloor. It also is an excellent example of how to analyze the major features of a region.

A What Is the Present Setting of South America?

The perspective view below shows South America and the surrounding oceans. Observe the topography of the continent, its margins, and the adjacent seafloor. See if you can find mid-ocean ridges, transform faults, ocean trenches, oceanic fracture zones, and other plate-tectonic features. From these features, infer the locations of plate boundaries in the oceans, and predict what type of motion (divergent, convergent, or transform) is likely along each boundary. Make your observations and predictions before reading the accompanying text.

1. The Galápagos Islands are located in the Pacific Ocean, west of South America. They consist of a cluster of about 20 volcanic islands, flanked by seamounts. Some of the islands are volcanically active and are interpreted to be over a hot spot. A mid-ocean ridge is also nearby.

9. The center of the South American continent has low, subdued topography because it is away from any plate boundaries. It is a relatively stable region that has no large volcanoes and few significant earthquakes. It is not tectonically active.

8. In the South Atlantic, the Mid-Atlantic Ridge is a divergent boundary between the South American and African plates. Seafloor spreading creates new oceanic lithosphere and moves the continents farther apart at a rate of 3 cm/yr.

2. The Andes mountain range follows the west coast of the continent and is the site of many dangerous earthquakes and volcanoes. A deep ocean trench along the edge of the continent marks where an oceanic plate subducts eastward beneath South America.

7. The eastern side of South America has a continental shelf that slopes gently toward the adjacent seafloor. There is neither a trench, nor any significant tectonic activity, nor any other evidence for a plate boundary. Instead, the continent and adjacent seafloor to the east are part of the same plate, and this edge of the continent is a passive margin.

3. The Pacific seafloor contains mid-ocean ridges with the characteristic zigzag pattern of a divergent boundary with offsets along transform faults. The Nazca plate lies north of this ridge, and the Antarctic plate is to the south.

6. The curved Scotia Island arc is related to a trench to the east and the westward subduction of the oceanic part of the South American plate.

10.17.a1

4. Many oceanic fracture zones cross the seafloor and were formed along transform faults, but they are no longer plate boundaries.

5. The southern edge of the continent is very abrupt and has a curving "tail" extending to the east. This edge of the South American plate is a transform boundary where South America is moving west relative to oceanic plates to the south.

B What Is the Geometry of the South American Plate and Its Neighbors?

This cross section shows how geoscientists interpret the configuration of plates beneath South America and the adjacent oceans. Compare this cross section with the plate boundaries you inferred in part A.

1. At the Mid-Atlantic Ridge, new oceanic lithosphere is added to the African and South American plates as they move apart. As this occurs, the oceanic part of the South American plate gets wider.

2. Along the eastern edge of South America, continental and oceanic parts of the plate are simply joined together along a passive margin. There is neither subduction, nor seafloor spreading, nor any type of plate boundary. As a result, the eastern continental margin of South America lacks volcanoes, earthquakes, and mountains.

10.17.b1

3. A subduction zone dips under western South America, carrying oceanic lithosphere beneath the continent. Subduction causes large earthquakes and produces magma that feeds dangerous volcanoes in the Andes.

C How Did South America Develop Its Present Plate-Tectonic Configuration?

If South America is on a moving plate, where was it in the past? When did it become a separate continent, and when did its current plate boundaries develop? Here is one commonly agreed-upon interpretation.

Around 140 Ma, Africa and South America were part of a single large supercontinent called Gondwana. At about this time, a continental rift developed, starting to split South America away from the rest of Gondwana and causing it to become a separate continent.

By 100 Ma, Africa and South America were completely separated by the South Atlantic Ocean. Spreading along the Mid-Atlantic Ridge moved the two continents farther apart with time. While the Atlantic Ocean was opening, oceanic plates in the Pacific were subducting beneath western South America. This subduction thickened the crust by compressing it horizontally and by adding magma, resulting in the formation and rise of the Andes mountain range.

Today, Africa and South America are still moving apart at a rate of several centimeters per year. As spreading along the mid-ocean ridge continues, the Atlantic Ocean gets wider. Earth, however, is not growing through time, and the expanding Atlantic Ocean is balanced by shrinking of the Pacific Ocean, whose oceanic lithosphere disappears into subduction zones along the Pacific Ring of Fire.

The photographs below contrast the rugged Patagonian Andes of western South America with landscapes farther east that have more gentle relief and are not tectonically active. The tectonics also significantly control the regional climates, with the Andes receiving abundant precipitation due to orographic effects, while the eastern lowlands are dry because they are downwind—in the rain shadow—of the Andes.

140 Ma

100 Ma

Today

10.17.c1–3

10.17.c4 Cuernos del Paine, Chile

10.17.c5 Central Argentina

Before You Leave This Page Be Able To

✓ Sketch and describe the present plate-tectonic setting of South America, and explain the main features on the continent and adjacent seafloor.

✓ Discuss the plate-tectonic evolution of South America over the last 140 million years.

10.17

11.4 What Are Composite Volcanoes and Volcanic Domes?

COMPOSITE VOLCANOES FORM STEEP, CONICAL MOUNTAINS that are hard to mistake for anything other than a volcano. They are common above subduction zones, especially along the Pacific Ring of Fire. Composite volcanoes are extremely dangerous. Volcanic domes consist of smaller, dome-shaped masses of highly viscous lava that accumulated over and close to the associated volcanic vent—they are also very dangerous.

A What Are Some Characteristics of a Composite Volcano?

Composite volcanoes are constructed of interlayered volcanic material formed by lava flows, pyroclastic flows, falling ash, and volcano-related mudflows and landslides. Composite volcanoes, also called stratovolcanoes, erupt over long time periods, which explains their large size and complex internal structure.

11.04.a2 Mount St. Helens, WA

◀ **1.** *Eruption Column*—Composite volcanoes produce a distinctive column of pumice, ash, and gas that rises upward many tens of kilometers into the atmosphere. Coarser pieces settle around the volcano, but finer particles (volcanic ash) can drift hundreds of kilometers in the prevailing winds.

▶ **2.** *Pyroclastic Flows*— These are the most violent eruptions from the volcano. They form when the eruption column collapses downward as a dense, swirling cloud of hot gases, volcanic ash, and angular rocks. Pyroclastic flows are one of the main mechanisms by which these volcanoes are constructed.

11.04.a3 Mount Mayon, Philippines

6. *Shape*—Composite volcanoes display the classic volcano shape because most material erupts out of a central vent and then settles nearby. They have steep slopes because they form from small eruptions of viscous lava flows that pile up on the flanks of the volcano and help protect pyroclastic material from erosion. The shape represents one snapshot in a series of stacked volcanic mountains that have been built over time.

5. *Lava Domes and Flows*— Lava domes and flows can erupt from any level of a composite volcano. Lava may erupt from the summit crater or escape through vents on the volcano's sides or base. Lavas associated with composite volcanoes are moderately to highly viscous, and so they move slowly and with difficulty. The lava may break into blocks that fall, slide, or roll downhill, forming a tongue or apron (▼) of jumbled pieces from the lava.

11.04.a1

3. *Landslides and Mudflows*—Composite volcanoes can be large mountains that collect rain or snow. Rain and snowmelt mix with loose ash and rocks on the volcano's flanks, causing a volcano-derived mudflow called a *lahar*. There is a high hazard for landslides and lahars (▼) because of the steep slopes, loose rocks, and abundant slippery clay minerals produced when hot water interacts with the volcanic rocks.

11.04.a4 Augustine volcano, AK

4. *Rocks*—Composite volcanoes consist of alternating layers of pyroclastic flows, lava flows, and deposits from landslides and mudflows. Most rocks are intermediate to felsic in composition. The volcanoes we see today, formed during eruptions from long-lived vents, are built on and around earlier versions of the volcanoes. A large composite volcano can be constructed in tens of thousands of years, or a few hundred thousand years.

11.04.a5 Mount St. Helens, WA

B   What Are Some Characteristics of a Volcanic Dome?

Volcanic domes form when viscous lava mounds around a vent. Some volcanic domes have a nearly symmetrical dome shape, but most have an irregular shape because some parts of the dome have grown more than other parts or because one side of the dome has collapsed. Domes may be hundreds of meters high and one or several kilometers across, but they can be much smaller, too.

1. This rubble-covered dome (▲) formed near the end of the 1912 eruption in the Valley of Ten Thousand Smokes in Alaska. Volcanic domes commonly have this type of rubbly appearance because their outer surface consists of angular, broken pieces of the dome.

Growth of a Dome

2. Domes mostly grow from the inside as magma injects into the interior of the dome. This new material causes the dome to expand upward and outward, fracturing the partially solidified outer crust of the dome. This process creates the blocks of rubbly, solidified lava that coat the outside of the dome. The blocks and other pieces, which are as small as several centimeters across to as large as houses, commonly tumble or slide down the steep slopes on the side of a dome.

50 m

11.04.b2

3. Domes also grow as magma breaks through to the surface and flows outward as thick, slow-moving lava. As the magma advances, the front of the flow cools, solidifies, and can collapse into angular blocks and ash. Most domes display both modes of growth.

Collapse or Destruction of a Dome

4. Domes can be partially destroyed when steep flanks of the dome collapse and break into a jumble of blocks and ash that flow downhill as small-scale pyroclastic flows (▶).

11.04.b3

11.04.b4

5. Domes can also be destroyed by explosions originating within the dome (◀). These typically occur when magma solidifies in the conduit and traps gases that build up until the pressure can no longer be held.

Deadly Collapse of a Dome at Mount Unzen, Japan

Mount Unzen towers above a small city in southern Japan. The top of the mountain contains a steep volcanic dome that formed and collapsed repeatedly between 1990 and 1995. The collapsing domes unleashed more than 10,000 small pyroclastic flows (top photograph) toward the city below. In 1991, the opportunity to observe and film these small pyroclastic flows attracted volcanologists and other onlookers to the mountain. Unfortunately, partial collapse of the dome caused a pyroclastic flow larger than had occurred previously. This larger flow killed 43 journalists and volcanologists and left a path of destruction through the valley (lower photograph). Note that damage was concentrated along valleys that drain the mountain.

11.04.t1 Mount Unzen, Japan

11.04.t2 Mount Unzen, Japan

Before You Leave This Page Be Able To

✓ Describe or sketch the characteristics of a composite volcano and how such a volcano forms.

✓ Sketch and explain the characteristics of a volcanic dome and the two ways by which a volcanic dome can grow.

✓ Explain or sketch how a volcanic dome can collapse or be destroyed by an explosion.

11.4

11.5 What Are Calderas?

CALDERA ERUPTIONS ARE AMONG NATURE'S most violent phenomena. They can spread volcanic ash over huge areas, and the largest erupt more than one thousand cubic kilometers of magma. As the magma vacates the magma chamber during an eruption, the roof of the chamber collapses to form a depression tens of kilometers across. The depression may fill with ash, lava flows, and sediment. On these two pages, we focus on large calderas, but smaller calderas can form on composite volcanoes or when magma withdraws beneath a shield volcano.

A What Is a Caldera?

A *caldera* is a large, basin-shaped volcanic depression, which typically has a low central part surrounded by a topographic escarpment, referred to as the *wall* of the caldera. The Valles Caldera of New Mexico nicely illustrates the important features of a caldera. It is relatively young and uneroded because it formed only two million years ago.

1. This image shows satellite data superimposed on topography. The circular Valles Caldera contains a central depression, about 22 km (14 mi) across, surrounded by steep walls. The caldera formed when a huge volume of magma erupted from a shallow magma chamber, producing a large eruption column and pyroclastic flows.

2. Internally, the caldera contains a series of faulted blocks that have been down-dropped relative to rocks outside the caldera. Faulting and ground subsidence occurred at the same time as the main eruption, so relatively thick amounts of ash and other materials accumulated inside the caldera.

Valles Caldera

11.05.a1

3. Small, rounded volcanic domes formed within the caldera after the main pyroclastic eruption. The domes were fed by fissures that tapped leftover magma. Some of this magma has solidified at depth, forming granite, but some may still be molten. The area has been explored for geothermal energy associated with the hot rocks at depth.

B How Does a Caldera Form?

The formation of a caldera and the associated eruption occur simultaneously—the caldera subsides in response to rapid removal of magma from the underlying chamber. The largest caldera eruptions produced volcanic ash layers more than 1,000 m thick.

5 km

11.05.b1

11.05.b2

1. The first stage is the generation of felsic magma. The magma rises and accumulates in magma chambers that can be kilometers thick and tens of kilometers across. The chambers may be within several kilometers of the surface.

2. Next, magma reaches the surface (at which point it is called lava) and eruptions begin. As the magma chamber loses material, the roof of the chamber subsides to occupy the space that is being vacated. Curved fractures allow crustal blocks to drop, outlining the edges of the caldera and providing many conduits for magma to reach the surface.

11.05.b3

3. The eruption forms columns and flows of pyroclastic material, much of which falls back into the caldera, creating a thick pile of ash. Landslides from steep caldera walls produce large blocks and clasts that become part of the caldera deposit. Some ash escapes the caldera and covers surrounding areas.

11.05.b4

4. As the eruption subsides, magma rises through fissures along the edges or in the caldera, erupting onto the surface as volcanic domes.

C Did the Eruption of a Caldera Destroy a Civilization Near Greece?

Santorini, east of the Greek mainland, is a group of volcanic islands with geology that records a major caldera collapse only 3,500 years ago, about the time of the collapse of the Minoan civilization on the island of Crete.

11.05.c1 Santorini, Greece

11.05.c2 Santorini, Greece

◄ **1.** This satellite view shows the islands of Santorini, including *Thera*, the largest island. Thera and the other islands encircle a submerged caldera that formed when the center of a larger volcanic island collapsed, leaving the modern islands as remnants. The steep cliffs around the caldera are eroded segments of the original wall of the caldera. The curving cliffs (►) expose volcanic layers, products of explosive volcanism that began one to two million years ago. Islands in the middle of the caldera were constructed by more recent eruptions.

2. The main eruption produced an ash column perhaps 40 km (25 mi) high, followed by pyroclastic flows. The erupted ash buried towns, now excavated (►), with up to 50 m of pumice and ash. Caldera collapse occurred as the eruption of ash emptied a large magma chamber.

3. The collapse of the caldera evidently unleashed a large destructive wave that traveled southward across the sea, probably helping lead to the downfall of the Minoan civilization on the island of Crete. This destruction of the civilization on Santorini and collapse of the volcanic island into the sea may have started legends about the sinking of a landmass and city (Atlantis) into the sea.

11.05.c3 Akrotiri, Santorini, Greece

D Why Is Yellowstone Caldera Considered to Be Dangerous?

Yellowstone is one of the world's largest active volcanic areas, occurring directly over a continental hot spot. Abundant geysers, hot springs, and other hydrothermal activity are leftovers from its recent volcanic history. During the last two million years, the Yellowstone region experienced three huge, caldera-forming eruptions. If a similar eruption occurs again, it would rain destructive ash over the Rocky Mountains and onto the Great Plains.

11.05.d1

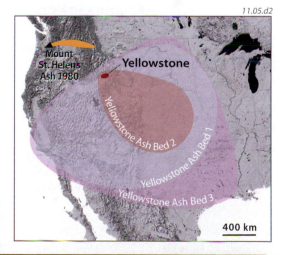

11.05.d2

◄ **1.** This image shows the outline of the youngest Yellowstone caldera, which formed 640,000 years ago. The boundaries of the caldera have been partially obscured by erosion, deposition of sediment, and lava flows that erupted after the caldera formed. Several areas within the caldera have been experiencing uplift. Magma and hot rocks at relatively shallow depth heat groundwater to form the spectacular geysers and hot springs for which Yellowstone is famous.

► **2.** Ash from the three Yellowstone eruptions was carried by the wind and deposited over a huge area that extends from northern Mexico to southern Canada and as far east as the Mississippi River. A repeat of such an eruption could devastate the region around Yellowstone and cause extensive crop loss in the farmlands of the Great Plains and Midwest.

Before You Leave This Page Be Able To

✓ Sketch and explain the characteristics of a caldera and the four stages by which one forms.

✓ Explain how a volcanic eruption destroyed Santorini.

✓ Describe what type of volcanic feature Yellowstone is, how it forms, and how a future eruption could impact the Great Plains.

11.6 What Hazards Are Associated with Volcanoes?

VOLCANIC ERUPTIONS can affect areas very close to the vent or can spread damaging volcanic ash over huge regions. The relative hazard depends on the type of eruption. Eruptions of pyroclastic flows have killed more than 30,000 people at a time, all within minutes. Surprisingly, volcanic eruptions can also cause huge floods.

A What Is Meant by a Hazard and a Risk?

The terms *hazard* and *risk* may seem more appropriate for a lesson about insurance, but geographers frequently apply these terms when discussing the effects that natural (and human-caused) events can have on humans and society. What is the difference between a hazard and a risk?

11.06.a2 Mount Rainier, WA

◄ A *hazard* is the existence of a potentially dangerous situation or event, such as a potential landslide of a steep slope or a lava flow erupting from a volcano. The hazard in this photograph was a basaltic lava flow.

Risk is an assessment of whether the hazard might have some *societal impact*, such as loss of life, damage to property, loss of employment, destruction of fields and forests, or implications for local or global climates. Remnants of destroyed houses, cars, and roads demonstrate that this area had a high risk for volcanic hazards.

11.06.a1 Kilauea, HI

The people living around Mount Rainier, including those in the cities of Tacoma and Seattle (▲), have a high risk from activity on the dangerous, composite volcano, especially from lahars. Some suburbs of Tacoma are built on or among deposits from lahars that flowed off the volcano within the last 1,000 years. If no people were living near this volcano, a hazard would still exist but there would be essentially no risk.

B What Hazards Are Associated with Eruptions of Cinder, Ash, and Gas?

Basaltic eruptions can be deadly and destructive, but mostly to nearby areas. They can hurl lava and solid rock into the air and spew out dangerous gases. Fine ash ejected high into the air can cause damage that is more widespread.

Falling Objects

11.06.b1 Hawaii

During the formation of lava fountains (◄), most cinder falls back to Earth near the vent and piles up in a cinder cone. Hazards include being struck and burned by cinders and being struck by blobs of lava and other projectiles. Larger ejected pieces, called *volcanic bombs* (▼), pose a severe hazard for anyone close to the erupting cone.

11.06.b2 Flagstaff, AZ

Volcanic Ash from Cinder Cones

Sand-sized cinders and finer particles of ash can bury nearby structures, and may cause breathing problems for people and livestock. In March and April of 2010, a shield volcano in Iceland, called Eyjafjallajökull, erupted large amounts of volcanic ash that drifted over Europe, shutting down most air travel and stranding hundreds of thousands of passengers. Eruption columns from cinder cones typically reach lesser heights, of several kilometers, and mostly impact areas nearby.

11.06.b3 Iceland

Gases

Volcanic gases are a significant hazard associated with many types of volcanoes. Gases such as carbon dioxide (CO_2) cause asphyxiation if concentrated. Other gases, including hydrogen sulfide (H_2S), cause death by paralysis. Gaseous sulfur dioxide (SO_2), hydrochloric acid (HCl), sulfuric acid (H_2SO_4), and fluorine compounds expelled during eruptions can destroy crops, kill livestock, and poison drinking water for people and animals. Gas eruptions can kill hundreds of people.

11.06.b4 Krafla, Iceland

C What Are the Hazards Associated with Pyroclastic Eruptions?

Lava flows usually move slowly enough that people can get out of the way, but such flows can completely destroy any structures in their path. In contrast, pyroclastic flows race downhill and across the landscape at hundreds of kilometers per hour, incinerating and smothering any creature in its path. They can cause massive destruction. Eruption columns spread their damage over wider areas and over longer time periods.

1. On May 18, 1980, at 8:32 a.m., an earthquake triggered a massive avalanche on the north side of Mount St. Helens in southwestern Washington state. The landslide reduced pressure on an underlying magma chamber, triggering a northward-directed blast that knocked over millions of trees and unleashed a pyroclastic flow. The pyroclastic flow swept downhill and across the landscape, burying and killing dozens of people and countless other living things in its path. This was followed immediately by a huge column of volcanic ash that rose 25 km (15 mi) into the atmosphere (◄). The ash was carried eastward by the wind and blocked sunlight as it settled back to Earth across a large area of Washington, Idaho, and Montana.

11.06.c1 Mount St. Helens, WA

2. After the main eruption, pyroclastic flows (◄) and other eruptions continued on Mount St. Helens until 1986. After a lull of several years, the volcano built, destroyed, and rebuilt several new volcanic domes. The domes would grow at some times and collapse at others, unleashing smaller pyroclastic flows, like the one shown here. Activity subsided in 2008, but the volcano is still being closely monitored by instruments designed to detect changes in temperature, gas emissions, or steepness of slopes.

11.06.c2 Mount St. Helens, WA

3. Vesuvius is an active composite volcano near the city of Naples in southwestern Italy. In A.D. 79, a series of pyroclastic flows moved down the flank of the volcano, destroying the coastal towns of Pompeii and Herculaneum and killing more than 25,000 people. This image (►) is an artist's conception of an explosive Vesuvius eruption striking the city of Naples, which covers most of the region shown. Naples and the surrounding area currently are home to over three million people. The area also has a significant hazard for lahars and landslides coming off the volcano, which looms over the city.

Herculaneum

Limit of Pyroclastic Flows

Pompeii

10 km

11.06.c3

4. The dashed red line marks the outward limit of pyroclastic flows from Vesuvius, but material from the eruption column covered a wider area. Note how much of the present city of Naples is within the area devastated by the eruption of 79 A.D. Evidence indicates that the catastrophe began with earthquakes and the formation of an eruption column that deposited a layer of ash over Pompeii, killing some inhabitants. This was immediately followed by pyroclastic flows that raced down the mountainside and hit Pompeii, killing the inhabitants and destroying the buildings. People were smothered, suffocated, subjected to thermal shock, or crushed.

Floods of Water, Ice, and Debris

Iceland is a land of ice and fire, due to its location near the Arctic Circle and its position on top of a mid-ocean ridge and an oceanic hot spot. It has many glaciers (the Icelandic term for glacier is *jökull*), including a large ice sheet that covers 25% of the country. Beneath the ice are a half dozen shield volcanoes, several of which have erupted in the last several decades.

In 1996, a volcanic eruption beneath an ice sheet melted the ice, releasing a catastrophic flood (►) of meltwater (called a *jökulhlaup* in Icelandic). The huge flood carried blocks of ice, rock, and other debris, causing widespread damage by destroying bridges and houses and by covering vast areas with sediment. In the photograph below, note a red plane for scale.

11.06.t1 Southern Iceland

Before You Leave This Page Be Able To

✓ Explain how risk differs from hazard, and provide an example of each.

✓ Describe and compare the hazards associated with lava flows, lava fountains, volcanic gases, and various types of pyroclastic eruptions.

✓ Describe the eruptions that occurred at Mount St. Helens and Vesuvius, and what caused most deaths.

✓ Explain eruption-caused floods.

11.6

11.7

What Areas Have the Highest Potential for Volcanic Hazards?

IN SOME PLACES, THE RISK POSED by volcanic hazards is great. In others, it is inconsequential. Volcanic eruptions are more likely in Indonesia than in Nebraska. Additionally, different types of volcanoes have different eruptive styles, so some volcanoes are more dangerous than others. What factors should we consider when determining which areas are the most dangerous and which are the safest?

A How Do We Assess the Danger Posed by a Volcano?

Potential hazards of a volcano depend on the type of volcano, which we infer from its shape and rock types, and on its history. Examine the volcano below for clues about what type of volcano is present and how it might erupt.

11.07.a1 Augustine volcano, AK

Shape—The shape of a volcano provides important clues about how dangerous the volcano might be. Volcanoes with steep slopes, such as composite volcanoes, are more dangerous because they form from potentially explosive, viscous magma and also are prone to landslides. Volcanoes that have relatively gentle slopes, such as most shield volcanoes, result from less explosive basaltic eruptions.

Types of Volcanic Materials—The types of materials on a volcano indicate the magma composition and style of eruption. A volcano is very dangerous if it contains deposits formed by pyroclastic flows. A volcano that erupted viscous or gas-rich magma is more dangerous than one that erupted less viscous magma that formed basalt. The volcano shown above is a steep, composite volcano composed of materials erupted as viscous lavas and pyroclastic flows.

Age and History—The age of a volcano is essential information. If the volcano has not erupted for a long time, maybe it is dormant (not active). The shape of a volcano, especially whether it still has a fresh-looking volcano shape or has been eroded, is one indicator of a volcano's age. Important clues are also provided by a volcano's history, if recorded, including oral histories from nearby people. Isotopic measurements on volcanic units can provide an accurate indication of a volcano's age. Studies of the sequence of volcanic layers, combined with isotopic ages, provide insight into how often eruptions recur. The volcano above clearly has recent activity, as expressed by the recent dark deposits and the steam and other gases escaping from the summit dome.

B What Areas Around a Volcano Have the Highest Risk?

Once we have determined the type of volcano that is present, we consider other factors that help identify which areas near the volcano have the highest potential risk.

11.07.b1

1. *Proximity*—The biggest factor determining potential risk is proximity, closeness to the volcano. The most hazardous place is inside an active crater. The potential risk decreases with increasing distance away from the volcano.

2. *Valleys*—Lava flows, small pyroclastic flows, and mudflows (lahars) are channeled into valleys carved into the volcano and surrounding areas. Such valleys are more dangerous than nearby ridges.

3. *Wind Direction*—Volcanic ash and pumice that are thrown from the volcano are carried farthest in the direction that the wind is blowing at the time of the eruption. Most regions have a prevailing wind direction, so a greater hazard of falling material exists in this direction from a volcano.

4. *Particulars*—Each volcano has its own peculiarities, and these influence which part of the volcano is most dangerous. Steeper parts of a volcano pose special risks, and one side of a volcano may contain a dome that could collapse and form pyroclastic flows. This image shows three small villages around a volcano. Is one village at greater risk than the others? Which one is in the least hazardous place, and what ideas led you to this conclusion?

C What Regions Have the Highest Risk for Volcanic Eruptions?

We can think on a broader regional scale about which regions are most dangerous. In North America, volcanoes are relatively common along the west coast and virtually absent east of the Rocky Mountains. *Tectonic setting*, especially proximity to certain types of plate boundaries, is the major factor making some places more prone to volcanic hazards than others. This map shows the locations of volcanoes (orange triangles) that have been active in the last several million years.

The largest concentration of composite volcanoes is along the Pacific Ring of Fire. The volcanoes form above subduction zones, in island arcs, and in mountain ranges along continental margins. Some subduction-zone volcanoes erupt so vigorously that they form calderas.

Much fluid basaltic lava erupts on the seafloor at mid-ocean ridges. Such eruptions pose little risk to humans because almost all of these occur at the bottom of the ocean. The island of Iceland, where a mid-ocean ridge coincides with a hot spot, is an exception.

11.07.c1

Many shield volcanoes occur along lines of islands and submarine mountains in the Pacific and other oceans. Most of these linear island chains, and a few other clusters of islands, formed above hot spots. Hawaii and the Galápagos Islands are good examples. Shield volcanoes also occur in other settings, including on continents.

Some volcanic features, including basalt flows, cinder cones, and composite volcanoes, are in the middle of continents. Most of these form over hot spots or in continental rifts, such as the East African Rift.

Forecasts, Policy, and Publicity

Predicting volcanic eruptions is currently an imprecise science. There have been some fabulous successes and disappointing failures. Volcanologists (scientists who study volcanoes) have successfully predicted some eruptions by studying clusters of small earthquakes generated as magma rises through the crust, by measuring changes in the amount of gas released by volcanoes, and through other types of investigations. Some predictions led to evacuations that saved lives because government officials and citizens acted on the scientific evidence. Some predictions have been unsuccessful because an eruption that was considered possible or even

likely did not occur. In other cases, predictions, policy, and publicity interacted in a bad way, with deadly results. In 1985, volcanologists working on Nevado del Ruiz, a composite volcano in the Colombian Andes, warned of an impending eruption. The city of Armero, with an estimated 29,000 inhabitants, lay in a valley that drained the steep volcano. Local government officials downplayed the risk and assured the citizens that there was no danger. A pyroclastic eruption occurred at night, melting snow and ice on the volcano and unleashing a mudflow that moved at hundreds of kilometers per hour, engulfing most of Armero and killing more than 20,000 people.

Before You Leave This Page Be Able To

☑ Summarize ways to assess the potential danger of a volcano based on its characteristics.

☑ Describe ways to identify which areas around a volcano have the highest potential hazard.

☑ Describe how the plate tectonic setting of a region influences its potential for volcanic hazards.

11.7

11.10 How Are Fractures Expressed in Landscapes?

FRACTURES ARE THE MOST COMMON geologic structures. They range from countless small cracks visible in an outcrop (an exposure of rock) to huge faults hundreds or thousands of kilometers long. What are the different types of fractures, how do the different types of fractures form, and how are fractures expressed in landscapes?

A In What Different Ways Do Rocks Fracture?

There are two main types of fractures: *joints* and *faults*. Joints and faults both result from stress but have different kinds and amounts of movement across the fracture.

Joints

11.10.a1

◄ **1.** Most fractures form as simple cracks representing places where the rock has pulled apart by a small amount. These cracks are called *joints* and are the most common type of fracture. They form under tension, where the rock is pulled apart.

2. These sandstone ledges (►) are cut by a series of near-vertical joints. The layers are not offset by the joints but are simply pulled apart by a very small amount.

11.10.a2 San Juan River, UT

Faults

► **3.** A *fault* is a fracture where rocks have slipped past one another. Rocks across a fault can slip up and down, as shown here, or they can slip sideways or at some other angle. A fault displaces the rocks on one side relative to the other side.

11.10.a4 Moab, UT

11.10.a3

◄ **4.** The long fracture in the center of this photograph cuts across and offsets the rock layers. That is, the layers across the fault have been displaced relative to each other.

B What Stresses Form Joints?

11.10.b1

The stresses that form joints arise from many sources, but they are mostly due to burial and to tectonic forces. Tectonic forces may push, pull, or shear the rock. These volcanic rocks (▼) are cut by vertical joints formed by tectonic stresses.

11.10.b3

Stresses build up as rocks get warmer or cooler. As some igneous rocks cool, they contract into polygon-shaped columns bounded by joints that commonly meet at 120° angles. The photograph below (▼) shows an example of such *columnar joints*.

11.10.b5

Stresses also arise during uplift of buried rocks, causing rocks to fracture due to reduced pressure. These joints, called *unloading joints*, form parallel to the surface and slice off thin sheets of rock (▼).

11.10.b2 South Fork, CO

11.10.b4 Iceland

11.10.b6 Yosemite, CA

C What Are the Main Types of Faults?

We classify faults based on the motion of one block relative to the other. There are three main types of faults: *normal faults, reverse faults,* and *strike-slip faults*. Black arrows show relative movement, and blue arrows show stress.

11.10.c1

Normal Fault—In a fault, if the block above the fault moves down relative to the block below the fault (◀), the fault is a *normal fault*. A normal fault forms when the rock units are pulled apart and lengthened, as for example by tension. The direction in which a fault (or layer) is inclined is called its *dip*. This normal fault dips to the right.

11.10.c2

Reverse Fault—If the block above the fault moves *up* relative to the block below the fault (◀), the fault is a *reverse fault*. A reverse fault forms as a result of *horizontal compression* and shortens the rock units in a horizontal direction. This reverse fault dips to the right.

11.10.c3

Thrust Fault—A reverse fault that has a gentle dip (◀) is a *thrust fault*. The rock above the fault is called a *thrust sheet* and is pushed up and over the block below the fault. On maps and other figures, teeth indicate that a fault is a thrust.

11.10.c4

Strike-Slip Fault—When rocks along a fault move with a side-to-side motion, parallel to the fault surface, the fault is a *strike-slip fault*. Relative motion is horizontal, offsetting the blocks laterally in one direction or the other.

D How Are Faults Expressed in Landscapes?

11.10.d1 Borah Peak, ID

11.10.d2 Marble Canyon, AZ

11.10.d3 Buckskin Mtns., AZ

1. When fault movement offsets Earth's surface, as during an earthquake (▲), it can cause a step in the landscape, called a *fault scarp*. If a fault forms and later erosion wears away one side more than the other, the step in topography is called a *fault-line scarp*.

2. We commonly recognize faults because of *offsets* or abrupt *terminations* of layers. Also, rocks along faults are highly fractured and easily eroded, so they erode into linear topographic notches (▲). This fault truncates layers and forms a linear notch.

3. Up close, faults have other distinctive characteristics (▲). They can fold the truncated layers, juxtapose two very different rocks against one another, or display shattered and scratched rocks along the fault zone. This fault has all these diagnostic features.

4. Faults commonly break the crust into large, fault-bounded blocks. If normal faults downdrop one *fault block* relative to blocks on either side, the resulting feature is called a *graben*. This term graben can refer to the downdropped block or to the resulting valley.

11.10.d4

5. A block that is uplifted relative to blocks on either side is called a *horst*. Parts of the American Southwest, including the aptly named *Basin and Range Province*, have fault-block horsts (mountains) and grabens (valleys).

Before You Leave This Page Be Able To

✓ Sketch and describe the formation and landscape expression of joints and faults.

✓ Sketch and describe three types of faults, showing the relative displacement and the type of stress.

✓ Sketch a horst and a graben.

11.10

What Happened During the Great Alaskan Earthquake of 1964?

THE SOUTHERN COAST OF ALASKA experienced one of the world's largest earthquakes in 1964. The moment magnitude 9.2 earthquake, which is the strongest to have ever struck North America, destroyed buildings, triggered massive landslides, and unleashed a tsunami that caused damage and deaths from Alaska to California. The nearby region also has active volcanoes, further evidence that this is a tectonically active area.

A What Types of Damage Did the Earthquake Cause?

The earthquake occurred along the southern coast, but was felt throughout Alaska, except for the far north coast. Ground shaking destroyed buildings and generated huge landslides of rock and soil. This dark, rocky landslide (▼) covered parts of the white Sherman Glacier.

The blue line on the map marks the limit of shaking-caused cracking of the ground and ice during the earthquake. The red line closer to the epicenter outlines the region where property damage occurred.

11.18.a2 Southern Alaska

11.18.a1

200 km

The epicenter of the earthquake was along the southern coast of Alaska, between the cities of Anchorage and Valdez. The earthquake began at depths of 20 to 30 km (12 to 19 mi). Based on the wide distribution of about 600 aftershocks, seismologists estimate that the earthquake ruptured a fault surface that was over 900 km (560 mi) long and 250 km (160 mi) wide. The earthquake occurred on a thrust fault that dips from the Aleutian trench gently north and northwestward beneath Alaska.

11.18.a3 Anchorage, AK

◄ Parts of downtown Anchorage were demolished when shaking caused the underlying land to slip and collapse. Some buildings sank so much that their second stories were level with the ground. Severe damage occurred in the Turnagain Heights area of Anchorage, where a layer of weak clay liquefied, carrying away shattered houses (▶).

11.18.a4 Anchorage, AK

B What Happened in the Sea During the Earthquake?

Because it occurred along the coast, the earthquake also caused (1) faulting and uplift of the seafloor, (2) huge waves from landslides, and (3) a tsunami that struck the coasts of Alaska, British Columbia, Washington, Oregon, California, Hawaii, and Japan.

11.18.b1 Southern Alaska

▶ The main fault that caused the earthquake did not break the land surface, but two subsidiary faults did. One fault cut a notch into a mountain and uplifted the seafloor 4 to 5 m (15 ft). The white material on the uplifted (left) side of the fault consists of calcareous marine organisms that were below sea level before the earthquake. The maximum observed uplift was 11.5 m (38 ft). Other areas subsided as much as 6 m (20 ft) during the earthquake, flooding docks, oil tanks, and buildings along the coast.

11.18.b2 Kodiak Island, AK

▲ Faulting uplifted a large area of seafloor off the southern coast of Alaska, sending a large tsunami out across the sea and up the many bays and inlets along the coast. The highest tsunami recorded was 67 m (220 ft), in a bay near Valdez. The photograph above shows damage done to Kodiak Island by a wave 6 m (20 ft) high. The tsunami killed 106 people in Alaska and 17 more in Oregon and California.

C What Are Some Other Manifestations of Tectonics?

1. Immediately after the earthquake, a team of scientists and surveyors investigated the coastline, measuring uplift and subsidence at hundreds of sites. They plotted and contoured the elevation change measurements on a detailed map, which they then generalized into a summary map (▶) that identifies broad zones of uplift and subsidence. As a result of the earthquake, a region of more than 250,000 km² (100,000 mi²) changed elevation, either up or down. Repetitions of the earthquake, hundreds of times over millions of years, could uplift a new mountain range where the sea is now, and form a basin where there is currently a ridge. Earth's elevations and landscapes are dynamic, always changing.

11.18.c1

11.18.c3 Mount Redoubt, AK

3. Southern Alaska also has a number of active volcanoes, like Redoubt volcano (▲), and these continue out across the ocean to form the Aleutian Islands, an active island arc. The volcanoes display large, somewhat symmetrical shapes with steep slopes—they are dangerous composite volcanoes. They are related to subduction of an oceanic plate beneath Alaska, the same process that caused the Great Alaskan Earthquake of 1964.

2. The surveying and other studies along the coast led to the recognition that the earthquake occurred along a thrust fault beneath the southern shoreline of Alaska. This study interpreted the earthquake as resulting from oceanic material being pushed beneath Alaska (▶),

11.18.c2

along what today we call a megathrust. These conclusions, in the 1960s, were drawn before development of the theory of plate tectonics. In fact, the studies of this earthquake, using a variety of geographic tools, like surveying, preparation of contour maps, and surveys of changes to the landscape, was an important step in the development of plate tectonics. Today, geographers, geologists, and other scientists would also use GPS devices, GIS, and satellite data.

Before You Leave This Page Be Able To

☑ Summarize events associated with the Alaskan earthquake, including effects on land and sea, and how studies of this area helped lead to the theory of plate tectonics.

11.18

12.2 How Does Chemical Weathering Affect Earth's Surface?

CHEMICAL WEATHERING alters and decomposes rocks and minerals, principally through chemical reactions involving water. When physical and chemical weathering processes interact to break down and alter rocks, new minerals, called secondary minerals, are created that are more stable in surface conditions than the original minerals. What are some processes by which chemical weathering occurs?

A How Does Changing a Rock's Environment Promote Weathering?

Many rocks and minerals form deep within Earth. When they are brought near the surface by uplift and erosion, these rocks encounter conditions very different from those in which they originally form and so they may become unstable.

Minerals that crystallize in high-temperature magmas are generally unstable when subjected to the low-temperature conditions that characterize Earth's surface. Most magma temperatures are above 700°C, whereas surface temperatures range from −40°C to 45°C (minus 40°F to plus 122°F).

During metamorphism, some minerals crystallize beneath the surface in high-pressure and high-temperature environments. Once such rocks reach Earth's low-pressure and low-temperature surface, they can change to different minerals that are stable at the new, wetter, low-pressure and low-temperature conditions.

12.02.a1

Oxygen (O_2) is abundant in the atmosphere and as a dissolved component in rain and most surface water. This oxygen chemically reacts with rocks, causing some minerals to oxidize (rust).

Liquid water is more abundant on and near Earth's surface than at depth in the crust. Water, especially when it is slightly acidic, is a chemically active substance that can break the bonds in many minerals. It increases the rate of chemical weathering. The availability of water largely depends on the climate of an area, so for this and other reasons climate is a key factor in weathering.

B What Happens When Rocks Dissolve?

The main agents for chemical weathering are water and weak acids formed in water, such as carbonic acid (H_2CO_3). These agents dissolve some rocks, loosen mineral grains, form clay minerals, and widen joints.

Limestone (below) and other rocks rich in calcium carbonate or magnesium carbonate are soluble in water and in acids. They dissolve and form pits and cavities (▼). On a large scale, the solution of limestone forms distinctive landforms of *sinkholes* and caves called *karst*.

12.02.b1 Capitol Reef NP, UT

Over time, the pits deepen, widen, and may interconnect, forming furrows (small troughs).

12.02.b2

Fractures can widen as water flows through them and dissolves material from the walls of the joints, as in the limestone outcrop below (▼). Caves can form by dissolution of limestone and other soluble rocks at depth.

12.02.b3 Austrian Alps

One Way That Calcite Chemically Dissolves in Water

Limestone is a relatively soluble rock because it is composed of calcite, which is soluble in weak acids. The most common weak acid in surface water is carbonic acid, produced when rainwater reacts with carbon dioxide (CO_2) in the atmosphere, soil, and rocks. The chemical reaction for the dissolution of calcite in carbonic acid is:

$CaCO_3$	+	H_2CO_3	$\longrightarrow$	Ca^{2+}	+	$2(HCO_3)^-$
Calcite		Carbonic acid		Calcium ion		Bicarbonate ion in solution

Acids in water produce unbonded H^+ ions, each of which is a proton without a balancing electron and is available to make other chemical bonds. H^+ ions are small and can easily enter crystal structures, releasing other ions, such as calcium, into the water. Rainwater is a weak acid, formed by CO_2 that was acquired from the atmosphere, to make carbonic acid.

C What Happens When Rocks Oxidize Near Earth's Surface?

Oxygen (O_2) is common near Earth's surface and reacts with some minerals to change the oxidation state of an ion. This is common in iron-bearing minerals because iron (Fe) has several oxidation states.

12.02.c1 Wilson Cliffs, NV

Many mafic igneous rocks contain dark, iron-bearing minerals, such as pyroxene. Iron in pyroxene can become oxidized, producing iron oxide minerals, as shown in the equation below.

Hematite consists only of iron and oxygen and is more stable than pyroxene under oxygen-rich conditions. It commonly forms during oxidation and gives oxidized rocks a reddish color.

If iron-bearing rocks become oxidized, such rocks generally take on a red color from the iron oxide mineral hematite. This oxidation process is called rusting and creates less stable by-products. Reddish rocks can lose this reddish color if interacting with fluids that have less oxygen (▶).

$$4FeSiO_3 \ + \ O_2 \longrightarrow 2Fe_2O_3 \ + \ 4SiO_2$$

Pyroxene   Oxygen   Hematite   Silica (in water)

D What Happens When Minerals Chemically React with Water?

When some minerals react with water they undergo a chemical reaction where the mineral combines with water to form a new mineral. This reaction, called *hydrolysis*, converts some minerals to clay, an important soil constituent.

12.02.d1 Crete

One kind of feldspar, containing potassium (K), is called K-feldspar. When this mineral reacts with acids (waters that have free H^+ ions), it can be converted into clay minerals by hydrolysis.

During the reaction, the H^+ ion moves into the crystalline structure of feldspar, expelling the K^+ ion and some silica, which both get carried away in the water.

If exposed to wet conditions, many rocks convert into clay minerals. The gray limestone shown here contained impurities that weathered into clay minerals and reddish hematite that accumulated between the blocks.

$$4KAlSi_3O_8 \ + \ 4H^+ \ + \ 2H_2O \longrightarrow 4K^+ \ + \ Al_4Si_4O_{10}(OH)_8 \ + \ 8SiO_2$$

K-feldspar   Hydrogen   Water   Potassium   Kaolinite   Silica
             ion   (in water)   (clay)   (in water)

E How Does Weathering Make the Ocean Salty?

Have you ever wondered why the ocean is salty or where the salt comes from? It turns out that most of the ocean's salts are derived from weathering and the dissolution of rocks on the land.

1. Rock, sediment, and soil on and near Earth's surface are exposed to water and oxygen in the atmosphere. Water that reacts with rocks and minerals can come from several sources, including rain and other forms of precipitation.

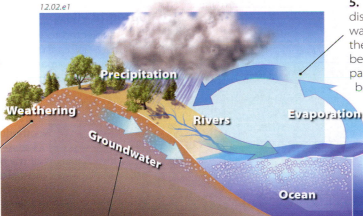
12.02.e1

Precipitation
Weathering
Groundwater
Rivers
Evaporation
Ocean

5. When seawater in the oceans evaporates, the dissolved salts remain in the seawater, increasing seawater's salt content (*salinity*). Such evaporation causes the seas to be saltier than the streams. The seas would be even saltier if salt were not removed from some parts of the sea by deposition of salt beds. If the salt beds, once formed, are uplifted and exposed on land, they can also contribute dissolved salt to streams and ultimately back to the sea.

2. Some water infiltrates into the subsurface, where it may chemically react with the materials. During weathering, hydrolysis reactions commonly produce clay minerals and also drive out positive ions (cations) from the pre-existing mineral structure. The dominant cations in feldspar, a very abundant mineral, are sodium and potassium, which can form common salts.

3. The dissolved cations, along with negative ions (anions) like chlorine, are carried by moving water, either in streams or in the subsurface by groundwater. Much of this water eventually finds its way to the oceans, where it contributes its salts and other ions to seawater.

4. Modern oceans typically contain about 3.5% dissolved salt, and are much saltier than stream water. Streams contain some dissolved salts but only a small amount, so they are considered to be freshwater, not salt water. If the oceans got their salt from the streams, why are the oceans saltier?

Before You Leave This Page Be Able To
✓ Describe several reasons why minerals formed at depth may not be stable at the surface.
✓ Summarize how limestone dissolves and what features are formed by dissolution.
✓ Briefly summarize the processes of oxidation and hydrolysis.
✓ Explain why oceans are salty.

12.2

12.8 What Controls the Stability of Slopes?

WITH MATERIAL LOOSENED BY WEATHERING, gravity pulls it downhill. Downward movement of material on slopes under the force of gravity is called *mass wasting*, and it occurs to some degree on all slopes. Mass wasting can proceed very slowly or very quickly, sometimes with disastrous results. Mass wasting is an important part of the erosional process, moving material downslope from higher to lower elevations, and feeding sediment from hillslopes to streams, beaches, and glaciers.

A How Does Gravity Affect Slope Stability?

The main force responsible for mass wasting is *gravity*. The force of gravity acts everywhere within and on the surface of Earth, tending to pull everything toward Earth's center.

12.08.a1

1. On a flat surface, the force of gravity acts on a block by pulling it vertically down against the base of the block. The block will not move under this force.

2. On a slope, gravity acts at an angle to the base of the block. Part of the force pulls the block against the slope and another part pulls the block down the slope. These two parts of the force are referred to as *components*.

3. The part of the force pulling the block against the slope is the *normal component*.

4. The other component acts parallel to the slope, trying to shear the block down the slope. It is the *shear component*.

5. As the angle of slope becomes steeper, the shear component becomes larger while the normal component becomes smaller. As the slope angle steepens, the shear component becomes enough to overcome friction, and it causes the block to slide. The effectiveness of the shear component is influenced by the strength of the slope material to resist deformation. If the material is very weak, like mud, it will internally shear down the hill rather than slide as a rigid block.

B What Makes Some Slopes Stable?

The steepest angle at which a pile of unconsolidated grains remains stable is called the *angle of repose*. This angle is controlled by frictional contact between grains. In general, loose, dry material has an angle of repose between 30° and 37°. This angle is somewhat higher for coarser material, for more angular grains, for material that is slightly wet, and for material that is partly consolidated. It is lower for material with flakes or rounded grains and for material that contains so much water that adjacent grains lose contact.

Dry Sand — Angle of Repose

12.08.b1

◀ 1. Dry, unconsolidated sand grains form a pile, and the angle of the resulting slope is at the angle of repose. If more sand is added, the pile becomes wider and higher, but the angle of repose remains the same. If part of the pile is undercut and removed, the grains slide downhill until the pile returns to a stable slope at the angle of repose. If sand is slightly wet, surface tension between the grains and a thin coating of water enables the sand to be stable on steeper slopes.

▶ 2. Loose rocks and other loose material (talus) accumulate on some slopes and at the bases of cliffs. Such talus material commonly forms slopes that are at the angle of repose for the particular sediment. The smooth *talus slope* shown here became too steep in places and so locally slid downward, forming a small pile of debris at the base of the slope.

12.08.b2 Tibet

▼ 3. Most slopes of sand dunes exhibit the angle of repose for dry sand. Slopes can be more gentle than the angle of repose, but if they begin to exceed the angle of repose then the slope fails, slumping downhill. Walking up a sand dune causes the barely stable sand to slide beneath your feet.

12.08.b3 Morocco

▶ 4. The slopes of a cinder cone reveal the angle of repose because they are typically composed of loose, volcanic cinders. The angle of repose will be steeper for coarser cinders and for material that partially fused together during the eruption.

12.08.b4 Northern AZ

C   What Resisting Forces Control Slope Stability?

The main control on slope stability is the angle of repose for the material. Intact rock can form cliffs or steep slopes, but soil, sediment, and strongly fractured rock form slopes as steep as their angle of repose.

The addition of minor amounts of water increases the strength of soil, but oversaturation pushes grains apart and weakens the soil. Materials with high clay-mineral content can flow downhill when they become wet.

Joints, cleavage, and bedding reduce the overall mechanical strength of a rock, and may allow rocks to slip downhill. In this illustration, rock layers oriented parallel to the slope allow material to slide.

12.08.c1

12.08.c2

12.08.c3

D   What Driving Forces Trigger Slope Failure?

Slope failure occurs when a slope is too steep for its material to resist the pull of gravity. Precipitation, load, oversteepening, shaking, and other factors are examples of driving forces. Some slopes slide or creep downhill continuously, but others fail because some event caused a previously stable slope to fail.

Precipitation can saturate sediment, increasing pore pressure and decreasing normal stress, allowing grains to slide past one another more easily. A slope that was stable under dry conditions may fail when wet. Slopes can also fail after wildfires, which destroy plants that help bind and stabilize the soil.

Hillslopes can fail when the load on the surface exceeds a slope's ability to resist movement. Humans sometimes build heavy structures on slopes, overloading the slope and causing it to fail. Areas with gentle slopes, such as near this town, are less prone to slope failures.

Modification of a slope by humans or natural causes can increase a slope's steepness so that it becomes unstable. Erosion along river banks, as shown here, or wave action along coasts can undercut a slope, making it unsafe.

Volcanic eruptions can shake, fracture, and tilt the ground, unleashing landslides from oversteepened slopes. Eruptions can cover an area with hot ash and other loose material, causing melting of ice and snow. Melting can rapidly release large amounts of water and mobilize volcanic material in destructive debris flows.

A sudden shaking, such as tremors caused by an earthquake along this fault scarp, may trigger slope instability. Minor shocks from heavy trucks or human-caused explosions can also trigger slope failure.

Oversteepening of cliffs or hillslopes during road construction can cause them to fail, especially if joints or layers are inclined toward the road.

12.08.d1

Slope Stability in Cold Climates

In cold climates, water is frozen much of the year, and ice, although solid, can flow. Freeze-thaw cycles, where ice freezes and then thaws repeatedly, cause ice to flow and can contribute to slope failure.

When water-saturated soil freezes, it expands, displacing rocks and slabs of frozen soil. When the soil thaws, the material moves down again. This process, called *frost heaving,* is a large contributor to the downslope movement of material in cold climates. In addition, when the upper layers of soil thaw during the warmer months, the water-saturated soil may move downslope more easily. Frost heaving can form polygon-shaped outlines in the soil, called *patterned ground* (▼).

12.08.t1 Kongakut, AK

Before You Leave This Page Be Able To

✓ Describe or sketch the role that gravity plays in slope stability.

✓ Describe the concept of the angle of repose and its landscape expressions.

✓ Describe some factors that control slope stability, and events that trigger slope failure.

12.8

12.9 How Do Slopes Fail?

THE RAPID DOWNSLOPE MOVEMENT of material, whether bedrock, soil, or a mixture of both, is commonly called a *landslide*. Movement during slope failure can occur by falling, sliding, rolling, slumping, or flowing, and movement of material (mass) downslope by any of these mechanisms is commonly called *mass wasting*, signifying that mass is moving downhill. We classify slope failures — types of mass wasting — by how the material moves and the type of material involved.

A What Are Some Ways that Slopes Fail?

Most people have seen evidence of slope failure when hiking, driving past a roadcut, or watching television news, nature shows, or movies. Slope failure can be as subtle as a small pile of rocks at the base of a hill or as spectacular as a mudflow that has destroyed a neighborhood. The photographs below show images of various types and sizes of slope failures.

12.09.a1 Yampa River, UT

12.09.a2 Denali region, AK

◄ **2.** During an earthquake, brown masses of rock and soil slid down these steep slopes in Alaska, smashed apart, and flowed as avalanches of rock, soil, and ice across a white glacier in the valley below. Parts of the avalanche flowed across the valley and partway up hillsides on the other side of the valley.

12.09.a3 Tibet

◄ **3.** On this hillside, millions of rock pieces slid off steep outcrops. Some pieces accumulated on a high talus slope, which is at the angle of repose for this material. Several scars indicate where the talus was remobilized and flowed downhill, constructing a debris fan at the base of the hill.

▲ **1.** This rocky cliff failed after being undercut by a river. Large sandstone blocks, one the size of a building, collapsed downward. The falling block detached along a prominent joint surface, which has since accumulated a brown rock varnish.

12.09.a4 El Salvador, Central America

◄ **4.** This landslide in El Salvador flowed down a steep, unstable slope and cut a swath of destruction across a neighborhood. Adjacent slopes on this hill appear to be just as steep as the part that failed, and pose a hazard to the remaining homes.

▶ **5.** Undercutting of hillsides by coastal erosion and highway construction has made many slopes steep and unstable. Along the California coast, rocks and soil slumped downward, covering and blocking the highway along the Pacific Palisades, California. Such slope failures commonly occur during or after intense rainstorms.

12.09.a5 Pacific Palisades, CA

B How Are Slope Failures Classified?

Classification of slope failure is imprecise because the processes commonly grade into one another, and more than one mechanism of movement can occur during a single slope-failure event. All classifications consider *how* the material moves, what *types of material* move, and the *rate of movement* of the material.

12.09.b1 Grand Junction, CO

Mechanism of Movement

◄ Geomorphologists classify slope failures primarily by how the material moved. Rocks and other material can *fall* off cliffs, can *slip* along joints, cleavage, or bedding planes, can *topple* over, or can do all three.

► Other slope failures involve the slow *creep* of the uppermost soil cover or the *flow* of material, as during turbulent flows of mud, rocks, and other debris. This brown mud flowed during an earthquake.

12.09.b2 Fort Funston, CA

12.09.b3 Monument Valley, UT

Type of Material

◄ Some slope failures involve slabs of *solid rock* or large pieces of broken rock derived from cliffs and rocky hillslopes. Such rocks can further break apart after they begin to move.

► Many slope failures mobilize *unconsolidated material* that is stripped from hillsides. Material can include soil, loose sediment, pieces of wood and other plant parts, boulders, and other types of loose debris.

12.09.b4 Grand Canyon, AZ

12.09.b5 Venezuela

Rate of Movement

◄ Fast rates—Another important factor is the *rate of movement* of the material. Some slope failures start in an instant and send material downhill at hundreds of kilometers per hour, or at least too fast for people to outrun.

► Slow rates—Other mass movements are more gradual and move downhill at rates that are imperceptible to an observer. This slow-moving mudflow carries trees, some tilted, along for the ride.

12.09.b6 Slumgullion, CO

Submarine Slope Failures

Slope failure is not restricted to the land; it can also occur on steep or even gentle slopes on the seafloor. Such *submarine slope failures* can be caused by overloading of sediment on a slope or in a submarine canyon. They can also be triggered by shaking during an earthquake, volcanic eruption, or storm. Various types of slope failure occurred off the southwestern coast of the Big Island of Hawaii, forming the large mass of debris shown here in green.

Big Island of Hawaii

Landslide

50 km

12.09.t1

Before You Leave This Page Be Able To

✓ Describe slope failures and some ways they are expressed in the landscape.

✓ Summarize the classification of slope failures, and describe the different types of movement, types of material, and rates of movement.

12.9

How Does Material on Slopes Fall and Slide?

SOME SLOPE FAILURES involve materials falling off a cliff or sliding down a slope. These mechanisms commonly involve rock and pieces of rock, but they can also involve materials that are less consolidated. Rocky cliffs and slopes might appear to be immune to slope failure because they consist of hard bedrock, but they can fail spectacularly and catastrophically.

A   What Happens When Rocks Fall or Slide?

Rocks and other material can fall from cliffs or can slide on joints or other weak planes. Falling and sliding rock masses may begin as relatively intact blocks, but they commonly break apart as they begin moving or when they hit the bottom of a cliff. Some slides rotate as they move. Others simply slide down the hill. One type of rock failure can lead to another because a rock fall may remove some support, causing higher parts of a cliff to slide.

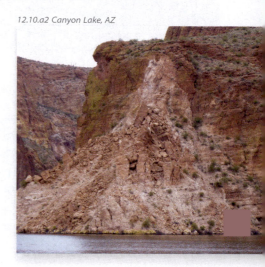
12.10.a2 Canyon Lake, AZ

Rock Falls and Debris Falls

In a *rock fall*, large blocks or smaller pieces of bedrock detach from a cliff face and fall to the ground. Rock falls can be triggered by rain, frost wedging, thawing of ice that had held rocks to the cliff, an earthquake, stream erosion, or human construction that undercuts a cliff. Less consolidated debris, including loose sand, can also fall off a cliff. Part of the cliff in this photograph (▶) has collapsed, producing a rock fall composed of large, angular blocks. As they fell, and during impact with the ground, the rocks fractured and broke apart, with most remaining as a large mass of shattered rocks. Some loose blocks fell, rolled, and slid downhill.

12.10.a1

12.10.a4 Naxos, Greece

Rock Slides

In a *rock slide*, a slab of relatively intact rock detaches from bedrock along a bedding surface, preexisting fault, joint, or other discontinuity that is inclined downslope. As it slides, parts of the slab typically shatter into angular fragments of all sizes, but large blocks can remain relatively intact. In the rock slide shown to the right (▶), road construction undercut these sedimentary rocks, which had layers and joints that were inclined downslope. At some point after construction, the rocks slid along the layers and into the road.

12.10.a3

12.10.a6 Fairbanks, AK

Rotational Slides (Slumps)

In some slides, rock layers and other material rotate backward as they slide. Such *rotational slides* move along one or more curved slip surfaces. This type of slide, also called a *slump*, can occur in bedrock or less consolidated material. Individual slices can remain relatively intact or can break and spread apart. The rotational slides in this photograph (▶) offset and tilted the land surface, and houses, as a result of shaking and loss of cohesion during the 1964 Alaskan earthquake. As the slide masses slumped downhill, they rotated, tilting back, opposite to the direction of movement.

12.10.a5

B What Is the Geometry of a Rock Slide?

A combination of geological circumstances is required to detach a slab of rock and create a rock slide. It requires sufficiently steep slopes along with bedding planes, joints, or other flaws that are inclined downslope.

1. Many rock slides occur in bedrock with discrete layers that differ in rock type and therefore in strength. Such rock layers are most common in sedimentary rocks, but they are also present in many volcanic and metamorphic sequences. In this figure, a sequence of different sedimentary layers has beds inclined downhill, toward a small stream valley.

2. At their upper end, most rock slides detach along a series of preexisting joint or fault surfaces. In other cases, as in this example, the stresses that build up in the rock before it slides are enough to form new joints, allowing the rock slide to detach from bedrock.

3. Detachment of the base of the rock slide from underlying bedrock commonly occurs along a layer of weak rock, such as shale, mudstone, or salt. This weak layer may allow the overlying slab to slide fairly easily and remain partially intact as it moves downhill.

4. To be able to slide down the dip of the layers, the upper layer must have space downhill in which to slide. That is, the layers will probably not slide if they are supported by more rocks in a down-dip direction. In this example, a stream has eroded a low area, giving the sliding rock slab somewhere to go.

5. Although not shown here, rock slides can slip along joints, fault surfaces, cleavage, or some other discontinuity, rather than bedding surfaces. Preexisting faults of the proper orientation are especially susceptible to rock slides because they are planar, fairly continuous, and structurally weak.

7. Most slides leave a linear or curved scar, or *scarp*, on the hillslope, marking where the slide pulled away from the rest of the hill. This upper end is also called the *head* of the slide.

6. The leading edge of a slide, the *toe*, can overrun the land surface in front of the slide.

12.10.b1

The Vaiont Disaster, Italy

In 1960 a dam was built across the Vaiont Valley in northeastern Italy. As seen in the figure below, this valley runs along the bottom of a syncline, where the rocks have been folded downward and dip into the valley from both sides. The rocks are mainly limestone but with interlayered thin beds of shale and sandstone. Some of the limestone beds contain caverns formed when groundwater dissolved the rock. Joints in the rocks run both parallel and perpendicular to the bedding planes.

During August and September 1963, three years after the dam was completed, heavy rain fell in the area. One day in October, the south wall of the valley failed and slid into the reservoir behind the dam. The slide was 1.8 km high and 16 km wide with a volume of 240 million cubic meters. The slide moved along shale layers that parallel the bedding planes in the limestone.

As the slide moved into the reservoir it displaced an equal volume of water, forcing a surge of water 240 m above lake level onto the village of Casso on the northern side of the valley. Waves within the reservoir killed 1,000 people. Waves 100 m high swept over the dam. Although the dam did not fail, the water that overtopped the dam killed 2,000 people living in villages below the dam.

Landslide
Block

12.10.t1

Before You Leave This Page Be Able To

✓ Sketch and describe a rock or debris fall, a slide, and a rotational slide.

✓ Sketch how the geometry of layers, faults, and other features could allow a rock slide to begin.

✓ Describe the Vaiont landslide disaster and factors that caused it to happen.

12.10

12.11 How Does Material Flow Down Slopes?

SLOPE FAILURE CAN MOBILIZE WEAK MATERIAL, including soil, sediment, broken rock, and loose debris. Material flows downhill if it is poorly attached to a hillside, is internally too weak to resist the downward pull of gravity, and the slope is steep enough to allow flow. If the material incorporates some water, it becomes weaker and better able to flow. Such material can move rapidly, as in a debris flow, or can creep down the hill at a nearly imperceptible rate. All these types of slope failure involve the flow of material.

Creep and Solifluction

12.11.a1

Creep and *solifluction* are very slow, continuous movement of soil, regolith, or weathered rock down a slope. Both processes occur on almost all slopes. At the surface, evidence for creep is expressed in bent or leaning trees, warps in roads and fences, and leaning utility poles. At roadcuts, creep commonly causes bedding and other layers in the subsurface to bend downhill (▶). A type of creep at high latitudes or elevations is called solifluction (sometimes called *gelifluction* in polar regions). Here, poorly drained ground freezes for most of the year (permafrost) but near the surface, ground thaws (called the active layer) during the summer. This layer of saturated ground can move downslope in irregular, often overlapping lobes as pictured here (▶).

12.11.a2 Colorado Springs, CO

12.11.a3 Norway

Debris Slides

12.11.a4

Soil, weathered sediment, or other unconsolidated material can move downslope as a *debris slide*. Debris slides are usually less than 10 m thick and leave behind a low *scarp*. A debris slide moves downhill partly as a sliding, coherent mass and partly by internal shearing and flow. The debris slide in this photograph (▶) occurred in clay-rich glacial sediment and displaced pieces of road down the mountain.

12.11.a5 British Columbia

Earth (Mud) Flows

12.11.a7 Thunder Mtn., ID

Earth flows are flowing masses of weak, mostly fine-grained material, especially mud and soil. The material moves like thick, wet concrete, generally slowly enough to outrun, but it contains enough water to be slightly fluid. Earth flows have more mud and other fine-grained material than rocks, and also can be called *mudflows*. The earth flow in this photograph (▶) mobilized clay-rich altered volcanic and sedimentary rocks, as well as a surface veneer of angular rocks. The earth flow moved downhill until it reached and dammed a larger valley and completely flooded a small village.

12.11.a6

Debris Flows

Debris flows are wet, downhill-flowing slurries of loose mud, soil, volcanic ash, rocks, and other objects picked up along the way. Some contain only a little water, whereas others are water rich and flow like a thick soup. Debris flows, especially their mud-dominated varieties, are called mudflows by the media, and can move rapidly. They often result from heavy rains that saturate the soil and other loose materials. Debris flows are thick and more dense than water and so can support and carry large boulders or even houses. In this photograph (▶), debris-flow deposits from the 1999 Venezuelan disaster, discussed in the opening pages of this chapter, consist of angular blocks, some several meters across. During transport, the rocks were enclosed in a matrix of wet mud and other debris.

12.11.a8

12.11.a9 Venezuela

Rock and Debris Avalanches

Debris avalanches are high-velocity flows of soil, sediment, and rock that result from the collapse of steep mountain slopes. A debris avalanche moves down the slope, in many cases traveling considerable distances down valleys and across relatively gentle slopes. A *rock avalanche* occurs when a rock mass falls off a cliff face and shatters on contact, sending a turbulent jumble of rock fragments, some bigger than cars, flowing downhill. They can flow at over a hundred kilometers per hour, fast enough to continue flowing uphill when they encounter a topographic obstacle. Both types of avalanches are often triggered by earthquakes and volcanic eruptions, and can kill thousands of people at a time. In 1970, an earthquake shook loose a large chunk of an Andean mountain, causing a rock and debris avalanche (▲) that buried a town (not visible in this photograph) and 18,000 of its inhabitants.

12.11.a10

12.11.a11 Huascaran, Peru

Classification of Slope Failures

WET

Solifluction

Debris Slide

Debris Flow

Rotational Slide

Earthflow

DRY

Creep

Rock Slide

Rock Fall and Debris Avalanche

SLOW   FAST
12.11.a11

◀ This diagram summarizes the classification of slope failures based on the type of material involved and the mechanism of movement, but viewed in the context of the *rates* of movement and whether movement occurred under *dry* or *wet* conditions. In this table, fast means you cannot outrun it, moderate means you have a chance to escape, and slow means you might have to stay for a while to see anything happen. For the slowest rates, you would not be able to see the movement in real time, only the results of months, years, or decades of accumulated movement.

Before You Leave This Page Be Able To

☑ Sketch and describe what happens during the following: creep, debris slide, earth flow, debris flow, and debris avalanche. Compare how each of these features flows.

☑ Draw a table summarizing different types of slope failures.

12.12 Where Do Slope Failures Occur in the U.S.?

LANDSLIDES AND OTHER SLOPE FAILURES have destroyed large parts of cities in some countries and have killed more than 30,000 people in a single event. Each year in the U.S., landslides cause 25 to 50 deaths and result in more than $3 billion in damage. What slope-failure events have affected the U.S.? What is the likelihood they will occur again? Which areas are most at risk in the future?

A Where and How Have Large Slope Failures Affected the U.S.?

There have been countless landslides, debris flows, and other slope failures. The landslide that accompanied the 1980 eruption of Mount St. Helens was the largest landslide in U.S. history. It generated enough debris to fill 250 million dump trucks. Other slope failures are described below.

12.12.a1 Denali Region, AK

A large (7.9 magnitude) earthquake struck the Denali (Mount McKinley) region of south-central Alaska in 2002. Ground shaking associated with the earthquake caused huge slope failures off the region's steep mountains. Spectacular rock falls, rock slides, and debris avalanches slid down the steep slopes and flowed across and buried the Black Rapid Glacier. The region is prone to slope failures partly because tectonics has rapidly uplifted the mountains and formed an array of joints, faults, and tilted rock layers.

12.12.a2 Pacific Palisades, CA

Landslides and debris flows are common in southern California (◄), where some houses have been built on risky sites that have been undercut by waves, streams, and roads, causing unstable slopes to collapse downhill.

12.12.a3 Tully Valley, NY

The largest landslide to affect New York in 75 years moved through the Tully Valley (◄), near Syracuse, in 1993. The earth flow mobilized over a million cubic meters of weak clays that had been deposited in a glacial lake. Similar glacial lakebeds occur in many parts of the region.

12.12.a4 Gros Ventre, WY

The Gros Ventre slide (◄) is one of North America's largest historic landslides. During this 1925 rock slide, a slab of limestone more than 1.5 km (1 mi) long broke loose along weak bedding planes that dipped toward, and were undercut by, the Gros Ventre River of Wyoming.

12.12.a5 Thistle, UT

The most costly landslide in U.S. history devastated the small Utah town of Thistle in Spanish Fork Canyon in 1983. The landslide (◄) moved slowly downhill, reaching a maximum rate of only 1 m (3 ft) per hour. It severed railroad service between Denver and Salt Lake City, flooded two major highways, and formed a new lake where the town had been. A railroad tunnel now cuts through the light-colored landslide material.

B What Is the Potential for Landslides and Debris Flows in the U.S.?

All states receive some damage from landslides and debris flows, but not all areas have the same potential hazard. The potential for landslides is highest near steep cliffs, hills, and mountains, such as in the Rocky and Appalachian mountains. Landslides are most common in areas with weak, heavily weathered materials or recent tectonic activity.

Potential for Landslides and Debris Flows in the U.S.

This USGS map portrays the landslide hazards of the lower 48 states. Red areas represent the greatest hazards, followed by yellow and then green. Other areas have less potential for landslides or are unstudied. What landslide potential exists where you live or visit?

Many parts in the Pacific Northwest experience landslides because of the many steep mountains, heavy rainfall, and rainfall that melts snow cover. The region also contains areas with high potential for debris flows, especially along slopes and valleys connected to the active Cascade volcanoes. An eruption on a snow-capped peak can unleash large debris flows.

A high potential for landslides and debris flows occurs in the Rocky Mountains and along the Wasatch Front, the steep mountain front that flanks Salt Lake City, Utah.

Landslide hazards in the central U.S. are mostly along steep bluffs that flank the rivers, or in areas, such as the northern Great Plains, underlain by weak materials.

12.12.b1

The coastal parts of central and southern California have high landslide potential because of steep mountains, earthquakes, and coastal erosion that undercuts weak material along hillslopes overlooking the shoreline. Debris flows are also common, especially in the high mountains flanking Los Angeles, California. These recently uplifted mountains have very steep slopes and receive locally intense rainfall. Wildfires worsen the situation.

In the east, landslides are common in the Appalachian Mountains, where landslides mobilize soil or occur along weaknesses in folded, faulted, and weathered rock layers. Shales, because of their inherently weak character, are especially prone to landslides.

Florida and the coastal plain of the southeast Atlantic seaboard have some of the lowest potential for landslides because the region lacks steep slopes.

250 km

Landslides and La Conchita

12.12.t1

The coastal community of La Conchita in southern California was partially overrun by a landslide in 2005, which was a repeat of one in 1995 (right). The landslide mobilized poorly consolidated sediment along the steep bluffs overlooking the town. The material flowed down into the community, burying and destroying a number of houses. Although the 1995 landslide had previously destroyed part of the town, houses were rebuilt in the area after 1995, only to be destroyed by the 2005 landslide. The 2005 landslide remobilized parts of the 1995 landslide and also incorporated new material farther back into the cliff. Would this be a good place to rebuild again? How much should hazard insurance cost people who live here?

Before You Leave This Page Be Able To

- ☑ Briefly describe factors involved in landslides in the U.S.
- ☑ Summarize some factors that make some areas of the U.S. have high risks for landslides or debris flows.
- ☑ Identify whether you live in an area with a high potential for landslides, and list some possible factors why your area has this potential.
- ☑ Describe what happened at La Conchita. Was it a good idea to build or rebuild there?

12.12

How Do We Study Slope Failures and Assess the Risk for Future Events?

HOW CAN WE LIMIT DAMAGE CAUSED BY SLOPE FAILURES? Geomorphologists observe and monitor active slope failures to better understand how such systems operate. They then use this information to recognize areas that either have experienced past slope failure or are susceptible to future slope failures. The most hazardous sites have dangerous combinations of geomorphic factors, such as steep slopes and weak materials.

A How Do We Recognize and Monitor Active or Recent Slope Failures?

To limit property damage and the potential for loss of life, geomorphologists study and monitor active and recently active slope failures. The resulting observations and measurements provide a solid basis for investigating the processes, causes, and hazard assessment of destructive slope failures.

12.13.a1 Venezuela

After a slope-failure event, geomorphologists conduct field studies to document the distribution, size, and other characteristics of the event, largely to understand what happened and to assess the potential for future events. This area (◄) was destroyed by debris flows.

12.13.a2 Mount Rainier, WA

Geomorphologists use surveying equipment to document topography, especially changes in slope, of hazardous areas. From such measurements, they construct maps and cross sections showing where slope failures are most likely. Finally, geomorphologists help devise plans to minimize losses from slope failure.

B How Do We Recognize Prehistoric Slope Failures?

Geomorphologists identify prehistoric slope failures by observing the topography, surface and subsurface distribution of rock types, condition of the rocks, and geometry of layers and other geologic structures.

12.13.b1 Hat Creek, British Columbia, Canada

Landslides and other slope failures can disrupt or even overrun existing topography, forming a random-looking assemblage of humps and pits (◄). We call this type of landscape *hummocky topography*, and it is often indicative of some type of slope failure.

12.13.b2 Venezuela

Each slope failure leaves behind characteristic deposits. Geomorphologists observe modern deposits and then use these characteristics to recognize deposits of past events that were similar to the modern one. These deposits (◄) were left by debris flows.

12.13.b3 Mount Hood, OR

This flank of Mount Hood (◄), a Cascade volcano, contains large deposits composed of broken, angular fragments of rock. These deposits formed when parts of the summit collapsed, slid, and tumbled downward, at some time in the recent past.

12.13.b4 Mount St. Helens, WA

Debris flows, debris avalanches, and landslides are sometimes recognized because they carried distinctive rock types into areas where such rocks do not otherwise occur. Rocks in this debris pile (◄) were derived from the distant Mount St. Helens, Washington.

C How Do We Assess an Area's Potential for Slope Failure?

Geomorphologists try to assess the likelihood that a location will suffer the disastrous consequences of slope failure by examining evidence of past slope failures in the area or in adjacent areas that have a similar setting. They also evaluate the steepness of slopes, including any recent changes, and any other factors leading to slope failure.

12.13.c1 Nepal

Evidence of Past Slope Failures

One of the best indications for potential slope failure is evidence of past failures. The more recent the failure, the more likely such an event will recur in the near future. This part of the Himalaya (◄), with steep hillslopes scarred with slope failures, looks risky.

12.13.c2 Durango, CO

Situated in Area with Known Problems

The slope angle of hillsides and mountains (►) is clearly a key factor in the potential for slope failure but must be evaluated in the context of the types of materials and subsurface structures present.

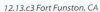

Steepness of Slopes
12.13.c3 Fort Funston, CA

A site may be at risk if its geologic setting is similar to other slopes that have failed. Part of this cliff (◄) collapsed during an earthquake, and nearby parts have the same steep slopes, weak rocks, and position that can be undercut by the ocean waves.

12.13.c4 Zermatt, Switzerland

Recent Changes in Slope

Natural processes, over time, act to adjust slopes to the appropriate, stable angle, but this equilibrium can be upset if the slope is steepened or undercut by natural or human activities. This rock slide (►) failed in 1991, burying 31 houses.

12.13.c5 Capitol Reef NP, UT

Conditions of Material

Another factor is the nature of material on a slope (◄): whether material is loose sediment or solid bedrock, and is resistant to erosion or relatively weak. The presence, spacing, and orientation of geologic structures can weaken rocks and facilitate downslope slippage of materials.

12.13.c6 Augustine volcano, AK

Potential Triggers

If other factors are equal, an area is at higher risk for slope failure if there are frequent events, such as volcanic eruptions or earthquakes, that could trigger slope failure. Steep slopes with loose material on an active, shaky volcano are trouble (►).

The Blackhawk Landslide

A huge lobe of shattered rock lies in a valley north of the San Bernardino Mountains, northeast of Los Angeles, California. This mass has hummocky topography and large, shattered pieces of rock that are different than rocks in adjacent areas. This feature, known as the *Blackhawk Landslide*, formed in prehistoric times when a large segment of the mountains collapsed, shattered, and flowed as debris avalanches more than 10 km (6 mi) out into the valley.

12.13.t1

Landslide

1 km

Before You Leave This Page Be Able To

✔ Describe some ways that geomorphologists investigate slope failures.

✔ Summarize characteristics used to identify prehistoric slope failures.

✔ Summarize some aspects that might indicate that an area has a high potential for slope failure.

12.13

What Is Happening with the Slumgullion Landslide in Colorado?

THE SCENIC SAN JUAN MOUNTAINS OF COLORADO contain the Slumgullion landslide, one of the best-studied landslides in the world. The landslide has been moving for more than 1,000 years, and part of it is still moving, allowing geomorphologists to examine up close the processes of a relatively slow slope failure.

A What Is the Setting and Morphology of the Slumgullion Landslide?

12.14.a2 Slumgullion, CO

1. The Slumgullion landslide is a conspicuous feature in the landscape of the San Juan Mountains of southwestern Colorado. It begins among high peaks of altered volcanic rocks, snakes downhill, and spreads out when it reaches the valley bottom. The landslide shows as a tan-fringed mass in this photograph (◄).

3. The head (top) of the landslide is a steep landslide scarp within volcanic rocks. Interactions with water altered the volcanic rocks into a weak, clay-rich material. This material is too weak to support the steep slope, and so it collapsed downward, starting the landslide.

1 km

12.14.a1

2. The flowing material in the landslide is mostly derived from the clay-rich, altered volcanic rocks and so is weak and fine grained; it could be called an earth flow. The abundant water in the landslide reduces its strength, allowing it to flow down relatively gentle slopes. Steeper slopes at the scarp, combined with weak materials and abundant water, produced a setting favorable for starting the landslide.

4. When it reached the valley, the landslide blocked the Lake Fork of the Gunnison River, forming Lake San Cristobal, Colorado's second-largest natural lake. The river flows from right to left in this view.

Parts of the Landslide

12.14.a3

Youngest Unit
Middle Unit
Oldest Unit
○ Radiocarbon Sample Locality

Lake Fork

107°15

38°

149

149

Lake San Cristobal

1 km

N

12.14.a4 Slumgullion, CO

5. When USGS geomorphologists investigated the landslide, they discovered that it actually consisted of three different ages of material (◄): a young, central part that was still active, and two older parts. In the active, center part, continued movement has tipped over trees and caused damage to a road that by necessity crosses the flow (►).

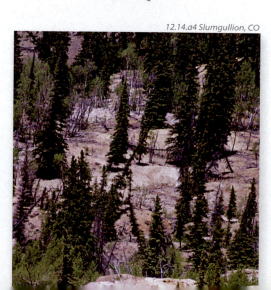

B What Features Are Associated with the Landslide?

As the landslide moves down the hill, shearing and flow form different features along different parts. USGS scientists documented and interpreted the various features and how they formed, in terms of where they were on the landslide.

12.14.b1 Slumgullion, CO

12.14.b2 Slumgullion, CO

12.14.b3 Slumgullion, CO

The upper part of the landslide contains fissures—*scarps*—where the landslide pulled away from the adjacent rocks or where there was internal stretching of the flowing mass. This photograph shows a still-active scarp.

The sides and middle of the landslide contain linear to gently curved zones of shearing, where the fastest-flowing parts of the landslide move past slower moving or stationary material, such as this trough on the right side.

The toe of the landslide commonly overruns areas downhill, as occurred here, where a slab of soil, bushes, and other material has been thrust over land that has not slipped or not slipped as much as the material on top.

C What Other Studies Did Geomorphologists Do to Understand the Landslide?

To study the landslide, the scientists combined surveying and aerial photographs to construct a very detailed topographic map. From various types of data, the geomorphologists calculated that the weak materials in the landslide are moving down a slope of 7° to 10°, much less than a typical angle of repose for normal, dry materials. The landslide is moving quite fast and continues to travel downhill as scientists study it.

1. Geomorphologists investigated the Slumgullion landslide not only to understand this particular landslide but to learn how landslides in general operate. They examined temperature and precipitation records to see if the landslide behaved differently during warm or wet times than during cold or dry periods. Velocity of the landslide decreased in the winter, when the formation of ice reduced the amount of water available to the landslide. The landslide also moved more slowly when conditions were drier.

2. Geomorphologists used GPS and other surveying methods to measure how fast different parts of the landslide were moving. On this map (►), longer red lines indicate faster rates of downhill flow.

3. The fastest rates of movement (the longest lines on this map) are in the middle segment of the landslide, where the landslide is moving as fast as 7.8 m (26 ft) per year.

4. The upper part of the landslide is moving more slowly, locally less than one meter per year.

Contours in Meters

3400
3350
3300
3250
3200
3150
3100
3050
3000

12.14.c1

0   30 m   Direction and Amount of Displacement

500 m   Map Scale

5. Geomorphologists also tested physical properties, including strength and moisture content, of the landslide material. They then used these data and computers to model how the landslide is moving.

Youngest Unit / Middle Unit / Oldest Unit

DEPTH(cm)
0
10
20
30

C
A C
A C
A C
A E B C
A E B C

12.14.c2

6. To constrain the three ages of landslide material that were identified and mapped, scientists compared the soil development in the three parts. As we discuss in another chapter, soil develops layers, called *horizons*, over time, with the A horizon on top, B horizon in the middle, and C horizon near the base. The oldest part of the landslide had well-developed A and B horizons, whereas the youngest part was too young to have developed either an A or B horizon. Carbon-14 ages obtained from samples of charcoal and wood within different parts of the landslide indicate that the landslide first formed about 1,000 years ago, yet it is still very active today.

Before You Leave This Page Be Able To

✓ Describe the setting and morphology of the Slumgullion landslide.

✓ Summarize the types of studies geomorphologists conducted on the landslide and what each type of study revealed about what was happening with the landslide.

12.14

What Happens in the Headwaters of Streams?

STREAMS CAN BEGIN in almost any setting, from a small hill in a pasture to a large, snowy mountain. In either case, the place where a stream starts is its *headwaters*. In its headwaters, a stream is fed by rain, snowmelt, groundwater, or some combination of these. The headwaters of many large rivers are in mountains, where streams are steep and actively erode the land with turbulent, fast-moving water, producing distinctive landforms.

A How Do Streams Start?

A stream does not start with a fully formed channel full of water, but instead it grows incrementally as surface runoff becomes concentrated into channels—is *channelized*. Smaller channels join others until a stream forms.

Channel Formation

13.06.a1

◄ Rainwater causes *splash erosion* as it hits the ground, and water flowing over the surface as *overland flow* causes subtle *sheet erosion*. Water tends to accumulate in natural cracks and low spots, such as these small channels, rather than spreading uniformly across the land.

Concentrated flow erodes or dissolves materials, especially those that are weak or loose, eventually carving a small channel or *gully*.

13.06.a2

Once formed, a channel accommodates runoff within its small drainage basin. The increased flow causes further *gully erosion* and channel deepening. *Headward erosion* is the cutting into the area between the channels (i.e., toward the divide) at the gully's headwater and can lead to *stream capture*, the natural diversion of water from one stream into another.

Channels occur at all scales. Microscopic channels feed into small channels that feed into larger ones, ultimately forming a stream. The spacing and geometry of the channels are influenced by the steepness of the slope, type of material in the slope, type and density of vegetation, and other factors.

B What Landforms Characterize the Headwaters of Mountain Streams?

Mountain streams begin in bedrock-dominated areas with relatively high relief and, in many cases, high elevation. The energetically moving water wears rock down and sculpts the bedrock into steep landforms with moderate to high relief.

1. The headwaters of many streams are in high mountainous areas, where they derive their water from rainfall, melting ice and snow, or mountain springs. Other streams originate in lower, flatter areas and are supplied by precipitation, lakes, springs, or the joining of small, local channels.

13.06.b1

13.06.b6 Ice Lake, Silverton, CO

2. The area below (▼) is the headwaters of several streams. It is near the top of a mountain, receives abundant snow and rain, and has several lakes left behind when glaciers melted away fairly recently.

3. Streams in mountains typically have steep gradients that result in fast-moving, turbulent flow (▼). Many have eroded down, or *incised*, into bedrock.

4. Steep and energetic streams can carry large materials. At times, this stream (▼) has a *competence* to carry large boulders.

5. With time, mountain streams begin to widen their valleys and deposit a thin, winding ribbon of stream gravels along which they flow (▼).

6. Lakes are common in mountains where water is impounded by some obstruction, such as a landslide, or where water fills a natural low spot (▲). If a lake is created by a human-constructed dam, it is a *reservoir*.

b2 Beartooth Plateau, WY and MT

13.06.b3 Boulder Canyon, MT

13.06.b4 Central CO

13.06.b5 ANWR, AK

C How Do Rapids and Waterfalls Form?

Streams that flow in an area of mountains and hills commonly have *rapids*, a segment of rough, turbulent water along a stream. Some segments can be so steep that water cascades through the air, forming a *waterfall*.

13.06.c1 Grand Canyon, AZ

13.06.c2 Grand Canyon, AZ

13.06.c3 Gulfoss, Iceland

Most rapids (▲) develop when the gradient of a stream steepens or the channel is constricted by narrow bedrock walls, large rocks, or other debris that partially blocks the channel. Many rapids form where tributaries have deposited fans of debris that crowd or clog the main channel. These obstructions cause water to flow chaotically over and around obstacles, creating extreme turbulence.

The obstruction forming a rapid generally causes water upstream to slow down and "pile up," producing a relatively smooth and slow-moving segment called a *pool*. The photograph above, looking downstream, shows a pool upstream of a rapid. Many streams have this characteristic alternation of rapids and pools, and so are called *pool-and-riffle streams*.

Most waterfalls are awe-inspiring, as water cascades over a ledge or cliff (▲). Where a stream has an abrupt change in gradient, such as at a waterfall, it is called a *nickpoint*. A nickpoint can be distinct or more subtle, and generally indicates some difference in erosion above and below the nickpoint. With continued erosion, the nickpoint, and in this case a waterfall, migrates upstream over time.

D What Features Develop When a Stream Exits from the Mountains?

As mountain streams flow toward lower elevations, they interact with tributaries and commonly decrease in gradient as they pass through foothills or mountain fronts. In response, they form other types of landforms.

Where a steep, narrow drainage enters a broader, more gentle valley, coarse sediment carried by running water or by muddy debris flows piles up just below the edge of the slope at the decrease in gradient, forming a fan-shaped feature called an *alluvial fan* (▶).

13.06.d1 Death Valley NP, CA

13.06.d2 Waiapu River, New Zealand

When a mountain stream reaches less confined spaces, it commonly spreads out in a network of sediment-filled braided channels (◀). These channels are not strongly incised, so the stream spreads out. As it does, the stream deposits sediment along its channel and over a broad floodplain.

How Do Mountain Streams Get Sediment?

Mountain streams are energetic primarily because their channels have steep gradients. Erosion dominates over deposition, forming deep *V-shaped valleys* with waterfalls and rapids. Steep valley walls promote landslides and other types of slope failure that widen the canyon and deliver abundant material to the stream for removal (▶). Erosion of materials to the side of a stream is *lateral erosion*. Soil and other loose materials on hillslopes can slide downhill toward the drainage. Tributaries carry debris flows and floods that scour their channels, providing even more sediment. Sediment in mountain streams ranges from bus-sized or car-sized boulders down to silt and clay. Larger clasts start out angular but begin to round as they are transported, or as they are struck by stones and other sediment within the turbulent waters.

13.06.t1 Tibet-Nepal Border Region

Before You Leave This Page Be Able To

☑ Describe how stream channels form.

☑ Describe some of the landforms associated with the headwaters of mountain streams.

☑ Describe why sediment is deposited along mountain fronts in alluvial fans.

☑ Describe how mountain streams get their sediment.

13.6

Glaciers and Glacial Landforms

GLACIERS ARE A POWERFUL FORCE in high-latitude and high-elevation areas, but in the past they also covered most of Canada, the northern parts of the U.S., and large areas of Europe and Asia. Erosion and deposition associated with glaciers reshape the underlying landscape and form many distinctive landforms, some of which indicate that ice once covered an area, even if none remains today. What features do glaciers form, and how can we use these features to reconstruct the distribution and flow directions of ice thousands or millions of years ago?

The northeastern U.S. and adjacent Canada, shown in this shaded relief map, have a striking collection of features on land and along the coast. The map extends from North and South Dakota (in the northwestern corner of the map) across the Great Lakes as far east as Maine and Virginia. The northern part of the map includes southeastern Canada.

Huge, smooth troughs (each labeled with a **T** on the map) cut across the landscape. Examples are near the northwestern corner of the map, in western Minnesota and the Dakotas, and southwest of Lake Erie, in Ohio and Indiana.

What caused these smooth areas of the landscape, and is the process still occurring?

Curiously curved ridges (labeled with an **R** on the map) cross some of the smooth areas, and are especially noticeable southwest of Lake Erie and Lake Michigan.

What are these ridges, how did they form, and what do they tell us about the geologic history of this region?

14.00.a1

TOPICS IN THIS CHAPTER

North of the Great Lakes, the landscape of Canada is rough on a small scale, containing many lakes, more than in most parts of the U.S.

Why does this region have so many lakes, and when and how did this landscape form?

The Finger Lakes of Upstate New York (▼) are elongated, like fingers, in a north-south direction and are among the deepest lakes in the U.S. They were carved by glaciers that no longer remain in the region.

How do glaciers form lakes, and how do we recognize a glacially formed landscape?

14.00.a2

Long Island, N.Y., largely consists of loose sedimentary materials that form two main ridges that define the island. From their character, these materials were deposited along the front of a glacier and from streams that emerged from the glaciers.

Did glaciers really get this far south, why did they grow so large, and what caused them to go away?

Ice Ages, Glaciers, and Lakes

Landscapes in the Great Lakes region contain evidence that huge ice sheets once flowed across this part of the continent—in the recent geologic past. This conclusion arises from comparing the distinctive landscape features and their associated sedimentary deposits with those observed today near currently active glaciers. With glaciers, as with most geologic features, the present is the key to interpreting the past.

During the last two million years, Earth has been experiencing an *Ice Age,* during which large regions of the Northern Hemisphere, as well as Antarctica, were covered year-round with ice and snow. Where the ice was thick enough, or rested on a steep enough slope, it moved downhill as a mass of flowing ice called a *glacier.* Some glaciers were small and restricted to mountain areas, whereas others covered large parts of the continents, forming *continental ice sheets.* Continental ice sheets flowed southward from Canada and smoothed off and carved grooves into the underlying landscape by grinding ice, rocks, and sand against the bedrock. The smooth troughs on this map were carved by these continental glaciers. Southward-flowing parts of the glaciers carved the valleys that hold the Finger Lakes of New York and some other lakes in the region. These glaciers advanced down the Hudson River, between the Finger Lakes and Long Island, and covered wide areas on both sides of the valley.

As the climate warmed over the past 20,000 years, the ice sheets and glaciers melted and covered less area. Rocks and other sediment carried in the ice were dropped along the front of the melting glaciers, forming a series of curved ridges south of the Great Lakes. The ice also left piles of glacially derived sediment on Long Island and Cape Cod. As the glaciers melted back, they uncovered a region with numerous topographic lows scoured by the ice. Many of these areas also lacked an established drainage network, because they had recently been covered with ice. For these reasons, the low spots became the countless lakes of Canada, Minnesota, Wisconsin, Michigan, and nearby states.

Water released from melting ice filled the glacially carved troughs and other low areas, resulting in the Great Lakes. The waters carved new river valleys or modified preexisting valleys and flowed into the sea, causing a rise in sea level.

CANADA

Finger Lakes

Cape Cod

Long Island

Delaware Bay

50 km

10 km

14.0

Coasts and Changing Sea Levels

PEOPLE LIVE ALONG COASTS for a variety of good reasons. Oceans moderate climate, supply fish and other sea resources, provide transportation for people and goods, and make available a range of recreation options. Living along a coast also presents challenges. Coasts are not static, and they can change dramatically as the result of a single storm. Coasts also experience waves, tides, and shifting beach materials. Past sea levels have been more than 200 m (660 ft) higher than today and about 120 m (390 ft) lower, flooding the continents or exposing continental shelves. The most rapid sea level changes are related to changes in glaciers and continental ice sheets.

15.00.a2 Cape Fear, NC

The East Coast of the U.S. (◄) has mostly gentle topography along the length of the coastline with a variety of different features found at different segments. In the figure below, observe the features along each part of the coast, from Georgia on the left to Maine on the right (▼).

What types of features are common along coasts, and why does a coastline vary in the features?

The Pacific Coast has a much different appearance (►). Instead of gentle topography along the coast, the Pacific Coast commonly has cliffs that drop off into the ocean.

What causes some coastlines to have subdued topography and others to have cliffs and steep slopes along the beach?

15.00.a3 Big Sur, CA

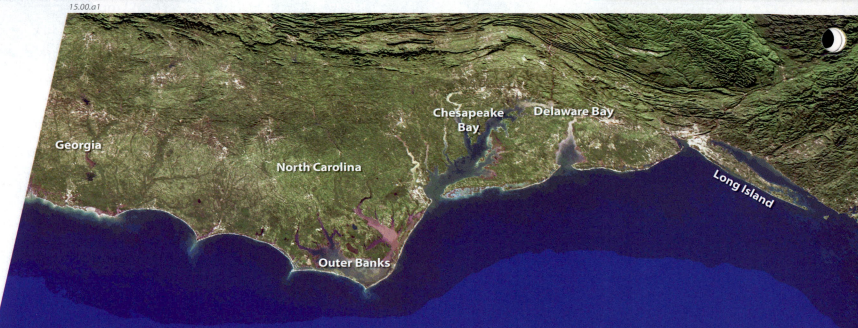

15.00.a1

Georgia

North Carolina

Outer Banks

Chesapeake Bay

Delaware Bay

Long Island

The Outer Banks of North Carolina, including Cape Hatteras, consists of a chain of linear islands offshore of the main coast. These islands help shield the main coastline from storms, but are themselves very vulnerable to hurricanes and other storms coming from the ocean.

What types of features are common along coasts, and why does a coastline produce these smooth areas of the landscape? Is the process still occurring?

Chesapeake Bay and Delaware Bay are large *embayments*, places where the coastline curves inward toward the land. In their outlines, they resemble stream valleys with tributaries joining from the sides. Between them, the land juts out toward the sea, forming a peninsula.

How do features such as Chesapeake Bay and Delaware Bay form, and why do they have this branching shape?

Long Island, the lengthy island extending eastward from New York City, and the adjacent shorelines of New Jersey, were recently damaged by waves associated with Hurricane Sandy. Houses and other buildings along the coast were completely destroyed by waves and storm waters that surged over the beach.

What processes cause damage along coastlines, and what factors make one coast more vulnerable than another?

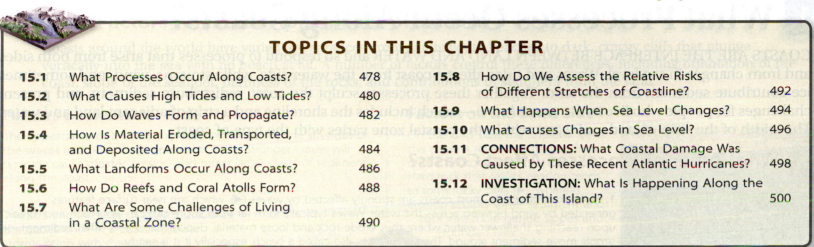

TOPICS IN THIS CHAPTER

15.00.a4 Acadia NP, ME

The beautiful Maine coastline is very irregular, with bays that curve into the coast and rocky headlands that protrude out into the sea (◄).

How did this irregular coastline form, and what does such a coastline indicate about its history?

Cape Cod protrudes into the ocean as a flexed arm or a boot with curled toes (▼). Offshore are a variety of islands, for which the region is well known.

How did Cape Cod form and what coastal processes caused this shape?

15.00.a5 Cape Cod, MA

Sea levels have been rising for at least the past century, and there are concerns that climate change may cause this rise to accelerate. Farther back In the past, sea levels have been more than 100 meters higher and lower than at present.

What causes sea level to rise and fall, and what happens to a coast when sea level changes by a significant amount?

Changing Sea Levels and Coasts

Features along the East Coast of the U.S. and Canada respond to the rising and falling of sea level relative to the land. As the climate warmed over the past 20,000 years, it caused melting of ice sheets and glaciers that had covered much of the northern U.S. and nearly all of Canada. Water released from melting ice carved new stream valleys as it flowed toward the sea. The shapes of Chesapeake Bay and Delaware Bay originated as stream valleys, with tributaries joining from the sides. As the meltwaters reached the sea, global sea level rose. The rising seas flooded coastlines and stream valleys, turning Chesapeake Bay from a stream valley to an embayment filled by the sea. The rising sea levels flooded similar features up and down the coastline, forming the many inlets and bays in Maine, Canada, and elsewhere along the Atlantic coast.

Other features along the East Coast likewise indicate rising and falling sea level over the last several million years. The Outer Banks region of North Carolina consists of long, sandy islands that form a barrier offshore of the main coastline. For this reason, this type of island is called a *barrier island*. The Outer Banks is very low, barely above sea level, juts eastward into the Atlantic, and lies within frequent tracks for hurricanes and other tropical storms. For these reasons, the Outer Banks is one of the most vulnerable stretches of coast in the world. The area is further threatened as sea level continues to rise.

Long Island and Cape Cod both contain materials deposited along the front of glaciers during the last glacial advance. As sea level rose from melting of the glaciers, glacial deposits were reworked and transported by waves and ocean currents. In Cape Cod, the prevailing currents move sand and other materials northward along the coast, curling around the north end to form Cape Cod's distinctive curved shape.

15.0

What Is Happening Along the Coast of This Island?

THIS PREVIOUSLY UNKNOWN ISLAND is being considered as a possible site for a small settlement and seaport to resupply ocean travelers with water and other supplies as they pass by. To avoid any missteps in this process, you are asked to observe the coastline, identify the various coastal features, infer what processes are occurring, and identify possible sites for the port and nearby settlement. Your considerations should include any issues that could affect the operation of the port, such as avoiding natural hazards and keeping shipping channels from becoming blocked with sand. The area does experience hurricanes.

Goals of This Exercise:

- Observe the three-dimensional perspective of the newly discovered island in order to identify the coastal features and processes.
- Fully consider any factors that would impact the location of a port and small settlement for people operating the port.
- Propose an acceptable location for the port and settlement and be prepared to defend your choices.

Geographic Setting

Listed below is some geographic information about the island.

1. The island, surrounded by the ocean, is in the Northern Hemisphere, located at approximately 30° N latitude.

2. The island is somewhat elongated and nearly 10 km in its longest direction. It reaches an elevation of 1,000 m above sea level in two mountains.

3. The climate of the island is warm during the summer, when it sometimes rains. It is wetter during the winter, when it receives enough rainfall to provide ships with water, as long as the water resources are managed intelligently. The ocean moderates the climate, keeping it a little cooler during the day and warmer at night relative to a place that is not next to the sea. The air is fairly humid so that the mountains sometimes cause short summer rains.

4. There is a moderate amount of vegetation on the island, except on the rocky mountains. The most dense vegetation is near a delta on the north side of the island and on a flat area on the western (left) side. Both sites are suitable for growing crops.

15.12.a2

1. Along the western side of the island are cliffs and steep slopes that rise out of the sea, with only a narrow beach along the shoreline (◄). The beach is composed of sand but it also contains many rounded pebbles and larger stones plus some large angular rocks that match the types of rocks in the cliff. There are also shells and fossils of other marine creatures.

15.12.a3

2. Above the cliffs is a relatively flat area, like a shelf in the landscape (▲). This type of feature is only observed on this western side of the island. Scattered about on this flat surface are rounded pebbles and larger stones, along with abundant loose sand. On some parts of the surface there is a layer that contains fossil shells (▼) and fossils of other marine organisms identical to those that occur on the beach.

15.12.a4

15.12.a1

Procedures

Follow the steps below, entering your answers for each step in the appropriate place on the worksheet or online.

1. Observe the overall character of the island, noting what is along each part of the coast and the locations of streams (a source of fresh water). Read the information about the geographic setting of the island (on the left-hand page).

2. Observe the photographs and read the accompanying text describing different parts of the island. Use what you know about coasts to interpret what type of feature is present, what processes are typical for this type of feature, and what the significance of each feature is for a port and small settlement.

3. Synthesize all the information and decide on the best locations for the settlement and port, which should be fairly close to one another (no more than 1 km apart). Be able to logically support your choices. There is not a single correct answer or best site.

4. OPTIONAL: Your instructor may have you write a short report or prepare a presentation describing all the factors you considered, why you chose the sites you did, or an interpretation of aspects of the natural history of the island.

3. Along the north shore of the island is a delta where the largest streams in the area reach the sea. This area has abundant trees (▶) and other plants, and is easily accessed from the east by walking along the beach.

15.12.a6

15.12.a5

4. The north shore of the island has gentle relief along the shore and has a well-developed sandy beach (▶). The beach slopes gently into the water, allowing people to wade quite a distance out into the sea. The sand is constantly in motion, moving up and back on the beach and also moving laterally along the coast. Lying on the beach are pieces of wood, similar in size to the trees observed near the delta.

5. Cut into the eastern end of the island are a series of embayments, or bays, that branch farther inland. In the largest bay, the land slopes into the bay at a moderate angle, gentle enough to walk up. The bay projects toward two mountains that have some well-developed vegetation on their flanks. The mountains are not too steep to walk up. Small streams flow on every side of the mountains but have more water in the summer. The mountains are not volcanoes.

15.12.a7

6. Extending off the land from the southern side of the island is a low ridge of relatively loose sand (▶). Some of the high parts of the ridge are clearly sand dunes. Some lack any vegetation, but others are anchored in place by grasses and other plants. To the east is a similar long island, again composed of sand, including dunes.

15.12

16.11 What Types of Soils Form Under Temperate and Polar Conditions?

FOUR OTHER SOIL ORDERS form in climatic settings ranging from the relatively warm, semi-arid lands of Africa to the coldest polar regions. These four soil orders include much of the agricultural heartland of the central U.S. and south-central Canada, and most of Europe. They also form large swaths across central and northern Asia, tracking the distribution of different types of climate, such as the Humid Continental climate.

16.11.a1

Alfisols

Alfisols are usually fertile soils with ample accumulation of organic matter in the A horizon. A yearly moisture surplus assists in clay eluviation into the B horizon, where the clays retain moisture and nutrients for plant growth. Alfisols have a natural tendency to become acidic, but this can be counter-acted by regular lime applications to raise soil pH. Only Mollisols (see below) are more productive agriculturally.

16.11.a2   16.11.a3   16.11.a4

Alfisols show one of the widest distributions of any of the soil orders, which indicates that they form under a variety of climatic conditions. Patches of Alfisols are scattered across North and South America, from the Great Lakes and Canada into southernmost Mexico. A large east-west belt of Alfisols crosses the southern part of western Africa, largely coinciding with a sub-arid region called the Sahel, which lies south of the Sahara. A dense concentration of Alfisols occupies western Russia and eastern India.

16.11.a5

Mollisols

Mollisols are the most agriculturally produc-tive soils, rich in calcium carbonates, which form in the process of calcification, and organic material, with little leaching, and a humus-rich A horizon. These soils of the mid-latitudes experience a period of moderate moisture deficit that prevents nutrients from being leached out of the root zone. Mollisols are generally found beneath tall and short prairie grass.

16.11.a6   16.11.a7   16.11.a8

Mollisols cover much of the agriculturally fertile Great Plains region of central North America and also cover relatively large parts of Mexico and southern South Amer-ica (e.g., Argentina). The distribution of Mollisols forms an east-west belt across southern Russia and central Asia, mostly to the south of the belt of Alfisols. As with Alfisols and the remaining two soil orders on the next page, Mollisols are relatively sparse in the region of Australia, Indonesia, and southeast Asia.

Spodosols

16.11.a9

Spodosols are formed under acidic forest litter in cooler portions of the mid-latitudes and in high latitudes. Acids deriving from the A horizon leach the soil of nutrients and organic matter, leaving a bleached, grayish subsurface E horizon. Because of the high acidity and short growing season, Spodosols tend to be fairly unproductive.

16.11.a10   16.11.a11   16.11.a12

Spodosols are largely restricted to the northern parts of North America, Europe, and Asia, which shows their development in colder climates. These soils are particularly prominent in eastern Canada, and they also occur along the western coast of Canada. Spodosols also cover much of Scandinavia, including most of Norway and Sweden. This soil order is mostly absent in the Southern Hemisphere due to the sparseness of land at high southern latitudes. A few small patches are barely visible in southernmost South America.

Gelisols

16.11.a13

Gelisols are soils above permafrost, having frozen soil for at least a portion of the year. Often waterlogged, these soils are rich in organic matter and sensitive to disturbance brought about through buildings, construction, and roads. Heat radiated from building foundations or sun-heated, dark-albedo roads can melt the top of these permafrost soils, causing the foundations of the building or road to sag and eventually collapse.

16.11.a14   16.11.a15   16.11.a16

If someone showed you the three globes above and asked you what types of soils form there, you would probably reply "very cold-climate ones." Since Gelisols form on ground that is frozen for part or all of the year, they are restricted to the polar regions or to very high elevations. They cover nearly all the polar regions of northernmost Canada and Alaska and northernmost Asia (Siberia). A few small patches occur in the western U.S., marking the highest and coldest peaks of the Rocky Mountains. A thin belt of Gelisols down the mountainous spine of South America likewise marks the highest and coldest part of the Andes. The large area of Gelisols in southern Asia is in the very high elevations of Tibet and the Himalaya.

Before You Leave This Page Be Able To

☑ Describe the main characteristics of Alfisols and Mollisols, where they occur, and why they are so important.

☑ Describe the character of Spodosols and Gelisols, and why each occurs mostly in the northernmost parts of the Northern Hemisphere.

16.11

Where Do the Various Soil Types Exist?

DISTRIBUTION OF THE 12 SOIL ORDERS across the planet can be viewed with globes, as done in the previous pages, or with a more traditional map, as done here. Globes have the advantage that they more properly show the latitudinal setting of each soil order and do not have the distortions inherent in any projection of a sphere to a flat map. Their main disadvantage is that only half of the world is visible at once. Examine this global map of soils to compare the distributions of the 12 soil orders, but pay most attention to the broad patterns, not the local ones.

Entisols and Inceptisols are characterized as being relatively poorly developed soils. Prospects for bringing this land into agricultural production can be limited by climate, terrain, inert parent material, and steep slopes, such as these mountainous slopes in the northern Rockies (◀).

16.12.a2 Yoho NP, British Columbia, Canada

In addition to topography, parent material, and climate, the type of overlying vegetation can greatly influence the type of soil. For example, different types of soils occur beneath grasslands (◀) than beneath forests.

16.12.a3, Badlands NP, SD

Aridisols are the main soil in many arid regions, like in this photograph from the deserts of the American Southwest. The relative lack of rainfall and warm temperatures result in somewhat sparse vegetation, and so these soils have poorly developed O horizons. Aridisols are relatively infertile because of a lack of organic matter. Desert soils also have distinctive soil-related features like caliche and desert pavement (◀).

16.12.a4 Big Maria Mtns., CA

16.12.a5 Amazon Basin, South America

Slash-and-burn agriculture has been a tradi-tional method of clearing land for cultivation in the tropics, especially in the world's rain forests, as shown here in a tropical rain forest (◀). This approach releases nutrients from trees to the relatively infertile Oxisols beneath the crops, but is incredibly destructive of the local ecosystem and can lead to soil erosion and eventually abandonment of the cleared land.

16.12.a1

Mollisols	Vertisols
Inceptisols	Spodosols
Alfisols	Oxisols
Entisols	Ultisols

Gelisols, limited to high latitudes or very high elevations, are permafrost soils and can melt, becoming unstable, disrupting the surface (▶) and damaging roads and buildings.

16.12.a6

Mollisols support extensive wheat (▶) and corn production. These soils feature an ideal balance between leaching and retention of nutrients. This soil order's high organic matter also contributes to fertility. With proper stewardship and sufficient rainfall, Mollisols are very productive soils.

16.12.a7

16.12.a8 Southeast Asia

Ultisols are generally infertile and need a heavy input of fertilizer to make them productive. This flooded rice paddy in southeast Asia (▶) has been made fertile through regular manure applications.

16.12.a9 SW Australia

	Aridisols		Ice
	Andisols		Rock
	Histosols		Shifting Sand
	Gelisols		Unclassified

Many soils, such as these of southwest Australia (▶), build up salt over time if they are not properly drained or do not receive enough rainfall to flush the salts farther down and out of the system. If these soils build up enough salt deposits, further agriculture is inhibited.

Before You Leave This Page Be Able To

☑ Describe some of the main patterns of distribution of some common soil types, and provide examples of how the climate or other conditions result in this type of soil.

16.12

16.14 How Does Soil Impact the Way We Use Land?

THE MOST COMMON way we think of soil is as a medium for plant growth, but soil is also a key component for construction, remediating pollution, and regulating and purifying our drinking water supply. The physical properties of soil can cause problems, but soil also has some uses that might surprise you.

A How Do We Engineer Problem Soils?

The suitability of a site for a building or road depends on soil. In addition to being a valuable resource, soil can cause problems because of its low strength and how it behaves when shaken, wetted, dried, or compacted. Problematic soil can be recognized by builders and home buyers so that building on, or buying, such risky sites can be avoided. In some cases, the soil can be compacted properly before building on it.

16.14.a1

16.14.a2

16.14.a3

Liquefaction occurs when loose sediment becomes saturated with water and individual grains lose grain-to-grain contact as water squeezes between them. *Quicksand* is an example of liquefaction. Liquefaction is especially common when loose, water-saturated sediment is shaken during an earthquake. The houses below, destroyed during the 1989 Loma Prieta earthquake, sank into artificial fill that liquefied during shaking.

Vertisols are a type of soil that contains a high proportion of certain clay minerals, called *swelling clays*, which increase in volume when wetted, expanding upward or sideways. As these clays dry out, they decrease in volume, causing the soil to shrink or compact. Repeated *expansion* and *compaction* during wet-dry cycles can crack foundations, make buildings unsafe, and ruin roads (▼).

In some soil, clay minerals start out arranged randomly, with much pore space between individual grains. When wheel traffic runs over clay-rich soils, or a building is constructed on top, pore space is reduced as clay minerals begin to lie flat, diminishing open spaces and therefore compacting the soil. Such *soil compaction* typically does not occur uniformly, because some parts of the soil have more clay than others. Differential compaction can crack walls (▼), foundations, and roads.

16.14.a4 California

16.14.a5 Boulder, CO

16.14.a6

B What Can Soils Tell Us About Watersheds?

Soils play a key role in the hydrologic cycle. The type of soil determines the fate of precipitation, whether it will infiltrate into the soil and percolate to recharge groundwater or run off at the surface, producing soil erosion and possibly flooding.

16.14.b1 Baraboo, WI

16.14.b2 Johnstown, PA

◄ The health of soils is a critical factor in protecting watersheds. Soils beneath a forested watershed will slowly release water into the watershed by reducing surface runoff. Here, soils act similar to a sponge, reducing the chance of floods in response to heavy rain.

◄ Suburbanization brings an increase in impermeable surfaces, as open space is converted into houses, streets, and sidewalks. The new impermeable surfaces promote more surface runoff, which contributes to urban flooding.

C   How Do Soils Interact with Certain Types of Pollution?

Soil is a resource that allows many pollutants to become less harmful—for this reason, it is called the *grand recycler*. Without soil acting as a decomposer and recycler, the depth of all the dead organisms that ever lived would be overwhelming. Soil also purifies groundwater, where billions of bacteria contained in soil can either detoxify or immobilize many harmful contaminants, including those contained in fertilizers, pesticides, and gasoline.

16.14.c1

Soil is best at a pH level of around 6.5 to 7.0 (close to neutral). When soils become increasingly acidic, lead, arsenic, and other heavy metals become mobile and more likely to be transported through and off the soil.

16.14.c2 Maryland

The ideal soil keeps surface pollutants from draining too quickly and contaminating groundwater. Conversely, the soil should have a coarse enough texture that it drains fast enough to prevent ponding.

16.14.c3 Eastern Washington

Soil can become polluted near farms that use pesticides, herbicides, fungicides, and fuel oils. Soil can also become contaminated by salt from irrigation water that has acquired a high salinity due to high rates of evaporation.

16.14.c4 Arizona

Silt- and clay-sized soils have a density low enough that these tiny sediments can become entrained into wind currents, depositing nutrients and even living bacteria thousands of kilometers from their source region.

D   What Are Some Unusual Uses of Soil?

Soil and water are the two most important natural resources on the planet. In addition to the importance of soil in agriculture and overall health of an ecosystem, soil is used in other ways.

Scientists have discovered that some clays, like the one in this soil pit (▼), have healing properties for certain skin infections. Flanking the central photograph below are microscopic images of dangerous bacteria. On the left are bacteria cells before being exposed to the clays, and on the right are bacteria cells that have been successfully damaged by exposure to the clays.

16.14.d1

16.14.d2

Baseball diamonds use soil that is the proper mixture of sand, silt, and clay, giving the field good traction and preventing excessive muddiness that might slow down the game. Likewise, we construct running tracks and clay-court tennis facilities using various mixtures of clay and other materials. The clay is derived from soils. The color of the track or court might tell us what type of soil was used.

16.14.d3

Soil is used in *forensic science* to investigate crimes. Footprints, soil on shoes, and the soil surrounding the crime scene can help identify the guilty party. Many criminals have been brought to justice as a result of clues left in soil (▶), as applied by soil scientists.

Before You Leave This Page Be Able To

✓ List the factors responsible for soil liquefaction, shrink-swell, and compaction.

✓ Discuss some practices that affect runoff versus groundwater recharge.

✓ Discuss the best texture, pH, and structure to contain pollution.

✓ Describe some unusual uses of soil.

16.14

Where Do These Soils Come From?

SOIL FORMATION INVOLVES a complex interplay of climate, vegetation, topography, parent material, and time. Another way to think of these variables is with the acronym ClORPT—Climate (Cl), Organic Matter/Vegetation (O), relief/topography (R), Parent Material (P), and Time (T). In this investigation you will use clues provided by ClORPT to locate these soil profiles correctly and provide a justification for your choice.

Goals of This Exercise:

1. Apply information from soil horizons to identify the soil order, and from the order infer the location.
2. Identify relevant ClORPT variables acting on each soil profile.
3. Synthesize ClORPT variables, specifically climate and vegetation patterns, to explain observed soil-horizon characteristics.

16.15.a2

Soil Profile 1—These agriculturally productive soils have a rich A horizon, a distinct E horizon where eluviation occurs, and a B horizon where a darker iron-rich layer is present.

16.15.a3

Soil Profile 2—A barely visible A horizon is present. This soil has a colluvium parent material that is derived from a bouldery glacial till. Soil processes have begun to act on this profile, but there is still no discernible B horizon.

16.15.a4

Soil Profile 3—Little organic matter can accumulate here and there is a weak A horizon. The dominance of evaporation has caused soluble calcium carbonate to accumulate in the B horizon.

16.15.a1

Procedures

This shaded-relief map shows elevations for North America and beyond. Light purple and brown colors depict areas of high elevation, whereas shades of green represent areas of lower elevation. Consider the elevation of an area and its impact on soil development when completing the steps in this investigation. For each step below, use the worksheet provided by your instructor or answer the questions in this investigation online.

1. For each soil profile, list the approximate depth and color of each horizon.

2. Based on clues provided for each pedon and a review of previous sections in the chapter, identify the soil order for each soil profile. Using the soil order as a clue, locate the position of the soil profile on the map. There are six possible locations on the map, one for each soil profile.

3. For each soil profile, create a table and write a short statement discussing how each variable of ClORPT likely has influenced the soil-profile patterns.

4. List the opportunities and challenges offered by each soil.

5. OPTIONAL PROCEDURE: Your Instructor may ask you to consult the climate chapter and biome chapter to list the climate and biome type associated with each of these soil profiles. Write a short paragraph discussing how climate and vegetation patterns would influence the origin of each soil order you identified.

16.15.a5

16.15.a6

16.15.a7

Soil Profile 4—No true humus exists in the A horizon because most of the time the entire soil profile is frozen. There is a brief summer period, however, when the A horizon will form a spongy substrate unsuitable for engineering purposes.

Soil Profile 5—A weak A horizon displays rapid decomposition of organic matter. An E horizon is discernible, where clays and iron-bearing minerals have been leached and moved downward to a B horizon.

Soil Profile 6—A rich A horizon extends far into the soil profile. The A horizon, with its soft granular structure, supports many earthworms. The wind blown loess has formed a blocky clay structure in the B horizon.

16.15

Ecosystems and Biogeochemical Cycles

THE BIOSPHERE IS THE REALM OF LIFE. *Biogeography* is the branch of physical geography that explores the biosphere and the Earth systems that support organisms. Adequate water, energy, and nutrients are necessary to sustain life, and their availability is controlled by climate and other aspects of physical geography. This chapter examines how organisms interact with each other and with their physical environment—*ecology*. The provisions and constraints of the physical environment, along with the entire community of interacting organisms (plants, animals, and others), in an area is an *ecosystem*. *Biogeographers* study ecosystems from a broad perspective that considers spatial and temporal variations of physical and chemical processes in the *biosphere* and how the biosphere impacts Earth's systems, from local to global scales.

Costa Rica provides a wonderful example of ecology, ecosystems, and biogeography. Earlier in this textbook, we explored the region's weather and climatic patterns, always an excellent place to start when considering the biogeography of a region. The large perspective view here (▶) is a satellite image draped over the topography of the land.

Land, Water, and Atmosphere

17.00.a2

The backbone of the country is a line of volcanoes, such as Volcán Arenal (◀), which has nearly continuous low-level eruptions of molten lava and ash. The volcanoes are formed by plate-tectonic activity (specifically subduction) along the west coast. The eruptions of Arenal are sufficiently continuous that the volcano has become a major site for *ecotourism*. The volcano interacts with the moist air in this tropical place, so it is often partially obscured by clouds and volcanic steam.

How do interactions between the land and atmosphere affect life, including humans?

Nicaragua

Costa Rica

Volcanoes

Pacific Ocean

17.00.a3

17.00.a4

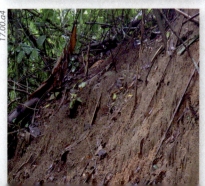

Eruptions of lava and ash flow down the mountain and bury anything on the surface (◀). The initially barren lava flows are colonized by certain types of plants and animals that can establish themselves on such "new land," and these are known as *pioneer communities*. Once the pioneer communities are established, an entire succession of other types of life can colonize the area. Over time, weathering begins to form soils (◀), which in this volcanic terrain are mostly Andisols, and the new land begins to blend into the landscape.

How do life and land interact chemically, interfacing through the soil?

17.00.a1

17.00.a5

Abundant tropical moisture sustains a dense growth of tropical rain forest, and the rain forest releases some of this moisture during transpiration. In addition, plants remove carbon dioxide (CO_2) from the air to produce carbon-rich wood, leaves, roots, and other material.

How does carbon cycle back and forth in the environment?

17.00.a6

Tropical rains, combined with locally impermeable tropical soils, result in plentiful runoff in streams. These corridors of running water attract many animals, forming a small ecosystem within the larger tropical ecosystem.

What are the main components of an ecosystem, and what role does water play in a healthy ecosystem?

TOPICS IN THIS CHAPTER

Caribbean Sea

Costa Rica

Life

17.00.a7

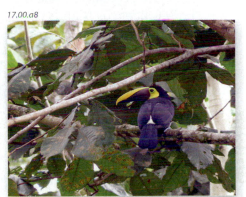
17.00.a8

Various forms of life thrive in the tropical and mountain habitats, including coatimundi (◄), a mammal in the raccoon family, and colorful and interesting tropical birds, such as a toucan (►). The lush vegetation hosts other classes of animals, for example reptiles, including this eyelash snake, a venomous pit viper (►).

17.00.a9

How do different types of animals interact, and what happens to their populations if they compete with one another or depend on one another?

17.00.a10

The rain forest is home to other small creatures, like this "poison-dart frog" (►), an amphibian, and countless types of insects, including ants that recycle vegetation and other materials that have fallen on the ground (photograph on the bottom right). The number of species in an ecosystem is one measure of biodiversity.

17.00.a11

The rain forest earns its name from the thick vegetation (the four photographs below) that forms a canopy and shades lower levels of the ecosystem. Vines and other plants climb or drape on the trees, trying to stay off the ground and gain better access to the light. Plants grow brightly colored flowers to better attract insects that, in turn, pollinate the plants, a symbiotic relationship that helps the plants and the insects.

What does it mean for a relationship to be called symbiotic, and what other types of relationships exist between living organisms?

What do we mean by biodiversity, and what can decrease biodiversity?

17.00.a15

17.00.a14

17.00.a13

17.00.a12

17.0

17.1 How Is the Ecosystem Approach Useful in Understanding the Biosphere?

IN AN ECOSYSTEM, organisms interact with individuals of the same species and with other types of organisms. The ecosystem is supported by various types of energy, matter, and processes, some that involve living organisms and others that involve nonliving components, like rocks, soil, and water.

A What Is Ecology?

Ecology is the study of how organisms and populations of organisms interact with one another and the nonliving components of their environment. These interactions occur from the local to regional scale—the ecosystem. The intricate array of interactions within an ecosystem evolves in a way that promotes increasing efficiency in energy exchanges and nutrient cycling, because efficiency promotes survival. *Ecologists* study ecosystems, and *biogeographers* use ecological principles to explain the distribution of life—the realm of *biogeography*. Photographs below explore the Galápagos Islands (west of South America), one of the world's truly unique and fragile ecosystems.

17.01.a1

Individual—Most living organisms exist as individuals that can function somewhat independently, at least for a while, such as this marine iguana, a type of lizard (reptile) that started on land but forages in shallow waters.

17.01.a2

Population—Although many individuals spend significant time as solitary creatures, they are always part of a population of individuals of the same species. The population can go up or down as the ecosystem changes.

17.01.a3

Community—Not only do organisms interact with others of their kind, but individuals from different species interact, as part of a community.

17.01.a4

Biotic Components—Living plants and animals, along with materials such as digestive wastes, discarded plant parts, and decaying remains of creatures, are the *biotic components* of an ecosystem.

17.01.a5

Abiotic Components—Components of an ecosystem not directly produced by living organisms are *abiotic components*. These include the air, rocks, soils, and water that provide a home for the plants and other organisms.

17.01.a6

Energy—All ecosystems require a source of energy, in most cases ultimately derived from the Sun. Insolation warms water and air, drives ocean currents, wind, and weather, and provides energy for plant photosynthesis.

Habitat—Some species live on land, such as this iguana, whereas others live in the water or air. The physical environment in which a species lives, such as a rocky coast where it finds its particular mix of food sources, is its *habitat*.

17.01.a7

17.01.a8

Ecological Niches—The unique position of a species in the ecosystem, especially its role relative to the other species, is called its *ecological niche*. How would you characterize the ecological niche of these Galápagos fur seals?

B What Are the Structures and Functions of Ecosystems?

All ecosystems consist of the storage and flow of matter and energy, and interactions between organisms and populations of organisms. The components of an ecosystem can be categorized into *structures* and *functions*. Ecosystems also have a variety of spaces for a wide array of organisms to exist, both supporting and being supported by the ecosystem.

Ecosystem Structures

1. *Ecosystem structures* consist of the biotic, abiotic, and energy components of the ecosystem. In most cases, these are the tangible, observable aspects of an environment. Observe the ecosystem structures shown in this diagram.

2. In the site depicted here, the most obvious *abiotic* components are the land, water, and air. For the land, there is the surface as well as the materials, like soil and rocks, that are beneath the surface. Less obvious abiotic structures would include gases and small particles in the air, the energy contained in sunlight, nutrients in the soil, and chemicals dissolved in the water.

17.01.b1

3. The *biotic* components include the cow, the trees, the low bushes and grass, and the roots of the plants. Other biotic components that are less visible, but certain to be present, include insects, birds, burrowing animals, bacteria in the soil, fish, algae, and other organisms in the water.

4. To understand this ecosystem, one of the first things we would want to do is inventory the ecosystem structures. How many trees and cows are there, how thick is the soil, and what is the pH of the water? Geographers and ecologists do similar inventories and quantitative measurements of structures in an ecosystem.

Ecosystem Functions

5. *Ecosystem functions* are the dynamic processes that occur in the environment to support the ecosystem. This figure shows a few of the processes that are occurring in this scene.

6. Some ecosystem functions occur on or near the land surface. These include the growth and other changes in the plants, including the sprouting of new individuals, shedding of leaves, cycling (movement) of nutrients, the uptake and release of water, and their eventual demise and return of their components to the soil or other sites.

17.01.b2

7. Other ecosystem processes involve the atmosphere, such as changes in the amount of sunlight from day to night and season to season, the constant exchange of gases between the atmosphere and organisms (plants, animals, and microbes), soils, and water. Other processes include the wind, evaporation, and at times precipitation, which provides fresh water to the system and causes the processes of runoff and infiltration.

8. A number of ecosystem functions are also important but not so obvious. These include the capturing and converting of sunlight for photosynthesis (called *fixation*), decomposition of organisms, and chemical and physical weathering that slowly forms soil.

The Aral Sea — An Example of a Dying Ecosystem

In the 1960s, the Russians diverted flow in the Amu Darya and Syr Darya Rivers to provide irrigation for agriculture, especially cotton. Today, Uzbekistan is one of the leading exporters of cotton, but at what price? When the streams were diverted, the Aral Sea, which at the time was the world's fourth largest lake (left image), lost most of its inflow and became about one-tenth of its former surface area (right image). The lack of inflow sharply increased salinity in

17.01.t1

the sea, and the entire aquatic ecosystem died. Toxic dust from the fertilizer and pesticides left behind as the Aral Sea evaporated has caused increasing instances of respiratory diseases and other ailments. This created impacts on the land surface that contributed to the decline of terrestrial ecosystems. The local temperature has become more severe in winter and summer, exacerbating risks for drought and causing other severe human impacts.

Before You Leave This Page Be Able To

✓ Explain what ecology and an ecosystem are.

✓ Describe the difference between habitat and niche.

✓ Sketch and explain an example that contrasts ecosystem structures and ecosystem functions.

✓ Discuss recent impacts on the Aral Sea ecosystem.

17.1

17.7 How Do Ecosystems React to Disturbance?

DISTURBANCES SUCH AS FIRE, storms, or human interference are common to most ecosystems. Following the disturbance, an ecosystem may remain largely unaffected, try to rebound from the disturbance, or start over. As an ecosystem recovers from a disturbance or tries to colonize new land, a succession of organisms may populate the area. While it is tempting to think of disturbance as endangering or destabilizing an ecosystem, disturbances can promote the stability of some ecosystems.

A How Does Disturbance Affect Ecosystem Stability?

Ecosystem stability refers to the degree of fluctuation in population of species over time. *Buffering mechanisms* help species avoid stress and thrive. For example, animals may hibernate during times of the year when food is scarce, or plants may evolve to tolerate a wider range of conditions. When these buffering mechanisms allow populations of all species within the ecosystem to stabilize, the ecosystem is said to be in balance, or *stable*.

1. This graph shows how the populations of three species, in three different ecosystems, change over time. The top curve represents the population of an abundant species. The number of individuals varies only slightly over time, so the ecosystem, as measured by this species, is *stable*. This species and its ecosystem can withstand environmental stresses without experiencing drastic changes in populations.

2. The bottom curve depicts a species with fewer individuals. The population of this species is also stable.

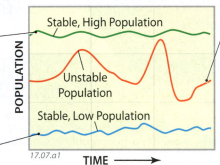
17.07.a1

3. The middle curve shows a species whose population varies widely with time, including some abrupt increases and decreases. This ecosystem is *unstable* and vulnerable to a population crash due to some event or change in conditions. This generally occurs in ecosystems that are considered to be less mature.

4. Some ecologists argue that energy constraints prohibit the stabilization of ecosystems, such as when there is not enough energy to withstand the addition of a new predator species.

5. Organisms generally have a limited range of conditions (▶), such as temperature and moisture, in which they can survive. The levels at which these conditions become too high or low for that organism to survive are known as the *lethal limits*.

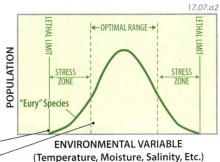
17.07.a2

6. Within the lethal limits exists an *optimal range* of conditions, under which the organism tends to thrive, resulting in a relatively large population. If conditions fall outside the optimal range but inside the lethal limits (the "stress zone"), the organism can survive but does not prosper, so the population will be low under these conditions.

7. Some species have wider optimal ranges and stress zones than others (▶). An organism that can tolerate a wide range of conditions for a given abiotic factor (e.g., temperature) is described with the prefix of "eury-" before a root word that refers to the factor, as in *euryhaline* for a creature that can tolerate a wide range of salinities. On the curve to the right, a "eury" species would be expressed by a broad curve of tolerances, represented by the green curve.

17.07.a3

8. In contrast, some species can only tolerate a narrow range of conditions for a given environmental factor, as represented by the black curve (▲). They are given a prefix of "steno," as in a *stenohaline* species that cannot tolerate changes in salinity or *stenothermal* for a species with a narrow range of tolerable temperatures.

Inertia and Resilience

9. There are two other important components of ecosystem stability. One is the ability of an ecosystem to resist change after a stress is applied—*inertia*. If an ecosystem is relatively stable, it may not respond or may respond slowly to an imposed change, like a decrease in rainfall at the start of a drought. It has inertia.

10. The other is *resilience*, the rate at which an ecosystem recovers following stress. For example, the area around Mount St. Helens was devastated by a volcanic eruption in 1980. Since that time, the vegetation has recovered quite rapidly (▶).

17.07.a4 Mount St. Helens, WA

11. Over the last 200 years, many scientists have viewed the natural world in the context of *uniformitarianism*—the concept that the natural world has evolved in the past by processes and rates similar to those we observe today. At the opposite end from uniformitarianism is *catastrophism*, which emphasizes the importance of sudden events in creating change. The eruption of Mount St. Helens was a perfect example of an abrupt catastrophe to the ecosystem.

B   What Is the Process of Succession?

Disturbances tend to cause great fluctuations in populations, as stresses affect species in different ways. The ecosystem typically responds with a predictable change in species composition, called *succession*. If the successional process begins with the formation of new land or with so much disturbance that little life is able to survive the disturbance (as in glaciations), the recovery process is a *primary succession* and begins with a *pioneer community*. *Secondary succession* occurs when soil is present in the landscape from the outset, as occurs in an area affected by a forest fire — the vegetation is destroyed but the soil is largely unaffected. Succession can lead to a stable community.

1. This figure shows the stages of recovery in a primary succession. The underlying graph shows a typical sequence of species replacement as the ecosystem evolves.

2. A pioneer community contains mostly r-selected species, including mosses and lichens.

3. With time, K-selected species, such as tree species, become possible and begin to dominate the ecosystem.

4. The final stable community, called a *climax community*, develops and includes an intricate food web and efficient nutrient cycling, with the diversity and productivity determined by the abundance or lack of necessary abiotic factors, specifically sunlight, water, and nutrients.

5. The concept of *patch dynamics* refers to the idea that subareas within an ecosystem may be influenced by different processes and are at different successional stages at the same time. Over time, a climax community may be disturbed and then reestablish a new climax.

Lichens   Mosses   Grasses   Shrubs   Trees

17.07.b1

6. Biodiversity increases in the early successional stages, but eventually it decreases as nutrient cycling becomes more efficient and competition weeds out less robust species. Different species have differing abilities to deal with disturbance, even in the same ecosystem.

SHANNON INDEX

CLIMAX

17.07.b2   TIME ➞

7. The resulting distribution of species (◀) after this steady-state "endpoint" is achieved (some time after the ecosystem was disturbed) is the climax community. K-selected species become increasingly favored, and r-selected species are at an increasing disadvantage at greater times after the disturbance.

8. Before establishment of a climax community, more energy is fixed by autotrophs than is consumed by respiration in the ecosystem.

Gross Primary Productivity

AMOUNT OF ENERGY

CLIMAX

Respiration

17.07.b3   TIME ➞

9. At the time when the climax community is established, the total energy fixed (gross primary productivity) will nearly balance energy expended in respiration.

10. After the climax community, nearly all of the energy available is used in respiration to break down food.

C   How Does Disturbance Affect Biogeochemical Cycles?

An important aspect of how ecosystems become established, evolve, and die is the concept of a *biogeochemical cycle*. A biogeochemical cycle is a way to conceptualize chemical connections between various components of an ecosystem — land, groundwater and surface water, oceans, ice, air, and organisms.

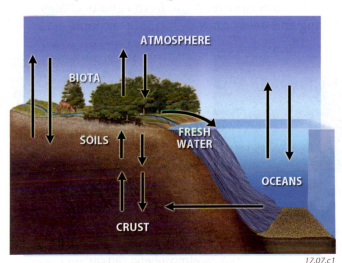

ATMOSPHERE

BIOTA

SOILS

FRESH WATER

OCEANS

CRUST

17.07.c1

This figure illustrates the main aspects of a biogeochemical cycle. Arrows indicate how some chemical substance, for example carbon, flows between various components of an ecosystem (a flux), including the soils, biota, atmosphere, and oceans. Typically, the amounts of that chemical substance are listed next to each arrow to convey the magnitude of the flux and how some change in the ecosystem could affect the ecological balance. Ecosystems store and cycle nutrients, so any changes in that ecosystem will be felt in the biogeochemical cycles involving them. This, in turn, can affect other ecosystems. The following pages have biogeochemical cycles for carbon, nitrogen, phosphorus, and sulfur, but others exist for many other chemical elements and compounds.

Before You Leave This Page Be Able To

✔ Explain the concept of ecosystem stability.

✔ Give examples of primary and secondary succession resulting from a disturbance of the land.

✔ Describe a biogeochemical cycle.

17.7

17.11 What Is the Role of Sulfur in the Environment?

SULFUR IS ANOTHER ESSENTIAL NUTRIENT for life. Sulfur (S) is present in two amino acids. S is necessary for plant growth, especially for legumes like peas and beans, which return N to the soil as they grow. S is also present in the molecules that protect plants and animals from bacterial infections. S is also used as the energy source in *chemosynthesis* to make food for deep-ocean organisms and for bacteria in oxygen-poor environments. The S cycle is not very well understood, largely because of the complicated chemical changes involving different forms of S, such as sulfur dioxide (SO_2), sulfate ions (SO_4^{2-}), and hydrogen sulfide (H_2S).

A How Does the Global Sulfur Cycle Operate?

The sulfur cycle looks similar to the phosphorus cycle in that it is almost entirely sedimentary, involving only minor interaction with the atmosphere. Again, the red arrow indicates the flow linked to the human-produced (anthropogenic) part of the cycle.

1. This figure depicts the global sulfur cycle, identifying the reservoirs of sulfur and the fluxes of sulfur between those reservoirs. The numbers indicate the size of the reservoir or the size of the flux in 10^{12} moles. The largest reservoirs are in the lithosphere, including sediments in the ocean. Large amounts of sulfur also occur in sedimentary rocks that were deposited in the ocean but are now on land and in rocks that have been affected by hot (hydrothermal) waters.

17.11.a1

ATMOSPHERE (1)

Fossil Fuel Emissions

Volcanoes (0.033)

Wind Erosion (0.6)

Deposition (1.2)

Deposition (1.9)

Seaspray (2.4)

BIOTA

FRESH WATER

SOILS (3,000)

Wastes and Decay

Biological Uptake

Runoff and Erosion (1-8)

OCEANS (50 X 10⁷)

Very Slow Geological Processes

Oxidized Sediments (2.0 x 10⁸)

Reduced Sediments (1.5 x 10⁸)

5. Although the effects of acid precipitation are most obvious outside the oceans, such as damage to land plants, most acid rain actually falls into the oceans, causing damage to aquatic animals.

4. Some sulfur goes from the ocean to atmosphere, and sulfur also goes from the atmosphere to the ocean and to the land, in the form of precipitation. In the atmosphere, sulfur oxides (SO_x) react with water to form sulfuric acid (H_2SO_4), which becomes incorporated into the precipitation, causing *acid rain* (or more generally acid precipitation).

2. As sulfur shifts between reservoirs, it moves from the land to the atmosphere from burning coal and petroleum, both of which contain variable amounts of sulfur. Sulfur is also released from volcanoes, and an increased release of sulfur may indicate that a volcanic eruption is imminent. Sulfur is also picked up by wind.

3. Sulfur is relatively mobile in water, so it enters streams and groundwater when sulfur-bearing rocks are weathered. Streams then carry this sulfur to the ocean. When sediment accumulates on the seafloor, it incorporates significant amounts of sulfur, essentially removing it from circulation through the system for a while.

Sulfur in the Land and Water

17.11.a2 Kidd Creek Mine, Ontario, Canada

17.11.a3 Silverton, CO

17.11.a4 Animas River, Silverton, CO

6. Sulfur occurs naturally in rocks, including these (▲) that are almost all sulfide minerals. They are exposed underground in a copper mine with a rock hammer for scale.

7. Sulfide minerals are not stable when exposed to atmospheric oxygen in the process of weathering. As a result, they convert to red and orange oxide and sulfate minerals.

8. The reddish rocks in the center photograph (to the left) drain into this river, causing the water and channel to be acidic and stained by the sulfur and iron.

B What Impacts Do Humans Have on the Sulfur Cycle?

The human impact on the S cycle is miniscule on a global scale, but it can be significant at the regional and local scales. About half of the atmospheric emissions come from humans, but much more than half is released in industrial areas. SO_x emissions, especially SO_2, are dangerous because they destroy plant hormones, stunt plant growth, reduce plant reproduction rates, and cause respiratory illnesses in humans.

17.11.b1 Fruitland, NM

1. The burning of coal (◄), especially coal that is high in sulfur, is the prime human source of atmospheric S. The sulfur falls back toward the surface in *acid precipitation*, like acid rain. The term *acid deposition* is also used because dust can be coated with acid and then fall to the surface. Later, when moisture comes in contact with this acidified dust, anything in contact with it, including raindrops, will become acidified.

2. Only a few counties in the U.S. (▶) are in noncompliance with the clean air standard for SO_2 as of December 2010. Most of these counties have smelters (furnaces in which to heat rocks) that process sulfur-rich rocks.

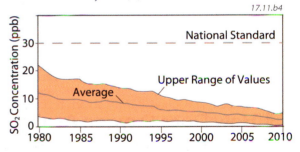

COUNTIES IN NON-ATTAINMENT FOR SO_2 EMISSIONS

Lewis and Clark (MT)
Yellowstone (MT)
Warren (NJ)
Salt Lake (UT)
Armstrong (PA)
Pinal (AZ)
17.11.b2
Piti and Tanguisson (Guam)

Spatial and Temporal Variations in Human-Generated Sulfur

3. This map depicts the amount of sulfur emitted in the continental U.S. The green circles show the magnitude of emissions from the largest sources and the colors represent the observed SO_4^{2-}. Examine this map and consider what might be causing the observed patterns.

17.11.b3

SO_4^{2-} ($\mu g/m^3$)

4. The largest emissions are in the eastern half of the country, around the industrial centers, including at coal-fired power plants.

5. The impact of acid deposition is most pronounced where bedrock compositions cannot buffer the acid deposition and where rainfall is orographically concentrated, for example in the Adirondack Mountains of upstate New York and the Appalachian Mountains. The good news is that there has been an improvement in acid deposition levels in the U.S. over the last 30 years (▼).

17.11.b4

Graph: SO₂ Concentration (ppb) vs. year (1980–2010)
National Standard
Upper Range of Values
Average

6. This unusual-looking map, called a *cartogram*, shows the size of each country in proportion to the amount of sulfur it releases into the atmosphere. Note that the industrialized countries, which look like big balloons, are responsible for most of the SO_2 emissions.

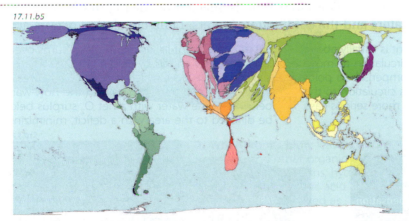

17.11.b5

7. Emissions elsewhere in the world are extremely variable. They are decreasing in many locations, such as in European Union countries, where emissions have fallen by 76% from 1990 to 2009. Emissions are increasing in the developing world, such as in China, which has been building coal-fired power plants at the rate of several per week! Scientists have tracked the increased emissions from China all the way across the Pacific and into the Pacific Northwest.

Before You Leave This Page Be Able To

✓ Sketch and explain the main reservoirs and fluxes involved in the sulfur cycle.

✓ Explain the role of humans in the sulfur cycle.

✓ Describe the spatial and temporal trends in sulfur dioxide emissions.

17.11

18.4 How Does Topography Influence Biomes?

DIFFERENCES IN TOPOGRAPHY, from valleys to mountain peaks, and differences in the directions slopes face, result in large local variations in biomes. The wide range of environmental conditions on mountain slopes supports many biomes in close proximity to one another that are normally thousands of kilometers apart. Topography also controls the flow of water, from peaks toward valleys.

A What Are Riparian Zones?

18.04.a1 San Juan River, UT

18.04.a2 Las Cruces, Baja California, Mexico

18.04.a3 Jackson Hole, WY

Riparian zones consist of trees, shrubs, and grasses that form along streams (▲), creating a local environment that is more lush with vegetation than the surrounding land. Due to the presence of flowing water for all or part of the year, riparian zones can host a very different biome than the surrounding region.

Riparian zones occur in many settings, such as grasslands and various forests. They are best expressed, however, in deserts and other arid lands, where the contrast between dry uplands and lush valleys can be sharp. They can form even where water is not flowing but is just below the surface (▲).

Riparian zones allow habitats for birds, shade-loving vegetation, and other organisms not otherwise common away from the stream valley. The water and vegetation in turn attract diverse types of animals, including frogs and other amphibians, snakes and other reptiles, and mammals, like moose (▲).

B What Factors Cause Conditions to Vary on Different Slopes?

Differences in elevation, slope orientation, precipitation, humidity, insolation, and wind intensity and direction cause mountain environments to differ greatly from one side to the other.

Clouds, Precipitation, and Humidity

1. One of the most significant factors causing local to regional variations in conditions is a mountain, which can cause orographic effects. The side of the slope facing upwind at any given time is the windward slope. Air blowing toward the mountain is forced upslope on this side, where it cools, condenses into clouds, and precipitates. The increased rain on this side of the mountain results in an increase in vegetation and other effects.

Windward Side — Cooling — Rainy (Wet)
Leeward Side — Warming — Rain Shadow (Dry)

18.04.b1

2. The opposite side, facing downwind, is the leeward slope. This side of a mountain tends to be drier because the air flowing down the leeward side has lost water to precipitation on the other side. It also compresses, heats up, and becomes less humid as it descends. The leeward side is considered to be in the rain shadow of the mountain and will have less vegetation, less soil, and drier ecosystems.

Insolation and Slope

3. Except near the equator, south-facing and north-facing slopes receive different amounts (▼) of insolation, when averaged over the year. A slope that receives more direct Sun is warmer and drier and is called an *adret slope*. The opposite side is colder, wetter, and more vegetated and is called a *ubac slope*.

Elevation

4. Elevation affects temperature and the amount and type of precipitation. If the mountain is high enough, there is an upper limit to trees—the *tree line* (▼). The tree line is a transition, not an abrupt line. The tree line typically is farther downslope on the colder, shaded (ubac) side of the mountain and higher on the warmer, sunlit (adret) side. The *snow line* occurs at the elevation where snow and ice cover the ground year-round.

Wind Variation

5. Deformed and stunted trees called *krummholz* occur near the tree line, where they must be able to survive the intense cold, dry winds. Note the branches taking advantage of the shelter offered by the trunk itself, growing only on the leeward side. Such a krummholz would be termed a *flag* or *banner tree* and indicates the direction of prevailing winds.

SUNLIGHT
Low Angle
High Angle
Ubac Slope
Adret Slope
18.04.b2 Ketchum, ID

18.04.b3 Galena, ID

18.04.b4 Patagonia, Chile

C How Do Biomes Change with Elevation on a Mountain?

On mountain slopes, environmental conditions change abruptly with changes in elevation. As a result, a number of biomes can occur over very short distances upslope and downslope.

18.04.c1 Mount Kilimanjaro from Kenya

1. As conditions vary with elevation up a mountain, so do the biomes, changing from warmer and commonly drier conditions near the base to colder and wetter settings upward. Mount Kilimanjaro (▲), the highest mountain in Africa at 5,895 meters (19,341 ft) above sea level, is located near the equator. From the base of the mountain, the vegetation progresses upward from savanna through several types of forests to alpine conditions with a glacier near the top. Remember that the global average *environmental lapse rate* is about 6.5 C°/km, so during the 3 km climb to the top, temperatures would drop more than 20 C° (36 F°). Nearly 80% of the glacier has melted since the early 1900s.

2. Another example of the effects of elevation is the Santa Catalina Mountains (▶) near Tucson, Arizona. The top of the mountain is at an elevation of 2,791 m (9,157 ft) and the base is about 2,000 m (about 6,600 ft) lower. Observe the different types of vegetation as one progresses from the base to the peak, beginning with the photograph below.

18.04.c2 Santa Catalina Mountains, AZ

18.04.c6

6. The highest peaks are in forests of large ponderosa pine and other high-altitude trees, and there is an alpine ski area.

18.04.c3 Tucson, AZ

18.04.c4

18.04.c5

3. The Tucson basin below the mountain consists of cacti and other heat-adapted plants of the Sonoran Desert.

4. One-third of the way up the mountain, the cacti give way to scrub and scattered trees, including piñon and juniper.

5. Two-thirds of the way up, the piñon and juniper become larger and closer, and are joined by some ponderosa pine trees.

D How Are Highland Environments Changing?

These vertical progressions of closely spaced ecosystems are particularly sensitive to climate change, in part because each is adapted to a particular range of elevation. Warming associated with climate change is expected to reduce snowpack, change the timing and magnitude of snowmelt, and alter stream discharge. When superimposed on effects of land use changes from resource and recreational exploitation, highland environments may change abruptly.

Humans and animals, like these guanacos (▶) in the Patagonian Andes of South America, have adapted to various mountain biomes. *Transhumance* is a type of seasonal nomadism in which people move to the optimal elevation to experience the best temperatures and animal fodder at a particular time of year—generally upslope in summer and downslope in winter. If the climate changes, plants and many animals cannot find their new niche quickly enough to survive.

Global or regional warming or cooling can shift mountain biomes to higher elevations, impacting some species. In many cases it is difficult to predict how the range of a species will change, because of local variations in soil, hydrology, microclimate, and topography. The net impact of migrating mountain biomes is likely to be increased habitat fragmentation and more endangered species, though there are likely to be new opportunities for some now-threatened populations. The details are not fully known.

18.04.d1 Patagonia, Chile

18.04.d2 Coal Bank Pass, CO

Before You Leave This Page Be Able To

✓ Describe what a riparian environment is, how you could recognize one, and why one attracts animals in the area.

✓ Sketch and explain how prevailing winds and Sun angle cause different conditions on opposite-facing slopes.

✓ Describe how changes in conditions up a mountain are reflected by changes in vegetation, providing at least one example.

✓ Explain why the effects of global climate change are magnified in mountainside environments.

18.4

18.8 How Do Desert Organisms Survive?

LIFE EXISTS EVEN IN THE HARSHEST conditions on Earth. In such environments, ecosystems and biomes are fragile, as growth is usually slow. Species that have built-in mechanisms that make them tolerant to arid environments are called *xerophytic*. What characteristics do xerophytic vegetation and animals possess?

A What Strategies Involve the Efficient Use of Water in Plants?

Plants have evolved numerous strategies to deal with scarce water, high temperatures, little shade, and drought usually associated with deserts. What are some of the active strategies employed by plants to survive harsh desert conditions?

Gathering Water

18.08.a1 Chandler, AZ

Tamarisk, an invasive species, dominate some riverbanks with roots that are over 20 m deep. They can out-compete native flora for limited water. Such a plant, with a deep taproot system, is termed a *phreatophyte*. Mesquite trees of the Desert Southwest have a similarly deep-rooted strategy.

Minimizing Water Loss

18.08.a2 Phoenix, AZ

Small, hard, waxy leaves on the creosote bush lose less water through transpiration than wider, softer leaves. In dry times, creosote leaves fold in half (▲) to conserve water, or the plant becomes deciduous by dropping its leaves. Some plants take advantage of cooler night temperatures to flower.

Strategic Positioning

18.08.a3 Namibia

Plants may be concentrated near oases or streams, or in places where groundwater is fairly close to the surface. Desert plants can have leaves that point downwind from the prevailing direction to minimize evaporation. In some cases, buds only develop on the leeward side of a branch.

Water Strategies of Cacti

Cacti and other desert plants have various strategies (▼) to maximize how much rain they collect, how much water they store, how much water they lose, and how much sunlight they receive. For example, some cacti, like the tall Giant Saguaro and barrel cactus, have accordion-like ribs that expand to store more water and contract during drier times to reduce their surface area and reduce the amount of water loss.

Other cacti, like agave, have sturdy, trough-shaped leaves that open upward, catching precipitation and funneling it toward the center of the plant. Some agave conserve water and energy by only blooming once, at the end of their life-cycle, sending up a tall stem crowned with flowers.

Many types of cacti, such as cholla and prickly pear, have spines, which discourage browsing by animals, help collect some moisture, and provide some shade from the hot desert sun.

18.08.a4 Tucson, AZ

CAM Photosynthesis

Some desert plants, like cacti and agave (▼), use a special type of photosynthesis to minimize water loss. In crassulacean acid metabolism (CAM), the *CAM plant* collects CO_2, converts it to an acid, and stores it at night rather than being active during daytime. This allows the leaf pores through which CO_2 (and water) passes—*stomates*—to remain closed during daytime hours, thus minimizing water losses during transpiration. A CAM plant can also leave its stomates closed during both night and day in extremely dry times. Photosynthesis continues over such intervals by recycling the CO_2 released during respiration. This "CAM-idling" allows desert plants to withstand dry periods and recover quickly when water becomes available.

18.08.a5 Tucson, AZ

B Which Survival Strategies Do Not Directly Involve Water Use?

Some methods for surviving searing heat and lack of rainfall do not directly involve water conservation. Rather, these adaptations involve rapid reproduction, spreading progeny through hitchhiking, and keeping plant neighbors at a distance. What is the nature of these adaptations?

Rapid Germination/Flowering

18.08.b1 Tucson, AZ

Despite active methods to conserve water, such as waxy leaves or deep taproots, a major survival strategy involves just waiting for the right moment to reproduce. Such a moment involves rapid germination and flowering following a significant rainfall event, either after winter rain or snow or following a summer monsoon. Deserts can be lush with spring flowers after a winter where rainfall was sufficient and somewhat evenly spread over several months.

Animal-Assisted Dispersal

18.08.b2 Copper Creek, AZ

The cholla, commonly called a "jumping" cactus, has spine-covered stem segments that detach easily (and painfully) when brushed by a passing hiker or animal. This serves not only as a defense mechanism against hungry herbivores but also as a means of opportunistic reproduction. Tiny barbs on the cholla spines assure dispersal. A new cholla easily grows where a stem segment was discarded. Note the many new cholla growing beneath the larger, older cholla.

Allelopathy

18.08.b3 Phoenix, AZ

Some desert plants employ a survival strategy of poisoning one's neighbors—*allelopathy*. Examples include the creosote bush (▲), which keeps away large plants by poisoning the adjacent soil with its roots. Spring grass avoids the poison. Another example is African rue, an aggressive and invasive species that out-competes native grasses by germinating earlier in the spring and by putting toxins into the soil via its root systems to prevent other plants from growing nearby.

C What Survival Strategies Do Desert Fauna Employ?

Mammals' and birds' body temperatures remain constant as long as they aren't overexposed to the hot Sun. Accordingly, these creatures seek shade during the hottest times of the day. Reptiles, amphibians, and insects cannot regulate their body temperatures, so their body temperatures mirror desert heat and cold.

Light colors give animals like this scorpion (▼) a high albedo. Sunlight reflects from their bodies and keeps them cooler.

18.08.c1

Camouflage can be important to survival. One of the best disguises ever is this grasshopper in Namibia (▼) that looks exactly like the rough-textured stones around it.

18.08.c2 Namibia

Desert fauna are typically small, so body heat can escape easily. The shape of body parts, such as the long ears of a desert hare (►), radiate body heat to the atmosphere. Note the ear's blood vessels that facilitate this process. Arctic hares, by contrast, have short ears in order to conserve heat and avoid frostbite.

Ephemeral (seasonal) behavior like estivation (summer dormancy) is used by species like the North American desert tortoise (▼). By day, many desert creatures seek the comfort of underground burrows, choosing instead to carry out most of their activity at night.

18.08.c4 Arizona

18.08.c3

Before You Leave This Page Be Able To

☑ Explain some strategies that desert plants use to collect water.

☑ Give other examples of survival strategies that do not involve water directly.

☑ Provide several examples of evolutionary adaptive strategies for desert fauna.

18.8

18.16 How Do Animal Realms Overlap with Biomes?

PLANTS CANNOT RAPIDLY move in response to environmental stress, so biomes provide a good indication of environmental conditions in a given place. The geographic range and migration of major animal types also reveal environmental changes across space, but faunal realms are separated more directly by physical, rather than climatic, barriers. These barriers include bodies of water, deserts, and mountain ranges that are difficult for plants and land-bound animals to cross. *Zoogeographers* study the factors that cause animal distributions and migrations.

A What and Where Are the Major Zoogeographic Realms?

1. Earth's surface can be subdivided based on differences in how the assemblage of plants and animals evolved over time instead of simply using modern-day distributions. Each of the resulting large, continental-sized areas, shown here on the central map, is called a *zoogeographic realm* or *ecozone;* each encompasses many biomes.

18.16.a2

2. The North American continent, called the *Nearctic realm,* has been mostly isolated from other continents for the past 70 million years, except for intermittent connections, as with Asia during the last glacial maximum. The harsh, seasonal environment has left this realm with limited richness and biodiversity, other than fishes and reptiles. However, canine-type animals, raccoons (▲), and horses are *endemic*—native—to this region.

3. The Pacific Islands are too small and isolated to have developed a rich natural fauna, but the *Pacific realm* is home to whales, dolphins, walrus, and seals.

18.16.a3

4. The largest zone, the *Palearctic,* encompasses all of Europe and most of Asia and North Africa. It has many similarities to the Nearctic region, because the two zones were once connected in a supercontinent called *Laurasia.* More recently a land bridge connected the two continents during the last glacial maximum, further allowing an exchange of fauna, including humans. The red panda (▲) is one of the few unique, endemic animals in the Palearctic.

18.16.a4

5. The *Paleotropic realm,* covering most of Africa, has great variety, with more mammalian families than any other region. It includes giraffe (▲) and many endemic bird and freshwater fish species.

6. The *Indomalayan realm* includes the Indian subcontinent, which was once joined with southern Africa as part of a southern supercontinent called *Gondwana.* As a result, the two regions share some faunal characteristics. India was not connected to Asia until much later, 45 million years ago. The realm also includes Southeast Asia and part of Indonesia.

18.16.a5 Ancient Mammal

18.16.a1

18.16.a6

7. Until relatively recently, North and South America were not connected and so evolved very different plants and creatures. South America constitutes the *Neotropic realm,* which contains unique mammals and colorful birds. Once the Panama isthmus connected the Americas (in the last 3 million years), fauna from two realms intermixed somewhat, with *Neotropic* fauna, like opossum, migrating into North America. Some Nearctic fauna migrated into South America, including large cats.

8. The unforgiving environment restricts fauna of Antarctica to birds, including penguins, but there is abundant evidence of extinct animals there.

18.16.a7 Antarctica

9. Known for its primitive primates, such as the lemur (▼) and other fauna that differ widely from other realms, the *Madagascan realm* is the smallest and most distinctive region.

18.16.a8 Madagascar

10. Lengthy isolation has made *Australian* fauna unique, particularly its pouched mammals—*marsupials*—like the kangaroo (▼) and wallaby. With little fresh water in these dry environments, amphibians and fish are rare. Other animals were able to migrate in from Asia.

18.16.a9 Anglesea, Australia

11. Birds, especially flightless ones like the kiwi (pictured in the New Zealand highway sign above), comprise an unusually large share of the fauna in the *New Zealand realm.* Otherwise, large animal life is sparse, with no native mammals.

18.16.a10 New Zealand

Map labels: NEARCTIC, PALEARCTIC, PALEOTROPIC (ETHIOPIAN), INDOMALAYAN (ORIENTAL), PACIFIC, NEOTROPIC, PACIFIC, MADAGASCAN, AUSTRALIAN, NEW ZEALAND, ANTARCTIC

B What Changes in Zoogeographic Realms Have Occurred in Geologic Time?

Different zoogeographic realms support different faunal assemblages, suggesting that animals evolved in response to the physical conditions (climate, geology, soils, vegetation) imposed by the biogeographic realm at the time. Mountains—and especially oceans—were barriers to dispersal. Such barriers, however, were mobile and temporary, at least in geologic time.

Before 250 Mya

18.16.b1

1. Prior to 250 million years ago (Mya), all of the Earth's major landmasses were joined in a supercontinent known as *Pangaea*. North America was connected on three sides to Europe, Africa, and South America.

2. Antarctica joined Africa, India, and Australia. At that time, animals could migrate between all of these areas, which today are separate continents. Sedimentary rocks deposited at that time contain some of the same animal and plant fossils, documenting this former linkage. East of the supercontinent was a large, wedge-shaped area of ocean.

After 250 Mya

18.16.b2

3. After 250 Mya, diversification of mammals and other animals began to increase sharply, just as Pangaea began to rift apart, dispersing the continents in various directions. The sea between the northern and southern continents was a barrier to faunal migration. For example, flightless birds of South America and Australia only had access to continental regions that were part of *Gondwana* (South America, Africa, India, Australia, and Antarctica).

4. Eventually, India became a separate, small continent that would carry its "southern" flora and fauna northward, until it collided with Asia 45 Mya. New land produced by volcanoes linked South America with Central America 3 Mya.

A Land Bridge Between Zoogeographic Realms

Humans have also migrated in response to changing environmental conditions. For hundreds of thousands of years they were restricted to parts of the world with dry-land connections. Around 20,000 years ago, during the *Pleistocene Epoch*, a broad land bridge called *Beringia* formed between Asia and Alaska. With sea level lower, due to water being locked up in glaciers and continental ice sheets, they could walk to Alaska and colonize the New World.

18.16.b3

18.16.b4

Animals like the woolly mammoth (◀), caribou, and saber-toothed cat also made the trip across the Beringia land bridge to join the fauna already in North America. Horses and camels from the Americas ventured westward to Asia. Later, humans also ventured across the land bridge, with their arrival largely coinciding with the extinction of many of the large animals (called *megafauna*), but whether this arrival was a cause or was coincidental with the megafauna extinction is being debated.

18.16.b5

The Panama Connection

18.16.b6

Before 5 Mya, the Atlantic flowed into the Pacific through the Central American Seaway, but this connection was closed with the Panama Land Bridge. In addition to allowing the migration of armadillos, opossums, and porcupines northward and deer, horses, and large cats southward, the Gulf Stream strengthened and may have triggered the Pleistocene glaciation. Off the coast of South America the Humboldt Current created a rich upwelling zone for nutrients. ENSO events may have also begun with the formation of the land bridge.

Before You Leave This Page Be Able To

☑ Explain the concept of zoogeographic realms, using several examples.

☑ Describe how the changing connections and positions of the continents affected the distribution of fauna and flora.

☑ Explain the importance of land bridges in affecting animal distributions and migrations.

18.16

18.17 What Is Sustainability?

SUSTAINABILITY, in the context of ecosystems and biomes, refers to whether the environment and its various components operate in a manner that allows them to continue operating without degrading the environment. What is required for sustainability, what threatens sustainability, and what are some opportunities for us to support sustainability? Like exploring other complicated issues, one way to examine sustainability is with a SWOT analysis—analyzing Strengths, Weaknesses, Opportunities, and Threats.

A What Issues Are Involved With Sustainability?

Sustainability can be defined as effective management that will allow long-term endurance and success, and it can refer to many different aspects of our world. The notion of sustainability implies the existence of responsibility, wisdom, foresight, cooperation, persistence, and balance. Here, we examine sustainability in the context of a biome and its ecosystems.

1. Sustainability requires the consideration of three partially overlapping aspects—*environmental*, *economic*, and *social*, as illustrated in the diagram below. The first step in achieving sustainability is to gather data and ideas about how to maximize the benefit to each of these components, without harming the others.

18.17.a5

18.17.a2

18.17.a1

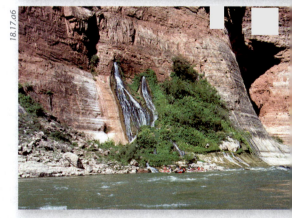
18.17.a6

2. Consider the management of a national park. The images on this page are all from Grand Canyon National Park in Arizona, mostly taken during hikes up the many side canyons that branch off from the Colorado River, which flows through and carved the canyon. Examine each photograph and think about the balance that should be achieved to ensure that the park is healthy environmentally, economically, and socially. What if nobody uses the park? What if it is overused? What if it isn't accessible to some segments of the population? What is the optimum approach that balances all of these considerations? Will such a plan for sustainability need to be revisited over time?

18.17.a3

18.17.a4

18.17.a7

3. Can you think of any drawbacks to trying to achieve sustainability in the national park example? What about in our management of the biosphere?

B What Are Some Threats and Opportunities for the Sustainability of Biomes?

The often-delicate interactions within an ecosystem—or within a regional biome—can be easily disturbed. Threats are things that might prevent us from achieving sustainability. Awareness of threats is a prerequisite for conquering them.

1. This map shows the amount of land that has been taken over by humans (domesticated) and the biomes of those regions that have not. Observe this map and consider all the factors—climate, biome, soil, water availability, suitability for agriculture, and livability, to name a few—that probably influenced whether land was domesticated or not. There are many other factors, including those brought up earlier in this chapter or in other chapters.

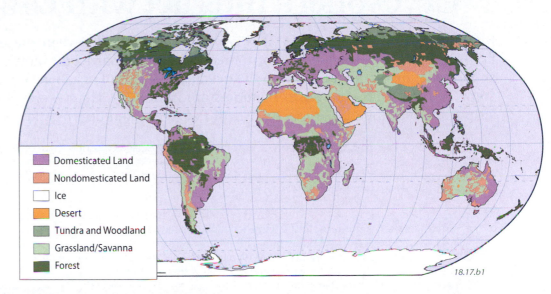

18.17.b1

Legend:
- Domesticated Land
- Nondomesticated Land
- Ice
- Desert
- Tundra and Woodland
- Grassland/Savanna
- Forest

2. As this map reveals, most of the land area has not yet been domesticated. Is this what you expected? Consider all the factors that determine whether the nondomesticated land in this map is used wisely, and remember that even nondomesticated lands are influenced by humans, sometimes heavily. Should only the local inhabitants decide what happens in their home? Should government officials make these decisions instead? The answer may be different for a national park versus a small plot of tundra.

18.17.b2

18.17.b3

18.17.b4

3. Rain forests are among the most threatened biomes, mostly due to incursions and development by humans. Threatening activities include clear-cutting rain forests for commercial logging and commercial or subsistence agriculture. Sometimes, the most apparent and quick path out of poverty involves exploiting such natural resources.

4. Marine and freshwater ecosystems are likewise vulnerable. Marine ecosystems are damaged by overfishing, pollution dumped into the ocean, and potentially by ocean acidification, where the pH of the ocean becomes more acidic due to increased amounts of CO_2 that move from the atmosphere to the water.

5. Certain large land animals have had their populations decimated by human activities. The most damaging activity is usually encroachment of human developments into the habitat of such creatures. Poaching, such as occurs for rhinoceros (▲), has nearly caused some large mammals to become extinct. Some animals, such as mountain gorillas, remain highly threatened, but the situation has recently improved for others.

18.17.b5 Kilimanjaro, Kenya

18.17.b6 Monterey Bay Aquarium, CA

6. There are opportunities to preserve threatened ecosystems. Some species have been greatly aided by conservation efforts, but far too many remain on the endangered list. Conservation efforts include a mix of education, enforcement, and empowerment.

7. Education, like at aquariums (▲), is a key in many situations, enabling people to take other pathways to prosperity and avoiding inadvertent damage. Science is showing that sustainability makes sense, and it has a history of, eventually, driving human behavior.

Before You Leave This Page Be Able To

✓ Describe the concept of sustainability and its components as it pertains to ecosystems and biomes.

✓ Describe some threats to sustainability and some opportunities to preserve the diversity in biomes.

18.17

18.18 How Do the Atmosphere, Hydrosphere, and Cryosphere Interact with Land to Form Biomes?

BIOMES ARE THE CULMINATION of the interactions between the atmosphere, hydrosphere, cryosphere, and lithosphere as they intersect to form the biosphere. A particular biome arises because of regional and local atmospheric regimes, its distance from the ocean, the kinds of ocean currents nearby, whether it is cold enough to have snow and ice, and what Earth materials are underfoot. Here, we explore these connections using a longitudinal strip from near the equator to the North Pole, centered on the Atlantic Ocean.

Observe this large diagram, the main part of which shows large-scale atmospheric circulation, semipermanent high-pressure and low-pressure areas, warm and cold ocean currents, and the lands flanking the Atlantic Ocean. The colors of the land are derived from satellite images, with green representing vegetation and tan and brown indicating sand, soil, and bare rocks. Examine how the patterns of vegetation on land correspond to the large-scale patterns in the atmosphere and ocean.

Tropics—Near the equator, warm and moist equatorial (E) air masses rise because this part of the Earth receives more intense insolation than latitudes farther north and south. The rising air causes the region to have low atmospheric pressure. As the air rises, it cools and results in heavy and regular precipitation over the tropics. The abundant water allows luxuriant and diverse plant growth, with many K-selected species, but also causes severe leaching of soluble components of the soil, resulting in reddish, leached *Oxisols*. The combination of high temperatures, lack of freezing, and abundant rainfall is characteristic of a *Tropical Rain Forest climate* (*Köppen Group A*) and produces a *tropical rain forest biome* with its characteristic flora and fauna.

Subtropics—In the subtropics, near 30° N latitude, descending air from the northern limb of the Hadley cell causes the underlying region to have high atmospheric pressure (subtropical highs, in this case the Bermuda-Azores High) and to be exceptionally dry. The dry, descending air stifles the formation of clouds and precipitation. Evaporation exceeds precipitation in these "biological desert" oceanic areas, increasing the salinity of the water. The lack of high-discharge rivers flowing into the North Atlantic reinforces the relatively high salinity, and this region becomes part of the global thermohaline circulation. The clear skies allow unrestricted insolation to the land and ocean, which warms and heats the overlying air. The land warms and cools faster than the water because it has a lower heat capacity (less thermal inertia). In the ocean offshore of Africa, the south-flowing Canary Current, part of the North Atlantic gyre, brings cold water southward along the coast, further stabilizing the atmosphere by reducing the environmental lapse rate and in some cases causing a temperature inversion, which in turn limits precipitation. The dry conditions from continental tropical (cT) air masses result in *Aridisols*, with their limited amount of organic material and calcium carbonate–rich B-horizon. The combination of descending dry air, sunny skies, and limited humidity and precipitation results in an *Arid climate* (*Köppen Group B*) and a *desert biome*. Plants and animals in this biome are adapted to the high temperatures and limited amount of water.

18.18.a1

Tropical Rain Forest

18.18.a2 Cambodia

18.18.a3 Costa Rica

Desert

18.18.a4 Namibia

18.18.a5 Zion NP, UT

Mid-latitudes—The mid-latitudes, near 45 to 60° N latitude, coincide with Great Britain east of the Atlantic Ocean (east is toward the bottom of the page) and eastern Canada west of the Atlantic. These areas are beneath rising air at the polar front, which also forms part of the polar air-circulation cell. The rising air causes semipermanent low pressure, in this case the Icelandic Low. As the moist air rises, its water vapor condenses into clouds and precipitation. The air is cold enough during the winter months to cause snow. Cold air flowing along the surface from the north into the Icelandic Low has an apparent deflection to the right, becoming polar easterlies. The west and southwest-moving air collides with air flowing from the southwest along the polar front, generating mid-latitude cyclones and trailing cold and warm fronts. This often involves the clash between maritime tropical (mT) air masses and continental polar (cP) air masses. The region is also under the influence of the polar front jet stream, which is embedded in the Rossby waves with its ridges and troughs that modulate the surface weather patterns. The combination of rising and converging air, low pressure, and the overlying jet stream results in an overall wet climate. Under the warming influence of the Gulf Stream, Great Britain maintains a *Marine West Coast climate* (Köppen Group C). Seasonal shifts in the strength and position of the Icelandic Low and Bermuda-Azores High (the North Atlantic Oscillation) cause climate variability. Great Britain has mostly *Alfisols* and *Inceptisols* and a *temperate forest biome*.

Polar Regions—In very high latitudes, at and adjacent to the North Pole, the land receives insolation at a very low angle of incidence in summer and is dark for months at a time during the winter. As a result, the land and water do not warm much and the air stays cold much of the year. Added to the weak insolation is cold air descending over the pole, the downward limb of the polar cell—the Polar High. The cold temperatures, low insolation, and semipermanent high pressure of these arctic (A) air masses limit the amount of water vapor in the air and precipitation, which in these cold latitudes is generally snow. The high albedo of snow reflects much of the paltry amount of insolation on it, in a positive feedback mechanism. The climate is *Polar* (Köppen Group E) or *Subarctic* (Köppen Group D). Permanently frozen ground (permafrost) at shallow depths—overlain by *Gelisols*—limits tree growth but allows ground-hugging cold-tolerant plants, the signature of the *tundra biome*. The harsh conditions limit the biodiversity, but allow polar bears, caribou, and other polar creatures to survive.

Polar Cell

Gulf Stream

Labrador Current

L Icelandic Low

H Polar High

Gulf Stream

Azores Current

Differences Across the Atlantic—The two sides of the Atlantic both fit this general pattern of tropical, subtropical, temperate, and polar climates, moving from equator to pole, but with some clear differences. Note for example that in the subtropics, Africa is clearly a desert (the tan colors characteristic of very limited vegetation), but areas at the same latitude on the Americas side of the Atlantic look fairly green. These areas, like Florida, are at the right latitude to be subtropical deserts, but they are bathed in a flow of moist air from the Caribbean, Gulf of Mexico, and Gulf Stream. If you move farther west, into the interior of the continent, at these same latitudes you reach southern Arizona, definitely a desert. Also, eastern Canada is subjected to a combination of colder air from the continent and moist air from the Atlantic, and so it has a *Humid Continental climate* (Köppen Group D); it is substantially colder than Great Britain, even though they are at equivalent latitudes. Great Britain experiences the relatively warm waters of the Gulf Stream, whereas northeastern Canada is adjacent to southward-flowing cold currents west of Greenland. Farther north in Canada, even colder temperatures cause the region to be a Subarctic climate with *Spodosols* and a *boreal forest biome*. Even farther north exists a Polar climate, *Gelisols*, and a *tundra biome*.

Tundra

18.18.a6 Quebec, Canada

18.18.a7

18.18.a8 ANWR, AK

18.18.a9 Churchill, Canada

18.19 What Factors Might Control the Biomes in This Location?

THE INVESTIGATION at the end of chapter 1 asked about possible impacts of logging of a mountainside, the same area shown here. Now that you have acquired a more heightened sense of the Earth-ocean-atmosphere-cryosphere-biosphere system, let us revisit the problem—and the entire environment—from a wide array of perspectives.

Goals of This Exercise:

- Review the list that you created in chapter 1 of the many possible impacts of deforestation in this area and elaborate on, refine, or dismiss the relevancy of each point.
- Examine the area and identify all the aspects that helped shape the observed landscape.
- Identify five areas or processes that might be a concern from the standpoint of sustainability. For each, conduct a SWOT analysis, factoring in all the relevant aspects presented in this book.
- Your instructor may have you develop a land-use plan to make this place as sustainable as possible.

This three-dimensional perspective shows an area with a variety of terrains and climates. Make as many observations as you can about the landscape and record these on a sheet of paper or on the worksheet. Complete each step in the "Procedures" on the facing page. Listed around the figure are some of the issues you should consider, but there are others not listed. Think broadly and be creative, using the knowledge and observational skills you have gained through the geographic approach.

Patterns of local atmospheric circulation, including prevailing wind direction and precipitation patterns:

Rock types:

Soil types:

Slope stability:

Weathering and erosion:

Local climate:

Types of vegetation:

Natural space versus developed space:

Procedures

1. Examine the scene and think about all the factors that are important with regard to the issue about deforestation and environmental threats to sustainability. Consider all the factors indicated by labels arranged around the figure, and think of other factors that are not listed. To help you organize your thoughts, list these factors. Compare your ideas with those you produced when you did the Investigation in chapter 1. Elaborate, refine, or replace any item based on what you can now surmise about this area (for example, what is the likely climate, soil, ecosystems, and biome).

2. Think broadly about the entire scene and each of its components. For each of the main components, sketch and explain the main processes involved in each system (e.g., the glacier and snow), including what conditions were required to have that system develop, what processes occur within that system (e.g., precipitation as snow), and what factors could influence how that system operates in the future (e.g., weather and climatic patterns that cause more or less snow). Complete one annotated sketch for at least five systems within this area. There are more systems than this, so choose the ones you think are most important or most interesting. Be sure to consider all aspects by reviewing the title of each chapter in the book to remind yourself of all the different avenues that might be important.

Flora and fauna on the land and in the freshwater biome, including birds:

Large-scale patterns of atmospheric circulation:

Possible seasonality:

Water supplies from the streams and lake:

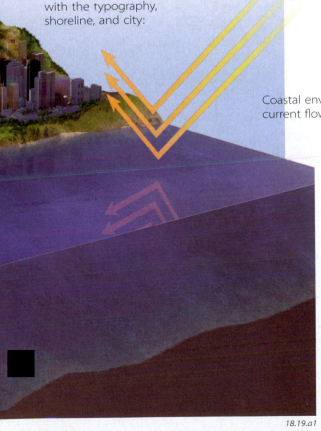

Microclimates associated with the typography, shoreline, and city:

Impact on solar incidence and storage of solar-derived energy:

Coastal environment, which features a current flowing parallel to the coast:

Vertical currents in the ocean and their effect on the marine ecosystems:

Plants and creatures in the upper part of the ocean (pelagic) and near the seafloor (benthic):

Oscillations in sea temperature and air pressure:

Optional Activities

Your instructor may have you complete the additional activities listed below. For these activities, type your answers into a word-processing document and submit this to your instructor in the manner in which she or he indicates (paper copy versus electronic copy).

3. For the system you sketched and explained in step 2 above, conduct a SWOT analysis to consider any factors that would affect the sustainability of that system.

4. Produce a land-use planning document that considers each system sketched in step 2, to develop a plan about how each part of the area would best be used. Your instructor will give you guidance about the length and format of the document.

18.19.a1

18.19

GLOSSARY

A

aa A blocky type of lava with a rough, jagged surface, or a lava flow characterized by jumbled, angular blocks of hardened lava. (11.3)

abiotic component The part of an ecosystem not directly produced by living organisms, including the air, rocks, soils, and water that provide a home for the plants and other organisms. (17.1, 18.3)

abrasion The mechanical wearing away of rock surfaces by contact with sand and other solid rock particles transported by running water, waves, wind, ice, or gravity. (14.4)

absolute humidity The mass of water vapor present per cubic meter of air. (4.2)

absolute zero The temperature at which all molecular motion theoretically ceases (i.e., there is no internal kinetic energy); occurs at 0 Kelvin, or –273°C. (2.3)

absorption The concentration of electromagnetic energy within the molecular structure of matter. (2.1)

abyssal plain A relatively flat, smooth, and old region of the deep ocean floor. (10.2)

abyssal waters Any water near the ocean floor, generally circulating toward the equator. (6.7)

acidic A material, such as water or soil, having a high concentration of hydrogen ions. (16.4)

acid rain An acidic secondary pollutant in the atmosphere formed when sulfur dioxide or nitrogen oxides interact with water in precipitation, either before or after the precipitation falls to the surface. (7.10, 17.11)

acre-foot The volume of water required to cover an acre of land to a height of one foot; equivalent to about 326,000 gallons or 1.23 million liters. (8.5)

active dune A dune of any type that is actively moving and being reshaped by the wind. (9.11)

active fraction The component of a soil consisting of decomposing organic matter. (16.1)

active layer In tundra, the zone between the surface and the permafrost that undergoes a freeze-thaw cycle each year. (18.13)

active remote sensing system A type of remote sensing in which the environment is monitored as the sensor emits energy and then detects properties of that energy after it interacts with matter in the atmosphere or near the surface and returns to the platform. (1.11)

adenosine triphosphate (ATP) A molecule produced as energy is released by "excited" electrons during the light reaction of photosynthesis. (17.5)

adiabatic A condition of a system in which energy does not enter or leave the system. (4.5)

adret slope The side of a slope that receives more direct sunlight, and is therefore warmer and drier. (18.4)

advection fog A type of fog that occurs when warmer air flows laterally over a colder surface, causing it to chill from beneath to the dew-point temperature or frost-point temperature. (4.9)

advection The horizontal transfer of energy through a gas, liquid, or weak solid. (2.2)

aeolian Any process relating to the wind's ability to shape the landscape or a geomorphic feature, primarily caused by wind deposition. (9.11, 16.5)

aerosol Any tiny solid or liquid particle suspended in the air. (2.1)

African easterly jet A west-flowing, fast stream of air in the northeast trade winds over or near northwestern Africa. (5.13)

aftershock A smaller earthquake that occurs after the main earthquake, in the same area. (11.16)

agent of transport Any phenomenon, including gravity, streams, glaciers, waves, and wind, that moves material, which can later be deposited to create landforms. (9.10)

agricultural lime Pulverized limestone used to improve fertility in acidic soils. (16.4)

A horizon Topsoil composed of dark gray, brown, or black organic material mixed with mineral grains. (16.7)

air mass A body of air that is relatively uniform in its meteorological characteristics over areas of thousands of square kilometers. (5.1)

air pollution A gaseous, liquid, or solid particle suspended above the surface, and usually introduced into the atmosphere by human activities and having effects that are harmful to life or property. (7.9)

Alaska Current A south-to-north-flowing, warm surface ocean current in the eastern North Pacific Ocean. (6.2)

albedo The percentage of insolation that is reflected by an object. (2.13, 14.9)

Alberta clipper A type of mid-latitude cyclone that forms on the leeward side of the Canadian Rockies and tracks generally east-northeastward. (5.3)

Aleutian Low A prominent area of semipermanent low atmospheric pressure (cyclone), particularly strong in winter, in the subpolar North Pacific Ocean. (3.6)

Alfisol A fertile type of soil, typically with ample accumulation of organic matter in the A horizon and clay eluviation into the B horizon. (16.11)

alkaline A material, such as water or soil, having a low concentration of hydrogen ions; also known as basic. (16.4)

allelopathy A survival strategy of desert vegetation in which a poison is injected from the roots into the soil to prevent other plants from growing nearby. (18.8)

alluvial fan A low, gently sloping mass of sediment, shaped like an open fan or a segment of a cone; typically deposited by a stream where it exits a mountain range or joins a main stream. (9.10, 13.6)

alluvial soil Any soil formed from gravels, sand, silt, and clay that was deposited by streams. (16.5)

alluvium Sedimentary material that was transported and deposited by a stream. (16.5)

alpine glacier A glacier that flows down a valley and tends to be narrow; also known as a valley glacier. (14.1)

altocumulus cloud A type of mid-level, vertically oriented cloud that contains both liquid water and ice. (4.8)

altostratus cloud A type of mid-level cloud that is wide and layered, and composed of both liquid water and ice. (4.8)

amphibole A group of silicate minerals characterized by double chains of silicon-oxygen bonds; most common in igneous and metamorphic rocks. (9.1)

amplitude The height between the trough and the crest of a wave, including electromagnetic, ocean, and seismic waves. (2.5, 15.3)

anabatic wind A local-to-regional-scale wind that blows from a valley uphill to the peak of a hill or mountain; also known as a valley breeze. (3.4)

analysis products Maps based on spatial-interpolation techniques that represent the observed and inferred state of the atmosphere at a given instant or interval of time at the Earth's surface or at a specified vertical constant pressure level. (5.15)

anastomosing channel A low-gradient stream that has semipermanent islands between several strands of the channel that split and rejoin. (13.8)

andesite An intermediate-composition igneous rock that is the fine-grained equivalent of diorite; typically gray or greenish gray, commonly with visible crystals of cream-colored feldspar or dark amphibole. (9.5)

Andisol A site-dominated soil type formed from volcanic ash and other volcanic material. (16.9)

angle of incidence The angle between the direction of incoming energy and a surface. (2.8)

angle of repose The steepest angle at which a pile of unconsolidated grains remains stable. (12.8)

anion A negatively charged ion. (6.4)

annular drainage pattern A drainage network in which the pattern of streams generally encircles a central area; characteristic of terraced areas around a central high or low point. (13.1)

Antarctic Bottom Waters The moderately saline, very cold, and very dense ocean waters in the middle to high latitudes of the Southern Hemisphere. (6.5)

Antarctic Circle The 66.5° parallel of latitude south of the equator; the most northerly latitude in the Southern Hemisphere that experiences 24 continuous hours of daylight (on the December Solstice) and 24 continuous hours of darkness (on the June Solstice). (1.7)

Antarctic Circumpolar Current A west-to-east-flowing, surface ocean current in the high latitudes of the Southern Hemisphere; also known as the West Wind Drift. (6.2)

Antarctic Intermediate Waters A tongue of less saline ocean waters from near Antarctica that forms a wedge below saline waters near the surface in the Southern Hemisphere, centered around 40° S. (6.5)

antecedent stream A stream that predates formation of a geologic structure through which it flows. (13.11)

anticline A fold, generally concave downward (in the shape of an *A*), with the oldest rocks in the center. (11.11)

anticyclone An enclosed area of high atmospheric pressure. (3.9)

aphelion The date of maximum distance between Earth and the Sun during the course of Earth's orbit, at present corresponding to a day on or near July 4. (2.7)

aphotic zone The part of a lentic system that does not have access to sunlight. (18.14)

aquiclude An impermeable surface that completely blocks flow of a liquid through it. (8.6)

aquifer A saturated body of permeable material through which groundwater can flow and that yields significant volumes of water to wells and springs. (8.6)

Arctic (or Antarctic - A) air mass A body of air that is relatively uniform in its very cold and very dry characteristics for thousands of square kilometers, originating in high-latitude source regions that are either inland or over sea ice. (5.1)

Arctic Circle The 66.5° parallel of latitude north of the equator; represents the most southerly latitude in the Northern Hemisphere that experiences 24 continuous hours of daylight (on the June Solstice) and 24 continuous hours of darkness (on the December Solstice). (1.7)

arête A hard, jagged bedrock ridge flanked by two or more cirques. (14.5)

arid (B) climate In the Köppen climate classification, a broad group of climatic types characterized by inadequate moisture availability, in which potential evapotranspiration typically exceeds precipitation. (7.5)

Aridisol A type of soil found in arid regions, characterized by a lack of organic matter, little leaching of nutrients, and limited horizonation; has a propensity for salinization and formation of a caliche layer. (16.10)

artesian Groundwater that is in a confined aquifer and is under enough water pressure to rise above the level of the aquifer. (8.6)

ash flow A fast-moving cloud of hot volcanic gases, ash, pumice, and rock fragments that generally travel down the flanks of a volcano; also known as a pyroclastic flow. (11.2)

asthenosphere The area of mantle beneath the lithosphere that is solid, but hotter than the rock above it, and can flow under pressure; functions as a soft, weak zone over which the lithosphere may move. (10.1)

Atlantic hurricane season The time of year when Atlantic tropical cyclones are most likely to occur—from June 1 through November 30. (5.13)

Atlantic Multidecadal Oscillation (AMO) An oscillation in sea-surface temperatures in the Atlantic Ocean, in which temperatures tend to be warmer than normal in much of the basin for many years and then colder than normal for many years. (7.12)

atmosphere The gaseous envelope that surrounds the solid Earth. (1.3, 2.1)

atmospheric attenuation The process by which electromagnetic energy is absorbed or scattered in the atmosphere. (2.8)

atmospheric window The part of the electromagnetic spectrum between 8 and 13 micrometers, in which atmospheric absorption of longwave radiation is minimal, except when water vapor is abundant. (2.14)

atoll A curved to roughly circular reef that encloses a shallow, inner lagoon or a circular arrangement of low, coral islands. (15.6)

atomic weight The sum of the weight of the subatomic particles in an average atom of an element, given in atomic mass units. (9.13)

aurora A light that originates from interactions of solar energy and energetic gas molecules, primarily in the thermosphere. (2.1)

autotroph An organism that acquires what it needs from the soil, water, air, and energy of the environment, without consuming other organisms; also known as a primary producer. (17.2)

B

backshore The highest part of the beach, only seldom covered by water; situated between the foreshore and the dune. (15.1)

backwash The flow of water back out to sea after a wave comes ashore. (15.3, 15.4)

badlands A distinctive type of topography with a soft, rounded appearance, usually composed of shale and other easily eroded rocks. (9.7)

bank-full The maximum stream flow without causing a flood. (13.12)

banner tree A wind-shaped tree, commonly near the tree line, that has branches growing only on the downwind (leeward) side of a trunk; also known as a flag tree. (18.4)

barchan dune A crescent-shaped dune of windblown sand with its tails pointing in the direction of the prevailing wind; also known as a crescent dune. (9.11)

barometer An instrument used to measure atmospheric pressure. (3.2)

barrier island A long, narrow island adjacent to a coastline and commonly containing sandy areas, like dunes, marshes, and sandy beaches. (15.5)

barrier reef A long, commonly curved, coral reef that parallels the coast of an island or continent from which it is separated by a lagoon. (15.6)

bar scale A visual representation of the relationship between actual distance and a distance on a map. (1.10)

basal slip The relatively fast sliding of a glacier over bedrock, either in a smooth bedrock-glacier interface or, if the glacier contains water, from pressure melting. (14.3)

basalt A fine-grained, dark-colored mafic igneous rock, with or without vesicles and visible crystals of pyroxene, olivine, or feldspar. (9.5)

basaltic lava flow A lava flow composed of dark, basaltic material. (9.7)

base level The lowest level to which a stream can erode, commonly represented by the level of the sea, lake, or main river valley. (13.4)

base map A map or image upon which data are plotted. (1.10)

basic A material, such as water or soil, having a low concentration of hydrogen ions; also known as alkaline. (16.4)

basin A low-lying area or an area in which sediments or volcanic materials accumulate with no or a limited surface outlet; also can refer to a structural feature toward which rock strata are inclined in all directions. (10.6)

Basin and Range Province The physiographic region in the interior Southwest of the U.S., consisting of almost all of the state of Nevada and parts of adjacent states, and characterized by alternating valleys and mountain ranges, commonly the expression of horsts and graben. (11.10)

batholith One or more contiguous intrusions that cover more than 100 square kilometers. (11.8)

baymouth bar A ridge of sand or gravel deposited entirely or partly across the mouth of a bay. (15.5)

beach A stretch of coastline along which sediment, especially sand and stones, has accumulated. (9.2, 15.1)

beach drift The process by which sand and other sediment move in a zig-zag fashion, but in a general lateral direction along a coast. (15.4)

bed A distinct layer formed during deposition, generally in sediment or sedimentary rock; can be present in volcanic tuff and some metamorphic rocks. (9.8)

bed load Material, commonly sand and larger, that is transported along the bed of a stream. (13.2)

bedrock stream A stream that is carved into bedrock, commonly in mountainous areas. (13.2)

Benguela Current A south-to-north-flowing, cold surface ocean current in the eastern South Atlantic Ocean. (6.2)

benthic Pertaining to an organism in the saltwater biome that resides mainly on the ocean floor. (18.15)

benthic zone The part of the saltwater biome that contains habitats of the sea bottom, whether in coastal, continental shelf, or deep-sea environments. (18.15)

Bergeron process A process of precipitation formation that occurs by evaporation of liquid water, which then allows the additional water vapor to deposit onto solid ice particles, enlarging the snowflake, which can then melt, forming a raindrop, or remaining solid on descent. (4.11)

Bering Current A small north-to-south-flowing, cold surface ocean current into the North Pacific Ocean through the Bering Strait. (6.2)

berm A nearly flat platform in the backshore, caused by wave deposition during the strongest storms. (15.1)

Bermuda-Azores High A prominent area of semipermanent, high atmospheric pressure (anticyclone) in the subtropical North Atlantic Ocean; is particularly strong in summer. (3.6)

B horizon A zone in the soil characterized by the accumulation of material, including iron oxide, clay, and calcium carbonate, depending on the climate and parent materials; also known as the subsoil layer, zone of accumulation, or the zone of illuviation. (16.7)

biodiversity The number of species and the evenness of the populations of those species in an ecosystem. (17.4)

biogeochemical cycle The movement of matter throughout the atmosphere, hydrosphere, lithosphere, and biosphere by interaction with living and dead organisms and involvement in chemical and physical processes. (1.4)

biogeography The branch of physical geography that explores the biosphere and the Earth systems that support organisms. (17.0)

biological oxygen demand (BOD) The amount of oxygen needed by the bacteria in an aquatic ecosystem. (17.12)

biomass The mass of carbon contained in a living or dead organism or in a collection of organisms. (17.6)

biome A number of adjacent ecosystems defined by the predominant types of vegetation and characterized by the presence of organisms that are adapted to that environment. (7.1, 18.0)

biosphere The spherical zone that includes life and all of the places life can exist on, below, and above Earth's surface. (1.3, 17.0)

biotic component The part of an ecosystem consisting of living plants and animals, along with materials such as digestive wastes, discarded plant parts, and decaying remains of creatures. (17.1)

biotic weathering Any physical or chemical weathering process caused by organisms. (12.4)

birdfoot delta A nearly flat tract of land formed by deposition of sediment at the mouth of a stream, formed by distributaries and having separate "toes" spreading out at the mouth. (13.9)

blocky soil A soil structure in which the peds are shaped like irregular chunks, which may hinder root penetration. (16.2)

blue jet A dim, blue, smoke-like streak of light that bursts out above a thunderhead, arcs upward high into the atmosphere, and then fades away. (5.9)

bluff The steep slope separating adjacent terraces or a terrace from the floodplain; also known as a riser. (13.8)

bog A wetland that is fed mostly by precipitation rather than groundwater, surface runoff, or streams; usually nutrient-poor. (18.14)

boiling point The temperature at which a liquid boils; that is, the temperature at which the vapor pressure in a liquid increases to a value equal to the pressure of the gas adjacent to the liquid, thereby forcing the molecules in the liquid to leave the liquid spontaneously. (2.3)

bomb (volcanic) A large rock fragment representing either a large mass of magma or a solid angular block ejected during an explosive volcanic eruption. (11.1, 11.6)

boreal forest biome A collection of adjacent ecosystems that consists of a few hardy cone-bearing (coniferous) tree species that have needles instead of normal flat leaves, and local stands of deciduous species, similar to some species in temperate forests. (18.1)

boulder A clast or rock fragment with a diameter exceeding 0.25 m. (13.2)

bow echo The arc-shaped pattern on Doppler radar that is characteristic of a derecho. (5.12)

Bowen's reaction series An idealized sequence in which minerals could crystallize from a magma as it cools. (12.3)

Boyle's Law The inversely proportional relationship between pressure and volume of a gas, under constant temperature. (3.1)

brackish Water that is intermediate in salinity between normal fresh water and seawater. (8.1)

braided stream A stream with an interconnecting network of branching and reuniting shallow channels. (13.5)

Brazil Current A north-to-south-flowing, warm surface ocean current in the western South Atlantic Ocean. (6.2)

breakwater An offshore structure (such as a wall), typically parallel to the shore, that breaks the force of the waves to protect a shoreline or harbor. (15.7)

breccia A rock composed of large, angular fragments; typically formed in sedimentary environments, but can be formed by volcanic, hydrothermal, or tectonic processes. (9.4)

brine Highly saline water. (6.6)

brine exclusion The nonparticipation of salt in the freezing of seawater, which makes the surrounding water saltier. (6.6)

brittle deformation A type of structural deformation that is characterized by breaking of material, such as by jointing, faulting, and the formation of tectonic breccia; typical of shallow crustal conditions. (11.9)

broadleaf Any tree with leaves rather than needles. (18.11)

buffering mechanism A feature of an ecosystem that helps species avoid stress and thrive. (17.7)

C

calcification A pedogenic regime characterized by the upward movement of soil moisture toward dry air in arid and semiarid regions, depositing calcium carbonate (caliche) in the B horizon. (16.6, 16.8)

calcite A common rock-forming calcium carbonate ($CaCO_3$) mineral occurring in limestone, travertine, and a variety of water-related deposits. (9.1)

caldera A large volcanic depression that is typically circular to elongated in shape and formed by collapse of a magma chamber. (11.1, 11.5)

caliche A soil-related accumulation of calcareous material that cements sand, gravel, and other materials, commonly forming a hard layer in arid environments. (16.6, 16.8)

California Current A north-to-south-flowing, cold surface ocean current in the eastern North Pacific Ocean. (6.2)

calorie A unit used to measure energy input. (17.6)

calving The breaking off of a mass of ice from a glacier, iceberg, or ice shelf. (14.3)

Cambrian explosion The widespread, relatively rapid appearance of diverse types of hard-shelled organisms near the beginning of the Cambrian Period. (9.13)

CAM plant Any plant that conducts C_4 photosynthesis at some times and C_3 at others, in an effort to enhance its chance of survival. (17.5, 18.8)

Canadian Shield The region of North America, mostly in Canada, that consists of relatively old crystalline metamorphic and igneous rocks. (10.13)

Canary Current A north-to-south-flowing, cold surface ocean current in the eastern North Atlantic Ocean. (6.2)

canopy The second-highest layer of life in a rain forest, typically forming an umbrella over the forest floor. (18.5)

capacity (stream) The amount of sediment, at a given flow rate, that a stream is capable of transporting. (13.2)

capillary action The ability of water, aided by surface tension, to travel upward against the force of gravity. (4.1)

capillary water Subsurface water that is within pore spaces between the grains but not adhering to the soil particles, and therefore is available to move upward through soil toward the surface. (16.3)

carbonate minerals A group of minerals containing a significant amount of the carbon-oxygen combination called carbonate. (9.1)

carbon cycle The movement of carbon between the biosphere, atmosphere, hydrosphere, and lithosphere. (1.4)

carbonic acid A weak acid (H_2CO_3) formed in water and serving as a major agent of chemical weathering. (12.2)

carnivore Any heterotroph that exclusively or mainly feeds on animals. (17.2)

carrying capacity The theoretical maximum population of a given species that an ecosystem can support. (17.3)

cartogram A type of map that distorts areal features (e.g., states, countries, or other political unit) so that their size conforms to the magnitude of the variable associated with that state, country, or other political unit being mapped. (17.11)

cartographer A person who makes maps. (1.9)

catastrophism The concept that emphasizes the importance of sudden events in creating change to a landscape, ecology, or other system. (17.7)

cation A positively charged ion. (6.4, 16.4)

cation exchange capacity The ease with which a cation in soil solutions is exchanged with the surface of organic matter or clay particles; high cation exchange capacity is generally associated with agriculturally productive soils. (16.4)

cave A natural underground chamber that is generally large enough for a person to enter. (12.2)

Celsius scale A system for indexing temperature calibrated such that water at sea level freezes at 0°C and water at sea level boils at 100°C. (2.3)

cementation The precipitation of a binding material (the cement) around grains in sedimentary rocks. (9.4)

centripetal drainage pattern A drainage network in which the pattern of streams is generally inward toward a central point. (13.1)

chalk A type of soft, very fine-grained limestone that forms from the accumulation of calcium carbonate remains of microscopic organisms that float in the sea. (9.4)

channelized Flowing water that is confined to a channel, either naturally or through artificial means. (13.6)

chaparral The distinguishing type of vegetation in the subtropical scrub and woodland biome, characterized by scrubby trees with thick bark and short, gnarly shrubs, thickets, and grasses with deep root systems; also known as maquis. (18.10)

Charles's Law The proportional relationship between temperature and volume in a gas at a constant pressure. (3.1)

chemical weathering Chemical reactions that affect a rock or other material by breaking down minerals and removing soluble material from the rock. (9.9, 12.1)

chemosynthesis A process by which inorganic compounds, such as hydrogen sulfide or diatomic hydrogen, make energy available for synthesis of organic molecules when combined with oxygen. (17.5, 17.11)

chert A silica-rich, nonclastic sedimentary rock that typically forms from the accumulation and compaction of tiny, silica-rich plankton shells that fall to the ocean bottom, or also as layers and irregular masses within limestone that form from the mixing of different types of water. (9.4)

Chile Current A local name for the Humboldt Current. (6.2)

Chinook wind A warm, dry wind that descends the leeward side of mountain ranges in western North America. (3.5)

chlorofluorocarbon A type of chemical released from aerosol cans, air conditioning units, refrigerators, and polystyrene, which contains halogens that can destroy ozone molecules by attracting one of the oxygen atoms away from them. (2.12)

chloroplast A specialized organelle, on a plant cell or in some algae, that captures insolation during photosynthesis. (17.5)

C horizon A zone in the soil composed of weathered bedrock or sediment, grading downward into unweathered bedrock or sediment. (16.7)

cinder A loose black or red, pebble-sized material, often a type of basalt, that composes a cinder cone. (11.1)

cinder cone A relatively small type of volcano that is cone shaped and mostly composed of volcanic cinders (scoria); also known as a scoria cone. (11.1)

circle of illumination The circle that separates the half of the Earth that is experiencing daylight from the half that is experiencing darkness at a given time. (2.10)

circulation cell A large, somewhat rectangular, vertically oriented circulation system in which air rises, moves laterally, sinks, and then moves laterally to replace the air that rose. (3.9)

cirque An open-sided, bowl-shaped depression formed in a mountain, commonly at the head of a glacial valley; produced by erosion of a glacier. (14.5)

cirque glacier The smallest type of alpine glacier; forms in a glacial depression that is somewhat sheltered from the Sun. (14.1)

cirriform cloud A family of cloud types that is "feathery" or wispy and composed of ice crystals, usually high in the troposphere. (4.8)

cirrocumulus cloud A type of high-level cloud that is partly cumuliform (lumpy) and partly cirriform (wispy) and almost entirely composed of ice. (4.8)

cirrostratus cloud A type of high cloud that is somewhat wispy but with a continuous, sheetlike appearance. (4.8)

cirrus cloud A type of high-level cloud with a wispy appearance, composed of ice particles. (2.13, 4.8)

clast An individual grain or fragment of rock, produced by the physical breakdown of a larger rock mass. (9.4)

clastic rock Rock consisting of pieces (clasts) derived from preexisting rocks; usually formed on Earth's surface in low-temperature environments. (9.4)

clay Any fine-grained sedimentary particle with a diameter less than 1/256 mm; also, a family of finely crystalline, hydrous-silicate minerals with a two- or three-layered crystal structure. (9.1, 16.1)

Clean Air Act of 1970 The (U.S.) federal legislation that mandated the formation of the federal Environmental Protection Agency along with other similar agencies at the state level, for the purpose of monitoring certain widespread pollutants nationwide and keeping the concentrations of each below a "threshold" level deemed hazardous. (7.9)

clear-cutting Deforestation in which all the trees in an area are removed. (18.6)

climate The long-term pattern of weather at a place, including not just the average condition but also the variability and extremes of weather. (7.0)

climatic water balance An accounting of the amount of water in each store and its rate of movement through the hydrologic cycle. (8.3)

climax community The culmination of succession, leading to an ecosystem with an intricate food web, efficient nutrient cycling, and biodiversity and productivity determined by the abundance or lack of necessary abiotic factors. (17.7)

climograph A graph that shows average monthly temperature (with a line) and precipitation (with bars) for a location. (7.4)

closed system Any interconnected set of matter, energy, and processes that does not allow material and energy to be exchanged outside itself. (1.3)

cloud-to-cloud lightning The electrical charges that are exchanged suddenly but only between two clouds; comprises the majority of lightning strikes. (5.9)

cloud-to-ground lightning Electrical charges exchanged suddenly between a cloud and the surface; comprises about 25 percent of lightning strikes. (5.9)

coal A natural, brown to black rock derived from peat and other plant materials that have been buried, compacted, and heated; used for heating and to generate electricity. (9.4)

coastal plain A plain that is barely above sea level and adjacent to the coast. (10.13)

coastal zone The area including the shoreline and a strip of adjacent land and water; specifically, the part of the saltwater biome nearest to the shore, characterized by wave action, temperature variations, and changes in substrate; also known as the intertidal zone. (15.1, 18.15)

cobble A clast or rock fragment larger than a pebble and smaller than a boulder, having a diameter in the range of 64 to 256 mm. (13.2)

col A low point on an arête, which can provide a pathway from one side of the ridge to the other. (14.5)

Cold Desert (BWk) climate In the Köppen climate classification, a climate type characterized by inadequate moisture availability, in which potential evapotranspiration typically exceeds precipitation by a large amount, and a mean annual temperature below 12°C (54°F). (7.5)

cold front The sloping boundary surface between an advancing mass of relatively cold air and a warmer air mass. (5.2)

cold phase (ENSO) A periodic cooling of the eastern equatorial Pacific surface ocean waters, and reduction in precipitation adjacent to South America, as the Southern Hemisphere's summer approaches; also known as La Niña. (6.10)

cold sector The part of a mid-latitude cyclone that is behind the cold front, characterized by cold air. (5.2)

Cold Steppe (BSk) climate In the Köppen climate classification, a climate type characterized by inadequate moisture availability, in which potential evapotranspiration typically exceeds precipitation by a small or moderate amount, and a mean annual temperature below 12°C (54°F). (7.5)

collision-coalescence A process of precipitation formation by the repeated merger of tiny liquid water droplets. (4.11)

collision regime The scenario of coastal inundation during which the height of the swash zone exceeds the base of coastal dunes but does not exceed the top of coastal dunes. (15.11)

colluvium Sedimentary material that was transported to a location after being loosened and moved downhill via mass-wasting processes. (16.5)

columnar joint A distinctive fracture-bounded column of rock formed when hot but solid igneous rock contracts as it cools. (9.7, 11.10)

columnar soil A soil structure in which the peds are taller than they are wide, commonly forming in arid climates where soil dries out and cracks. (16.2)

commensalism A type of symbiosis in which one species benefits from the interaction but the other is largely unaffected. (17.3)

compaction Process by which soil, sediment, and volcanic materials lose pore space in response to the weight of overlying material. (9.4)

compensation point The level within a body of water where the production of oxygen through photosynthesis is balanced by the use of oxygen by respiration or fermentation. (17.12)

competence The largest size of particle that a stream can carry. (13.2)

competition A situation in which members of two or more species, or two or more individuals of the same species, are seeking the same scarce resources. (17.3)

composite volcano A common type of volcano constructed of alternating layers of lava, pyroclastic deposits, and mass-wasting deposits, including mudflows; also known as a stratovolcano. (11.1)

compression The type of differential stress that occurs when forces push in on a rock. (11.9)

condensation A change in state from a gas to a liquid, which releases latent energy to the environment. (2.4)

condensation nuclei The solid aerosols that facilitate the conversion of water vapor to liquid water around them. (4.11)

conditional instability (atmosphere) A situation in which the atmosphere is stable if unsaturated but unstable if saturated; the environmental lapse rate is between the unsaturated adiabatic lapse rate and the saturated adiabatic lapse rate. (4.5)

conduction The transfer of thermal energy by direct contact, primarily between two solids. (2.2)

cone of depression A depression in the water table that has the shape of a downward-narrowing cone; develops around a well that is pumped, especially one that is overpumped. (8.8)

confined aquifer An aquifer that is separated from Earth's surface by materials with low permeability. (8.6)

conformal projection Any type of mathematical algorithm for representing the three-dimensional Earth on a flat map that accurately preserves the shapes of features depicted. (1.9)

conglomerate A coarse-grained clastic sedimentary rock composed of rounded to subrounded clasts (pebbles, cobbles, and boulders) in a fine-grained matrix of sand and mud. (9.4)

conical projection A family of map projections based on the premise that the globe is projected onto a cone, with the tangent line having no distortion; other locations are neither conformal nor equal area, but retain some advantages of both conformal and equal area maps when depicting small areas. (1.9)

coniferous A cone-bearing tree with needles rather than broad leaves. (18.1)

coniferous evergreen forest A forest occurring in the mid-latitudes, dominated by trees that do not lose their needles seasonally. (18.11)

coning plume A distribution of smoke or steam from a smokestack characterized by similar dispersion upward and downward as the smoke or steam moves laterally downwind. (4.16)

consequent stream A stream that continues to follow the gradient of the landscape that existed when it formed. (13.11)

contact effects Evidence of baking, passage of hot fluids, or some other manifestation of the thermal and chemical effects of a magma chamber or a hot volcanic unit, as expressed in changes to adjacent wall rocks. (9.12)

contact metamorphism Metamorphism that principally involves heating of the rocks next to a magma. (9.6)

continental collision A plate-tectonic boundary where two continental masses collide; also known as a continent-continent convergent boundary. (10.7)

continental crust The type of Earth's crust that underlies the continents and the continental shelves; has an average composition of granite, but includes diverse types of material. (10.1)

continental drift The concept of the movement of continents and other landmasses across the surface of the Earth. (10.3)

continental glacier An ice sheet that covers a large region. (14.1)

Continental Monsoon (Dwa or Dwb) climate In the Köppen climate classification, a climate type characterized by adequate moisture in summer but not winter, a mean temperature of the coldest month below −3°C, and a hot or warm summer. (7.7)

continental platform A broad region that surrounds the continental shield and typically exposes horizontal to gently dipping sedimentary rocks. (10.13)

continental polar (cP) air mass A body of air that is relatively uniform in its cold and dry characteristics for thousands of square kilometers, originating in inland, subarctic source regions. (5.1)

continental rift A low trough or series of troughs bounded by normal faults, especially where two parts of a continent begin to rift apart. (10.6)

continental shelf A gently sloping, relatively shallow area of seafloor that flanks a continent, is underlain by thinned continental crust, and is covered by sedimentary layers. (10.2)

continental shield A central region of many continents, consisting of relatively old metamorphic and igneous rocks, commonly of Precambrian age. (10.13)

continental tropical (cT) air mass A body of air that is relatively uniform in its hot and dry characteristics for thousands of square kilometers, originating in inland, subtropical regions that are isolated from moisture sources. (5.1)

continent-continent convergent boundary A plate-tectonic boundary where two continental masses collide; also known as a continental collision. (10.7)

contour A line on a map or chart connecting points of equal value, generally elevation. (1.6)

contrail A condensation trail; a linear streak of clouds left behind by a jet. (4.8)

convection The vertical transfer of energy through a gas, liquid, or weak solid. (2.2)

convection cell A circulation system in a fluid that includes rising motion where the fluid is relatively hot, sinking motion where the fluid is relatively cold, and lateral movement between the areas of rising and sinking. (2.2)

convection current A flowing liquid or solid material that transfers heat from hotter regions to cooler ones, commonly involving movement of material in a loop. (10.10)

convergence (atmospheric) The horizontal piling up of air, either from two air streams colliding as they move laterally, or from an air current slowing down as it moves laterally. (3.3)

convergence (tectonic) The relative movement of two tectonic plates toward each other. (10.5)

convergent boundary A plate-tectonic boundary where two plates have relative movement toward each other. (10.5)

cool sector The part of a mid-latitude cyclone that is ahead of the warm front, characterized by air that is intermediate in temperature, as compared to that ahead of and behind the cold front. (5.2)

core The inner layer of Earth, below the mantle; interpreted to be divided into a molten outer core and a solid inner core. (10.1)

Coriolis effect An apparent force resulting from Earth's rotation that causes moving objects to appear to be deflected to the right of their intended path in the Northern Hemisphere and to the left in the Southern Hemisphere. (3.7)

counter-radiation The longwave radiation emitted from the atmosphere down to the surface. (2.14)

C_3 plant Any plant that creates sets of two 3-carbon-based molecules in photosynthesis; these include most green plants. (17.5)

C_4 plant Any plant that creates sets of 4-carbon-based molecules in photosynthesis; such plants tend to thrive in hotter and drier environments because water loss through the plant's stomates is reduced in this process. (17.5)

crater A typically bowl-shaped, steep-sided pit or depression, generally formed by a volcanic eruption or meteorite impact. (11.1)

creep The slow, continuous movement of material, such as soil and other weak materials, down a slope. (12.9, 16.13)

crescent dune A crescent-shaped dune of windblown sand with its tails pointing in the direction of the prevailing wind; also known as a barchan dune. (9.11)

crest The highest point of any type of wave. (2.5, 15.3)

crevasse A steep fracture in a glacier caused by internal stresses generated by the glacier's movement. (14.1)

cross bed A series of beds inclined at an angle to the main layers or beds. (9.11)

crosscutting relations The principle that a geologic unit or feature is older than a rock or feature that crosscuts it. (9.12)

crown fire A wildfire that leaps from treetop to treetop. (18.10)

crust The outermost layer of the solid Earth, consisting of continental and oceanic crust. (10.1)

crystalline A mineral that has an ordered internal structure due to its atoms being arranged in a regular, repeating way. (9.1)

crystalline rock A rock composed of interlocking minerals that grew together; usually formed in high-temperature environments by crystallization of magma, by metamorphism, or by precipitation from hot water. (10.13)

crystallization The formation of crystals during the slow cooling and solidification of rock, usually at depth. (9.3)

cuesta Any ridge with a sharp summit and one slope inclined approximately parallel to the dip of layers, resembling in outline the back of a hog; also known as a hogback. (11.11)

cumuliform cloud A family of puffy cloud types that is taller than it is wide. (4.8)

cumulonimbus cloud A cumuliform cloud type that is actively precipitating and has an extensive vertical extent, composed of liquid water near the base and ice near the top; commonly associated with severe weather. (4.8)

cumulus cloud A type of cumuliform cloud that occurs in fair weather, without extensive vertical extent, but under free-convection conditions. (4.8)

cumulus stage The first stage in a single-cell thunderstorm's life cycle, characterized by updrafts that form and grow cumuliform clouds, but without precipitation. (5.6)

cutbank A steep cut or slope formed by lateral erosion of a stream, especially on the outside bend of a channel. (13.5, 18.14)

cutoff meander A meander cut off from the main channel by a new channel that formed when a stream cuts through the neck of a meander. (13.5)

cyclogenesis The formation stage in the life cycle of a cyclone, particularly a mid-latitude cyclone. (5.4)

cyclone An enclosed area of low atmospheric pressure; can also refer to the most intense type of storm of tropical origin in the Indian Ocean basin. (3.9)

cylindrical projection A family of conformal map projections based on the premise that the globe is projected onto a cylinder, with distortion increasing with distance from the tangent at which the globe touches the cylinder. (1.9)

D

dart leader A second flow of electrical energy after the first lightning stroke, which flows through approximately the same pathway as the original stroke between the cloud and the ground. (5.9)

daughter product The element produced by radioactive decay of a parent atom; also known as daughter atom. (9.13)

Daylight Savings Time A one-hour offset of standard time used in much of the United States for the purpose of skewing daylight hours toward evening hours on the clock. (1.13)

debris avalanche A high-velocity flow of soil, sediment, and rock, commonly from the collapse of a steep mountainside. (11.6, 12.11)

debris flow Downhill-flowing slurries of loose rock, mud, and other materials, and the resulting landform and sedimentary deposit; also known as a mudflow. (9.10, 12.11)

debris slide The downslope movement of soil, weathered sediment, or other unconsolidated material, partly as a sliding, coherent mass and partly by internal shearing and flow. (12.11)

December Solstice The day, on or near December 21, when the North Pole (marking where the Earth's axis of rotation intersects the Earth's surface in the Northern Hemisphere) is pointed most directly away from the Sun. (2.9)

deciduous Any tree that loses its leaves or needles seasonally. (18.11)

deciduous temperate forest A forest occurring in the mid-latitudes, dominated by trees that lose their leaves seasonally. (18.11)

decomposition The breakdown of organic materials from dead organisms into inorganic nutrients, primarily by bacteria and fungi. (17.3)

deep-sea thermal vents A type of productive ecosystem within the saltwater biome on the ocean floor where the local energy source is chemosynthesis. (18.15)

deficit The amount of water (in centimeters or inches of precipitation equivalent) that could have evaporated if it were available; equivalent to the difference between potential evapotranspiration and actual evapotranspiration; a measure of drought severity. (8.3)

deforestation The clearing of a forested landscape. (4.6)

deformation Processes that cause a rock body to change position, orientation, size, or shape, such as by folding, faulting, and shearing. (1.4, 9.3, 11.9)

delta A nearly flat tract of land formed by deposition of sediment at the mouth of a stream. (13.9)

dendritic drainage pattern A drainage network in which the pattern of streams is treelike; usually an indicator that underlying rocks have similar resistance to erosion or that the streams have been flowing for a long time. (13.1)

denitrification The movement of nitrogen from the land to the atmosphere when microbes convert other compounds of nitrogen into nitrogen gas, and during combustion of fossil fuels. (17.9)

density An object's mass divided by the volume it occupies. (2.2)

denudation The wearing away of the surface after erosion and transportation remove material that may be loosened by weathering. (9.9)

deposition (atmospheric) A change in state from a gas to a solid, which releases latent energy to the environment. (2.4)

deposition (geomorphology) The dropping of sediment that occurs when the transporting agent can no longer carry the sediment, such as when a flowing stream slows down. (1.4, 9.3)

depth The distance of an object below the surface or below sea level. (1.6)

derecho A regional windstorm characterized by nonrotating winds that can advance over huge areas quickly. (5.12)

desert biome A collection of adjacent ecosystems that consists of vegetation that is especially adapted to survive in places where moisture deficits are typical year-round. (18.1)

desert pavement A natural concentration of pebbles and other rock fragments that mantle a desert surface of low relief. (9.11)

desert varnish A thin, dark film or coating of iron and manganese oxides, silica, and other materials; formed by prolonged exposure at the surface; also known as rock varnish. (9.11)

desertification The process by which a climate becomes increasingly arid, either through natural or anthropogenic causes. (7.13, 18.7)

detritivore Any organism that consumes the remains of decomposing (nonliving) plants and animals in the soil. (17.2)

dew A condensed liquid water drop on a solid surface. (4.4)

dew-point depression The difference between the current air temperature and the dew-point temperature. (4.4)

dew-point temperature The temperature to which a volume of air must be cooled for it to become saturated with water vapor, thereby causing condensation. (4.4)

diabatic A condition of a system in which energy enters or leaves the system. (4.5)

differential erosion The varying rate of wearing away of soil, sediment, and rock in a landscape caused by varying composition of rock types in that landscape. (11.12)

Differential GPS An enhanced GPS that uses a correction signal to check each GPS satellite's signal independently. (1.11)

differential weathering The breakdown of rock at differing rates because of differences in mineral composition, hardness, or other resistance to weathering. (12.3)

dike A sheetlike intrusion that cuts across any layers in a host rock, commonly formed with a steep orientation. (9.12, 11.8)

diorite A medium- to coarse-grained, intermediate-composition igneous rock; equivalent in composition to andesite. (9.5)

dip The angle that a layer or structural surface makes with the horizontal. (11.10)

dip slope A slope caused by the erosion of hard and soft layers parallel to the dipping layers. (11.11)

disappearing stream A stream that vanishes suddenly from the surface as it begins to flow beneath the surface, often captured by pits in karst terrain. (12.7)

discharge The volume of water flowing through some stretch of a stream per unit of time. (13.1)

displacement The movement of a body in response to stress, either from one place to another, or manifested as a change in shape of the body. (11.9)

dissipating stage The final stage in a single-cell thunderstorm's life cycle, characterized by downdrafts, an end of precipitation, and decay of the system. (5.6)

dissolution The process by which a material is dissolved. (9.9, 12.6, 13.2)

dissolved load Chemically soluble ions, such as sodium and chlorine, that are dissolved in and transported by moving water, as in a stream. (13.2)

distributary system The branching drainage pattern formed when a stream branches and spreads out into a series of smaller channels. (13.9)

divergence (atmospheric) The horizontal spreading apart of air, either from two air streams moving laterally in opposite directions, or from an air current accelerating as it moves laterally. (3.3)

divergence (tectonic) The relative movement of two tectonic plates away from each other. (10.5)

divergent boundary A plate-tectonic boundary in which two plates move apart (diverge) relative to one another. (10.5)

diversity (ecology) The variety of life in an area; biodiversity. (17.4)

Dixie Alley The zone of high tornado frequency across the United States Gulf Coast region. (5.11)

doldrums The belt of weak winds along the Intertropical Convergence Zone, between the northeast trade winds and the southeast trade winds. (6.1)

dolomite A carbonate mineral containing calcium and magnesium. (9.1)

Doppler radar A form of remote sensing that emits a microwave signal and relates changes in frequency or amplitude of that wave to the rate of movement of an object that intercepts the wave; used to detect precipitation, storm movement, and motion within a storm. (5.15)

downcutting The process by which a stream incises (cuts) downward as it erodes into its bedrock. (13.2)

downdraft A local downward flow of air, usually on very short timescales. (3.3)

downstream The direction of flow of water from higher elevations to lower ones. (13.3)

drainage basin An area in which all drainages merge into a single stream or other body of water. (8.6, 13.1)

drainage network The configuration or arrangement of streams within a drainage basin; also known as stream system. (13.1)

drapery A thin sheet of travertine formed by the evaporation of water dripping from the ceiling of a cave. (12.6)

driving force A force that causes a tectonic plate to move. (10.9)

drought The dry conditions caused by excessive water-balance deficits. (8.3)

drumlin A commonly teardrop-shaped hill formed when a glacier reshapes glacial deposits. (14.6)

drumlin field A group of drumlins in the same area. (14.6)

dry line The boundary between a warm, dry (cT) air mass and a warm, moist (mT) air mass, in the warm sector of a mid-latitude cyclone. (5.6)

ductile deformation A type of structural deformation that is characterized by the material flowing as a solid, without fracturing; typical of the crust below about 15 kilometers and of the asthenosphere. (11.9)

dust devil A small-scale, rotating dust storm. (5.12)

dynamic equilibrium (streams) A condition in which a stream has neither a net eroding nor depositing capability. (13.4)

dynamic equilibrium (systems) A condition of a system in approximate steady state maintained by equal quantities of matter and energy flowing into and out of a system. (1.3)

dynamic system Any interconnected set of matter, energy, and processes in which matter, energy, or both are constantly changing their position, amounts, or form. (1.3)

dynamo An electrical generator. (10.10)

E

earth flow A flowing mass of weak, mostly fine-grained material, especially mud and soil; also commonly called a mudflow. (12.11)

East Australian Current A north-to-south-flowing, warm surface ocean current in the western South Pacific Ocean. (6.2)

easterly wave A bend in the trade-wind flow, where relatively low atmospheric pressure from the Intertropical Convergence Zone protrudes farther poleward than elsewhere. (5.13)

East Pacific Rise A mid-ocean ridge (divergent boundary) in the South Pacific Ocean that crosses the eastern Pacific and approaches North America. (10.2)

ebb tide The local falling elevation of the sea surface between high tide and low tide. (15.2)

eccentricity A change in the degree to which Earth's orbit deviates from being perfectly circular. (14.9)

ecological niche The unique position of a species, especially its role relative to the other species, in an ecosystem. (17.1)

ecologist A scientist who studies how organisms interact with each other and with their physical environment. (17.1)

ecology The complex set of relations between living organisms and their environment. (17.0)

ecosystem The provisions and constraints of the physical environment, along with the entire community of interacting organisms (plants, animals, and others) in an area. (17.0)

ecosystem functions Dynamic processes that occur in the environment to support an ecosystem. (17.1)

ecosystem stability The degree of fluctuation of the population of species in an ecosystem over time. (17.7)

ecosystem structure A biotic, abiotic, or energy component of an ecosystem. (17.1)

ecotone A broad transition zone between two zones of natural vegetation (biomes), and therefore between two Köppen climate types. (7.1, 18.2)

eddy A swirl in a fluid's current that can form in both horizontal and vertical directions, characteristic of turbulent flow, as in streams or the atmosphere. (13.2)

E horizon A light-colored, leached zone of soil, lacking clay and organic matter; also known as the zone of eluviation. (16.7)

Ekman layer The layer of the atmosphere characterized by an increased turning of the winds with height; from about 100 meters to about 1–2 kilometers above the surface. (3.8)

Ekman spiral The change in the direction of flow from the surface of the ocean downward to the null point, which turns from parallel to the wind direction at the surface to an increasingly rightward (in the Northern Hemisphere) deflection with depth. (6.1)

Ekman transport The apparent deflection of currents in the saltwater biome with depth; the ocean's Coriolis effect. (18.15)

electromagnetic spectrum The full assemblage of energy of all possible wavelengths, resulting from electrical and magnetic fields. (2.6)

electron shell A discrete "level" away from a nucleus in an atom, on which a maximum number of electrons can exist, with successively higher energy content existing on more outward levels. (2.5)

elevation The vertical distance of an object above or below a reference datum (usually mean sea level); generally the height of a ground point above sea level. (1.6)

El Niño A periodic warming of eastern equatorial Pacific surface ocean waters, accompanied by excessive precipitation and flooding adjacent to South America, as the Southern Hemisphere's summer approaches; also known as a warm phase of ENSO. (6.9)

El Niño–Southern Oscillation (ENSO) The combined alterations of oceanic temperatures, atmospheric pressures, and resulting oceanic and atmospheric circulations in the equatorial Pacific Ocean. (6.10)

eluviation The separation or dissolution of soluble constituents from a rock, sediment, or soil by percolation of water; also known as leaching. (16.6)

elve A flat disk of dim reddish light that forms high in the atmosphere after lightning strikes. (5.9)

embayment A place where the coastline curves inward toward the land. (15.0)

emergent coast A coast that forms where the sea has retreated from the land due to falling sea level or due to uplift of the land relative to the sea. (15.9)

emergent layer The highest layer of life in a rain forest, consisting of a few trees that tower up to 70 meters above the forest floor but without forming a closed canopy at that level. (18.5)

endemic Native to an area. (18.16)

energy The ability to do work. (1.3, 1.4)

englacial debris Material transported within the ice of a glacier. (14.4)

Entisol A type of recently developed soil, characterized by no B horizon and little or no profile development. (16.9)

entrainment The mixing of drier air with more humid air inside a cloud, characterized by the evaporation of cloud water and the decay of a cloud. (5.14)

entrenched meander A curved canyon that represents a meander carved into the land surface. (13.11)

environmental lapse rate The rate of change of temperature with height in a static atmosphere. (4.5)

ephemeral Seasonal. (18.8)

ephemeral stream A stream that only flows during some times of the year. (13.3)

epicenter The point on Earth's surface directly above where an earthquake occurs (directly above the focus or hypocenter). (11.13)

epipelagic zone The part of the saltwater biome located seaward of the coastal zone's low-tide mark and containing the vast open waters of the ocean, at shallow depth, above the mesopelagic zone, where light penetrates and separates into colors. (18.15)

epiphyte A plant that lives suspended on other plants and derives its nutrients and water without contacting soil. (18.5)

equal-area projection Any type of mathematical algorithm for representing the three-dimensional Earth on a flat map that accurately preserves areas of features depicted. (1.9)

equator The imaginary line of latitude, equidistant between the North and South Poles, that divides the Earth into halves, and is defined as the 0° parallel. (1.7)

equatorial (E) air mass A body of air that is relatively uniform in its warm and very humid characteristics for thousands of square kilometers, originating in oceanic, equatorial source regions. (5.1)

equilibrium The state of a system characterized by an equal movement of atoms and molecules between two states of matter; also used in ecosystems and other contexts. (4.1)

equilibrium line The zone across a glacier where the losses of ice and snow balance the accumulation of ice and snow; also known as firn line. (14.3)

equilibrium profile A profile reflecting a stream in an approximate steady state where deposition of sediment is balanced by erosion. (13.4)

equitability The evenness of populations of different species in an ecosystem. (17.4)

era A broad division of geologic time, including the Paleozoic, Mesozoic, and Cenozoic, with each divided into periods. (9.13)

erosion The removal of soil, sediment, and rock by running water, waves, currents, ice, wind, and gravity. (1.4, 9.3)

erosional remnant A mountain or hill that remains when adjacent areas have eroded to lower levels. (11.12)

eruption column A rising column of hot gases, pyroclastic material, and rock fragments ejected high into the atmosphere during a volcanic eruption. (11.2)

esker A long, narrow, sinuous ridge composed of sediment deposited by a stream flowing within or beneath a glacier. (14.6)

estuary A channel where fresh water from the land interacts with salt water from the sea and commonly is affected by ocean tides. (15.9, 18.15)

Ethiopian realm The zoogeographic realm consisting of most of Africa. (18.16)

euryhaline Capable of tolerating a wide range of salinity conditions. (17.7)

eutrophic An aquatic ecosystem with high nutrient concentrations and high net primary productivity. (17.10, 18.14)

eutrophication The addition of so many nutrients, especially phosphates, into an aquatic ecosystem that the ecosystem is harmed; also known as hypertrophication. (16.13, 17.10)

evaporation A change in state from a liquid to a gas, requiring an input of energy from the environment and the absorption of that energy at the molecular level as latent energy. (2.4)

evaporation fog A type of fog that occurs when cold air overlies a warm water surface, destabilizing the atmosphere and encouraging evaporation into the colder air. (4.9)

evaporite Any type of rock, such as rock salt, that forms from the evaporation of water. (12.7)

evapotranspiration The combined effect of evaporation and transpiration in moving liquid water from the surface up to the atmosphere in its gaseous form. (4.0, 16.3)

evergreen Any tree that does not lose its leaves or needles seasonally, even though individual leaves or needles fall and are replaced by new growth. (18.11)

evergreen temperate forest A forest occurring in the mid-latitudes, dominated by trees that do not lose their leaves seasonally. (18.11)

exfoliation The processes by which a rock sheds concentric sheets, as occurs due to the release of pressure during exposure. (12.1)

exfoliation dome A curved rock face left behind after an exfoliation joint shaves off thin, curved slices of rock parallel to the surface. (12.1)

exfoliation joint A joint that forms when a rock expands near the surface, shedding concentric plates, mimicking the topography. (12.1)

expansion joint A joint that forms as a result of expansion due to cooling or to a release of pressure as rocks are uplifted to the surface. (12.1)

exponential growth curve A plot that represents an accelerating population growth rate over time. (17.3)

Extended Köppen Classification The categorization of climatic types over both land and ocean, by applying the algorithm for the Köppen method to climatic data collected over the ocean. (7.2)

extremophile An organism that lives in extreme environments, such as scalding hydrothermal vents at the very high pressures of the deep seafloor. (17.5)

eye The center of a tropical cyclone, characterized by sinking air and clear and calm conditions. (5.13)

eyewall The area of a tropical cyclone immediately surrounding the eye. (5.13)

F

Fahrenheit scale A system for indexing temperature calibrated such that water at sea level freezes at 32°F and water at sea level boils at 212°F. (2.3)

failed rift A narrow trough (rift) at a triple junction, which becomes less active and fails to completely break up a continent into separate plates. (10.12)

Fall Line An imaginary line connecting waterfalls on several adjacent rivers, especially along the boundary between the Piedmont and Coastal Plains in eastern North America. (13.4)

false echo A reflection of a radar image from ground clutter, such as buildings or birds; also known as noise. (5.16)

fanning plume A distribution of smoke or steam from a smokestack that spreads out much more in the horizontal than vertical direction. (4.16)

fault A fracture along which the adjacent rock surfaces are displaced parallel to the fracture. (10.8, 11.9, 11.10)

fault block A section of rock bounded on at least two sides by faults. (11.10)

fault scarp A step in the landscape caused when fault movement offsets Earth's surface. (9.14)

faunal succession The systematic change of fossils with age. (9.13)

feedback A reinforcement or inhibitor to a change in a system. (1.3)

feldspar A very common rock-forming silicate mineral that is abundant in most igneous and metamorphic rocks and some sedimentary rocks, with varieties containing different amounts of potassium, sodium, and calcium. (9.1)

felsic rock An igneous rock with a light color and generally abundant quartz and feldspar. (9.5)

fen A wetland that has neutral or alkaline, but not acidic, waters; usually fed by nutrient-rich groundwater and dominated by grasses and sedges. (18.14)

fermentation The breakdown of carbohydrates to carbon dioxide by yeasts and bacteria. (17.12)

Ferrel cell A hypothetical atmospheric circulation cell over the mid-latitudes in each hemisphere that appears only weakly if at all in the real world. (3.13)

fertilizer A supplement to soils containing nutrients, typically comprised of some combination of nitrogen, phosphorus, and potassium. (16.4)

field capacity The maximum amount of water (in centimeters or inches of precipitation equivalent) that a soil can hold. (8.3, 16.3)

firn Snowflakes that become compressed into small, irregular spheres of more dense snow; also known as neve. (14.3)

firn line The zone across a glacier where the losses of ice and snow balance the accumulation of ice and snow; also known as equilibrium line. (14.3)

First Law of Thermodynamics The fundamental scientific principle that energy and matter can be neither created nor destroyed, but only transferred from one form to another. (1.4, 17.5)

first-level heterotroph An organism that consumes a primary producer. (17.2)

first-order stream A stream with no tributaries. (13.1)

fissure (volcanic) A magma-filled fracture in the subsurface, which typically solidifies into a dike, or a linear volcanic vent erupting onto the land surface. (11.1)

fjord A long, narrow arm of the sea contained within a U-shaped valley, interpreted to be carved by a glacier and later invaded by the sea as the ice melted and sea level rose. (14.5, 15.9)

flag tree A wind-shaped tree with branches growing only on the downwind (leeward) side of a trunk; also known as a banner tree. (18.4)

flash flood A local and sudden flood of short duration, such as that which may follow a brief but heavy rainfall. (13.12)

flocculation The clumping together of clay particles as fresh water mixes with salt water. (13.9)

flood The result of water overfilling a channel and spilling out onto the floodplain or other adjacent land. (13.12)

floodplain An area of relatively smooth land adjacent to a stream channel that is intermittently flooded when the stream overflows its banks. (9.10)

flood stage The level at which the amount of discharge causes a stream to overtop its banks and spill out onto the floodplain. (13.12)

flood tide The local rising elevation of the sea surface between low tide and high tide. (15.2)

flowstone Any deposit of calcium carbonate or other mineral formed by flowing water on the walls or floor of a cave. (12.6)

flume A large, sand-filled tank, used to study stream behavior in a laboratory; also known as a stream table. (13.5)

focus The place where an earthquake is generated; also known as a hypocenter. (11.13)

fog A cloud at the surface of the Earth. (4.8)

folding A form of deformation in which horizontal rock layers are bent into a wavy pattern. (9.12)

foliation The planar arrangement of textural or structural features in metamorphic rocks and certain igneous rocks. (9.6)

food chain The sequencing of organisms from autotrophs to higher level hetero- trophs. (17.2)

food web A network of predator-prey relationships between different types of organisms within an ecosystem; typically more complex than a single, linear food chain. (17.2)

force A push or pull that causes, or tends to cause, change in the motion of a body, commonly expressed as the amount of acceleration experienced by a mass. (11.9)

forced convection The rising motion in the atmosphere caused by horizontal con- vergence, an orographic barrier, or a frontal boundary, rather than by unstable conditions. (4.7)

forecast products The maps of the forecasted state of the atmosphere at a given time into the future, at the surface or at a specified vertical constant pressure level. (5.15)

forensic science The use of technologies to address scientific questions needed for criminal investigations or civil litigation. (16.14)

foreshore The part of a beach submerged during high tides but exposed during low tides, located between the inshore (or swash zone) and the backshore. (15.1)

forest floor The lowest layer of life in a rain forest, where organic debris decom- poses rapidly. (18.5)

forest soil An informal name for soils that support temperate forests. (16.6)

formation A rock unit that is distinct, laterally traceable, and mappable; also used as an informal term for an eroded, perhaps unusually shaped, mass of rock. (12.6)

fossil Any remains, trace, or imprint of a plant or animal that has been preserved from some past geologic or prehistoric time. (9.13)

fracture A break or crack in a rock; subdivided into joints and faults. (9.8)

free convection The rising motion in the atmosphere caused by unstable condi- tions. (4.7)

freezing A change in state from a liquid to a solid, which releases latent energy to the environment. (2.4)

freezing point The temperature at which a substance changes its state from a liquid to a solid. (2.3)

freezing rain A form of precipitation in which a snowflake produced by the Bergeron process melts on its way down to the surface, but refreezes in a cold layer upon impact with the surface, causing it to stick to the surface. (4.12)

frequency The number of waves per second passing a point. (2.5)

freshwater biome A collection of adjacent ecosystems that includes lakes, streams, marshes, and other surface waters. (18.1)

friction layer The lowest one kilometer or so of the troposphere, where friction with the Earth's surface exerts an important effect on motion. (3.3)

fringing reef A reef that fringes the shoreline of an island or continent. (15.6)

front A narrow zone separating two different air masses. (5.2)

frontal lifting The rising motion in the atmosphere caused by a warmer, less dense air mass being pushed up over a colder, more dense one. (4.7)

frost heaving The uneven upward movement and distortion of soils and other mate- rials due to subsurface freezing of soil water. (12.8)

frost-point temperature The subfreezing temperature to which a volume of air must be cooled in order to become saturated with water vapor, thereby causing frost to form. (4.4)

frost wedging The process by which jointed rock is pried and dislodged by the expansion of ice during freezing. (12.1, 15.4)

full moon The position of the Moon in its orbit around Earth when it is directly on the opposite side of Earth from the Sun; results in the Moon being fully illumi- nated by the Sun. (15.2)

fumigating plume A distribution of smoke or steam from a smokestack that expands downward more than upward, as it moves laterally downwind; indicative of sta- ble air above the steam/smoke. (4.16)

funnel cloud A tornado-like circulation that does not reach the surface. (5.10)

fusion (nuclear) The joining together of protons and neutrons to produce larger particles, releasing electromagnetic radiation in the process. (2.5)

G

gabbro A medium- to coarse-grained mafic igneous rock, the compositional equiv- alent of basalt. (9.5)

gaining stream (river) A stream or part of a stream that receives water from the inflow of groundwater. (8.7)

gamma ray The electromagnetic radiation of the shortest wavelengths (less than 10^{-11} micrometers). (2.6)

gas A state of matter in which the energy content is high enough that molecules break apart from each other and move freely at high speeds. (2.2)

gauging station A site where measurements from a stream are collected in order to calculate discharge of that segment of the stream. (13.14)

Gelisol A type of soil situated above permafrost, characterized by waterlogged con- ditions and organic matter that is slow to decompose. (16.11)

general circulation model (GCM) A type of complicated, computer-based model that predicts global atmospheric circulation using physically based and statisti- cally based equations incorporating processes and feedbacks in the atmosphere, hydrosphere, cryosphere, lithosphere, and biosphere. (7.14)

geographic information system (GIS) A hardware and software system that allows a variety of spatial and nonspatial information to be combined quickly and effi- ciently to examine relationships among the different features. (1.12)

geologic timescale A chronologic subdivision of geologic time depicting the sequence of geologic events, including those represented by fossils; ages of boundaries are assigned through numeric dating of key rock units. (9.12)

geomagnetic polarity timescale A chronology based on the pattern and numeric ages of reversals of Earth's magnetic field. (10.10)

geomorphology The study of landforms. (9.0)

georeferencing The process of matching an image or map to standard coordinates; also known as rectification. (1.12)

geostrophic level The lowest height in the atmosphere at which airflow is parallel to isobars, neither toward nor away from lower atmospheric pressure. (3.8)

geostrophic wind A wind that blows parallel to the isobars, characterized by a bal- ance between the pressure gradient force and the Coriolis effect, above the level at which friction exerts an important impact on winds. (3.8)

geyser A type of hot spring that intermittently erupts a fountain or spray of hot water and steam. (8.7)

glacial drift Any sediment carried by ice, icebergs, or meltwater. (14.4)

glacial erratic A rock fragment carried by moving ice and deposited some distance from where it was derived. (14.4)

glacial groove A very large, deep scratch mark amid an otherwise smooth polished surface created by glacial abrasion. (14.4)

glacial period A time interval within an ice age when glaciers were abundant. (14.2)

glacial striation A scratch mark amid an otherwise smooth polished surface created by glacial abrasion. (14.4)

glacier A moving mass of ice, snow, rock, and other sediment on land. (1.4, 9.2, 14.1)

glaciofluvial deposition Dropping of sediment by glacial streams. (14.4)

gleization A pedogenic regime characterized by a dark, organic-rich A horizon produced in saturated and waterlogged soils where the rate of decay of organic material is slow. (16.8)

global cooling Decreasing global atmospheric and oceanic temperatures as mea- sured or inferred from some point in the past to the present. (7.12)

global outgoing longwave budget The apportioning of upward-directed longwave radiation to various components of the global system on an annual basis. (2.15)

Global Positioning System (GPS) A tool that provides locational information by receiving radio signals from orbiting satellites that relay the position and distance of each satellite. (1.11)

global shortwave radiation budget The apportioning of shortwave radiation to various components of the global system on an annual basis. (2.13)

global warming Increasing global atmospheric and oceanic temperatures as mea- sured or inferred from some point in the past to the present. (7.11)

Glossopteris A seed-bearing plant that lived before 250 million years ago, whose widespread fossils across many southern continents provides evidence support- ing the theory of continental drift. (10.3)

gneiss A metamorphic rock that contains a gneissic foliation defined by a preferred orientation of crystals and generally by alternating lighter and darker colored bands. (9.6)

Gondwana The inferred supercontinent consisting of many of the southern con- tinents. (10.3)

Goode projection A popular type of sinusoidal map projection. (1.9)

graben An elongate, downdropped crustal block that is bounded by faults on one or both sides. (11.10)

graded stream A stream in equilibrium, showing a balance between its capacity to transport sediment and the amount of sediment supplied to it, thus with no overall erosion or deposition of sediment. (13.4)

gradient The change in a variable, such as elevation or atmospheric pressure, across a particular horizontal distance, expressed in degrees; also referred to as slope. (1.6, 13.3)

gradient flow The airflow between higher and lower atmospheric pressure areas parallel to the curving isohypses; typically occurs above the friction layer. (3.8)

granite A coarse-grained, felsic igneous rock containing mostly feldspar and quartz. (9.5)

granular soil A soil structure in which the peds are shaped like small spherical grains less than 0.5 centimeters in diameter, a structure that is often ideal for root penetration. (16.2)

grassland biome A collection of adjacent ecosystems that has tall-grass and short-grass subtypes, related to different climatic conditions, but generally characterized by moisture deficits for part of the year. (18.1)

grassland soil An informal name for soils that support the mid-latitude grassland biome; generally corresponds to Mollisols. (16.6)

gravitational water Water within pore space in soil that is pulled by gravity deeper into the subsurface. (16.3)

gravity The force exerted between any two objects, such as that between the Sun, Earth, and Moon. (12.8)

great circle Any curved line formed by the intersection of a plane and a sphere that divides the sphere into two equal halves and therefore represents the shortest distance between any two points on the sphere. (1.7)

Great Pacific Climate Shift The reversal of the Pacific Decadal Oscillation (PDO) from a cool phase to a warm phase around 1977. (6.0)

greenhouse effect The phenomenon by which Earth's lower atmosphere is warmed by the presence of gases that absorb longwave radiation from the Earth and emit longwave radiation downward to the Earth. (2.14)

greenhouse gas An atmospheric gas that effectively absorbs infrared (longwave) radiation and emits it downward to the Earth, contributing to the warming of the lower atmosphere. (2.14)

Greenland Current A small, north-to-south-flowing, cold surface ocean current in the North Atlantic Ocean near the Greenland coast. (6.2)

groin A low wall built out into a body of water to affect the lateral transport of sand by waves and longshore currents. (15.7)

gross primary productivity The total amount of energy fixed for photosynthesis. (17.6)

ground clutter A non-weather object that intercepts a radar signal, giving the false impression that there is precipitation where none exists. (5.16)

ground heat The insolation absorbed at Earth's surface. (2.13)

ground layer The lower of the two layers in most boreal forests. (18.12)

ground-level ozone A harmful air pollutant, produced by sunlight striking hydrocarbons, such as car exhaust, in the lower troposphere; also known as tropospheric ozone. (2.12)

ground moraine Sediment deposited from the base of the ice in a glacier and scattered around a large area. (14.6)

groundwater Water that occurs in the pores, fractures, and cavities in the subsurface. (1.4, 8.0, 16.3)

groundwater divide A relatively high area of the water table, separating groundwater that flows in opposite directions. (8.6)

groundwater recharge The replenishment of water below the water table by percolation. (8.7)

growing degree day A variable indexing the length of the growing season and how warm it is above a threshold temperature required for the growth of a particular type of vegetation. (17.6)

Gulf low A type of mid-latitude cyclone that forms near the Gulf of Mexico coast of the United States mainly in winter and tracks generally northeastward. (5.3)

Gulf Stream A south-to-north-flowing, warm surface ocean current in the western subtropical North Atlantic Ocean. (6.2)

gully A small channel eroded into the land surface, formed from rills that coalesce. (13.6, 16.13)

gully erosion The wearing away of soil, sediment, and rock as increasing flow in a small but growing channel occurs. (13.6)

gypsum A common calcium sulfate mineral, generally formed by the evaporation of water. (9.1)

gyre Any circular system of currents near the ocean's surface, caused by effects of atmospheric flow around subtropical anticyclones or subpolar lows. (6.1)

H

habitat The physical environment in which a species lives. (17.1)

haboob A severe, downdraft-related, non-rotating dust storm. (5.12)

hadal zone The part of the saltwater biome located seaward of the coastal zone's low-tide mark and containing the vast open waters of the ocean, at the greatest depth, below the mesopelagic zone (i.e., part of the aphotic zone). (18.15)

Hadley cell An atmospheric circulation cell with rising motion at the Intertropical Convergence Zone, poleward flow aloft to the subtropics in each hemisphere, sinking motion at the subtropical high-pressure belt, and surface motion back to the Intertropical Convergence Zone, with one cell in each hemisphere and the most vigorous motion occurring at longitudes experiencing noon. (3.10)

hailstone A ball of ice that forms in subfreezing temperatures within a cumulonimbus cloud and that subsequently falls toward the Earth's surface. (5.8)

half-life In radioactive decay, the time required for one-half of the parent atoms in a sample to decay into the daughter product. (9.13)

halite A salt mineral (NaCl) generally formed by the evaporation of water; cleaves into cubes and has a distinctive salty taste. (9.1)

halogen Any of a group of five elements, including fluorine and chlorine, that easily bonds with another element or molecule. (2.12)

hand specimen A hand-sized piece of rock for study, sampling, or inclusion in a collection. (9.1)

hanging valley A glacial valley whose mouth is higher than the bottom of a larger glacially carved valley it joins. (14.5)

harsh mid-latitude (D) climate In the Köppen climate classification, a broad group of climatic types characterized by adequate moisture availability, the mean temperature of the coldest month less than −3°C, and the mean temperature of the warmest month above 10°C. (7.3)

Hatteras low A type of mid-latitude cyclone that forms near the mid-Atlantic coast of the United States, mainly in winter, and tracks generally northeastward. (5.3)

Hawaiian High A prominent area of semipermanent high atmospheric pressure (anticyclone) in the subtropical North Pacific Ocean; is particularly strong in summer. (3.6)

hazard The existence of a potentially dangerous situation or event. (11.6)

headward erosion The wearing away of soil, sediment, and rock at the headwaters of a stream, into the area between channels (the divide). (13.6)

headwaters The location or general area where a stream begins. (13.3)

heat capacity The amount of energy required to change the temperature of a volume by one Kelvin, expressed in joules per cubic meter per Kelvin. (2.17)

heat flux The transfer of thermal energy from higher temperature to lower temperature objects. (2.2)

heat The thermal energy transferred from one object to another. (2.3)

hematite An iron oxide (Fe_2O_3) metal that has a reddish streak and commonly forms under oxidizing conditions. (9.1, 12.2)

herbivore Any organism that consumes only plants. (17.2)

heterotroph An organism that consumes other organisms. (17.2)

hibernation The inactivity or dormancy of an organism or species, usually in winter. (18.11)

high tide The maximum height to which water in the ocean rises relative to the land in response to the gravitational pull of the Moon; also refers to the time when such high levels occur. (15.2)

hinge The part of a fold that is most sharply curved. (11.11)

Histosol A site-dominated soil type formed in wetlands without permafrost, influenced by anaerobic conditions and the accumulation of organic matter in a thick O horizon. (16.9)

hogback Any ridge with a sharp summit and one slope inclined approximately parallel to the dip of layers, resembling in outline the back of a hog; also known as a cuesta. (11.11)

hoodoo A tall, thin rock formation produced by erosion of highly jointed rock layers, with somewhat rounded sides. (12.5)

hook echo The curved pattern on Doppler radar characteristic of a tornado. (5.10)

horizon A soil layer that is distinct from adjacent layers, including differences in color, texture, content of minerals and organic matter, or other attributes. (12.14, 16.5)

horizonation The process of forming soil horizons. (16.9)

horn A glacial erosional feature formed when three or more cirques merge by headward erosion. (14.5)

horse latitudes The oceanic regions near 30° N and 30° S. (3.12)

horst An elongate, relatively uplifted crustal block that is bounded by faults on two sides. (11.10)

Hot Desert (BWh) climate In the Köppen climate classification, a climate type characterized by inadequate moisture availability, in which potential evapotranspiration typically exceeds precipitation by a large amount, and a mean annual temperature above 12°C (54°F). (7.5)

hot spot A volcanically active site interpreted to be above an unusually high-temperature region in the deep crust and upper mantle. (10.9)

Hot Steppe (BSh) climate In the Köppen climate classification, a climate type characterized by inadequate moisture availability, in which potential evapotranspiration typically exceeds precipitation by a small or moderate amount, and a mean annual temperature above 12°C (54°F). (7.5)

human geography The study of the spatial distribution of phenomena that are related to the activities of people. (7.10)

Humboldt Current A south-to-north-flowing, cold surface ocean current in the eastern South Pacific Ocean; also known locally as the Peru Current or the Chile Current. (6.2)

Humid Continental (Dfa or Dfb) climate In the Köppen climate classification, a climate type characterized by adequate moisture year-round, a mean temperature of the coldest month below −3°C, and a hot or warm summer. (7.7)

humidity The moisture in the atmosphere. (4.0)

Humid Subtropical (Cfa) climate In the Köppen climate classification, a climate type characterized by adequate moisture year-round, hot summers, and the mean temperature of the coldest month between −3°C and 18°C. (7.6)

hummocky topography A type of landscape characterized by randomly distributed humps and pits, commonly created by a landslide or less commonly by a pyroclastic eruption. (12.13)

humus A durable, partially decomposed form of organic matter. (16.1)

hundred-year flood The size of a flood that is predicted, from the existing data, to have a 1 in 100 chance (1%) of occurring in any given year. (13.14)

hurricane A regional term used in the Atlantic basin and eastern Pacific basin applied to the most intense type of storm of tropical oceanic origin. (5.13)

hydraulic conductivity A measure of the ability of a material to transmit a fluid; also known as permeability. (16.3)

hydraulic gradient The slope of energy content of subsurface water. (8.6)

hydrogen bond A weak bond in water that forms between one molecule's hydrogen atom and another molecule's oxygen atom. (4.1)

hydrograph A graph showing the change in the amount of flowing water (discharge) over time. (13.1)

hydrologic cycle The cycle representing the movement of water between the oceans, atmosphere, land, rivers and other surface water, groundwater, and organisms. (1.4, 16.3)

hydrolysis A form of chemical weathering involving the chemical reaction of silicate minerals with water, converting the original materials to clay minerals. (9.9, 12.2)

hydrophyte A plant species adapted to the constantly wet conditions of wetlands. (18.14)

hydrosphere The part of Earth characterized by the presence of water in all its expressions, including oceans, lakes, streams, wetlands, glaciers, groundwater, moisture in soil, water vapor, and drops and ice crystals in clouds and precipitation. (1.3)

hygroscopic water Water within pore space in soil that adheres to soil particles; typically so tightly held that it is unavailable to plants. (16.3)

hypertrophication The addition of so many nutrients, especially phosphates, into an aquatic ecosystem that the ecosystem is harmed; also known as eutrophication. (17.10)

hypocenter Place where an earthquake is generated; also known as the focus. (11.13)

hypothesis A proposition that is tentatively assumed, and then tested for validity by comparison with observed facts and by experimentation. (1.2)

hypsometric curve A graph that plots elevation versus the percentage of land or seafloor that is above that elevation (or depth). (10.13)

I

iceberg A massive piece of ice floating or grounded in the ocean or other body of water. (6.4)

Ice Cap (EF) climate In the Köppen climate classification, a climate type characterized by all 12 months having a climatological average temperature below freezing. (7.8)

ice sheet A mass of ice of considerable thickness and more than 50,000 square kilometers in area, forming a nearly continuous cover of ice and snow over a land surface. (14.1)

ice shelf An ice sheet or large glacier that flows into the ocean and floats on seawater. (14.3)

Icelandic Low A prominent area of semipermanent low atmospheric pressure (cyclone) in the subpolar North Atlantic Ocean. (3.6)

Ideal Gas Law The relationships between pressure, density, and temperature in certain gases, including the atmosphere. (3.1)

igneous rock A rock that formed by solidification of molten material (magma). (9.2)

illuviation The deposition and accumulation of materials in the lower levels of a soil. (16.6)

impermeable A property of a material or surface that completely blocks flow of a liquid through it; also known as an aquiclude. (8.6, 16.3)

Inceptisol A type of recently developed soil characterized by mild weathering and the beginning of a weak B horizon, exhibiting similar characteristics to Entisols but generally older and having better horizonation. (16.9)

incised Cut into bedrock by downward erosion. (13.6)

inclusion A fragment of older rock or material that is contained within another rock or material, as in a fragment of preexisting rock in a magma. (9.12)

index contour A dark line on a topographic map which helps emphasize the broader elevation patterns of an area and allows easier tracking of lines across the map; on most topographic maps, every fifth line is an index contour. (1.6)

Indian Ocean Dipole An atmospheric pressure oscillation between the eastern Indian Ocean and the western Indian Ocean. (6.13)

Indomalayan realm The zoogeographic realm consisting of the Indian subcontinent and adjacent parts of southern and southeastern Asia. (18.16)

inertia The ability of an ecosystem to resist change after a stress. (17.7)

infiltration Water and other fluids that seep into the ground through open pores, fractures, and cavities in soil and rocks. (8.1, 16.3)

infrared energy (IR) A form of electromagnetic energy with longer wavelengths than visible light (from about .7 to about 1,000 micrometers); much of the Sun's radiaton that reaches Earth's surface is ultimately converted into this type of energy. (2.6)

input (ecology) Any energy and materials, such as nutrients and water, that are available in a system, especially in an ecosystem. (17.5)

insectivore A carnivore that consumes insects. (17.2)

inshore The part of the coastal zone that is permanently submerged by shallow water where waves break; also known as the swash zone. (15.1)

insolation The energy transmitted from the Sun to Earth, as incoming solar radiation. (2.7)

interglacial period A time during an ice age when glaciers are melting, retreating, or diminished in extent. (14.2)

intermediate (igneous rock) An igneous rock with a medium color and generally containing both felsic and mafic minerals. (9.5)

intermittent stream A stream that flows every year but not during the entire year, typically flowing only during rainstorms and spring snowmelt. (13.3)

International Date Line The imaginary line that generally runs north-south but zigzags through the Pacific Ocean near the 180° meridian; locations west of this line are one day later in the calendar than locations east of this line. (1.7)

intertidal zone The area including the shoreline and a strip of adjacent land and water; specifically, the part of the saltwater biome nearest to the shore, characterized by wave action, temperature variations, and changes in substrate; also known as the coastal zone. (18.15)

Intertropical Convergence Zone (ITCZ) The belt of relatively low surface atmospheric pressure around the equatorial region, characterized by the collision of the northeast and southeast trade winds and rising motion. (3.9)

intrusion A subsurface magma body or the mass of igneous rock into which it solidifies; also known as a pluton. (11.8)

intrusive rock Any igneous rock that solidified from magma below Earth's surface; also known as a plutonic rock. (11.8)

inundation regime The scenario of coastal inundation during which the height of the swash zone exceeds the tops of coastal dunes. (15.11)

invasive species A non-native species that disrupts its new ecosystem by developing a broad geographical range and occupying a niche. (17.13)

invigorated vegetation The notion that higher temperatures may increase the rate of carbon uptake by plants under longer growing seasons with more favorable conditions for photosynthesis. (17.8)

ion A charged atom. (6.4)

iron formation A rock composed of millimeter- to centimeter-thick layers of iron-bearing minerals, especially iron oxide, commonly with quartz. (9.4)

island and seamount chain A series of islands and submarine peaks in a long, commonly linear belt. (10.2)

island arc A generally curved belt of volcanic islands above a subduction zone; also used as an adjective to refer to this setting. (10.2)

isobar A line connecting locations having the same atmospheric pressure; usually shown for sea-level-corrected pressures. (3.2)

isohyet A line on a map connecting points having the same precipitation total, either from a single event or for a climatological average. (7.1)

isohypse A line on an upper-level weather map connecting locations having the same height at which a particular atmospheric pressure value occurs. (3.8)

isostacy The concept that crust of different thickness will ride on the mantle at different elevations. (10.1)

isostatic rebound Uplifting caused by the removal of weight on top of the crust, as when an ice sheet melts away or when erosion strips material off the top of a thick crustal root of a mountain. (10.15, 13.10, 15.10)

isotherm A line on a map connecting locations having the same temperature, either from a single event or for a climatological average. (2.17)

isotope One of two or more species of the same chemical element but differing from one another by having a different number of neutrons. (9.13)

isotopic dating The determination of the age of a rock usually by chemical analysis of the products of natural radioactive decay. (9.12)

J

jet stream A fast-flowing current of air, often near the tropopause. (3.9)

jetty An engineering structure built from the shore into a body of water to redirect current or tide, for example, to protect a harbor. (15.7)

joint A fracture in a rock where the rock has been pulled apart slightly, without significant displacement parallel to the fracture. (11.10)

joule A unit of measurement of work or energy. (2.3)

June Solstice The day, on or near June 21, when the North Pole (marking where the Earth's axis of rotation intersects the Earth's surface in the Northern Hemisphere) is pointed most directly toward the Sun. (2.9)

K

kame A hill formed where meltwater in stagnant ice deposited sediment in crevasses or in the space between the glacier and the land surface. (14.6)

kame and kettle topography A terrain of hills, ponds, and poor drainage, characteristic of the effects of kames and kettles. (14.6)

karst A topography characterized by sinkholes, caves, limestone pillars, poorly organized drainage patterns, and disappearing streams; generally formed from the dissolution of limestone or other soluble rocks. (12.2)

katabatic wind A local-to-regional-scale wind that blows from the peak of a hill or mountain down to the valley; also known as a mountain breeze. (3.4)

Kelvin scale An absolute system for indexing temperature in which all molecular motion ceases at 0 K, water freezes at 273 K, and water at sea level boils at 373 K. (2.3)

Kelvin wave A subsurface equatorial wave in the Pacific Ocean that propagates from west to east along the thermocline, at the onset of an El Niño event. (6.12)

Kerguelen Plateau The second-largest oceanic plateau on Earth, in the southern Indian Ocean. (10.11)

kettle A pitlike depression in glacial deposits, commonly a lake or swamp; formed by the melting of a large block of ice that had been at least partly buried in the glacial deposits. (14.6, 18.14)

kettle lake A body of water occupying a kettle. (14.6)

keystone predator The species at the top of the food chain in an ecosystem. (17.3)

kinetic energy The energy possessed by a solid, liquid, or gas expressed by the motion of that object. (2.2, 8.2)

Köppen climate classification A scheme for categorizing climates based on the annual regime of temperature and precipitation, with the categories of climate types corresponding to the major realms of natural vegetation. (7.1)

krummholz A deformed, stunted tree near the treeline, such as banner trees. (18.4)

K-selected species Any species with a slow rate of reproduction and maturation, but with a strong and prolonged nurturing, leading to a relatively high likelihood of survival into adulthood. (17.2)

Kuroshio Current A south-to-north-flowing, warm surface ocean current in the western North Pacific Ocean. (6.2)

L

Labrador Current A small, north-to-south-flowing cold surface ocean current in the northern Atlantic Ocean between Greenland and eastern Canada. (6.2)

laccolith A bulge-shaped igneous body that has domed and tilted overlying layers and that is observed or interpreted to have a relatively flat floor. (11.8)

lagoon A shallow part of the sea between the shoreline and a protecting feature, such as a reef or barrier island, farther out to sea. (15.5)

lahar A mudflow mostly composed of volcanic-derived materials and generally formed on the flank of a volcano. (11.4)

lake-effect snow Intense snowfall events that occur near lakes, especially the Great Lakes, resulting from enhancement of moisture contributed by the lakes, in air that has been destabilized by very cold air overlying relatively warm lakes. (5.4)

laminar flow Flow characterized by smooth movement in parallel layers. (13.2)

land breeze A local-to-regional-scale wind that blows from land to sea, typically at night. (3.4)

land bridges Hypothesized ridges, now barely submerged in the oceans, proposed to explain the worldwide migration of animals from their geographic origin; most proposed ridges have been disproved by more recent data. (10.3)

land cover The natural or human-made features on the surface of the Earth, such as bare rock, water, soil, forest, or concrete. (4.6)

landfall The arrival of a tropical cyclone on land. (5.13)

landslide A general term for the rapid downslope movement of soil, sediment, bedrock, or a mixture of these; also the material or landform formed by this process. (11.4, 12.9)

La Niña A periodic cooling of the eastern equatorial Pacific surface ocean waters, and reduction in precipitation, adjacent to South America, as the Southern Hemisphere's summer approaches; also known as a cold phase of ENSO. (6.10)

lapse rate The change in temperature of an air parcel or the atmosphere in general, per unit change in elevation. (4.5)

large-scale map A map that shows great detail within a small geographic area. (1.10)

latent heat The energy stored or released as a consequence of a change between the solid, liquid, or gaseous states of matter. (2.4)

latent heat flux The flow, or transfer, of latent heat in the atmosphere by convection and advection. (2.15)

latent heat of fusion The energy absorbed as a consequence of a change in state from a solid to liquid state of matter (melting), or released as a consequence of a change in state from a liquid to solid state of matter (freezing). (2.4)

latent heat of sublimation The energy absorbed as a consequence of a change in state from a solid to gaseous state of matter (sublimation), or released as a consequence of a change in state from a gaseous to solid state (deposition). (2.4)

latent heat of vaporization The energy absorbed as a consequence of a change in state from a liquid to gaseous state of matter (evaporation), or released as a consequence of a change in state from a gaseous to liquid state of matter (condensation). (2.4)

lateral erosion The wearing away of soil, sediment, and rock along the side of a stream. (13.6)

lateral moraine Sediment carried in and deposited along the sides of a glacier. (14.5)

laterite A type of tropical soil rich in iron (Fe) and aluminum (Al) oxides, commonly giving the soil a deep red color. (16.6)

laterization A pedogenic regime characterized by rapid chemical weathering and extreme dissolution in hot, wet settings, producing red, leached soils, such as laterites. (16.8)

latitude A measure of position north or south of the equator; all locations sharing the same latitude in a hemisphere fall along an imaginary east-west line that encircles the Earth. (1.7)

Laurasia The inferred northern supercontinent that existed in the Mesozoic and included North America, Europe, and Asia. (10.14)

lava Magma that is erupted onto the surface, or the rock mass into which it solidifies. (9.2)

lava dome A dome-shaped mountain or hill of at least partly solidified lava, generally of felsic to intermediate composition. (11.2)

lava flow A volume of magma that erupts onto the surface and flows downhill from the vent of a volcano; also the solidified body of rock formed by this magma. (11.2)

lava fountain A shooting stream of molten lava propelled into the air by pressure and escaping gases during an explosive volcanic eruption. (11.2)

lava tube A long, tubular opening under a roof of solidified lava and representing an active or partially emptied subsurface channel of lava. (11.3, 12.6)

leaching The separation or dissolution of soluble constituents from a rock, sediment, or soil by percolation of water; also known as eluviation. (16.3)

leeward The downwind side of a slope. (3.5)

lentic system Any freshwater ecosystem that contains still water, such as lakes and ponds. (18.14)

lethal limit The threshold at which an environmental variable becomes too high or too low for a particular organism to survive. (17.7)

levee A long, low ridge of sediment deposited by a stream next to the channel, or built by humans to prevent floodwaters from spilling onto a floodplain. (13.15)

liana A vine in a rain forest that winds around the maze of branches and leaves in the canopy layer. (18.5)

lichens A symbiotic combination of algae and fungi. (12.4, 18.13)

lidar A type of active remote sensing in which electromagnetic radiation usually at near-infrared wavelengths is emitted, and the amount and timing of returning energy are used to determine distance and other characteristics of the land. (1.11)

lightning The bright flash observed at the instant that electrical energy is transferred through the atmosphere, after the return stroke is established. (5.9)

limb The planar or less curved parts of a fold on either side of the hinge. (11.11)

limestone A sedimentary rock composed predominantly of calcium carbonate, principally in the form of calcite, and which may also include chert, dolomite, and fine-grained clastic sediment. (9.4)

limnetic zone The upper zone of a lake in which more oxygen is generated from photosynthesis than is used by respiration or fermentation. (17.12, 18.14)

lineation A linear structure in rock, usually in a metamorphic rock. (9.6)

liquefaction A loss of cohesion when grains in water-saturated soil or sediment lose grain-to-grain contact, as when shaken during an earthquake. (11.16, 16.14)

liquid A state of matter in which a collection of mobile atoms and molecules more or less stay together but are not held in the rigid form of a solid. (2.2)

lithification The conversion of unconsolidated sediment or volcanic ash into a coherent, solid rock, involving compaction and cementation. (9.3)

lithosphere Earth's upper, rigid layer composed of the crust and uppermost mantle. (1.3, 10.1)

lithospheric mantle The part of the uppermost mantle that is in the lithosphere. (10.1)

litter Dead plant material on the ground. (16.4)

litter layer A layer of loose leaves, twigs, branches, pine needles, and other plant debris on top of soil. (16.7)

Little Ice Age A period from about 1450 to about 1850 A.D. in which global temperatures were relatively cold. (2.7, 7.11)

littoral zone The part of a lentic system or coastal zone that has access to sunlight throughout its entire depth; typically near the shore where the water is not too deep. (15.1, 18.14)

loading The process by which weight is added to the lithosphere. (15.10)

loam A soil texture consisting of a balance between particles of different sizes, enhancing the soil's ability to support agriculture. (16.2)

local climate The average atmospheric conditions, along with the extremes and variability of those conditions, at a local scale; also known as microclimate. (7.10)

location The setting of a place, both in terms of its absolute geographic coordinates and its position relative to other features. (1.2)

loess An essentially unconsolidated sediment consisting predominantly of silt, interpreted to be windblown dust, commonly of glacial origin. (9.11, 16.5)

lofting plume A distribution of smoke or steam from a smokestack that expands upward from the stack, but not downward, as it migrates laterally downwind. (4.16)

longitude An imaginary north-south-running line that passes through the North and South Poles, encircles the Earth, and indicates how far east or west of the Prime Meridian a location is. (1.7)

longitudinal dune A type of sand dune that is linear or only very gently curving for many kilometers, oriented parallel to prevailing winds. (9.11)

longitudinal profile A plot of the elevation of a stream versus distance from the source, typically showing a steeper gradient near the headwaters and a progressively less steep gradient downstream. (13.4)

longshore current A current, generally in an ocean or large lake, flowing more or less parallel to a coastline. (15.4)

longshore drift The process by which sand and other sediment move parallel to a coast with the prevailing longshore current. (15.4)

longwave radiation Electromagnetic energy at wavelengths greater than about 4 micrometers; corresponds to the wavelengths at which the Earth emits nearly all of its energy. (2.6)

looping plume A distribution of smoke or steam from a smokestack characterized by prominent up-and-down curves as it advances downwind, indicative of abundant turbulence and vertical movement. (4.16)

losing stream (river) A stream or part of a stream that loses water from outflow to the unsaturated zone beneath it; also known as a losing river. (8.7)

lotic system Any freshwater ecosystem that contains moving water, such as a stream. (18.14)

low tide The minimum height to which water in the ocean falls relative to the land in response to the gravitational pull of the Moon; also refers to the time when such low levels occur. (15.2)

M

Madagascan realm The zoogeographic realm consisting of Madagascar. (18.16)

mafic A group of generally dark-colored, silicate minerals with a high magnesium (Mg) and iron (Fe) content. (9.1)

mafic rock A generally dark-colored igneous rock with a mafic composition. (9.5)

magma Molten rock, which may or may not contain some crystals, solidified rock, and gas. (9.2)

magma chamber A large reservoir in the crust or mantle that is occupied by a body of magma. (9.2)

magmatism The melting of rock at high temperatures beneath the Earth's surface, causing magma to form. (9.3)

magnetic reversal A switch between normal polarity and reversed polarity of the Earth's magnetic field. (10.10)

magnetite An iron oxide (Fe_3O_4) mineral that is typically black and is strongly magnetic. (9.1)

magnetometer An instrument used to measure the direction and strength of magnetism in rocks and other materials. (10.10)

magnitude (seismic) A measure of the amount of energy released by an earthquake; used to compare sizes of earthquakes; can refer to Richter magnitude or moment magnitude. (11.16)

O horizon An upper, organic-rich soil horizon composed of dead leaves and other plant and animal remains. (16.7)

oligotrophic An aquatic ecosystem with low nutrient concentrations and low net primary productivity. (17.10, 18.14)

omnivore Any organism with a diverse diet of both meat and plants. (17.2)

Ontong Java Plateau The largest oceanic plateau on Earth, in the tropical western Pacific Ocean near Papua New Guinea. (10.11)

open system Any interconnected set of matter, energy, and processes that allows material and energy to be exchanged outside itself. (1.3, 17.5)

optimal range The thresholds of an environmental variable between which an organism thrives. (17.7)

orbital plane The imaginary plane in which the Earth revolves around the Sun. (2.9)

original horizontality The principle that most sediments and many volcanic units are deposited in layers that originally are more or less horizontal. (9.12)

orographic effect The rising motion and enhancement of precipitation on the windward slopes of a mountain or other topographic feature. (4.7)

outgoing longwave radiation (OLR) Electromagnetic energy at wavelengths longer than 4 micrometers that radiates into space after having been emitted by the Earth's surface and atmosphere. (2.14)

outwash plain An area in front of a glacier where streams deposit glacially produced sediment. (14.6)

overland flow Water flowing over the surface but not in a channel. (13.6, 16.3)

overstory The higher of the two layers in most boreal forests. (18.12)

overturning The mixing of waters from different depths in a large body of water; also known as turnover. (6.5)

overwash The process by which a wave or storm surge moves material from the side of a landmass nearest to the sea to an area farther from the sea. (15.11)

overwash regime The scenario of coastal inundation during which the height of the swash zone reaches the tops of coastal dunes. (15.11)

oxbow lake An isolated, curved lake formed when a cutoff meander is filled with water. (13.5)

oxidation The chemical process during which a material combines with oxygen; a form of chemical weathering. (9.9, 17.5)

oxide minerals A group of nonsilicate minerals that contains oxygen bonded with a metallic element, such as iron, manganese, or titanium. (9.1)

Oxisol A type of soil formed in wet tropical conditions and characterized by deep weathering, leaching, and oxidation, usually high in iron and aluminum and low in nutrients. (16.10)

ozone A molecule consisting of three oxygen atoms, concentrated in the stratosphere where it absorbs harmful ultraviolet radiation, but also existing near the Earth's surface where it is a pollutant. (2.12)

ozone layer The lower part of the stratosphere where ozone is formed and photodissociated quickly. (2.12)

P

Pacific Decadal Oscillation (PDO) An oscillation in sea-surface temperatures in the North Pacific Ocean, in which temperatures are warmer than normal either near the Alaska coast or over the central North Pacific, while they are simultaneously below normal at the other location. (6.0)

Pacific Decadal Oscillation (PDO) index A numerical indicator of the degree to which the Pacific Decadal Oscillation is in its warm phase, with warmer than normal sea-surface temperatures in the eastern North Pacific near the Alaska coast (when the index is positive), or in its cool phase, which is characterized by cooler than normal sea-surface temperatures in the eastern North Pacific near the Alaska coast (when the index is negative). (6.0)

Pacific realm The zoogeographic realm consisting of most of the Pacific islands. (18.16)

Pacific Ring of Fire The belt of frequent volcanic activity from the southwestern Pacific, through the Philippines, Japan, and Alaska, and then down the western coasts of North and South America, resulting from subduction of oceanic plates around the Pacific Ocean. (10.7)

pahoehoe A type of lava or lava flow that has a relatively smooth upper surface or folds that form a "ropy" texture. (11.3)

Palearctic realm The zoogeographic realm consisting of Europe and most of Asia and northern Africa. (18.16)

Palmer Drought Index A numerical indicator of the severity of drought conditions or of the excessiveness of water, customized to the place of interest at the time of year of interest. (4.14)

Pampas An agriculturally fertile plains region of South America, encompassing Uruguay, part of eastern Argentina, and extreme southern Brazil. (18.2)

Pangaea An inferred supercontinent that existed from about 300 million to about 200 million years ago and included most of Earth's continental crust. (10.14)

parallel Any line of latitude. (1.7)

parasitism A type of symbiosis in which one species benefits from the interaction and the other is harmed by the interaction. (17.3)

parent atom An atom before it undergoes radioactive decay; also known as parent isotope. (9.13)

parent material Any minerals and rocks upon which physical and chemical weathering acts, and from which soils are derived. (12.3, 16.5)

Pascal A unit of pressure (force per unit area) equivalent to one Newton per square meter. (3.2)

passive margin A continental edge that is not a plate boundary. (10.6)

passive remote sensing system A type of remote sensing in which the sensor detects whatever light, heat, or other energy is naturally coming from the region being monitored. (1.11)

patch dynamics The idea that subareas within an ecosystem may be influenced by different processes and are at different successional stages at the same time. (17.7)

paternoster lakes A series of tarns connected by a stream. (14.5)

patterned ground Polygon-shaped outlines in the soil of Arctic areas, caused by frost heaving. (12.8, 14.7)

peak discharge The maximum volume of water per unit time, shown on a hydrograph. (13.3)

peat An unconsolidated deposit of partially decayed plant matter. (16.1)

pebble A small stone between 6 and 64 mm in diameter. (13.7)

ped A group of soil particles clumped together. (16.2)

pediment A gently sloping, low-relief plain or erosion surface carved onto bedrock, commonly with a thin, discontinuous veneer of sediment. (9.9)

pediment dome A broad, gently sloping peak or ridge formed by erosion of bedrock over a long period of geologic time. (9.9)

pedogenic regime An environmental setting where certain physical, chemical, and biological conditions dominate; includes laterization, salinization, calcification, podzolization, and gleization. (16.8)

pedon The smallest unit within which soils are described; a three-dimensional body about 2 meters deep and large enough (1 to 10 square meters) to distinguish soil horizons. (16.8)

pegmatite An igneous rock containing very large crystals, which may be centimeters to meters long. (9.5)

pelagic Pertaining to an organism in a saltwater biome that resides mainly suspended in the water. (18.15)

pelagic zone The part of the saltwater biome located seaward of the coastal zone's low-tide mark and containing the vast open waters of the ocean, consisting of the epipelagic, mesopelagic, and hadal zones at various depths. (18.15)

perched water table Groundwater that sits above the main water table and generally is underlain by an impermeable layer or lens that blocks the downward flow of groundwater. (8.6)

percolation The deep downward movement of water from the unsaturated soil layer into the saturated layer. (8.7)

perennial stream A stream that flows all year. (13.3)

perihelion The date of closest approach of Earth to the Sun during the course of its orbit, at present corresponding to a day on or near January 4. (2.7)

period The largest subdivision of geologic time within an era, usually lasting tens of millions of years. (9.13)

permafrost A condition in which water in the uppermost part of the ground remains frozen all or most of the time. (8.1, 14.7, 18.12)

permeability A measure of the ability of a material to transmit a fluid; also known as hydraulic conductivity. (8.6, 16.3)

Peru Current A local name for the Humboldt Current. (6.2)

photic zone The topmost layer of an aquatic system that has access to sunlight, generally to a depth of 100 meters or less. (6.7, 18.14)

photodissociation The splitting apart of chemical bonds caused by exposure to radiation. (2.12)

photosynthesis The process by which plants produce carbohydrates, using water, light, and atmospheric carbon dioxide. (17.1)

phreatophyte Any type of plant with a deep taproot system, to enhance survival in a desert biome. (18.8)

phyllite A shiny, foliated, fine-grained metamorphic rock, intermediate in character between slate and schist. (9.6)

physical geography The study of spatial distributions of phenomena across the landscape, processes that created and changed those distributions, and implications of those distributions for people. (1.1)

physical weathering The physical breaking or disintegration of rocks when exposed to the environment. (9.9, 12.1)

phytoplankton A broad group of mostly microscopic photosynthesizing organisms. (17.6)

piedmont glacier A broad glacier that forms when an ice sheet or valley glacier spreads out as it moves into less confined topography. (14.1)

pillow A rounded, pillow-shaped structure that forms when lava erupts into water. (11.3)

pioneer community The first group of organisms to appear in an ecosystem following primary succession. (17.7)

planar projection A family of map projections based on the premise that the globe is projected onto a plane, with no distortion at the tangent point at which the globe touches the plane, and increasing distortion of both shape and area with distance from that point. (1.9)

plateau A broad, relatively flat region of land that has a high elevation. (10.2)

platform The instrument-carrying vehicle used in remote sensing. (1.11)

platy soil A soil structure in which the peds take the form of horizontal plates, commonly because the soils have been compacted by the weight of overlying materials. (16.2)

plucking The tearing away of rock by a glacier, particularly as a result of jointing caused by expansion during freeze-thaw action; common on the downflow side of a hill. (14.4)

plutonic rock Any igneous rock that solidified from magma below Earth's surface; also known as an intrusive rock. (11.8)

PM$_{10}$ Any solid aerosol (particulate matter) with a diameter of ≤ 10 micrometers. (7.10)

podzolization A pedogenic regime characterized by shallow acidic soils with a well-developed soil profile, created in environments where winters are long, cold, and subject to frost action, and chemical weathering is relatively slow. (16.7, 16.8)

point bar A series of low, arcuate ridges of sand and gravel deposited on the inside of a stream bend or meander. (13.5, 18.14)

point source A localized site which produces pollution. (7.9)

point-pattern analysis A type of spatial analysis in which the degree of clustering of a spatial distribution can be determined. (1.12)

polar (chemical property) A molecule, such as water, that has an uneven distribution of charges from one side to the other. (4.1)

polar cell An atmospheric circulation cell with sinking motion near the pole, equatorward surface flow to subpolar latitudes, rising motion at the subpolar low-pressure belt, and motion aloft back to the pole, with one cell in each hemisphere, and strong Coriolis deflection affecting these motions. (3.11)

polar (E) climate In the Köppen climate classification, a broad group of climatic types characterized by adequate moisture availability, the mean temperature of the coldest month less than $-3°C$, and the mean temperature of the warmest month below $10°C$. (7.8)

polar easterlies The prevailing surface general circulation feature that exists between the pole and about 60° of latitude in each hemisphere, characterized by flow from east to west. (3.9)

polar front jet stream The fast-flowing current of air near the tropopause in each hemisphere at the boundary between the polar cell and the mid-latitudes. (3.13)

polar high The large permanent areas of high surface atmospheric pressure centered near the North Pole and South Pole. (3.11)

polar stereographic projection A popular type of planar map projection. (1.9)

polyconical projection A family of map projections based on the premise that the globe is projected onto a cone that intersects it in two arcs, with the two tangent lines having no distortion; other locations are neither conformal nor equal area, but retain some advantages of both conformal and equal area maps when depicting small areas. (1.9)

polypedon Up to 10 pedons, if similar in properties. (16.8)

ponding Water that accumulates in a concentrated area on the surface when the intensity of precipitation exceeds the rate of infiltration. (16.3)

pool A slow-moving segment in a stream upstream from rapids. (13.5, 13.6)

pool-and-riffle stream A stream characterized by an alternation of pools with riffles and rapids. (13.6)

poorly drained soil A soil in which too much water remains between the grains for too long, or in which water cannot pass easily through it; usually results in the ineffective dispersion of nutrients throughout the root zone. (16.3)

pore space Any open space within rocks, sediment, or soil, including between grains in a sedimentary rock, within fractures, and in other cavities. (9.4, 16.1)

porosity The percentage of the volume of a rock, sediment, or soil that is open space (pore space). (16.1)

porphyritic An igneous rock having relatively larger crystals in a finer grained matrix; signifies that the magma cooled somewhat at depth but then rose and solidified on, or closer to, the surface where it cooled rapidly to solidify the finer grained matrix. (9.5)

positive feedback A response to a change in a system that reinforces and amplifies more of the same type of change. (1.3, 13.5)

positive Indian Ocean Dipole (IOD) The phase of the pressure oscillation between the eastern and western Indian Ocean characterized by lower atmospheric pressure and warmer water than normal in the western Indian Ocean and higher than normal atmospheric pressure with cooler water than normal in the eastern Indian Ocean. (6.13)

positive magnetic anomaly A strong signal from magnetic instruments, indicative of more magnetic rock or of a rock formed during a time of normal polarity. (10.10)

positive North Atlantic Oscillation index A condition characterized by simultaneous stronger than normal atmospheric pressures in the Bermuda-Azores High and Icelandic Low. (6.13)

potential energy The energy stored within the molecular structure of a solid, liquid, or gas as a consequence of its position relative to where it could possibly move. (2.2, 8.2)

potential evapotranspiration The maximum amount of water that could possibly be lost from the surface to the atmosphere over a given interval of time through the combined effect of evaporation and transpiration, if the water were abundantly available. (7.3)

pothole A bowl-shaped pit eroded into rock by swirling water and sediment. (13.2)

prairie A tall-grass subcategory of the grassland biome. (18.9)

Prairie Peninsula The narrow, eastward protrusion of the grassland biome into Iowa, Illinois, and Indiana. (18.2)

precession The long-term, cyclic process involving the change in direction of the Earth's axial tilt. (14.9)

precipitation (atmospheric) The water in solid or liquid form that falls from the atmosphere to the surface, including rain, freezing rain, sleet, snow, and hail. (1.4)

precipitation (chemical) The movement of a dissolved material out of the solution and into solid form. (9.4)

precipitation fog A type of fog that occurs when falling raindrops evaporate, increasing the humidity of the colder air beneath it, or when precipitation evaporates rapidly upon reaching the surface. (4.9)

predator An animal that consumes another animal. (17.3)

prescribed burn A fire that is intentionally started for beneficial reasons. (18.10)

pressure A force exerted on a unit area, expressed in a gas by the frequency of molecular collisions. (3.1)

pressure gradient The horizontal difference in pressure across some distance. (3.3)

pressure-gradient force The force that initiates movement of air from areas of higher atmospheric pressure to areas of lower pressure at right angles to the isobars. (3.3)

prey An animal that is consumed by another animal. (17.3)

pride A small family unit of lions. (18.9)

primary data Any new information, such as that generated when a new map is made based on original observations. (1.10)

primary mineral Minerals in soil, such as quartz, that are derived directly from the weathering and erosion of various types of rocks but undergo no chemical transformation during the soil-forming process. (16.1)

primary producer An organism that acquires what it needs from the soil, water, air, and energy of the environment, without consuming other organisms; also known as an autotroph. (17.2)

primary succession The predictable change in species composition in an ecosystem following stress which causes so much disturbance that little life is able to survive. (17.7)

primary wave (P-wave) A seismic body wave that involves particle motion, consisting of alternating compression and expansion, in the direction of propagation. (11.13)

Prime Meridian The line of longitude running through Greenwich, England, which serves as the zero degree of longitude. (1.7)

principle of superposition The concept that a sedimentary or volcanic layer is younger than any rock unit on which it is deposited. (9.12)

probability The statistical likelihood that an event will occur. (13.14)

profundal zone The lower zone of a lake in which more oxygen is used by respiration or fermentation than is generated from photosynthesis. (17.12)

promontory A local ridge of land that juts out into a body of water. (15.4)

proxy evidence A clue about past climatic conditions during times when direct measurements are not available. (7.11)

Public Land Survey System (PLSS) A coordinate system used in much of the United States that characterizes location based on the number of 6-mile by 6-mile squares north, south, east, or west of the intersection of a baseline with a principal meridian, with subdivisions within the square also described; also known as township-range system. (1.8)

pumice Volcanic rock, especially of felsic or intermediate composition, containing many vesicles (holes) formed by expanding gases in magma. (9.5)

pycnocline The level in the ocean at which density changes rapidly with depth. (6.7)

pyrite A common, pale bronze to brass yellow, iron sulfide mineral (FeS_2), commonly called "fool's gold." (9.1)

pyroclastic eruption A volcanic eruption in which hot fragments and magma are thrown into the air. (11.2)

pyroclastic flow A fast-moving cloud of hot volcanic gases, ash, pumice, and rock fragments that generally travels down the flanks of a volcano; also known as an ash flow. (9.5, 11.2)

pyroxene One of a group of mostly dark, silicate minerals, with a single chain of silicon-oxygen bonds. (9.1)

Q

qualitative data Data or observations that include descriptive words, labels, sketches, or other images. (1.2)

quantitative data A numeric measurement that is typically visualized and analyzed using data tables, calculations, equations, and graphs. (1.2)

quarter moon The position of the Moon at one-quarter and three-quarters of the distance through its orbit around Earth; characterized by a 90-degree angle formed between the position of the Moon, Earth, and Sun; results in the Moon being half illuminated by the Sun, with half of the illuminated side (that is, a quarter of the Moon) being visible from Earth. (15.2)

quartz A very common rock-forming silicate mineral, consisting of crystalline silica, with the chemical formula of SiO_2. (9.1)

quartzite A very hard rock consisting chiefly of quartz grains joined by secondary silica that causes the rock to break across rather than around the grains; formed by metamorphism or by silica cementation of a quartz sandstone and other quartz-rich rocks. (9.6)

quicksand Loose sand saturated with water and undergoing liquefaction, rendering it unable to support weight. (16.14)

R

radar A type of active remote sensing in which a microwave or radio wave is emitted, and the time and amount of energy returning at those wavelengths are used to map the surface. (1.10, 1.11)

radial drainage pattern A drainage network in which the pattern of streams is generally outward away from a central high point. (13.1)

radiation (electromagnetic) The energy emitted from charged particles and manifested as interacting electrical and magnetic fields called electromagnetic waves; can be transferred either through matter or through a vacuum. (2.2)

radiation balance The difference between incoming and outgoing radiant energy. (2.16)

radiation deficit A property of matter at a given place and time characterized by a greater magnitude of outgoing than incoming radiant energy. (2.16)

radiation fog A type of fog that occurs when nocturnal longwave radiation loss cools the surface to the dew-point or frost-point temperature. (4.9)

radiation surplus A property of matter at a given place and time characterized by a greater magnitude of incoming than outgoing radiant energy. (2.16)

radioactive decay The spontaneous disintegration and emission of particles from an unstable atom. (9.13)

radiosonde An instrument package attached to a weather balloon that measures a range of variables at various heights. (2.3)

radio wave The electromagnetic radiation of the longest wavelengths (10 meters to 100 kilometers). (2.6)

rain forest biome A collection of adjacent ecosystems that consists of evergreen trees requiring abundant water balance surpluses year-round. (18.1)

rain shadow A relatively dry region on the downwind side of a topographic obstacle, usually a mountain range; rainfall is noticeably less than on the windward side. (18.3)

rapid A segment of rough, turbulent water along a stream. (13.6)

recessional moraine A moraine that forms as the front of a glacier melts back and stagnates for some time in one location, depositing a pile of sediment. (14.6)

rectification The process of matching an image or map to standard coordinates; also known as georeferencing. (1.12)

recurrence interval The average length of time that would be expected to occur between events that equal or exceed a given magnitude; also known as a return period. (4.14)

red sprite Faint red lights that sometimes project up from the top of a thunderhead after an energetic bolt of lightning strikes the ground. (5.9)

reef A shallow, mostly submarine feature, built primarily by colonies of living marine organisms, including coral, sponges, and shellfish, or by the accumulation of shells and other debris. (15.4, 15.6)

reflection The change in direction of electromagnetic energy when it interacts with matter, with the energy remaining concentrated before and after interaction. (2.1)

regelation The refreezing of basal ice after it had melted by compression earlier. (14.3)

regional metamorphism Metamorphism affecting an extensive region and related mostly to burial, heating, and deformation of rocks. (9.6)

regolith A near-surface zone of weathered rock and soil, grading downward into unweathered bedrock or sediment. (12.1, 16.5)

relative dating The determination of the age of a rock relative to another rock, by using the principles of superposition, cross-cutting relationships, and other commonsense geological principles. (9.12)

relative humidity The ratio of the atmospheric vapor pressure to the water-vapor capacity, usually expressed as a percentage. (4.2)

relief The relative difference in elevation between two features, such as from a valley to a mountain peak. (1.6)

remediation The cleanup of a contaminated area. (8.10)

remote sensing Any technique used to collect data or images from a distance, including the processing and analysis of such data and the construction of maps using these techniques. (1.11)

representative fraction The ratio of the distance on a map to the actual distance (in the same units) on Earth; also known as scale. (1.10)

reproduction The production of offspring. (17.3)

reservoir A lake created by a human-constructed dam. (13.6)

residence time The average amount of time that an atom or molecule of a certain type of matter remains in a store. (8.2)

residual soil Soils that develop in place rather than being deposited into a place from elsewhere. (16.5)

residuum Parent material formed in place from bedrock beneath the soil. (16.5)

resilience The ability of an ecosystem to recover following stress. (17.7)

resisting force A force that resists the motion of an object, such as resisting the movement of tectonic plates. (10.9)

resource reservoir The storage of energy and/or matter in an ecosystem for later use. (17.5)

return period The average length of time between events that equal or exceed a given magnitude; also known as a recurrence interval. (4.14)

return stroke The positively charged particles that flow upward from the ground as negatively charged particles approach the ground from the cloud, after an electrical connection is established in cloud-to-ground lightning. (5.9)

reverse fault A fault in which the block above the fault moves up relative to the block below the fault. (11.10)

reversed polarity The orientation of the Earth's magnetic field during many times in the geologic past, when the field flowed from north to south, causing the magnetic ends of a compass needle to point toward the south; the opposite of the present-day polarity. (10.10)

rhyolite A mostly fine-grained, felsic igneous rock, generally of volcanic origin; can contain glass, volcanic ash, pieces of pumice, and variable amounts of visible crystals (phenocrysts). (9.5)

richness The number of species present in an ecosystem. (17.4)

Richter magnitude A numeric scale of earthquake magnitude, devised by the seismologist C. F. Richter. (11.16)

ridge (atmospheric) An elongated area of high atmospheric pressure. (3.2)

ridge push A plate-driving force that results from the tendency of an oceanic plate to slide down the sloping lithosphere-asthenosphere boundary near a mid-ocean ridge. (10.9)

riffle A fast-moving, shallow zone in a stream or other freshwater ecosystem. (18.14)

rift A narrow trough that runs along the axis of most mid-ocean ridges, formed when large blocks of crust slip down as spreading occurs. (10.6)

rill A small and shallow (2–4 centimeters deep) channel that forms when sheet flow begins to become concentrated in some places more than other adjacent places. (16.13)

riparian zone A relatively narrow strip of trees, shrubs, and grasses that form along streams. (18.4)

rip current A narrow, fast flow of water from the shoreline back out to sea. (15.3)

rip rap Large rocks and other debris that are dumped along a coast or stream channel for protection from erosion. (15.7)

riser The steep slope separating adjacent terraces or a terrace from the floodplain; also known as a bluff. (13.8)

risk An assessment of whether a hazard might have some societal impact. (11.6)

roaring forties The belt of particularly strong westerly winds aloft in the Southern Hemisphere between about 40° S and 50° S, where little land exists to provide friction to slow the winds. (3.12)

Robinson projection A popular type of map projection for depicting the entire Earth that blends aspects of cylindrical maps with other types of projections. (1.9)

roche moutonnée An asymmetrical glacial erosional landform caused by abrasion on the upflow side and plucking on the downflow side. (14.4)

rock cycle A conceptual framework presenting possible paths and processes to which an Earth material can be subjected as it moves from one place to another and between different depths within Earth. (1.4, 9.3)

rock fall A mass-wasting process whereby large rocks and smaller pieces of bedrock detach and fall onto the ground. (12.10)

rock glacier A feature consisting of angular rocks on the outside and slowly flowing ice on the inside. (14.1)

rock slide A slab of relatively intact rock that detaches from bedrock and slides downhill, shattering as it moves. (12.10)

rock varnish A thin, dark film or coating of iron and manganese oxides, silica, and other materials; formed by prolonged exposure at the surface; also known as desert varnish. (9.11)

Rodinia An inferred supercontinent, consisting of all the continents joined, that existed near the Precambrian-Paleozoic boundary. (10.14)

root wedging The process of plant roots extending into fractures and growing in length and diameter, expanding preexisting fractures. (12.1)

Rossby wave Upper-tropospheric, mid-latitude waves of motion that are primarily west-to-east in each hemisphere, but with meanders toward the poles (ridges) and toward the equator (troughs), displaying wavelengths of hundreds to thousands of kilometers. (3.13)

rotation (tectonic) The tilting or horizontal rotation of a body in response to an applied stress. (11.9)

rotational slide A slide in which shearing takes place on a well-defined, curved shear surface, concave upward, producing a backward rotation in the displaced mass; also known as a slump. (12.10)

r-selected species Any species that produces relatively frequent batches of numerous offspring, which mature faster but are less supervised and relatively unlikely to survive into adulthood. (17.2)

runoff Precipitation that flows on the surface, such as in streams. (8.1, 16.13)

S

Saffir-Simpson scale A classification system for the strength of Atlantic and eastern Pacific hurricanes, from Category 1 (weakest) to Category 5 (strongest). (5.14)

Sahel A semiarid, east-west belt in Africa south of the Sahara, characterized by a Tropical Savanna climate and a grassland biome. (18.7, 18.9)

salinization A pedogenic regime characterized by the upward movement and evaporation of moisture, leaving salts behind; typically occurs in arid or semiarid regions. (16.8)

saltation Transport of sediment in which particles are moved in a series of short, intermittent bounces on a bottom surface. (9.11, 13.2, 16.13)

salt minerals A group of soluble, nonsilicate minerals including halite (NaCl). (9.1)

saltwater biome A collection of adjacent ecosystems in saline waters, in the ocean or inland water bodies. (18.1)

saltwater incursion (intrusion) Displacement of fresh groundwater by the advance of salt water, usually in coastal areas. (8.8)

salt wedging The forced widening of pores in rock by growing salt crystals, primarily in coastal environments, which enhances the ability of wave action to erode rocks. (15.4)

sandbar A low, sandy feature, possibly submerged, offshore of a shoreline or within a sandy stream. (15.5)

sand dune An accumulation of loose sand piled up by the wind. (9.2)

sandstone A medium-grained, clastic sedimentary rock composed mostly of grains of sand, along with other material. (9.4)

Santa Ana wind A warm, dry regional wind of southern California that blows from the northeast, across desert areas, over mountains, and then warms as it descends the leeward slopes. (3.5)

saturated (atmosphere) A condition characterized by atmospheric vapor pressure that is equal to the water-vapor capacity; implies that no more water can evaporate at that temperature and that the relative humidity is 100 percent. (4.2)

saturated (soil) A condition of the soil in which all the pore spaces are filled with water. (16.3)

saturated adiabatic lapse rate The rate of change of temperature with height of an adiabatically rising air parcel with a relative humidity of 100 percent; the rate itself differs depending on the temperature but can be no more than 10 C°/km. (4.5)

saturated zone The area in the subsurface where groundwater fills the pore spaces. (8.3)

scale The ratio of the distance on a map to the actual distance (in the same units) on Earth; also known as representative fraction. (1.10)

scarp (fault-line) A step in the landscape caused when fault movement offsets Earth's surface. (9.14, 11.10)

scarp (landslide) A linear or curved scar left behind by a landslide on a hillslope, marking where the landslide pulled away from the rest of the hill. (12.10, 12.11)

scattering The change in direction of electromagnetic energy when it interacts with matter, characterized by the energy dispersing in different directions after interaction. (2.1)

schist A shiny, foliated, metamorphic rock generally containing abundant visible crystals of mica and other metamorphic minerals. (9.6)

schlerophyllic Pertaining to vegetation that has developed a tolerance to drought and fire. (18.10)

scientific method A logic-based framework for investigating and revisiting some types of problems, involving observation, hypothesis, prediction, testing, retesting, and reaching a conclusion. (1.2)

scoria cone A relatively small type of volcano that is cone shaped and mostly composed of volcanic cinders (scoria); also known as a cinder cone. (11.1)

sea arch An opening through a thin promontory of land that extends out into the ocean. (15.5)

sea breeze A local- to-regional-scale wind that blows onshore in a coastal area, typically by day. (3.4)

sea cave A cave at the base of a sea cliff, usually flooded by seawater. (15.5)

sea cliff A cliff or steep slope situated along the coast. (15.5)

seafloor spreading The process by which two oceanic plates move apart and new magmatic material is added between the plates. (10.3)

sea ice Ice that forms from the freezing of seawater. (6.4, 14.1)

seamount A submarine mountain, in some cases flat-topped, that rises above the seafloor. (10.2)

sea stack An isolated, pillar-like, rocky island or pinnacle near a rocky coastline. (15.5)

seawall A human-constructed wall or embankment of concrete, stone, or other materials along a shoreline, intended to prevent erosion by waves. (15.7)

secondary data Any information derived from preexisting information, such as a preexisting map that is used for providing the input for answering some other question. (1.10)

secondary mineral A mineral formed, generally in soils, after physical and chemical weathering processes interact to break down and alter rocks; usually more stable than the minerals composing the original rock. (12.2, 16.1)

secondary pollutant A contaminant that occurs in the atmosphere as a by-product of a chemical reaction between an emitted pollutant and sunlight, water, or another primary pollutant in the atmosphere. (7.10)

secondary succession The predictable change in species composition in an ecosystem following stress that does not remove the soil from the ecosystem. (17.7)

secondary wave (S-wave) A seismic body wave propagated by a shearing motion that involves movement of material perpendicular to the direction of propagation; an S-wave cannot travel through magma and other liquids. (11.13)

Second Law of Thermodynamics The fundamental scientific principle that energy and matter tend to become dispersed into a more uniform spatial distribution. (1.4, 17.5)

second-level heterotroph An organism that consumes a first-level heterotroph or products derived from a first-level heterotroph. (17.2)

second-order stream A stream that results from the confluence, or merging, of two first-order streams. (13.1)

sediment Grains and other fragments that originate from the weathering and transport of rocks, and the unconsolidated deposits that result from the deposition of this material. (1.4, 9.2)

sediment load The amount of sediment carried by a stream, comprised of suspended load and bed load, but not dissolved load. (13.3)

sedimentary rock A rock resulting from the consolidation of sediment. (9.2)

seep A small spring. (8.7)

seismic station The location of a scientific instrument (seismograph) that measures seismic vibrations. (11.13)

seismic wave A wave produced by earthquakes or generated artificially. (11.13)

seismicity Any of the Earth's movements, either on the surface or at depth, caused by earthquakes. (11.14)

senescence The old-age state of an organism or ecosystem. (18.11)

sensible heat A type of heat in a material detectable as its temperature. (2.3)

sensible heat flux The flow, or transfer, of sensible heat in the atmosphere by convection and advection. (2.15)

September Equinox The day, on or near September 21, when Earth's axis of rotation is pointed exactly sideways relative to the Sun, and the durations of daylight and darkness are equal throughout the world. (2.9)

shaded-relief map A representation of elevation of Earth's surface by using color tones or shading to mimic the slope's appearance in three dimensions. (1.6)

shale A fine-grained clastic sedimentary rock, formed by the consolidation of clay and other fine-grained material. (9.4)

Shannon index A measure of biodiversity used by biogeographers and ecologists, which incorporates both richness and equitability. (17.4)

shear The type of differential stress that occurs when stresses on the edge of a mass are applied in opposing directions. (11.9)

shear component The part of the gravitational force that pulls a mass parallel to the sloping surface on which it sits, and therefore may cause it to slide downhill. (12.8)

sheet erosion The detachment and transportation of soil, sediment, or rock in a thin layer of unchanneled water that runs off from the surface during and shortly after a precipitation event. (13.6, 16.13)

sheet lightning The electrical charges that are exchanged in less discrete mechanisms than by a single stroke or series of strokes, causing a broad area of the sky to be illuminated. (5.9)

sheetflow Any water that flows across the Earth's surface in a thin, continuous fashion rather than in a stream. (8.3)

sheetwash An unconfined flow of water in a thin layer downslope. (16.13)

shelf cloud A distinctive type of cloud that is characteristic of derechos. (5.12)

shield volcano A type of volcano that has broad, gently curved slopes constructed mostly of relatively fluid basaltic lava flows. (11.1)

shortwave radiation The electromagnetic energy at wavelengths less than about 4 micrometers; corresponds to the wavelengths at which the Sun emits nearly all of its energy. (2.6)

shrub layer The shaded, humid layer of life in a rain forest below the understory, experiencing little wind. (18.5)

Siberian High An area of surface high atmospheric pressure (anticyclone) over north-central Asia in winter, caused by the sinking of very cold, dense air. (3.6)

silicate A group of minerals usually consisting of one silicon atom bonded with four oxygen atoms; the most common mineral group on Earth. (9.1)

sill A tabular igneous intrusion that parallels layers or other planar structures of the surrounding rock and which usually has a subhorizontal orientation. (11.8)

silt A fine-grained rock fragment or clast, 1/256 to 1/16 mm in diameter. (9.10, 16.1)

single-cell thunderstorm The simplest category of thunderstorm, existing independently of other storm systems. (5.6)

single-grain soil Pertaining to a soil with no apparent soil structure, because the grains do not stick together into peds; also known as structureless soils. (16.2)

sinkhole A closed, circular depression, usually in a limestone karst area, resulting from the collapse of an underlying cave. (12.2)

sinuosity The amount that a stream channel curves for a given length. (13.5)

sinuosity index The ratio of a stream's channel length (including all the bends) to the down-valley length (i.e., the straight-line distance down the valley). (13.5)

sinusoidal projection A family of equal-area map projections in which map coverage is interrupted, showing only lobes where shape distortion is minimized. (1.9)

slab pull A plate-driving force generated by the sinking action of a relatively dense, subducted slab. (10.9)

slash-and-burn A type of mostly subsistence agriculture typical of the tropics in which an area is deforested, cultivated, and then abandoned after the soils harden into laterites and become unproductive. (18.6)

slate A compact, fine-grained, weakly metamorphosed rock that possesses slaty cleavage. (9.6)

sleet A form of precipitation in which a Bergeron process–produced snowflake melts on its way down to the surface, but refreezes in a cold layer before hitting the surface, resulting in a small ball of ice that bounces from the surface. (4.12)

slip face The steep side of an asymmetrical mound, such as a sand dune, which is also the side that is downwind (leeward) of the prevailing atmospheric circulation. (9.11)

slope The change in elevation across a particular horizontal distance, expressed in degrees; also referred to as gradient. (1.6)

slope aspect The direction (orientation) that a slope faces. (12.4, 18.3)

slope failure The sudden or gradual collapse of a slope that is too steep for its material to resist the pull of gravity. (12.8)

slump A slide in which shearing takes place on a well-defined, curved shear surface, concave upward, producing a backward rotation in the displaced mass; also known as rotational slide. (12.10, 15.4)

small circle Any curved line formed by the intersection of a plane and a sphere that does not divide the sphere into two equal halves. (1.7)

small-scale map A map that shows little detail but covers a large area. (1.10)

smog A type of hazy air pollution caused by interactions between sunlight, hydrocarbons, and nitrogen oxides. (2.12)

snowfield A large area covered with snow and ice that, unlike a glacier, does not move. (14.1)

snow line The elevation on a slope at which snow and ice cover the ground year-round. (18.4)

soil Unconsolidated material at and near the surface, produced by weathering; includes mineral matter, organic matter, air, and water, and is generally capable of supporting plant growth. (16.1)

soil color A property of soil that can reveal information about soils, such as the degree of aeration, moisture conditions, and mineral composition. (16.2)

soil compaction The reduction of pore space as clay particles begin to lie flat, often because of land uses such as overgrazing or passage of traffic. (16.14)

soil erosion The detachment of a soil particle from underlying material and subsequent transportation of the particle. (16.13)

soil profile A vertical section of the soil from the ground surface to parent material, usually around one to two meters in depth. (16.7)

soil structure The manner in which soil particles are clumped together, such as granular, blocky, columnar, or platy. (16.2)

Soil Taxonomy The most common soil classification system used in the United States. (16.8)

soil texture The average distribution of particle sizes in a soil, sand, silt, or clay. (16.2)

soil water Water trapped within the pore spaces in a soil, as hygroscopic water, capillary water, or gravitational water. (16.3)

soil-water recharge Any water that soaks into the soil. (8.3)

soil-water use The evapotranspiration of water stored in soil during times when potential evapotranspiration exceeds precipitation. (8.3)

solar constant The amount of energy emitted by the Sun each second and received at the "top" of the Earth's atmosphere, equivalent to 1,366 joules per square meter per second, or 1,366 watts per square meter. (2.7)

solar cycle An approximately 11-year cycle of sunspot activity. (2.7)

solar flare A burst of intense energy and matter directed from the Sun into space, which can affect wireless communication systems on Earth. (2.7)

solid A state of matter in which the constituent atoms and molecules are bound together tightly enough with low enough energy levels that the mass can withstand the vibrations and other motions without coming apart. (2.2)

solidification The process in which magma cools and hardens into solid rock, with or without the formation of crystals. (9.3)

solifluction The very slow, continuous movement of wet soil, regolith, or weathered rock down a slope. (12.11)

sonar A type of active remote sensing in which a sound wave is emitted, and the amount of energy at those wavelengths and the time required for it to return to the sensor are used to map the surface. (1.11)

source region An area of thousands of square kilometers with similar meteorological characteristics where an air mass forms and acquires its characteristics. (5.1)

South Atlantic High A semipermanent subtropical high atmospheric pressure system (anticyclone) in the Atlantic Ocean. (7.6)

southeast trade winds The prevailing surface atmospheric circulation feature that exists in the tropical latitudes of the Southern Hemisphere, characterized by flow from southeast to northwest. (3.9)

South Equatorial Current An east-to-west-flowing, surface ocean current south of the equator in the Pacific Ocean. (6.2)

Southern Oscillation (SO) The oscillation of surface atmospheric pressure between the area near Darwin, Australia, and the area near Tahiti; when pressure is higher at one of these places, it tends to be lower at the other. (6.9)

Southern Oscillation index (SOI) A numerical indicator of the degree to which atmospheric pressure patterns differ between the western equatorial Pacific Ocean, near Darwin, Australia, and the eastern equatorial Pacific Ocean, near Tahiti; when the index is positive, pressures are higher near Tahiti and lower near Darwin; when it is negative, pressures are higher near Darwin and lower near Tahiti. (6.9)

spatial interpolation A family of geostatistical methods of estimating a data value at any point based on values for nearby points. (1.12)

specific heat (capacity) The amount of energy required to change the temperature of a mass of one kilogram by one Kelvin, expressed in joules per kilogram per Kelvin. (2.17)

specific humidity The ratio of the mass of water vapor in a body of air to the total mass of that air, often expressed in grams per kilogram. (4.2)

spectral model A weather or climate forecasting model that represents atmospheric motion as a series of sine waves rather than relying on computations at discrete grid points. (5.15)

spheroidal weathering A form of mostly chemical weathering in which concentric or spherical shells of decayed rock are successively separated from a block of rock. (9.7, 12.5)

spit A small point or low ridge of sand or gravel projecting from the shore into a body of water. (15.5)

splash erosion The breakdown and transportation of soil, sediment, and rock as raindrops hit the ground. (13.6)

Spodosol An agriculturally unproductive type of soil formed under acidic forest litter in cooler portions of the mid-latitudes and in high latitudes. (16.11)

spreading center A divergent boundary where two oceanic plates move apart (diverge). (10.6)

spring A place where groundwater flows out of the ground onto the surface. (8.7)

spring tide Higher-than-average high tides and lower-than-average low tides caused when the Sun's gravity adds to the effects of the Moon's gravity. (15.2)

squall-line thunderstorm A series of organized thunderstorms oriented along the dryline in the warm sector of a mid-latitude cyclone. (5.6)

stabilized dune A dune of any type that is stationary or barely moving, commonly with vegetation holding the sand in place. (9.11)

stable (ecology) An ecosystem that is in balance, such that populations of all species within it remain fairly constant. (17.7)

stable condition (atmosphere) A situation in which sinking atmospheric motion is encouraged because an adiabatically moving parcel of air is colder than its surrounding environmental air. (4.5)

stalactite A conical or cylindrical cave formation that hangs from the ceiling of a cave and is composed mostly of calcium carbonate. (12.6)

stalagmite A conical, cylindrical, or mound-like cave formation that developed upward from the floor of a cave and is composed mostly of calcium carbonate. (12.6)

standard line The tangent line at which the globe touches a cylinder in a cylindrical projection, where no distortion exists. (1.9)

standing water Any water at the surface of the Earth that is not flowing. (8.3)

star dune A type of sand dune with an irregular, complex shape, and variably trending sand ridges radiating out from a central peak. (9.11)

State Plane Coordinate System (SPCS) A coordinate system used in the United States that divides a state into elongated zones and characterizes location based on distance from a central meridian or parallel that runs along the middle of the zone; map distortion is zero at the central meridian or parallel and expands away from it. (1.8)

states of matter The three major types of arrangements of atoms and molecules: solid, liquid, or gas. (2.2)

stationary front A narrow zone separating two air mass types in which neither air mass is displacing the other. (5.2)

Stefan-Boltzmann Law A relationship linking the Kelvin temperature of an object to the amount of electromagnetic radiation it emits. (2.5)

stenohaline Capable of tolerating only a narrow range of salinity conditions. (17.7)

stenothermal Capable of tolerating only a narrow range of temperature conditions. (17.7)

steppe A short-grass subcategory of the grassland biome. (18.9)

stepped leader An invisible channel of charged particles that starts and stops and zigzags between a cloud and the ground, immediately before a lightning strike. (5.9)

stereo pair A pair of aerial photographs that have overlapping coverage and slightly different perspectives. (1.10)

stereoscope An instrument that allows stereo pairs to be seen in three-dimensional form. (1.10)

stomate A pore on a plant's leaf through which water is lost to the atmosphere during transpiration. (17.5, 18.8)

store A part of a system that retains matter for a period of time rather than allowing that matter to flow through the system. (8.2)

storm surge The pileup of water from waves onshore as a severe storm, particularly a tropical cyclone, makes landfall. (5.13, 15.7)

storm track The path that a storm takes across the surface. (5.3)

strain A change in size or shape of a body as a result of stress. (11.9)

stratiform cloud A family of cloud types that is layered, and wider than tall. (4.8)

stratocumulus cloud A type of low, layered cloud that is composed solely of tiny drops of liquid water. (4.8)

stratopause The transition zone between the stratosphere and overlying mesosphere. (2.11)

stratosphere The second layer of the atmosphere above the surface, which contains most atmospheric ozone. (2.1)

stratovolcano A common type of volcano constructed of alternating layers of lava, pyroclastic deposits, and mass-wasting deposits, including mudflows; also known as a composite volcano. (11.4)

stratus cloud A type of low-level stratiform cloud that is not precipitating. (4.8)

streambed The zone of contact between the water in a stream and the earth material over which it is passing; formed by the coalescence of gullies. (16.13)

stream capture The natural diversion of water from one stream to another, often occurring as headward erosion removes the divide between two adjacent drainage basins. (13.6)

streamer The positively charged streams of energy that move upward from the surface to meet the stepped leader immediately prior to a lightning strike, in cloud-to-ground lightning. (5.9)

stream load The material carried by a stream, including the material chemically dissolved in solution; comprised of suspended load, dissolved load, and bed load. (13.2)

stream system The configuration or arrangement of streams within a drainage basin; also known as drainage network. (13.1)

stream table A large, sand-filled tank, used to study stream behavior in a laboratory; also known as a flume. (13.5)

stream terrace A relatively level or gently inclined surface, or bench, bounded on one edge by a steeper descending slope, above the floodplain that flanks a stream. (9.14)

stress The amount of force divided by the area where the force is applied; similar to pressure. (11.9)

stress zone The range of an environmental variable outside the optimal range but within the lethal limit, within which an organism can survive but not thrive. (17.7)

strike-slip fault A fault in which the relative movement between the material on either side is essentially horizontal. (11.10)

strip cropping The spatial alternating of different crops, each planted in a narrow strip often along the contours of the land, as a technique to minimize soil erosion. (16.13)

structureless Pertaining to a soil with no apparent soil structure, because the grains do not stick together into peds; also known as single-grain soils. (16.2)

Subarctic (Dfc or Dfd) climate In the Köppen climate classification, a climate type characterized by adequate moisture year-round, a mean temperature of the coldest month below −3°C, and a mild or cool summer. (7.8)

Subarctic Monsoon (Dwc or Dwd) climate In the Köppen climate classification, a climate type characterized by adequate moisture in summer but not winter, an average temperature of the coldest month below −3°C, and a mild or cool summer. (7.8)

subduction zone An area where subduction takes place, either referring to the actual downgoing slab and its surroundings, or to the region, including Earth's surface, above the subducting slab. (10.7)

subglacial debris Material transported at the base of the ice of a glacier. (14.4)

sublimation A change in state from a solid to a gas, requiring an input of energy from the environment and the absorption of that energy at the molecular level as latent energy. (2.4)

submarine slope failure Slope failure that occurs on the seafloor. (12.9)

submergent coast A coast that forms where land has been inundated by the sea because of a rise in sea level or subsidence of the land. (15.9)

subpolar low A large, permanent area of low surface atmospheric pressure centered over the oceans near 60° N and 60° S, including the Icelandic Low and Aleutian Low in the Northern Hemisphere. (3.11, 3.12)

subsequent stream A stream that changes its course to follow weaker rock units; resulting when tectonic uplift raises the land or base level drops. (13.11)

subsoil layer A zone in the soil characterized by the accumulation of material, including iron oxide, clay, and calcium carbonate, depending on the climate and starting materials; also known as the B horizon, the zone of accumulation, or the zone of illuviation. (16.7)

subtropical jet stream The fast-flowing current of air near the tropopause in each hemisphere at the boundary between the Hadley cell and the mid-latitudes. (3.13)

subtropical scrub and woodland biome A collection of adjacent ecosystems that has hardy bushes, grasses, and low trees to areas with larger trees but no closed canopy, which must tolerate a hot, dry summer. (18.1)

succession The predictable change in species composition in an ecosystem following stress. (17.7)

sulfide minerals A group of nonsilicate minerals containing sulfur (S) bonded with a metal. (9.1)

sunspot A dark spot on the Sun that is slightly cooler than the rest of the Sun but associated with increased overall total solar irradiance. (2.7)

sunspot activity The extent to which dark spots on the Sun are occurring at any given time. (14.9)

sunspot number The average number of sunspots present at a given time. (2.7)

supercell thunderstorm A very large, single thunderstorm characterized by an organized set of features that allow the system to sustain itself for many hours and produce very severe weather. (5.6)

superposed stream A stream that was superimposed on already existing geologic features. (13.11)

supersaturated A theoretical condition of the atmosphere characterized by vapor pressure that slightly exceeds the water-vapor capacity; implies that active condensation or deposition must be occurring. (4.2)

supraglacial debris Material transported on the surface of a glacier. (14.4)

surface boundary layer The lowest part of the friction layer in the atmosphere; the part of the atmosphere adjacent to the surface. (3.8)

surface tension The tendency for molecules, such as water, to coalesce with a discrete outer surface. (4.1, 16.3)

surface water Water that occurs in streams, rivers, lakes, oceans, and other settings on Earth's surface. (8.0)

surface wave A seismic wave that travels on Earth's surface. (11.13)

surplus Any "extra" water (in centimeters or inches of precipitation equivalent) on the surface after the soils are saturated. (8.3)

suspended load Fine particles, generally clay and silt, that are carried suspended in moving water. (13.2)

suspension A mode of sediment transport in which water or wind picks up and carries the sediment as floating particles. (9.11)

sustainability Effective management that will allow long-term endurance and success, in environmental, economic, and social aspects. (18.17)

swamp A forested wetland along a stream or floodplain. (18.14)

swash regime The scenario of coastal inundation during which the height of the swash zone remains below the bases of coastal dunes. (15.11)

swash zone The part of the coastal zone that is permanently submerged by shallow water where waves break; also known as the inshore. (15.1)

swelling clay Soils, especially Vertisols, composed of a high percentage of certain clay minerals; they increase in volume when wet. (16.14)

symbiosis Any type of intricate, life-long interaction between individuals of different species, including through mutualism, commensalism, or parasitism. (17.3)

syncline A fold, generally concave upward (in the shape of a *U*), with the youngest rocks in the center. (11.11)

system An interconnected set of matter, energy, and processes. (1.3)

T

talus A pile of angular rock fragments on a steep slope loosened by weathering processes, or referring to an accumulation of such fragments. (12.1)

talus slope A steep slope composed of rock fragments loosened by weathering processes (talus). (12.1)

tarn A small lake within a cirque, a glacially scoured depression. (14.5)

tectonic plate Any of the dozen or so fairly rigid blocks into which Earth's lithosphere is broken. (10.5)

tectonic terrane A fault-bounded body of rock that has a different geologic history than adjacent regions. (10.13)

tectonism Earthquakes, volcanoes, and other processes that deform Earth's crust and mantle. (9.14)

teleconnection An association in atmospheric features, usually referring to opposing pressure conditions, between two distant places. (6.11)

temperate forest biome A collection of adjacent ecosystems that have deciduous and evergreen trees, often intermixed, that thrive in humid areas with warm summers and cool to cold winters. (18.1)

temperate mid-latitude (C) climate In the Köppen climate classification, a broad group of climatic types characterized by adequate moisture availability, and the mean temperature of the coldest month between –3°C and 18°C. (7.6)

Temperate Monsoon (Cwa, Cwb, or Cwc) climate In the Köppen climate classification, a climate type characterized by adequate moisture in summer but dry winters, hot, warm, or mild summers, and the mean temperature of the coldest month between –3°C and 18°C. (7.6)

temperature A measure of the object's average internal kinetic energy—the energy contained within atoms or molecules that are moving. (2.3)

temperature inversion An atmospheric condition in which temperature increases as height increases. (2.11)

temporary base level Any base level, other than sea level, that limits the downward extent of erosion. (13.9)

tension The type of differential stress where stress is directed outward, pulling the material. (11.9)

terminal moraine Glacially carried sediment that accumulates at the terminus (end) of a glacier and a landform composed of such material; generally marks the glacier's farthest downhill or most equatorward extent. (14.5)

terminus The farthest extent, especially the downslope end, of a glacier. (14.4)

terrace A relatively flat bench that is perched above a stream and that was formed by past deposition or erosion by the stream. (13.8)

terrestrial biome Any biome that exists on land. (18.3)

Tethys Sea An inferred wedge-shaped ocean that separated Asia from landmasses farther to the south, in the late Paleozoic Era and later. (10.14)

thalweg The continuous line connecting the deepest parts of a stream channel. (13.5)

theory of convergent evolution The notion that, despite wide separation and unrelated lineage, organisms develop similar physical and behavioral characteristics if they develop in similar environments. (18.10)

theory of plate tectonics The theory, currently accepted by nearly all earth scientists, that Earth's lithosphere is broken into a number of fairly rigid plates that move relative to one another. (10.0)

thermal inertia The degree to which a particular type of matter remains at the same temperature when a change in energy is imposed on it. (4.6)

thermal infrared (IR) energy The part of the electromagnetic spectrum with wavelengths between about 3 and 20 micrometers. (1.11)

thermocline The level in a body of water at which temperature changes rapidly with depth. (6.7, 17.12)

thermohaline conveyor The globally interconnected, three-dimensional oceanic circulation system. (6.7)

thermopycnocline The level in a body of water where both the temperature and density change rapidly with depth; represents a correspondence between the thermocline and the pycnocline. (6.7)

thermosphere The upper most layer of the atmosphere, characterized by a temperature inversion. (2.1)

therophyte Any annual plant that completes its life cycle quickly by having its seeds lie dormant during the dry season; common in the subtropical scrub and woodland biome. (18.10)

thrust fault A reverse fault that has a gentle dip. (11.10)

thrust sheet The sheet of rock that has been displaced above a thrust fault. (11.10)

thunder The sound made when air rushes back into the channel where a lightning stroke passed. (5.9)

thunderhead A cumulonimbus cloud that is producing hail and lightning. (5.8)

Tibetan Low A prominent area of low atmospheric pressure in summer, centered over Tibet. (3.14)

tidal bulge The raised level of water associated with high tide, on the side of the Earth closest to the Moon and on the opposite side of the Earth. (15.2)

tidal range The difference in local sea elevation between high and low tide. (15.2)

tide A cyclic change in the height of the sea surface, generally measured at locations along the coast; caused by the pull of the Moon's gravity and to a lesser extent the Sun's gravity. (15.2)

till Unsorted, generally unlayered sediment, deposited directly by or underneath a glacier. (14.4)

tilth A property of soil indicating its amenability to being plowed. (16.2)

time step The determined period of time into the future (usually minutes for a weather forecasting model) for which an initial forecast is generated, with subsequent forecasts being made out farther into the future using this same time increment, repeatedly. (5.15)

topographic map A representation of the elevation of Earth's surface by using contours. (1.6)

topographic profile A collection of connected points used to depict the change in elevation across a feature. (1.6)

topsoil The soil found in the A horizon, composed of dark gray, brown, or black organic material mixed with mineral grains. (16.7)

tornado A violent, rapidly rotating, funnel-shaped column of air that extends to the ground. (5.10)

Tornado Alley The swath from south-central Canada to central Texas where tornado frequency is among the highest on Earth. (5.11)

total solar irradiance (TSI) A measure of the Sun's total output of energy. (2.7)

township-range survey system A coordinate system used in much of the United States that characterizes location based on the number of 6-mile by 6-mile squares north, south, east, or west of the intersection of a baseline with a principal meridian, with subdivisions within the square also described; also known as the Public Land Survey System. (1.8)

traction The process by which particles roll, slide, or otherwise move on the surface, by such transport agents as streams, wind, or waves. (13.2)

transform boundary A plate boundary in which two tectonic plates move horizontally past one another. (10.5)

transform fault A strike-slip fault that accommodates the horizontal movement of one tectonic plate past another. (10.8)

transhumance A type of seasonal nomadism in which people move to the optimal elevation to experience the best temperatures and animal fodder at a particular time of year—generally upslope in summer and downslope in winter. (18.4)

transmission The passing of energy through matter, such as air. (2.1)

transpiration The transfer of water from the soil to the atmosphere via plant roots, stems, and leaves. (4.0, 16.3)

transverse dune A type of sand dune that is oriented linearly or only very gently curving for many kilometers, perpendicular to prevailing winds. (9.11)

travertine A variety of limestone that is commonly concentrically banded and porous. (12.6)

trellis drainage pattern A drainage network in which the pattern of streams follows a layer or structure, such as along a fracture or folded area, and then cuts across a ridge to follow a different feature; results in a rectangular pattern of streams. (13.1)

triple junction The place where three tectonic plates and three plate-tectonic boundaries meet. (10.8)

Tropic of Cancer The 23.5° parallel of latitude north of the equator; represents the most northerly latitude that experiences the directly overhead rays of the Sun (on or near June 21). (1.7)

Tropic of Capricorn The 23.5° parallel of latitude south of the equator; represents the most southerly latitude that experiences the directly overhead rays of the Sun (on or near December 21). (1.7)

tropical (A) climate In the Köppen climate classification, a broad group of climatic types characterized by adequate moisture availability and the mean temperature of the coldest month above 18°C. (7.4)

tropical cyclone A well-organized, rotating storm that originates in the tropical part of the Earth, ranging in strength from weak storms to hurricanes. (5.10)

tropical depression An enclosed area of low atmospheric pressure of tropical origin, with at least one closed isobar. (5.14)

Tropical Monsoon (Am) climate In the Köppen climate classification, a climate type characterized by a mean temperature of the coldest month above 18°C, and by precipitation totals that exceed potential evapotranspiration in most months, but with enough water stored in the soils during the brief dry season to support the growth of forests. (7.4)

Tropical Rain Forest (Af) climate In the Köppen climate classification, a climate type characterized by a mean temperature of the coldest month above 18°C and precipitation totals that typically exceed potential evapotranspiration year-round. (7.4)

Tropical Savanna (Aw) climate In the Köppen climate classification, a climate type characterized by a mean temperature of the coldest month above 18°C, and by precipitation totals that exceed potential evapotranspiration for about half of the year, providing enough moisture for tropical grasslands and a few scattered trees, but not forests; also known as Tropical Wet-Dry climate. (7.4)

tropical storm A tropical cyclone with well-developed circulation features and wind speeds of 64 km/hr (39 mi/hr) to 118 km/hr (73 mi/hr). (5.14)

Tropical Wet-Dry (Aw) climate *See* Tropical Savanna (Aw) climate.

tropics The region between the Tropic of Cancer (23.5° N) and the Tropic of Capricorn (23.5° S). (2.9)

tropopause The transition zone between the troposphere and overlying stratosphere. (2.11)

troposphere The lowest layer of the atmosphere, which contains nearly all of the atmosphere's water. (2.1)

tropospheric ozone A harmful air pollutant, produced by sunlight striking hydrocarbons, such as car exhaust, in the lower troposphere; also known as ground-level ozone. (17.9)

trough (atmosphere) An elongated area of low atmospheric pressure. (3.2)

trough (waves) The lowest point of any type of wave. (2.5, 15.3)

tsunami A large sea wave produced by uplift, subsidence, or some other disturbance of the seafloor, especially by a shallow submarine earthquake. (11.16, 15.7)

tuff Volcanic rock composed of consolidated volcanic ash and other material, commonly including pumice, crystals, and rock fragments. (9.5)

tundra biome A collection of adjacent ecosystems that is treeless, but includes low-lying bushes, shrubs, mosses, grasses, and lichens (algae and moss) that can withstand bitter cold. (18.1)

Tundra (ET) climate In the Köppen climate classification, a climate type characterized by adequate moisture, the warmest month's temperature averaging below 10°C (50°F), and frosts possible in any month. (7.8)

turbidity The degree of murkiness of water. (18.14)

turbulent flow Flow characterized by chaotic motion. (13.2)

turnover The mixing of waters from different depths in a large body of water; also known as overturning. (6.5)

typhoon A regional term used in the western Pacific basin applied to the most intense type of storm of tropical oceanic origin. (5.13)

U

ubac slope The side of a slope that receives less direct Sun, and is therefore cooler and more damp. (18.4)

Ultisol A type of soil formed in the subtropics and warmer mid-latitudes that is less highly weathered than Oxisols, with the eluviation of clay minerals into the B horizon, and generally acidic. (16.10)

ultramafic A composition of magma and igneous rocks, generally expressed as a dark or greenish igneous rock composed chiefly of mafic minerals rich in magnesium and iron. (9.5)

ultraviolet light The electromagnetic radiation of wavelengths between about 10^{-8} and 0.4 micrometers. (2.6)

unconfined aquifer An aquifer where the water-bearing unit is open (not restricted by impermeable rocks) to Earth's surface and atmosphere. (8.6)

unconformity A boundary between underlying and overlying rock strata, representing a significant break or gap in the geologic record; an unconformity represents an interval of nondeposition or erosion, commonly accompanied by uplift. (8.7)

unconsolidated material Material sitting on top of solid rock, including soil, loose sediment, pieces of wood and other plant parts, boulders, and other types of loose debris. (12.9)

understory The shaded, humid layer of life in a rain forest above the shrub layer, with little wind. (18.5)

uniformitarianism The concept that the natural world has evolved in the past by processes and rates similar to those we observe today. (17.7)

Universal Transverse Mercator (UTM) System A coordinate system that divides Earth into zones 6° of longitude wide and 20° of latitude long, with eastings identifying the number of meters east or west of the central meridian within a zone and northings representing the number of meters north or south of the equator. (1.8)

unloading The process during which weight is removed from a landmass, such as by the melting of a glacier. (12.1, 15.10)

unloading joint A joint formed from stresses that arise during uplift of buried rocks and that cause rocks to fracture due to reduced pressure. (11.10)

unsaturated (atmosphere) A condition characterized by atmospheric vapor pressure that is less than the water-vapor capacity; implies that more water could evaporate at that temperature under the right conditions and that the relative humidity is less than 100 percent. (4.2)

unsaturated adiabatic lapse rate The rate of change of temperature with height of an adiabatically rising air parcel with a relative humidity less than 100 percent; always amounts to 10 C°/km. (4.5)

unsaturated zone The area below the surface but above the water table, in which much of the pore space is filled by air rather than water. (8.3)

unstable (ecology) An ecosystem that is not in balance, such that populations of all species within it fluctuate considerably. (17.7)

unstable condition (atmosphere) A situation in which rising atmospheric motion is encouraged because an adiabatically moving parcel of air is warmer than its surrounding environmental air. (4.5)

updraft A local upward flow of air, usually on very short timescales. (3.3)

uplift The raising of rock up to the surface by tectonic forces. (9.3)

upslope fog A type of fog that occurs when air cools to the dew-point temperature or frost-point temperature as it moves upslope. (4.9)

upwelling The flow of colder, deeper waters up toward the surface of the ocean, replacing surface ocean water that circulated elsewhere. (6.3)

urban canyon The area between buildings in an urban area, containing different atmospheric properties than the surrounding rural environment. (7.10)

urban heat island (UHI) The phenomenon of increased air temperatures over cities as compared to their surrounding rural areas. (3.4, 4.6, 7.10)

urbanization The replacement of natural land cover with built-up land cover associated with the growth of cities and towns. (4.6)

U-shaped valley A steep-sided, rounded valley carved by a glacier. (14.5)

V

Valley and Ridge Province A physiographic region in the Appalachian Mountains of the eastern U.S., characterized by linear or curved ridges and valleys. (11.11)

valley breeze A local-to-regional-scale wind that blows from a valley uphill to the peak of a hill or mountain; also known as an anabatic wind. (3.4)

valley fog A type of fog that occurs when cold nocturnal downslope winds trap atmospheric moisture in the valleys, causing the air to reach the dew-point temperature through the addition of humid air; or when cold winds move downslope into the valley, chilling the valley air to the dew-point or frost-point temperature. (4.9)

valley glacier A glacier that flows down a valley and tends to be narrow; also known as alpine glacier. (14.1)

valley train A valley in which glacial outwash is concentrated. (14.6)

vapor pressure The portion of the total atmospheric pressure exerted by water vapor molecules, usually measured in millibars. (4.2)

vein A generally tabular accumulation of minerals that filled a fracture or other discontinuity in a rock; formed by precipitation of material from fluids, especially hydrothermal fluids. (9.7)

vertical wind shear A change in horizontal wind speed or direction with height. (5.6)

Vertisol A type of clay-rich soil located in temperate and subtropical environments, characterized by swelling when wet and shrinking when dry. (16.10)

vesicle A small hole found in a volcanic rock, representing a gas bubble in a magma that was trapped when the lava solidified. (9.5)

virga Any precipitation that evaporates before reaching the ground. (5.12)

viscosity A measure of a material's resistance to flow. (11.2, 13.2)

visible light The part of the electromagnetic spectrum consisting of wavelengths between about 0.4 and 0.7 micrometers. (1.11, 2.6)

volatile organic compound A large family of carbon-based chemicals that evaporate easily at room temperatures, including acetone, benzene, formaldehyde, and many others. (7.10)

volcanic ash Sand-sized or smaller particles originating from volcanic pyroclastic eruptions, and accumulations of such material. (9.2, 11.2)

volcanic breccia A volcanic rock containing angular fragments in a matrix of finer material. (9.5)

volcanic dome A dome-shaped volcanic feature, largely composed of solidified lava of felsic to intermediate composition. (11.1)

volcanic glass A natural glass produced by the cooling and solidification of molten lava at a rate too rapid to permit crystallization. (9.5)

volcanic neck A steep, typically butte-shaped topographic feature composed of volcanic materials that formed in the conduit within or beneath a volcanic vent and that were more resistant to erosion than surrounding materials. (11.8)

volcanic rock An igneous rock that formed from a volcanic eruption, including lava, pumice, volcanic ash, and other volcanic materials. (11.1)

volcanic tremor A small earthquake caused by the movement of magma. (11.13)

volcano A vent in the surface of Earth through which magma and associated gases and ash erupt; also the form or structure constructed from magma erupted from the vent. (11.1)

volume The amount of three-dimensional space, measured in cubic units of length, that a mass occupies. (3.1)

vorticity Any clockwise or counterclockwise horizontal rotation (spin) of air. (5.3)

V-shaped valley A very narrow, steep valley typically found in environments in which erosion dominates over deposition, such as the headwaters of a mountain stream, where the slope is steep. (13.6)

W

Walker cell circulation Any east-west-oriented atmospheric circulation system in the equatorial part of the Earth characterized by relatively high pressure and subsiding air on one side, relatively low surface pressure and rising motion on the opposite side, east-to-west trade winds at the surface, and west-to-east flow of air aloft; often refers to the cell of this type over the equatorial Pacific Ocean. (6.8)

wall cloud A sharp, low-forming protrusion on a cumulonimbus cloud often associated with tornadoes and other severe weather. (5.10)

warm front The sloping boundary between an advancing relatively warm air mass and a colder one. (5.2)

warm phase (ENSO) A periodic warming of eastern equatorial Pacific surface ocean waters, accompanied by excessive precipitation and flooding adjacent to South America as the Southern Hemisphere's summer approaches; also known as El Niño. (6.10)

warm pool A large region of ocean water warmer than the water surrounding it. (6.6)

warm sector The part of a mid-latitude cyclone that is between the warm front and the cold front, characterized by warm air. (5.2)

warning An alert by the United States National Weather Service that a specific type of life-threatening severe weather, including severe thunderstorms, tornadoes, winter storms, strong winds, or tropical cyclones, has been observed and is approaching a specific area at a designated time. (5.16)

watch An alert by the United States National Weather Service that conditions are favorable in a specified area for some specific type of life-threatening severe weather, including thunderstorms, tornadoes, winter storms, strong winds, or tropical cyclones. (5.16)

waterfall A steep descent of water within a stream, such as the place where it crosses a cliff or steep ledge. (13.6)

waterspout A rotating, columnar vortex of air and small water droplets that is over a body of water, usually an ocean or a large lake. (5.12)

water table The surface between the unsaturated zone and the saturated zone, as in the top of groundwater in an unconfined aquifer. (8.3)

water-vapor capacity The maximum atmospheric vapor pressure that can exist at a given temperature. (4.2)

wave (ocean) An irregularity on the surface of a body of water usually generated by wind. (15.3)

wave base The depth, usually about half the wavelength, at which the action of a wave in an ocean or lake no longer has an effect. (15.3)

wave-cut notch An indentation produced in rocks or sediment by continued wave action at a specific level along a coast. (15.4)

wave-cut platform A gently sloping surface or bench produced by wave erosion in the coastal zone. (15.4)

wave height The vertical distance between the height of the crest and the height of the trough of a wave. (15.3)

wave interference The presence of multiple sets of waves propagating from different directions onto the same point. (15.3)

wavelength The distance between two adjacent wave crests or troughs. (2.5, 15.3)

wave of oscillation A type of wave motion in which the individual molecules are merely oscillating around a fixed location, rather than propagating forward with the wave. (15.3)

wave refraction The bending of any type of wave, such as when one side of an ocean wave encounters the sea bottom before the other side of the wave. (15.3)

weather The conditions in the atmosphere at some specific time and place. (5.0)

weathering The breaking apart of rock. (1.4, 12.1)

weathering rind A weathered, outer crust on a rock fragment or bedrock mass exposed to weathering. (12.5)

westerlies The prevailing surface and upper-tropospheric general circulation feature that exists in the middle latitudes in each hemisphere, characterized by flow from west to east. (3.9)

Western Atlantic Warm Pool A large region of warm ocean temperatures centered on the Caribbean Sea. (6.6)

Western Australian Current A south-to-north-flowing, cold surface ocean current in the eastern Indian Ocean. (6.2)

Western Pacific Warm Pool The broad region of the warmest ocean temperatures on Earth, from China to Australia, spanning the western Pacific and eastern Indian Oceans. (6.6)

West Wind Drift A west-to-east-flowing, surface ocean current in the high latitudes of the Southern Hemisphere from southern Australia to South America and south of Africa; also known as the Antarctic Circumpolar Current. (6.2)

wetland Any area of standing water with emergent vegetation. (18.14)

Wien's Law A relationship between the temperature of a body and the wavelength at which it emits the most electromagnetic energy. (2.6)

wilting point The minimum amount of water (in centimeters or inches of precipitation equivalent) that is necessary in the rooting zone to permit extraction by plants; any less water in the soil will adhere to rock particles and resist evapotranspiration. (8.3, 16.3)

wind barb A particular type of symbol on a weather map that shows the direction and speed that wind is blowing at a point. (3.8)

windbreak A technique to minimize soil erosion, whereby a row of trees is planted perpendicular to the prevailing wind direction with the intent of slowing down the wind and therefore minimizing transportation of soil particles. (16.13)

wind streak A thin, small ridge of sand formed when winds in arid areas are strong enough that sand is blown through, without accumulating in sizeable dunes, often on the downwind (leeward) side of small plants and stones. (9.11)

windward The upwind side of a slope. (3.5)

X

xerophytic Pertaining to a species that has built-in mechanisms that make it tolerant to arid environments. (18.8)

x-ray The electromagnetic radiation of wavelengths between about 10^{-11} and 10^{-8} micrometers. (2.6)

Y

yazoo stream A channel that runs parallel to the main channel of a low-gradient stream but cannot join it for many kilometers because the sediment deposited on the floodplain by the main channel is high enough to create a natural separating barrier. (13.8)

Z

zenith angle The angle formed by the Sun's position to the observer and then directly over the observer's head. (2.8)

zonal flow A wind that is primarily blowing from west to east or east to west. (4.15)

zone of ablation The lower or equatorward part of a glacier or ice sheet, where snow and ice are removed by melting and other processes faster than they can be added. (14.3)

zone of accumulation (glaciers) The upper or poleward part of a glacier or ice sheet, where snow and ice are added faster than they are removed by melting and other processes. (14.3)

zone of accumulation (soil) A zone in the soil characterized by the accumulation of material, including iron oxide, clay, and calcium carbonate, depending on the climate and starting materials; also known as B horizon, zone of illuviation, or subsoil layer. (16.7)

zone of eluviation A light-colored, leached zone of soil, lacking clay and organic matter; also known as E horizon. (16.7)

zone of illuviation *See* zone of accumulation (soil).

zoogeographer A physical geographer/biogeographer who studies the factors that cause animal distributions and migrations. (18.16)

CREDITS

Select Abbreviations Used in Photograph and Text/Line Art Credits

CPC: Climate Prediction Center
CSC: Coastal Services Center
EPA: Environmental Protection Agency
ERSL: Environmental Remote Sensing Laboratory (University of Nebraska, Lincoln)
FAO: Food and Agriculture Organization
GSFC: Goddard Space Flight Center
HPC: Hydrometeorological Prediction Center
NASA: National Aeronautic and Space Administration
NCAR: National Center for Atmospheric Research
NCEP: National Centers for Environmental Prediction
NCDC: National Climatic Data Center
NESDIS: NOAA Satellite and Information Service
NGA: National Geospatial-Intelligence Agency
NHC: National Hurricane Center
NOAA: National Oceanic and Atmospheric Administration
NWS: National Weather Service
SPC: Storm Prediction Center
UN: United Nations
USGS: United States Geological Survey

PHOTOGRAPHS

Unless otherwise credited: © Stephen Reynolds.

DESIGN ELEMENTS

Chapter 2 Banner: NASA.

CHAPTER 1

Figure 0100a1: NASA; 0100a3, 0100a5, 0101a3: Photo by Susanne Gillatt; 0102b1: Cyrus Read/U.S. Geological Survey; 0102b2: Kate Bull/U.S. Geological Survey; 0102b3: T.A. Plucinski/U.S. Geological Survey; 0103b1: Photo by Susanne Gillatt; 0103b2: NASA; 0105a1: Photo by Susanne Gillatt; 0105a2: Photo by M.E. Yount, USGS; 0105a3: Photo by Susanne Gillatt; 0105a5: Photo by Cynthia Shaw; 0105b1: © Design Pics/The Irish Image Collection RF; 0106a2: Wendell Duffield/U.S. Geological Survey; 0109b1: Photo by Cynthia Shaw; 0109b4: NASA; 0110b1: NASA/JPL/University of Arizona; 0111a1: National Air and Space Museum, Smithsonian Institution; 0111a2: NASA; 0111b1: © Thinkstock/Getty RF; TA 1 0111b2: Photo by Ramon Arrowsmith; 0111c1: NASA; 0111d1: USGS; 0111d2–d4: NASA; 0111d5: U.S. Geological Survey; 0113a1: © McGraw-Hill Education; 0113c2,c3: U.S. Geological Survey; 0114a1: NASA.

CHAPTER 2

Figure 0200a3: Photo by Susanne Gillatt; 0200a4: © Glow Images RF; 0200a5: © Alamy RF; 0200a7: © Photographer's Choice/Getty RF; 0200a8: © McGraw-Hill Education. Bear Dancer Studios/Mark Dierker, photographer; 0200a9: NASA; 0201a1: © Getty RF; 0201a2: © McGraw-Hill Education. Bear Dancer Studios/Mark Dierker, photographer; 0201b1: Photo by Cynthia Shaw; 0202c1: © McGraw-Hill Education. Kevin Cavanagh, photographer; 0202c2: © Getty RF; 0203b1: © Comstock/Getty RF; 203b2: Game McGimsey/Alaska Volcano Observatory/USGS; 0203b3: Courtesy of ICT International; 0203b4: NOAA; 0204d1: NASA/GSFC; 0204d2: © Photodisc/Getty RF; 0204d3: © Flickr/Getty RF; 0205a1: © Getty RF; 0205a2: © Corbis RF; 0206c2: © Jim Reed/Getty RF; 0207c1–c4: NASA; 0207c5: NASA/SDO; 0208b2: © Getty RF; 0208c1: Photo by Susanne Gillatt; 0208c3: Photo by Susanne Gillatt; 0211c2: U.S. Geological Survey; 0210c1: © Getty RF; 0211a3, 0212a2: © Getty RF; 0212a3: © Ingram Publishing RF; 0213a3: © Alamy RF; 0213b1: NASA; 0213c2: USGS; 0218a1, 0218a10: Photo by Susanne Gillat; 0218a14: © Getty RF; 0218a17: Photo by Susanne Gillatt.

CHAPTER 3

Figure 0301a1: Photo by Susanne Gillatt; 0302b2: NOAA; 0302c1: Photo by Cynthia Shaw; 0304a4: Photo by Susanne Gillatt; 0304c1: Courtesy of Chuck Carter; 0305b1, 0305b2: NASA; 0305c3: NPS Photo by R. G. Johnsson/U.S. National Park Service (NPS); 0305c4: NASA/MODIS Rapid Response; 0307d2: NASA; 0309a7: Jacques Desclotres, MODIS Rapid Response Team, NASA/GSFC; 0310a2: Photo by Susanne Gillatt; 0314a5, 0314a6: NASA; 0315a3: Library of Congress Prints and Photographs Division [LC-DIG-pga-00675]; 0315a4: Library of Congress Prints and Photographs Division [LC-D41-13]; 0315a5: Library of Congress Prints and Photographs Division [LC-USZ62-38479]; 0315a6: Photo by Susanne Gillatt; 0315a7: Transportation Collection, Division of Work & Industry, National Museum of American History, Smithsonian Institution; 0315a8: Library of Congress Prints and Photographs Division [LC-DIG-pga-00809].

CHAPTER 4

Figure 04000a2: Photo by Susanne Gillatt; 0400a3: Laura J. Hartley, USGS; 0400a4, 0401a2, 0404a1: Photo by Susanne Gillatt; 0406a1: © Digital Vision/PunchStock RF; 0406a2: NASA/JPL; 0408a2: NOAA/NASA; 0408a4: NASA; 0408a5: © Robert Rohli; 0408a8: Ralph F. Kresge/NOAA; 0409b1: © Ingram Publishing/SuperStock RF; 0409b2: Jeff Schmaltz/NASA; 0409c5: © Getty RF; 0412t1: NOAA/Warren Hornsby; 0413t1: Henry Reges/CoCoRaHS HQ; 0416b2: U.S. Fish & Wildlife Service.

CHAPTER 5

Figure 0500a1: © Getty RF; 0500a2, 0500a3, 0500a4, 0501a1: NOAA; 0506a4, 0506b4: © Getty RF; 0506b6: U.S. National Park Service (NPS)/Photo by J. Schmidt; 0509a1: © Getty RF; 0509a2: © McGraw-Hill Education. Bear Dancer Studios/Mark Dierker, photographer; 0509a3: Photo by Cynthia Shaw; 0509a4: Photo by Cynthia Shaw; 0509a5: Photo by Cynthia Shaw; 0510a1: © Corbis RF; 0510b2: NASA; 0510b3: National Severe Storms Lab/NOAA; 0510c1: © Willoughby Owen/Getty RF; 0510c2: © D. Falconer/PhotoLink/Getty RF; 0511b3: © Thomas Augustine RF; 0512a1: © Tim Osborn, NOAA Coast Survey; 0512a2: Melody Ovard, OMAO/NOAA Ship Nancy Foster, NOAA/Department of Commerce; 0512a6: Sean Waugh NOAA/NSSL; 0512a7: © Patrick Byrd/Alamy RF; 0512a10: NOAA; 0512a11: Photo courtesy of Bill Cartwright; 0512a12: NASA/U. of Michigan; 0513a2: NASA; 0514b1: Barbara Ambrose NOAA/NODC/NCDDC/Department of Commerce; 0514b2: Andrea Booher/FEMA; 0514b3: FEMA/Tim Burkitt; 0514b5: Federal Emergency Management Agency (FEMA); 0515a1: NOAA/ARM.gov; 0515a2: Eric Kurth, NOAA/NWS/ER/WFO/Sacramento, NOAA/Department of Commerce; 0515a3, a4: NOAA; 0515b1: NASA/NSF; 0516a2: NOAA/NASA; 0516b3: U.S. Fish & Wildlife Service/Steven R. Emmons; 0517t1: U.S. Geological Survey; 0517t2: Andrea Booher/FEMA; 0517t3: DoD/U.S. Air Force photo by Master Sgt. Mark C. Olsen.

CHAPTER 6

Figure 0600a4: John Scurlock/U.S. Geological Survey; 0604b6, 0606a4–a6: Photo by Susanne Gillatt; 0606b2: © Brand X Pictures/Getty RF; 0609c4–c6: Photo by Susanne Gillatt.

CHAPTER 7

Figure 0701c2, 0701c3, 0701c6, 07004c1, 0704c2, 0704c3: Photo by Susanne Gillatt; 0708c1: Photo by Cynthia Shaw; 0709a1: © Getty RF; 0709a2: Design Pics/Kelly Redinger RF; 0709a3: Comstock/Jupiterimages RF; 0709a4: © Thinkstock/Jupiterimages RF; 0709a5: © Ingram Publishing RF; 0709a6: © Digital Vision/Punchstock RF; 0709c1: © Getty RF; 0709c3: NOAA; 0710c2, 0710c3, 0713a2, 0713a3: NASA; 0713a1: Photo by Cynthia Shaw; 0716a3: © Vol. 36/PhotoDisc/Getty RF; 0716a6: © Vol. 31/PhotoDisc/Getty RF; 0716a7: © Rab Harling/Alamy RF; 0716a12: Photo by Susanne Gillatt; 0716a13: NASA.

CHAPTER 8

Figure 0800a3: Peg Owens/Department Tourism-Idaho; 0802d2: Photo by Susanne Gillatt; 0803a1:

USDA/NOAA; 0803a2: NASA; 0803a3: Western Regional Climate Center/Desert Research Institute; 0803c2: U.S. Department of Agriculture; 0804a13: Photo by Susanne Gillatt; 0805a1: Photo by David Walsh/Bureau of Reclamation; 0805a2: © Corbis RF; 0805a6, 0805a7, 0805b1, 0805b2, 0805b3: © PhotoDisc/Getty RF; 0805b4: Bureau of Reclamation; 0805b5: Photo by David Walsh/Bureau of Reclamation; 0807b3: © McGraw-Hill Education. John A. Karachewski, photographer; 0809a2: © Gyori Antoine/Corbis; 0809t2: U.S. Geological Survey International; 0810b1: © Colin Cuthbert/Newcastle University/Photo Researchers, Inc; 0810b2: Doug Bartlett/Clear Creek Associates.

CHAPTER 9

Figure 0900a4: © PhotoTake/Alamy; 0901a1: © Dr. Parvinder Sethi; 0901a2: Photo by Michael Ort; 0901a3: © Doug Sherman/Geofile RF; 0901a5: © PhotoDisc/Getty RF; 0901a7: © McGraw-Hill Education. Stephen Frisch, photographer; 0901b1: Doug Sherman/Geofile RF; 0901b2: © Robert Rutford/James Carter, photographer; 0901b3: © Charles E. Jones; 0901b4, 0901b5: Thomas Sharp/Arizona State University; 0901b6: © Doug Sherman/Geofile RF; 0901b8: © The McGraw-Hill Education. Doug Sherman, photographer; 0902a4: © Getty RF; 0902b2, 0902b3: U.S. Geological Survey; 0902b4: Photo by Ariel Anbar; 0910a3: Matthew Larsen/U.S. Geological Survey; 0911a2, 0911b1: Photo by Susanne Gillatt; 0911b2: NASA; 0911b6: © Julia Waterlow/Eye Ubiquitous/Corbis.

CHAPTER 10

Figure 1001a5: © Dr. Parvinder Sethi; 0008b2: Robert E. Wallace/U.S. Geological Survey; 1010b1: Gary Wilson-Paleomagnetic Research Facility, University of Otago; 1011c2: NASA; 1012a4: U.S. Geological Survey; 1014a1(all): Ron Blakey, Colorado Plateau Geosystems; 1016b1: Photo by Susanne Gillatt; 1016b2: U.S. Geological Survey; 1016b3: Game McGimsey/U.S. Geological Survey.

CHAPTER 11

Figure 1100a4-5, 1100a6: NASA; 1100a8: U.S. Geological Survey; 1101a1: J.D. Griggs/U.S. Geological Survey; 1101a2: E. Klett/U.S. Fish and Wildlife Services; 1102a1, 1102a2: J.D. Griggs/U.S. Geological Survey; 1102a3: Don Swanson/J.D. Griggs/U.S. Geological Survey; 1102a4: R. Clucas/J.D. Griggs/U.S. Geological Survey; 1102a5: U.S. Geological Survey; 1102a6: M.E. Yount/Alaska Volcano Observatory/U.S. Geological Survey; 1102c1: John Pallister/U.S. Geological Survey; 1102c2: Hawaii Volcano Observatory/U.S. Geological Survey; 1103a1: J.D. Griggs/U.S. Geological Survey; 1103a3: J.D. Griggs/U.S. Geological Survey; 1103a4: U.S. Geological

Survey; 1103a6, 1103b2, 1103b3: J.D. Griggs/U.S. Geological Survey; 1103c2: Photo by Cynthia Shaw; 1104a2: U.S. Geological Survey; 1104a3: C. Newhall/U.S. Geological Survey; 1104a4: Game McGimsey/Alaska Volcano Observatory/U.S. Geological Survey; 1104a5: T.A. Plucinski/Alaska Volcano Observatory/U.S. Geological Survey; 1104b1: T. Miller/U.S. Geological Survey; 1104t1: Mike Dukas/U.S. Geological Survey; 1104t2: U.S. Geological Survey; 1105c1: NASA; 1106a2: Lyn Topinka/U.S. Geological Survey; 1106b1: J.D. Griggs/U.S. Geological Survey; 1106b3, 1106b4: Courtesy of Henrik Thorburn; 1106c1: Donald A. Swanson/U.S. Geological Survey; 1106t1: © AP Images; 1107a1: G. McGimsey, Alaska Volcano Observatory; 1108a3: Photo by Steven Semken; 1108b6: Photo by George H. Davis; 1110b6: © Dean Conger/Corbis; 1112a2: Photo by Scott U. Johnson; 1112a3: © JRC, Inc./Alamy; 1116b1: © Visions of America, LLC/Alamy; 1116b2: U.S. Geological Survey; 1116b3: © Karl V. Steinbrugge Collection, Earthquake Engineering Research Center, University of California, Berkeley; 1116c4: © Lynette Cook/SPL/Photo Researchers, Inc.; 1117a2: © Thomson Reuters; 1117a3: NOAA; 1117b2: Logan Abassi/UN Development Programme; 1117b4: Commander Dennis J. Sigrist/International Tsunami Information Center/NZ Defence Force; 1118a2: A. Post/U.S. Geological Survey; 1118a3: U.S. Army Photo; 1118a4: W.R. Hansen/U.S. Geological Survey; 1118b1: Photo by G. Plafker/U.S. Geological Survey; 1118b2: U.S. Navy Photo; 1118c3: USGS, Alaska Volcano Observatory.

CHAPTER 12

Figure 1200a3, 1200a4, 1200a5: Matthew Larsen/U.S. Geological Survey; 1201b3: National Park Service; 1203b3: © Creata/PunchStock RF; 1204c2: Photo by Mark Franeck; 1204c3: U.S. Department of Agriculture; 1204d1, 1204d2: Photo by Mark Franeck; 1205c1: © Image Source/Corbis RF; 1205c2, 1205c3, 1205c4: Photo by Mark Franeck; 1205c6: © L. Clarke/Corbis RF; 1206a2: U.S. Geological Survey; 1206b1: Douglas Peebles Photography/Alamy RF; 1206b2: Courtesy of Emma Harrison; 1206c2: © Dr. Parvinder Sethi; 1206t1: D. Luchsinger/National Park Service; 1207a2: © Dr. Parvinder Sethi; 1207a3: U.S. Geological Survey; 1207a4: amana images/Getty RF; 1207a5,1208b3,1208t1: Photo by Susanne Gillatt; 1209a2: Photograph courtesy of USGS Earthquake Hazards Program; 1209a4: Edwin L. Harp/U.S. Geological Survey; 1209a5: Photograph by J.T. McGill, USGS Photo Library, Denver, CO; 1209b2: U.S. Geological Survey; 1209b5: Matthew Larsen/U.S. Geological Survey; 1210a6: T.L. Pewe/U.S. Geological Survey; 1211a3: Ole Humlum, Univ. of Oslo, Dept. of Geography and the Univ. Centre in Svalbard; 1211a5: Reproduced with the permission of

Natural Resources Canada 2012, courtesy of the Geological Survey of Canada (Photo 2002-585 by Réjean Couture); 1211a8: Matthew Larson/U.S. Geological Survey; 1211a10: © Lyod Cluff/Corbis; 1212a1: Peter Haeussler/U.S. Geological Survey; 1212a2: Photograph by J.T. McGill/U.S. Geological Survey; 1212a3: Gerald Wieczorek/U.S. Geological Survey; 1212a5: © McGraw-Hill Education. John A. Karachewski, photographer; 1212t1: R.L. Schuster/U.S. Geological Survey; 1213a1: Matthew Larsen/U.S. Geological Survey; 1213a2, 1213b1: Lyn Topinka/U.S. Geological Survey; 1213b2: Matthew Larsen/U.S. Geological Survey; 1213b4: U.S. Geological Survey; 1213c6: Jennifer Adleman/Alaska Volcano Observatory/U.S. Geological Survey; 1214a2: D.J. Varnes/U.S. Geological Survey; 1214b1, 1214b2, 1214b3: U.S. Geological Survey.

CHAPTER 13

Figure 1300a1: NASA; 1300a4: Joel Schmutz/U.S. Geological Survey; 1300a5: NASA; 1306a1: U.S. Department of Agriculture; 1306b5, 1306b6: Photo by Susanne Gillatt; 1306d1: © Doug Sherman/Geofile RF; 1306d2, 1307b1, 1307b2, 1307t1–t6: Michal Tal/National Center for Earth-surface Dynamics; 1308a2: © Alan Morgan RF; 1308a3: © Pixtal/AGE Fotostock RF; 1308a4, 1308a5: © Doug Sherman/Geofile RF; 1308a10: © Pixtal/AGE Fotostock RF; 1309a2: NASA; 1309b1: U.S. Geological Survey/EROS Data Center/NASA; 1309c2: Photo by Lynn Betts, USDA Natural Resources Conservation Service; 1310a4: © Aimin Tang/Getty RF; 1310b1: © Denise McCullough RF; 1311a1: Photo by William R. Dupre; 1312b2, 1312b4: U.S. Geological Survey; 1312b5: Lyn Topinka/U.S. Geological Survey; 1312b6: © AP Images/U.S. Bureau of Reclamation; 1312b7: © Dr. Parvinder Sethi; 1313b3, 1313b4: Photo by W.R. Hansen, U.S. Geological Survey; 1315a1: © Denise McCullough RF; 1315a4: Natural Resources Conservation Service; 1315a5: Bae Lee/Alamy RF; 1315a6: © Digital Vision/PunchStock RF; 1315a7: © Ocean/Corbis RF; 1315a8: Steve Bly/Alamy RF; 1315a10: © Creatas/Punchstock; 1315t1: Federal Emergency Management Agency.

CHAPTER 14

Figure 1401b1: Photo by Paul J. Morin; 1401t1: Photo by Susanne Gillatt; 1402b5: U.S. Geological Survey; 1404a2: Courtesy of Skye L Rodgers, www.skyewritercom.com; 1404a4: © John A. Karachewski RF; 1404a5: Dan Trimble; 1404a6: U.S. Geological Survey; 1404b4: Courtesy of Bixler McClure; 1404c1, 1404c3: U.S. Geological Survey; 1405a3: Photo by Susanne Gillatt; 1405b1: Austin Post/U.S. Geological Survey; 1405c3, 1405c6: U.S. Geological Survey; 1405c7: NASA/Michael Studinger; 1406a3: © Doug

Sherman/Geofile RF; 1406a5: © Thinkstock/Jupiter RF; 1406a8: U.S. Department of Agriculture; 1406a13: National Park Service; 1407a2: Pixtal/AGE Fotostock RF; 1407b3: Courtesy of Vladimir E. Romanovsky; 1407b4: Photo by Susanne Gillatt; 1407c2: P. Weiss/U.S. Geological Survey; 1407c4: © Dean Conger/Corbis; 1408a2: © Doug Sherman/Geofile; 1408a5: U.S. National Park Service; 1408a6: Photo by Gary Fleeger; 1408b2: Photo by Cynthia Shaw; 1408b5: © MedicalRF; 1410b1,1410b2: NASA.

CHAPTER 15

Figure 1500a2: © Digital Vision RF; 1500a5: NASA; 1501a2: © McGraw-Hill Education/John A. Karachewski, photographer; 1501a4: © PhotoDisc/Getty RF; 1501a6: © Miles Ertman/Masterfile; 1501a7: © Corbis RF; 1502t1, 1502t2: © Everett C Johnson - Stock Connection/Science Faction/Corbis; 1505b1: © Corbis RF; 1505b2: U.S. Geological Survey; 1506a2: Photo by Cynthia Shaw; 1506a3: © PhotoDisc/Getty RF; 1506a4: © Paul Souders/Corbis; 1506a5: International Space Sation/NASA; 1507a3: ©PhotoDisc/Getty RF; 1507a4,1507a5: U.S. Geological Survey; 1507b1: Photographed by Robert Morton/U.S. Geological Survey; 1507b2: U.S. Army Corps of Engineers, Portland District; 1507b3: U.S. Army Corps of Engineers; 1508a2: Goodshoot/Punchstock RF; 1508t1,1508t2,1508t3, 1508t4, 1508t5: U.S. Geological Survey; 1509a3: NASA; 1509a5: U.S. Geological Survey; 1509b3: © McGraw-Hill Education/John A. Karachewski, photographer; 1509b5: Photo by Susanne Gillatt; 1511a1: © Thinkstock/Alamy RF; 1511a6, 1511A7, 1511b1–b12: U.S. Geological Survey; 1512a5: Photo by Susanne Gillat; 1512a6: Ron Chapple Stock/Alamy RF; 1512a7: National Geographic/SuperStock RF.

CHAPTER 16

Figure 1600a2: © Doug Sherman/Geofile RF; 1600a3: © Alamy Creativity/Alamy RF; 1600a6: Centers for Disease Control and Prevention; 1600a7: Photo by Susanne Gillatt; 1600a8: U.S. Department of Defense photo; 1601a4: © Mark Schneider/Visuals Unlimited; 1601c2, 1601c3: © McGraw-Hill Education. Ken Karp, photographer; 1602a2: © Digital Vision/Getty RF; 1602c1: Photo by Hari Eswaran, USDA Natural Resources Conservation Service; 1602c2: U.S. Department of Agriculture; 1602c3: U.S. Geological Survey; 1604a2: © Jupiterimages RF; 1604a3: Design Pics/Con Tanasiuk RF; 1604c1: Photo by Susanne Gillat; 1604c2: © Ingram Publishing RF; 16 1604c3: © J. Luke/PhotoLink/Getty RF; 1604d2: © Getty RF; 1605a2: Photo courtesy of Patricia A. Trochlell/DNR; 1606b4: U.S. Department of Agriculture; 1605c4: Photo Courtesy of Illinois Geological Survey, Joel Dexter; 1606a3, 1606c2, 1606c3: U.S. Department of Agriculture;

1607a2: © Tinke Hamming/Ingram Publishing RF; 1607a4: © Kenneth Fink/Photo Researchers, Inc.; 1609a1, 1609a2, 1609a7, 1609a13, 1610a1, 1610a5, 1610a9, 1610a13, 1611a1, 1611a4, 1611a7, 1611a10: U.S. Geological Survey; 1612a5: © Digital Vision/PunchStock RF; 1612a6: Photo courtesy of Vladimir E. Romanovsky; 1612a7: © Pixtal/AGE Fotostock RF; 1612a8: © Corbis RF; 1612a9: U.S. Geological Survey; 1613a1, 1613a2, 1613a4, 1613b1: U.S. Department of Agriculture; 1613b3: Photo by Jeff Vanuga, USDA Natural Resources Conservation Service; 1613c1, 1613c2: U.S. Department of Agriculture; 1613d1: Photo by Tim McCabe, USDA-NRCS; 1613d2: U.S. Department of Agriculture; 1613d3: U.S. Department of Agriculture; 1613d4: Photo by Lynn Betts, USDA Natural Resources Conservation Service; 1614a4: C.E. Meyers/U.S. Geological Survey; 1614a5: Photograph courtesy USGS Photo Library, Denver, CO; 1614a6: © Victor de Schwanberg/Photo Researchers, Inc.; 1614b2: © Ilene MacDonald/Alamy RF; 1614c1: U.S. Army; 1614c2: Photo by Tim McCabe, USDA-NRCS; 1614c3: U.S. Department of Agriculture; 1614c4: Natural Resources Conservation Service; 1614c5: NASA; 1614d1: Antibacterial clay research funded by the U.S. National Science Foundation (EAR 1123931) grant to Lynda B. Williams. Photos by Sandra Carolina Londoño-Arias; 1614d2: © Ocean/Corbis RF; 1614d3: Aaron Roeth Photography RF; 1615a2–a7: U.S. Department of Agriculture.

CHAPTER 17

Figure 1700a2, 1700a5–a15, 1701a1–a8: Photo by Susanne Gillatt; 1701t2: NASA; 1702a4: © Tom Brakefield/Purestock/Superstock RF; 1702a5, 1702a6: Photo by Susanne Gillatt; 1702a7: © Ed Reschke/Getty RF; 1702a8: © Gallo Images/Anthony/Alamy RF; 1702a9: © Corbis RF; 1703a1: Photo by Susanne Gillatt; 1703a3: © Digital Vision; 1703a12–a15: Photo by Susanne Gillatt; 1703a16: Paparazzi by Appointment/Alamy RF; 1704c1: © Steven P. Lynch RF; 1704c9: © James L. Amos/Corbis; 1705t1: Image courtesy of New Zealand American Submarine Ring of Fire 2007 Exploration, NOAA Vents Program; 1707a4: © EdBookPhoto/Alamy RF; 1708b1: © Jupiterimages RF; 1708b3: © Kent Knudson/PhotoLink/Getty RF; 1708b4: © Getty RF; 1708d2: Photo by Susanne Gillatt; 1710b1: © PhotoDisc/Getty RF; 1712d2: © Doug Menuez/Getty RF; 1713a1: © Brand X Pictures/PunchStock RF; 1713a2: © Fuse/Getty RF; 1713a4: National Geographic RF; 1713b1: U.S. Fish & Wildlife Service; 1713b3, 1713b5: U.S. Geological Survey; 1714a3–a17: Photo by Susanne Gillatt.

CHAPTER 18

Figure 1800a2: © Corbis RF; 1800a3: National Park Service; 1800a5: Photo by Susanne Gillatt; 1800a6:

Photo by Cynthia Shaw; 1800a8: Photo by Susanne Gillatt; 1800a9: © Radius Images/Alamy RF; 1801a1, 1801a2: Photo by Susanne Gillatt; 1801a4: © Steven P. Lynch RF; 1801a5: © Author's Image/PunchStock RF; 1801a7: © Corbis RF; 1801a8: Photo by Susanne Gillatt; 1802a2: © Tetra Images/Tetra Images/Corbis RF; 1802a3: © Pixtal/AGE Fotostock RF; 1802a4: Regional Cabled Network Ocean Observatory Program University of Washington School of Oceanography; 1802a5: H. Fougere/Iconotec.com RF; 1802a6: © Design Pics/Natural Selection Jerry Whaley RF; 1802a7: Photo by Mark Francek; 1802a8: © David Ponton/Design Pics/Getty RF; A 18 1802a9: © Bill Westermeyer RF; 1803a2: © John A. Karachewski RF; 1803a3: Photo by Susanne Gillatt; 1803a4: John McColgan/BLM Alaska Fire Service; 1803b3: National Park Service; 1803c2: © Getty RF; 1804a3: © Brand X Pictures/PunchStock RF; 1804c1: © Fuse/Getty RF; 1804c3, 1804d1: Photo by Susanne Gillatt; 1805b1: NASA; 1805b2: Photo by Susanne Gillatt; 1805c2–c5: Photo by Susanne Gillatt; 1806b2: © Dr. Parvinder Sethi RF; 1806b3: © Digital Vision/PunchStock; 1806b4: © PhotoDisc/Getty; 1806c2, 1806c3: Photo by Susanne Gillatt; 1806c4, 1807d2: NASA; 1808a2: Julia Johnson; 1808a3,1808a5: Photo by Susanne Gillat;1808b3: Julia Johnson; 1808c1: © Goodshot/Alamy RF; 1808c2: Photo by Susanne Gillatt; 1808c3: National Park Service; 1808c4: © Design Pics/David Ponton RF; 1809b1: © Javier Perini/Image Source RF; 1809b2: © Corbis Premium/Alamy RF; 1809b3: National Fish and Wildlife Service; 1809c1: U.S. Fish and Wildlife Service. Photo by Michelle Gadd; 1809c2, 1809c4: Photo by Susanne Gillatt; 1809c5: © Comstock RF; 1809c6: Photo by Susanne Gillatt; 1810b1: U.S. Geological Survey; 1810b2: © Charles Smith/Digitalstock RF; 1810b3: EPA; 1810c1: National Park Service; 1811a4: © Radius Images/Alamy RF; 1811a5, 1811b2: Photo by Susanne Gillatt; 1811c2: © Paul E. Tessier/Getty RF; 1812b2: © Punchstock RF; 1812b5: © Jupiterimages RF; 1812c2: U.S. Fish and Wildlife Service; 1812d1: © Vetta/Getty RF; 1812d2: © John A. Rizzo/Getty RF; 1813a4: Photo by Susanne Gillatt; 1813b2: U.S. Fish & Wildlife Service/Fran Mauer; 1813b3, 1813c3, 1813c4: U.S. Geological Survey; 1813d1: Photo by Susanne Gillatt; 1813d3: © Corbis RF; 1813d4: U.S. Fish & Wildlife Service; 1814a6: © Design Pics RF; 1814b2: USDA; 1814b3: NOAA; 1814b4: U.S Fish and Wildlife Service/Bill Radke; 1814b6: © Gail Shotlander/Getty RF; 1814b7: © McGraw-Hill Education. Pat Watson, photographer; 1814b8: © Pixtal/AGE Fotostock; 1814b10: U.S. Fish & Wildlife Service; 1815a1: Photo by Susanne Gillatt; 1815a3: © Punchstock RF; 1815b1, 1815b3: Photo by Cynthia Shaw; 1815b4: © Diane Nelson RF; 1815c1: NOAA PMEL Vents Program; 1815c2: U.S. Fish & Wildlife Service; 1816a2: © Digital Stock/Corbis

RF; 1816a3: © Partha Pali/Getty RF; 1816a4: Photo by Susanne Gillatt; 1816a6: © Jeremy Woodhouse/ Getty RF; 1816a7–a9: Photo by Susanne Gillatt; 1816a10: © Brand X/Punchstock RF; 1816b1, 1816b2: Ron Blakey, Colorado Plateau Geosystems; 1816b4: "North American Pleistocene Landscape," copyright Karen Carr; 1816b5: National Park Service; 1817a3, 1817a4: Photo by Susanne Gillat; 1817a7: NPS Photo by Mark Lellouch; 1817b2: © McGraw-Hill Education. Barry Barker, photographer; 1817b3: Photo by Cynthia Shaw; 1817b4: Photo by Susanne Gillat; 1817b5: © DLILLC/Corbis RF; 1818a2–a9: Photo by Susanne Gillat.

TEXT AND LINE ART

CHAPTER 1

01.06.a4: USGS; 01.06.a5: USGS; 01.08.a3: NGA Office of GEOINT Sciences; 01.08.b1: http://www.gsd.harvard.edu/gis/manual/projections/fundamentals/zones.jpg; 01.10.a1: USGS; 01.10.d2: NOAA/NWS; 01.11.d1 NASA; 01.12.a3: NOAA/CSC; 01.12.b3: California Department of Conservation; 01.14.b1: Modified from NOAA; 01.14.b2: Modified from UCAR MedEd program.

CHAPTER 2

02.00.a6: NASA; 02.03.d1: NOAA; 02.04.d1: NASA; 02.07.c6–7: NASA; 02.11.c2: Some data from Sellers, W.D., 1965: *Physical Climatology*. University of Chicago Press; 02.12.a1: Adapted from NOAA/NCEP; 02.12.c1: NOAA/GSFC; 02.12.d1: NOAA; 02.12.d1: NOAA/GSFC; 02.12.d3: NOAA/GSFC; 02.13.a1: Modified from NASA/ERSL02.13.c1: Data from NASA; 02.13.c3: NASA/EOS; 02.14.b1: Adapted from Fleagle, R.G. and J.A. Businger, 1963: *An Introduction to Atmospheric Physics*. Academic Press; 02.15.a2: NASA/EOS; 02.16.a1: NCEP/NCAR Reanalysis dataset; 02.16.b1: NCEP/NCAR Reanalysis dataset; 02.16.c1: NOAA/NASA COMET Program; 02.16.c2: NCEP/NCAR Reanalysis dataset; 02.17.b2–b4: NCEP/NCAR Reanalysis dataset; 02.18.a3: http://www.worldclimate.com; 02.18.a6: http://www.worldclimate.com; 02.18.a9: http://www.worldclimate.com; 02.18.a12: http://www.worldclimate.com; 02.18.a16: http://www.worldclimate.com; 02.18.a19: http://www.worldclimate.com; 02.18.a22: http://www.worldclimate.com; 02.19.a3: Adapted from Neiburger, M., J.G. Edinger, and W.D. Bonner, 1982: *Understanding Our Atmospheric Environment*. W.H. Freeman and Company; 02.19.a4–a5: NCEP/NCAR Reanalysis dataset; 03.02.d1: NOAA/NWS.

CHAPTER 3

03.06.b1–b3: NCEP/NCAR Reanalysis dataset; 03.09.b2. NOAA/NHC; 03.12.a8–a12: NCEP/NCAR Reanalysis dataset; 03.13.D1: NCEP/NCAR Reanalysis dataset; 03.14.a3–a4: http://www.worldclimate.com; 03.14.b1–b2: NCEP/NCAR Reanalysis dataset;03.14.b3: http://www.worldclimate.com; 03.14.b4–b5: NCEP/NCAR Reanalysis dataset; 03.14.b6: http://www.worldclimate.com; 03.14.b7–b8: NCEP/NCAR Reanalysis dataset; 03.14.b9: http://www.worldclimate.com; 03.16.a2–a5: NCEP/NCAR Reanalysis dataset; 03.16.a6–a7: NCEP/NCAR Reanalysis dataset and NASA Earth Observatory; 04.03.a1–a2: NCEP/NCAR Reanalysis dataset.

CHAPTER 4

04.03.b1–b2: NCEP/NCAR Reanalysis dataset; 04.04.c1–c4: NCEP/NCAR Reanalysis dataset; 04.07.a2: NASA/NOAA; 04.07.c2: NOAA/NCEP; 04.10.c1–c5: National Climatic Data Center Climate Atlas; 04.12.c1: Data from Changnon, S.A. and T.R. Karl, 2003: Journal of Applied Meteorology 42: 1302–1315; 04.13.a1–a2: NCEP/NCAR Reanalysis dataset; 04.13.b1: NCEP/NCAR Reanalysis dataset; 04.14.a2: NCEP/NCAR Reanalysis dataset; 04.14.b1: NASA/USGS; 04.14.b2: NOAA/CPC; 04.14.c1–c3: USDA/Natural Resources Conservation Service; 04.15.a1–a4: NCEP/NCAR Reanalysis dataset; 04.15.b1–b2: NOAA/CPC.

CHAPTER 5

05.00.a6–a9: NCEP/NCAR Reanalysis dataset; 05.01.b1: NCEP/NCAR Reanalysis dataset; 05.02.a1, b1, c1, and c3: NOAA/NWS/HPC; 05.03.a4: NCEP/NCAR Reanalysis dataset; 05.03.b2: NCEP/NCAR Reanalysis dataset; 05.03.b3: NOAA/NWS/HPC; 05.03.b4: NCEP/NCAR Reanalysis dataset; 05.03.b5: NOAA/NWS/HPC; 05.04.a1: NOAA/NWS/HPC; 05.04.a2: NOAA/NWS; 05.04.a3: NOAA/NWS/HPC; 05.04.a4: NOAA/NWS; 05.04.a5: NOAA/NWS/HPC; 05.04.a6: NOAA/NWS; 05.05.a1: NCEP/NCAR Reanalysis dataset; 05.05.c2–c4: NCEP/NCAR Reanalysis dataset; 05.06.b3: NOAA/NWS/HPC; 05.07.a1–a3: Data from NASA; 0507.a4: Modified from NASA; 05.07.b1–b2: NOAA/NCDC/SPC/National Severe Weather Database; 05.08.b1–b3: NOAA/NCDC/SPC/National Severe Weather Database; 05.09.t1: NOAA/NWS; 05.10.b1: NOAA/NWS/HPC; 05.11.a1–a5, b2: NOAA/NWS/SPC/National Severe Weather Database; 05.12.a3: NOAA/NWS; 05.12.a4: NOAA; modified from Coniglio, M.C. and D.J. Stensrud, 2004: Weather and Forecasting 19: 595–605; 05.12.a5: NOAA/NWS; 05.13.b1: NASA; 05.13.c4: NOAA; 05.16.b2: NOAA; 05.16.c1: NOAA;

05.17.a3: NOAA/NWS; 05.17.a4: NOAA; 05.18.a1: NOAA/NWSNOAA/NWS; 05.18.a2: NOAA/NWS; 05.18.a3: NOAA/NWS/HPC; 05.18.a4: NOAA/NWS/HPC; 05.18.a5–a10: NCEP/NCAR Reanalysis dataset.

CHAPTER 6

06.00.a1–a2: NCEP/NCAR Reanalysis dataset; 06.00.a3: NOAA/Northwest Fisheries Science Center; 06.00.a5: D'Aleo, J.D. and D. Easterbrook. Multidecadal tendencies in ENSO and global temperatures related to multidecadal oscillations. Energy and Environment 21(5), 437–460; 06.00.a6–a7: NCEP/NCAR Reanalysis dataset; 06.01.c2: NASA/GSFC/Scientific Visualization Studio; 06.03.a1–a4: NCEP/NCAR Reanalysis dataset; 06.05.a1–a2: NOAA/National Ocean Data Center; 06.05.b1–b2: http://www.soes.soton.ac.uk/teaching/courses/oa631/hydro.html#alt; 06.05.c1: Modified from Webster, P.J., 1994: The role of hydrological processes in ocean-atmosphere interactions. *Reviews of Geophysics*, 32(4), 427–476; 06.06.a2–a3: NCEP/NCAR Reanalysis dataset; 06.06.b5: S. Adapted from Barker et al., 2011: 800,000 years of abrupt climate change. *Science* 334, 347–351; 06.07.b1: Argonne National Laboratory; 06.07.b2–b3: Modified from Durack, P.J., S.E. Wijffels, and R.J. Matear, 2012: Ocean salinities reveal strong global water cycle intensification during 1950 to 2000. Science 336, 455–458; 06.07.b4: Modified from NASA; 06.08.a2: NCEP/NCAR Reanalysis dataset; 06.08.t1: Adapted from International Research Institute for Climate and Society, Lamont-Doherty Earth Observatory; 06.09.a1: NASA; 06.09.a2: International Research Institute for Climate and Society, Lamont-Doherty Earth Observatory; 06.09.b1: NCEP/NCAR Reanalysis dataset; 06.09.b3: Climate Prediction Center/NOAA; 06.09.c1: Data from Walker, G.T., and E.W. Bliss, 1932: World Weather V., Memoir Royal Meteorological Society, 4, No. 36, 53–84; 06.09.c2: NCEP/NCAR Reanalysis dataset; 06.10.b1–b6: NCEP/NCAR Reanalysis dataset; 06.11.a2–a3: NCEP/NCAR Reanalysis dataset; 06.11.b1–b2: NOAA/CPC; 06.12.a4: NASA/TAO/Triton database; 06.12.a5–a7: NASA/TAO/Triton database; 06.12.b1–b4: NCEP/NCAR Reanalysis dataset; 06.12.t1: NCEP/NCAR Reanalysis dataset; 06.12.t2: NOAA/Climate Prediction Center; 06.13.a2: NCEP/NCAR Reanalysis dataset; 06.14.a6–a7: NCEP/NCAR Reanalysis dataset; 06.14.a8–a9: P. Waylen, unpublished.

CHAPTER 7

07.00a1–a2: NCEP/NCAR Reanalysis dataset; 07.00.a3: NCEP/NCAR Reanalysis dataset; 07.01.

b1: NCEP/NCAR Reanalysis dataset; 07.01.c1: Rohli and Vega, 2012: Climatology, Second Edition, Jones and Bartlett Learning; 07.02.00: NCEP/NCAR Reanalysis dataset; 07.03.a1–a4: NCEP/NCAR Reanalysis dataset; 07.04.a1–a3: NCEP/NCAR Reanalysis dataset; 07.04.c5–c7: http://www.worldclimate.com; 07.05.a1–a3: NCEP/NCAR Reanalysis dataset; 07.05.c1–c2: NCEP/NCAR Reanalysis dataset; 07.05.c4: NCEP/NCAR Reanalysis dataset; 07.05.d1–d4: http://www.worldclimate.com; 07.06.a1–a3: NCEP/NCAR Reanalysis dataset; 07.06.b2: NCEP/NCAR Reanalysis dataset; 07.06.c2: NCEP/NCAR Reanalysis dataset; 07.06.c3–c5: http://www.worldclimate.com; 07.07.a1–a3: NCEP/NCAR Reanalysis dataset; 07.07.b1–b2: NCEP/NCAR Reanalysis dataset; 07.07.b3–b5: http://www.worldclimate.com; 07.08.a1–a3: NCEP/NCAR Reanalysis dataset; 07.08.b1: http://reynolds.asu.edu/biosphere3d/bio3d_home.htm; 07.08.c4: National Snow and Ice Data Center, University of Colorado, Boulder; 07.08.c5: NCEP/NCAR Reanalysis dataset; 07.08.c6–c8: http://www.worldclimate.com; 07.09.b1–b2: U.S. EPA; 07.10.a1–2: Kaufmann, R.K. and C.J. Cleveland, 2008: *Environmental Science*, McGraw-Hill Higher Education; 07.10.c2–c3: NASA; 07.11.a1–a2: NOAA/NCDC/NESDIS; 07.11.a3–a6: National Research Council, 2006: Surface Temperature Reconstructions for the Last 2,000 Years, The National Academies Press; 07.11.a7: Dr. Roy Spencer, University of Alabama, Huntsville; 07.12.a3: NASA/Marshall Space Flight Center; 07.13.b2–b3: NOAA; 07.16.a13: NASA.

CHAPTER 8

08.01.a2: After USGS; 08.03.a4: NWS; 08.03.c1: NOAA/CPC; 08.04.a1: Adapted from: Willmott, C.J. and J.J. Feddema, 1992: A more rational climatic moisture index, *Professional Geographer*, 44, 84–87; 08.04.a2–a5: WebWIMP, University of Delaware; 08.04.a7–a8: Adapted from: Willmott, C.J. and J.J. Feddema, 1992: A more rational climatic moisture index, *Professional Geographer*, 44, 84–87; 08.04.a9–a12: WebWIMP, University of Delaware; 08.04.a14–a15: WebWIMP, University of Delaware; 08.04.a17: WebWIMP, University of Delaware; 08.05.a4: USGS; 08.11.a1: USGS/High Plains Regional Ground Water Study; 08.11.a2: Texas Water Development Board; 08.11.b1: USGS; 08.11.b2: Texas Water Development Board; 08.11.b3: USGS; 08.11.b4: USGS.

CHAPTER 10

10.04.a1: USGS; 10.04.b1: Centers for Disease Control and Prevention/USGS; 10.05.a1, c1: Data from Michelle K. Hall-Wallace; 10.09.b1, c1: Data from Michelle K. Hall-Wallace; 10.10.

b2: After J. Kious/USGS, 1996; 10.10.c1–c2: USGS; 10.11.c1: Don Anderson/Seismological Laboratory, California Institute of Technology; 10.13.c2: P. King, 1977: Princeton University Press; 10.13.t2: After E. Moores and others, 1999: Geological Society of America Special Paper 338.

CHAPTER 11

11.05.d1: USGS Fact Sheet 100-03; 11.05.d2: USGS Fact Sheet 2005-3024; 11.14.a1: NOAA/Paula Dunbar; 11.14.b1: NOAA/Paula Dunbar; 11.17.a3: NOAA; 11.18.a1: E. Eckel, 1970: USGS Professional Paper 546; 11.18.c1–c2: After G. Plafker, 1969: USGS Professional Paper 543.

CHAPTER 12

12.07.b1: http://web.env.auckland.ac.nz/our_research/karst/; 12.07.b2: USGS; 12.10.t1: After Kiersch, G., 1964: Civil Engineering; 12.12.b1: USGS Open-File Report 97-0289; 12.14.a3, c1–c2: Varnes, D. and W. Savage, 1996: USGS Bulletin 2130.

CHAPTER 13

13.03.b1: Adapted from Mount, J., 1995: *California Rivers and Streams: The Conflict between Fluvial Process and Land Use*, University of California Press; 13.04.a2: Bailey, C., 1998: William and Mary College; 13.10.a2–a3: Martin Jakobsson/Stockholm GeoVisualisation Laboratory; 13.13.a1: K. Wahl, USGS Circular 1120-B; 13.13.a2:C. Parett, USGS Circular 1120-A; 13.13.a5: USGS; 13.13.b2: Based on data from Maddox, R.A., L.R. Hoxit, C.F. Chappell, and F. Caracena, 1978: Comparison of meteorological aspects of the Big Thompson and Rapid City flash floods, 1977: *Monthly Weather Review* 106, 375–389; 13.14.b1–b2: Data from USGS/National Water Information System; 13.16.a6: After D. Topping, 2003: USGS Professional Paper 1677.

CHAPTER 14

14.06.a2: USGS; 14.06.a10: USGS; 14.06.a13: USGS; 14.07.b2: After European Science Foundation/PACE21 Network; 14.07.c1: After J. Feth, USGS Professional Paper 424B, 1961; 14.07.c6–c8: NOAA/Great Lakes Environmental Research Laboratory; 14.08.b3: University of British Columbia; 14.08.b4: Emi Ito, University of Minnesota; 14.10.d1: USGS Fact Sheet 076-00.

CHAPTER 15

15.01.a3: After U.S. Army Corps of Engineers, 1992: Engineer Manual EM1110-2-1502: "Coastal Littoral Transport"; 15.06.c1: Chris Jenkins/Institute of

Arctic and Alpine Research, University of Colorado at Boulder; 15.08.t1–5: USGS; 15.10.t1: After S. Dutch/University of Wisconsin, Green Bay; 15.11.a2–a5: USGS; 15.11.b6–b7: 15.11.b10–b11: St. Petersburg Coastal and Marine Science Center/USGS; 15.11.b10–b11: St. Petersburg Coastal and Marine Science Center / USGS; 15.11.b12: USGS.

CHAPTER 16

16.01.b2: http://www.terragis.bees.unsw.edu.au/terraGIS_soil/sp_particle_size_fractions.html; 16.09.a2–a4: UN/FAO; 16.09.a6–a8: UN/FAO; 16.09.b2–b4: UN/FAO; 16.09.b6–b8: UN/FAO; 16.10.a2–a4: UN/FAO; 16.10.a6–a8: UN/FAO; 16.10.a10–a12: UN/FAO; 16.10.a14–a16: UN/FAO; 16.11.a2–a4: UN/FAO; 16.11.a6–a8: UN/FAO; 16.11.a10–a12: UN/FAO; 16.11.a14–a16: UN/FAO; 16.12.a1: UN/FAO; 16.13.c3: USDA/NRCS.

CHAPTER 17

17.02.a2–a3: Kaufmann, R.K. and C.J. Cleveland, 2008: *Environmental Science*, McGraw-Hill Higher Education; 17.02.t1: Kaufmann, R.K. and C.J. Cleveland, 2008: *Environmental Science*, McGraw-Hill Higher Education; 17.04.b1: OpenStax College. (2013, June 20). *The Biodiversity Crisis*. Retrieved from the Connexions Web site: http://cnx.org/content/m44892/1.5/; 17.06.a1: NASA; 17.06.a2: Cunningham, W.P. and M.A. Cunningham, 2012: *Environmental Science: A Global Concern*. McGraw-Hill Higher Education; 17.06.B: Data from Turner, D.P., S. Urbanski, D. Bremer, S.C. Wofsy, T. Meyers, S.T. Gower, and M. Gregory, 2003: A cross-biome comparison of daily light use efficiency for gross primary production, Global Change Biology 9, 383–395; 17.06.b1–b3: NCEP/NCAR Reanalysis dataset; 17.06.b5: NASA; 17.07.c1: Kaufmann, R.K. and C.J. Cleveland, 2008: *Environmental Science*, McGraw-Hill Higher Education; 17.08.a1: Kaufmann, R.K. and C.J. Cleveland, 2008: *Environmental Science*, McGraw-Hill Higher Education; 17.09.b1, c1: Kaufmann, R.K. and C.J. Cleveland, 2008: *Environmental Science*, McGraw-Hill Higher Education; 17.09.d1: U.S. EPA; 17.10.a1: Kaufmann, R.K. and C.J. Cleveland, 2008: *Environmental Science*, McGraw-Hill Higher Education; 17.10.b2: Adapted from USGS; 17.10.b3: Swap, R.M. et al., 1992: Saharan Dust in the Amazon Basin, Tellus 44B, 133–149; 17.10.c2: Adapted from Cunningham, W.P. and M.A. Cunningham, 2012: *Environmental Science: A Global Concern*. McGraw-Hill Higher Education; 17.11.a1: Kaufmann, R.K. and C.J. Cleveland, 2008: *Environmental Science*, McGraw-Hill Higher Education; 17.11.b2–b4: U.S. EPA; 17.11.b5: http://www.worldmapper.org/; 17.12.c1: Kaufmann, R.K. and C.J. Cleveland, 2008:

Environmental Science, McGraw-Hill Higher Education; 17.13.a3: Adapted from USGS; 17.13.a5: University of Georgia, Center for Invasive Species and Ecosystem Health; 17.13.b2: Adapted from USGS; 17.13.b4: Adapted from USGS; 17.13.b6: Adapted from USGS.

CHAPTER 18

18.01.b1; 18.01.c1: Adapted from Cunningham, W.P. and M.A. Cunningham, 2012: *Environmental Science: A Global Concern*. McGraw-Hill Higher Education; 18.02.a1: Adapted from https://www.sciencebase.gov/catalog/item/508fece8e4b0a1b43c29ca22; 18.03.b1: NASA; 18.05.a1: Modified from NCAR; 18.06.a1: http://www.grida.no/graphicslib/detail/changing-global-forest-cover_b110; Philippe Rekacewicz assisted by Cecile Marin, Agnes Stienne, Guilio Frigieri, Riccardo Pravettoni, Laura Margueritte and Marion Lecoquierre); 18.06.c1: Modified from JPL/NASA; 18.07.b1–b3: Adapted from https://www.sciencebase.gov/catalog/item/508fece8e4b0a1b43c29ca22; 18.07.c1: Adapted from https://www.sciencebase.gov/catalog/item/508fece8e4b0a1b43c29ca22; 18.07.d1: Modified from USDA/Reich, P: World Soil Resources; 18.09.a1–a3, 18.10.a1–a3, 18.11.a1–a3: Adapted from https://www.sciencebase.gov/catalog/item/508fece8e4b0a1b43c29ca22; 18.11.c1: Adapted from Cunningham, W.P. and M.A. Cunningham, 2012: *Environmental Science: A Global Concern*. McGraw-Hill Higher Education; 18.12.a1–a3: Adapted from https://www.sciencebase.gov/catalog/item/508fece8e4b0a1b43c29ca22; 18.13.a1–a3: Adapted from https://www.sciencebase.gov/catalog/item/508fece8e4b0a1b43c29ca22; 18.13.b4: http://dec.alaska.gov/; 18.13.c1: NOAA; 18.13.d2: Cunningham, W.P. and M.A. Cunningham, 2012: *Environmental Science: A Global Concern*. McGraw-Hill Higher Education; 18.14.b9: Adapted from National Resources of Canada; 18.15.a2: Adapted from: Kaufmann, R.K. and C.J. Cleveland, 2008: *Environmental Science*, McGraw-Hill Higher Education; 18.15.b2: Adapted from Kaufmann, R.K. and C.J. Cleveland, 2008: *Environmental Science*, McGraw-Hill Higher Education; 18.16.b3: National Park Service; 18.16.b6: Woods Hole Oceanographic Institution; 18.17.a1: Adapted from: Adams, W.M., 2006: The future of sustainability: Rethinking environment and development in the twenty-first century. In Report of the IUCN Renowned Thinkers Meeting *The Future of Sustainability Re-thinking Environment and Development in the Twenty-first Century*, World Conservation Union; 18.17.b1: Cunningham, W.P. and M.A. Cunningham, 2012: *Environmental Science: A Global Concern*. McGraw-Hill Higher Education.

INDEX

Northern Alaska insolation, 70
Northern Hemisphere
    Coriolis effect, 88, 92, 168
    general air circulation
        patterns, 96, 97, 98, 99
    on globe of Earth, 16
    latent heat, 69
    seasons, 52, 53
northings, 18
North Pacific Current, 184
North Pole, 96, 97, 229
North Star, 52
Norway, glaciers in, 456
no-till farming, 502, 529
nuclear explosions, 371
nuclei (atomic), 45
null point, 182
nunataks, 455
nutria, 560
nutrients in biomes, 570, 571
Nuuk (Greenland) climate, 229

O

oasis effect, 233
observations, in scientific
        method, 7
occluded fronts, 151
occlusions, 151
ocean acidification, 237, 489
ocean-continent convergent
        boundaries, 320
ocean currents
    atmospheric effects, 121, 136
    effects on climate, 218,
        223, 237
    effects on coastlines, 478
    effects on sea-surface
        temperatures, 187
    formation, 182–183
    global patterns, 184–185
    impact on fog formation,
        127, 128
    impact on glaciations, 471
    oil spill dispersal, 31
ocean fertilization, 551

oceanic crust, 308
oceanic fracture zone, 310, 322
oceanic plateaus, 311, 328
oceanic trenches, 372, 373
ocean-ocean convergent
        boundaries, 320
oceans. See also seafloor; sea-
        surface temperatures
    acidification, 237, 489, 599
    as biomes, 564, 566, 594–595
    as carbon dioxide reservoir,
        237, 241
    climate, 213, 216, 218
    density, 188–189
    earthquakes, 372, 377, 378,
        381
    evaporation from, 189,
        249, 250
    glaciers meeting, 459
    humidity over, 114
    impact on fog formation,
        127, 128
    land and sea breezes, 82
    movement of matter and
        energy, 10, 12
    oil spills, 30–31
    as percentage of Earth's
        surface, 310
    as percentage of Earth's
        water, 248, 250
    primary productivity, 546
    role in cyclone formation,
        148, 168, 169, 170
    salt deposition in, 389
    sediments, 276
    surface area, 474
    thermal properties,
        68–69, 121
ocean trenches, 311, 314, 320
offshore breeze, 82
offshore flows, 103
Ogallala aquifer, 268–269
Ohio, cross section, 333
O horizon, 516, 517
oil fields, 4
oil reserves, 4

oil spills, 30–31, 264
Old Faithful geyser, 260
oligotrophic water bodies,
        555, 592
olivine, 275, 308, 390
Olympic Peninsula, 575
omnivores, 539
onshore breeze, 82
Ontong Java Plateau, 329
open systems, 8, 544
orbital plane, 52, 53
orbital plane of Earth, 52, 53
orbit of Earth
    climate change and,
        236, 470
    impact on insolation, 48,
        52, 53
orchards, 60
orders, stream, 419
organic matter in soils, 505,
        510, 511, 515
original horizontality
        principle, 296
orographic effects, 122, 157,
        570, 572. See also
        mountains
oryx, 563
oscillations
    climate change and, 237
    ENSO, 199, 200–205
    Indian Ocean, 206
    North Atlantic, 207
    of ocean waves, 482
    Pacific Decadal, 180–181
Outer Banks (North Carolina),
        476, 477, 499
outer core (Earth), 308
outgoing longwave radiation,
        63, 64–65, 66, 67, 73
outliers, 27
outwash plains, 466
overgrazing, 529, 579
overland flow, 428
overpumping, 262–263,
        268–269
overturning, 191

overwash regime, 498
oxalic acid, 393
oxbow lakes, 427, 432, 433
oxidation, 290, 389, 395
oxide minerals, 275
Oxisols, 519, 522, 526, 600
oxygen
    as agent of chemical
        weathering, 388, 389
    dissolved, 571
    ecosystem shortages,
        558–559
    as greenhouse gas, 63
    isotopes, 469
    ozone form, 58
    as percentage of
        atmosphere, 37, 56
    release by plants, 11, 577
ozone
    in Earth's atmosphere, 37,
        57, 58–59
    as greenhouse gas, 63
    as pollutant, 230, 232, 553
ozone "hole," 34, 59
ozone layer, 34, 58, 59

P

Pacific Coast, 157
Pacific Decadal Oscillation,
        180–181, 237
Pacific Northwest, 70, 129
Pacific Ocean
    coral reefs, 489
    currents, 184, 185
    seafloor features, 310, 323
    tectonic activity, 321, 323,
        328–329, 341
Pacific Palisades, 402
Pacific plate, 323
Pacific realm, 596
Pacific Ring of Fire, 321, 359
Pacific Walker cell, 197
pahoehoe lava, 350
Painted Desert, 302
Palau, 570

porphyritic texture, 282
Port Moresby (Papua New Guinea) water balance, 255
positive feedbacks
  defined, 9
  with fish kills, 559
  in global climate models, 240–241, 471
  in stream channels, 427
positive IOD, 206
positive magnetic anomalies, 327
positive NAO phase, 207
potassium, 510, 511
potassium feldspar. *See* K-feldspar
potential energy, 38, 251
potential evapotranspiration, 218, 222, 223, 252–253
potholes, 395, 420
Potomac River, 418, 436
power plants, 256, 257
prairie dogs, 582
prairies, 582
precession, 470
precipitation
  acidic, 232, 551, 553, 556, 557
  average annual amounts in U.S., 13
  biome distribution and, 570, 572
  climate change impacts on, 238
  cloud types and, 124–125
  in Costa Rica, 534
  distribution on Earth, 134–135, 212, 250
  effects on primary productivity, 547
  effects on soil formation, 514
  effects on weathering, 392
  erosion by, 428, 528
  extreme events, 137, 143
  flooding from, 441, 442, 443
  formation, 108, 123, 130–133

in hydrologic cycle, 10, 249, 253
  measuring, 135, 252
  in mid-latitude climates, 225, 226, 227
  in polar and subarctic climates, 228
  radar imaging, 174
  role in landslides, 384–385, 401
  soil water from, 508
  at stationary fronts, 147
  in urban areas, 233
precipitation fog, 126
precipitation of minerals, 269
predation, 540, 541
predictions, 7
prehistoric slope failures, 410, 411
prescribed burns, 585
pressure
  of gases, 76, 77, 78
  impact on states of matter, 43
  measuring for air, 78
  role in metamorphism, 284
pressure-gradient force, 81, 82, 90
pressure gradients
  causes in atmosphere, 81, 83
  defined, 80
  effect on wind direction, 90, 92
  local atmospheric circulation, 82–83
pressure melting, 460
prevailing winds, 74
prickly pear cactus, 580
prides, 583
primary data, maps as, 23
primary impacts of drought, 139
primary minerals, 504
primary producers, 538, 544
primary productivity, 546–547, 567

primary succession, 549
primary waves (P waves), 370
Prime Meridian, 17, 28
Principal Meridians, 19
principles of relative dating, 296–297
prisms, 44
probability, 445
probability of exceedance, 451
productivity in ecosystems, 546–547
profiles of streams, 424–425
profundal zone, 558
projections, 20–21
promontories, 484, 486
propagation of waves, 44, 482
proportional relationships, 76
proxy evidence, 234, 235
Prudhoe Bay oil field, 4
Public Land Survey System, 19
pumice, 283
pump-and-treat remediation, 267
Pu`u `Ō `ō volcano, 346
P-waves, 370
pycnocline, 194
Pyrenees Mountains, 456
pyrite, 275
pyroclastic eruptions, 348, 357
pyroclastic flows
  basic features, 282
  common to composite volcanoes, 352
  eruption columns versus, 348
  hazards, 357
  Mount Unzen disaster, 353
pyrophilic species, 585
pyroxenes, 275, 389, 390

Quaternary Period, 299
questions, in scientific method, 7
quicksand, 530

R

raccoons, 596
radar imaging
  applications, 25
  tornadoes, 142
  weather, 172, 174–175
radial drainage patterns, 419
radiation, 39
radiation balance, 67
radiation fog, 127, 128, 129
radioactive decay, 298
radiosondes, 40, 78
radio waves, 47
raindrops, 130
rain forests. *See also* tropical rain forests
  biomes, 564, 566, 567, 574–575
  climate of, 219, 220, 221
  precipitation patterns, 135
  threats to, 526, 576–577, 599
rain gauges, 135, 252
rain-shadow deserts, 579
rain shadows, 117, 223, 341. *See also* mountains
rainy seasons, 583
random distributions, 27
ranges, 19
Ranier, Mount, 356
rapid germination, 581
rapids, 429
rates, calculating, 29
rates of movement in slope failures, 407
recessional moraines, 465
recharge, groundwater, 260
recreational water use, 257
rectification, 26
recurrence intervals, 137
Redoubt volcano, 381

Shaded-Relief Map of the Conterminous United States and Adjacent Areas

This map simulates shading on the land surface as if lighted from the upper left.
It includes the southern part of Canada and the northern part of Mexico.